Lecture Notes in Computer Science 9642

Commenced Publication in 1973
Founding and Former Series Editors:
Gerhard Goos, Juris Hartmanis, and Jan van Leeuwen

Editorial Board

More information about this series at http://www.springer.com/series/7409

Shamkant B. Navathe · Weili Wu
Shashi Shekhar · Xiaoyong Du
X. Sean Wang · Hui Xiong (Eds.)

Database Systems
for Advanced Applications

21st International Conference, DASFAA 2016
Dallas, TX, USA, April 16–19, 2016
Proceedings, Part I

 Springer

Editors
Shamkant B. Navathe
Georgia Institute of Technology
Atlanta, GA
USA

Weili Wu
University of Texas at Dallas
Richardson, TX
USA

Shashi Shekhar
University of Minnesota
Minneapolis, MN
USA

Xiaoyong Du
Renmin University
Beijing
China

X. Sean Wang
Fudan University
Shanghai
China

Hui Xiong
Rutgers, The State University of New Jersey
New Brunswick, NJ
USA

ISSN 0302-9743   ISSN 1611-3349   (electronic)
Lecture Notes in Computer Science
ISBN 978-3-319-32024-3   ISBN 978-3-319-32025-0   (eBook)
DOI 10.1007/978-3-319-32025-0

Library of Congress Control Number: 2016934671

LNCS Sublibrary: SL3 – Information Systems and Applications, incl. Internet/Web, and HCI

Printed on acid-free paper

This Springer imprint is published by Springer Nature
The registered company is Springer International Publishing AG Switzerland

Preface

Welcome to the proceedings of the 21st International Conference on Database Systems for Advanced Applications (DASFAA)! The DASFAA conference is held in varying locations throughout the world, and the 2016 DASFAA conference was held in Dallas, Texas, USA. DASFAA is an annual international database conference, which showcases state-of-the-art R&D activities in database systems and their applications. It provides a forum for technical presentations and discussions among database researchers, developers, and users from academia, business, and industry.

The DASFAA conference is truly an international forum. During its 21-year history, the conference has been held in more than 12 countries around the world. This year's conference continued this global trend: Our Organizing and Program Committee (PC) members represent 10 countries, and authors submitted papers from 24 different countries/regions.

This year's conference was competitive. A total of 183 papers were submitted for review. Each paper was reviewed by at least three PC members (except for a few reviewed by two PC members) and the selection was made on the basis of discussion among the reviewers and the program co-chairs. This year, 61 papers were accepted for presentation, representing an acceptance rate of about 33 %. In keeping with the goal of advancing the state of the art in databases, paper topics span numerous active and emerging topic areas including big data, crowdsourcing, Web applications, cloud data management, data archive and digital library, data mining, data model and query language, data quality and credibility, data semantics and data integration, data streams and time-series data, data warehouse and OLAP, databases for emerging hardware, database usability and HCI, graph data management, index and storage systems, information extraction and summarization, multimedia databases, parallel, distributed and P2P systems, probabilistic and uncertain data, query processing and optimization, real-time data management, recommendation systems, search and information retrieval, security and privacy, Semantic Web and knowledge management, sensor data management, social network analytics, statistical and scientific databases, temporal and spatial databases, transaction management, Web information systems, and XML and semi-structured data.

Reviewing and selecting papers from such a large set of research groups required the coordinated effort of many individuals. We want to thank all 83 members of the Program Committee and 90 external reviewers, who provided insightful feedback to the authors and helped with this selection process. In addition to the technical presentations, our program also included two invited speakers: Aidong Zhang, Jian Pei and 10-year best paper "Probabilistic Similarity Join on Uncertain Data" which appeared in DASFAA 2006 (written by Hans-Peter Kriegel, Peter Kunath, Martin Pfeifle, Matthias Renz). In addition, a set of four workshops completed the program.

Organizing the DASFAA 2016 program required the time and expertise of numerous contributors. We are grateful for the tremendous help of Hong Gao, Jinho Kim, and Yasushi Sakurai, who organized the workshops, Shaojie Tang, who was this year's publication chair, Ming Wang and Jun Liang, who organized the indusrial/practitioners track, Latifur Khan, Lidong Wu, and Dingzhu Du, who served as local organization co-chairs, Sang Won Lee, Jin Soung Yoo, and Jiaofei Zhong, who served as publicity co-chairs, Jing Yuan, who served as the registration chair, and Lei Cui and Jing Yuan, who served as webmasters. In addition, the guidance of the Steering Committee liaison, Xiaofang Zhou, was invaluable throughout each step of the conference organization and we thank him for his tireless efforts as well.

Finally, we thank the DASFAA community for their support of this international conference. We hope you enjoy the DASFAA conference and that you are inspired by the ideas found in these papers.

January 2016

Shamkant B. Navathe
Weili Wu
Shashi Shekhar
Xiaoyong Du
X. Sean Wang
Hui Xiong

Organization

Organizing Committee

General Co-chairs

Shamkant B. Navathe   Georgia Tech, USA
Weili Wu   University of Texas at Dallas, USA
Shashi Shekhar   University of Minnesota, USA

Program Committee Co-chairs

Xiaoyong Du   Renmin University, China
X. Sean Wang   Fudan University, China
Hui Xiong   Rutgers, The State University of New Jersey, USA

Workshop Co-chairs

Hong Gao   Harbin Institute of Technology, China
Jinho Kim   Kangwon National University, South Korea
Yasushi Sakurai   Kumamoto University, Japan

Industrial/Practitioners Track Co-chairs

Ming Wang   Google, USA
Jun Liang   SAP, USA

Publication Co-chair

Shaojie Tang   University of Texas at Dallas, USA

Local Organization Co-chairs

Latifur Khan   University of Texas at Dallas, USA
Lidong Wu   University of Texas at Tyler, USA
Dingzhu Du   University of Texas at Dallas, USA

Publicity Co-chairs

Sang Won Lee	Sungkyunkwan University, Korea
Jin Soung Yoo	Indiana University - Purdue University Fort Wayne, USA
Jiaofei Zhong	California State University, USA

Registration Chair

Jing Yuan	University of Texas at Dallas, USA

Webmasters

Lei Cui	University of Texas at Dallas, USA
Jing Yuan	University of Texas at Dallas, USA

Steering Committee Liaison

Xiaofang Zhou	University of Queensland, Australia

Program Committee

Toshiyuki Amagasa	University of Tsukuba, Japan
Spiridon Bakiras	John Jay College, CUNY, USA
Zhifeng Bao	RMIT University, Australia
Boualem Benatallah	University of New South Wales, Australia
Zhipeng Cai	Georgia State University, USA
K. Selcuk Candan	Arizona State University, USA
Jianneng Cao	Institute for Infocomm Research, A* Singapore
Varun Chandola	State University of New York at Buffalo, USA
Lei Chen	Hong Kong University of Science and Technology, SAR China
Yi Chen	New Jersey Institute of Technology, USA
James Cheng	Chinese University of Hong Kong, SAR China
Reynold Cheng	The University of Hong Kong, SAR China
Bin Cui	Peking University, China
Ugur Demiryurek	University of Southern California, USA
Prasad Deshpande	IBM Research, India
Bowen Du	Beihang University, China
Hong Gao	Harbin Institute of Technology, China
Yunjun Gao	College of Computer Science, Zhejiang University, China
Yong Ge	The University of North Carolina at Charlotte, USA
Vikram Goyal	IIIT-Delhi, India
Le Gruenwald	The University of Oklahoma, USA
Ralf Hartmut Güting	Fernuniversität Hagen, Germany
Takahiro Hara	Graduate School of Information Science and Technology, Osaka University, Japan

Hongzhi Wang	Harbin Institute of Technology, China
Jianmin Wang	Tsinghua University, China
Jie Wang	Indiana University, USA
Li Wang	Taiyuan University of Technology, China
Peng Wang	Fudan University, China
Wei Wang	University of New South Wales, Australia
Jia Wu	University of Technology, Sydney, Australia
Keli Xiao	Stony Brook University, USA
Yanghua Xiao	Fudan University, China
Xike Xie	Aalborg University, Denmark
Jianliang Xu	Hong Kong Baptist University, Kowloon Tong, SAR China
Jeffery Xu Yu	Chinese University of Hong Kong, SAR China
Xiaochun Yang	University of California, Irvine, USA
Jian Yin	Sun Yat-sen University, China
Ge Yu	Northeastern University, China
Dayu Yuan	Google, USA
Zhongnan Zhang	Xiamen University, China
Fay Zhong	CSUEB, USA
Aoying Zhou	East China Normal University, China
Yuanchun Zhou	Computer Network Information Center, Chinese Academy of Sciences, China
Yuqing Zhu	California State University Los Angeles, USA
Roger Zimmermann	National University of Singapore
Lei Zou	Peking University, China

External Reviewers

Chunyu Ai	Shumo Chu	Zhipeng Huang
Ibrahim Almubark	Ananya Dass	Hui-Ju Hung
Daichi Amagata	Atreyee Dey	Atsushi Keyaki
Moshe Chai Barukh	Aggeliki Dimitriou	Huayu Li
Favyen Bastani	Zhaoan Dong	Jinfeng Li
Seyed-Mehdi-Reza	Yixiang Fang	Mingda Li
Beheshti	Xiaoyi Fu	Yafei Li
Hongzhi Chen	Chuancong Gao	Xinsheng Li
Jinchuan Chen	Li Gao	Yusan Lin
Lei Chen	Yanjun Gao	Huan Liu
Lu Chen	Yash Garg	Sicong Liu
Xilun Chen	Kazuo Goda	Yanchi Liu
Yueguo Chen	Xiaotian Hao	Zhi Liu
Ji Cheng	Juhua Hu	Wei Lu
Wenliang Chen	Shengyu Huang	Siqiang Luo
Lingyang Chu	Xiangdong Huang	Xiangbo Mao

Contents – Part I

Recommendation

Semantics Computing and Knowledge Base

Textual Data

Contents – Part II

Advanced Applications(1)

Crowdsourcing

Cost Minimization and Social Fairness for Spatial Crowdsourcing Tasks

Qing Liu[1]([✉]), Talel Abdessalem[2], Huayu Wu[3], Zihong Yuan[3], and Stéphane Bressan[1]

[1] School of Computing, National University of Singapore, Singapore, Singapore
liuqing@u.nus.edu, steph@nus.edu.sg
[2] LTCI/IPAL CNRS, Télécom ParisTech, Université Paris-Saclay, Paris, France
talel.abdessalem@telecom-paristech.fr
[3] Institute for Infocomm Research, A*STAR, Singapore, Singapore
{huwu,yuanzh}@i2r.a-star.edu.sg

Abstract. Spatial crowdsourcing is an activity consisting in outsourcing spatial tasks to a community of online, yet on-ground and mobile, workers. A spatial task is characterized by the requirement that workers must move from their current location to a specified location to accomplish the task. We study the assignment of spatial tasks to workers. A sequence of sets of spatial tasks is assigned to workers as they arrive. We want to minimize the cost incurred by the movement of the workers to perform the tasks. In the meanwhile, we are seeking solutions that are socially fair. We discuss the competitiveness in terms of competitive ratio and social fairness of the Work Function Algorithm, the Greedy Algorithm, and the Randomized versions of the Greedy Algorithm to solve this problem. These online algorithms are memory-less and are either inefficient or unfair. In this paper, we devise two Distribution Aware Algorithms that utilize the distribution information of the tasks and that assign tasks to workers on the basis of the learned distribution. With realistic and synthetic datasets, we empirically and comparatively evaluate the performance of the three baseline and two Distribution Aware Algorithms.

Keywords: Spatial crowdsourcing · Task assignment · Cost · Social fairness

1 Introduction

TaskRabbit[1] is one of a growing number of new crowdsourcing platforms where users can outsource various tasks to crowd workers in the physical world. Some of these task require, indeed, that a worker moves to a specified location.

This activity is referred to as spatial crowdsourcing. It consists in outsourcing spatial tasks to a community of online, yet on-ground and mobile, workers. A spatial task is characterized by the requirement that workers must move from their current location to a specified location to accomplish the task.

---

[1] www.taskrabbit.com.

© Springer International Publishing Switzerland 2016
S.B. Navathe et al. (Eds.): DASFAA 2016, Part I, LNCS 9642, pp. 3–17, 2016.
DOI: 10.1007/978-3-319-32025-0_1

For instance, in the aftermath of natural disasters (e.g., earthquakes, virus outbreaks), new platforms such as Ushahidi[2] and in STEDD[3] help request and orchestrate actions of volunteering members of the public and independent relief forces to gather information or provide assistance. TaskRabbits wait-for-delivery service[4] finds someone to wait for a delivery and sign for a package. The assigned worker moves to the mailing address to receive and sign for the package and replace the consignee. The service alleviates the costly delivery failure problem.

Workers of the spatial crowdsourcing platform complained that the time spent on commuting between different locations to perform different tasks is non-negligible and is unpaid [1]. Therefore, in this paper, we study the assignment of spatial tasks to workers with the objective of minimizing the "commuting cost" which can refer to time and transportation fee. The spatial task is known by its location and assignment time interval and needs to be assigned to workers as it arrives. Workers have an initial position and move to the locations of the tasks to which they have been assigned. We want to find an assignment policy that minimizes the commuting cost of the workers to reach the assigned tasks. Cost refers to the commuting cost in this paper. In addition to cost optimization, we are seeking solutions that are socially fair. Namely, we are looking for solutions that minimize the variance of the workload among workers. We want to assign a similar number of tasks with a similar total cost to each worker. No worker should, preferably, be overloaded or starved.

In many applications, the spatial tasks are not randomly distributed. The distribution follows the density distribution of requesters or resources, and it may evolve over time. For example, we observed that the delivery failures occur densely in residential areas during office hours, and sparsely in those areas in the evening when most people are back home. By considering the spatial temporal distribution of tasks that can be learned from the history, we can design better strategies for the assignment.

In this paper, we study the opportunity to leverage the clustering of tasks (e.g. calls for wait-for-delivery are clustered in residential areas; calls for assistance are clustered in the disaster area). While the actual mechanisms to learn the distribution of tasks and clusters of tasks are orthogonal to the main issue discussed here, we devise algorithms that assign tasks to workers on the basis of the distribution of tasks and its evolution. Our hypothesis is that knowledge of the distributions of tasks and of its evolution not only help minimize the cost but also provide a basis for the fair assignment of tasks and cost among the workers. We evaluate the efficiency, in terms of cost minimization, and effectiveness in terms of social fairness of our proposed algorithms with several datasets. We use a realistic dataset where the schedule of spatial tasks is constructed from the log of a real parcel delivery service for an application-level evaluation. We create three synthetic datasets for the micro evaluation of the various behaviours of the algorithms. We empirically and comparatively evaluate the cost for each

---

[2] www.ushahidi.com.

[3] instedd.org.

[4] www.taskrabbit.com/m/shopping-delivery/wait-for-delivery.

algorithm with each dataset. We compare the algorithms from the point of view of social fairness.

The remainder of this paper is organized as follows. Section 2 synthesizes related works. Section 3 gives a formal definition of the assignment and cost minimization problem. In Sect. 4, we presents the algorithms we propose. Thereafter, Sect. 5 presents the experimental results. Section 6 concludes the paper.

2 Related Work

Several variants of the problem of assigning spatial tasks to mobile workers have been studied under various sets of hypothesis [5, 7, 10–12, 16].

The authors of [11] devise a taxonomy of spatial crowdsourcing. In their taxonomy and in its vocabulary the problem that we consider is referred to as a single task server assignment. We consider that workers that are willing to accomplish the given tasks are identified. Therefore incentives and rewards are orthogonal issues. The crowdsourcing service does not publish the spatial tasks to the workers. Instead, any online worker sends his location to the crowdsourcing server. The crowdsourcing server assign tasks to the workers. In [11], Kazemi and Shahabi consider workers constrained by their location and workload: workers can only be assigned to tasks in their neighbourhood and they can be only be assigned up to a maximum number of tasks. Given a sequence of sets tasks, the authors define and propose approximate solutions to the problem of assigning tasks to workers while maximizing the number of tasks assigned.

In [7], workers can select the tasks they want to accomplish. Given a sequence of sets tasks and workers choices, the authors define and propose approximate solutions to the problem to the problem of assigning tasks to workers while maximizing the number of tasks assigned. In [12], Kazemi et al. stressed the validity of the results provided by the crowd workers. They studied the problem of maximizing the number of assigned tasks while ensuring that the quality of the answers reaches a confidence level. In [5], Chen et al. consider the assignment of moving workers to spatio-temporal tasks. Tasks can only be accomplished within a specified area and time interval. The authors consider moving workers who have known position, direction and speed, as well as confidence scores quantifying how reliably workers can accomplish assigned tasks. Diversity is obtained by assigning workers coming from different locations. The authors propose approximate solutions to the problem of assigning a set of workers to a set of tasks while maximizing reliability and diversity. The authors of [10, 16] propose a differential privacy model for protecting location privacy of workers participating in spatial crowdsourcing tasks, which is not the concern in our problem setting.

Unlike previous works, our proposal stresses both the cost incurred by the movement of worker and the fairness of the assignment among the workers. Our objective is to minimize the total cost after completing all the tasks and the variance of the number of task and total cost per worker.

3   Problem Formulation

3.1   Preliminaries

We consider finite metric two dimensional space $\mathcal{M}$, which contains finite number of locations, on which tasks are requested. For simplicity, we will name *spatial task* as *task* in the following of this paper. A task r_j is represented as a tuple $< r_j, l_j, b_j, e_j >$ where l_j is the location where r_j is requested, and $[b_j, e_j]$ is its assignment time interval. For simplicity, we assume that there is no difference between the tasks in terms of execution time (i.e., they all need the same amount of time for their execution). For a given time step t, we denote by $R^t = \{r_1^t, r_2^t, ..., r_{n^t}^t\}$ the set of spatial tasks that can be assigned at that step ($b_j < t < e_j$), n^t denotes the total number of tasks in time step t, and n the overall number of tasks.

A worker w_i is represented as a tuple $< w_i, l(w_i), v >$ where $l(w_i)$ is the location of the worker, and v its availability. Let $W = \{w_1, w_2, ..., w_m\}$ be a set of workers who are committed to perform tasks, m be the total number of workers.

The *assignment instance* is the fact that a batch of tasks $R^t = \{r_1^t, r_2^t, ..., r_{n^t}^t\}$ is assigned to a set of workers by the platform at time step t.

Worker w_i is required to move to the locations of the tasks assigned to her (i.e., $l(w_i)^{t+1} = l_j^t$). Considering geographical constraints, a worker cannot be assigned two different spatial tasks located at different locations in the same assignment instance. A batch of n^t tasks at time step t has to be assigned to n^t different workers. In our model, each task is assigned to exactly one worker since every copy of the task can be regarded as an independent task.

We assume that the platform does not know when and where the next batch of tasks is going to arrive. Thus, it has to make instant decisions to assign this batch of tasks before the coming of the next ones. However, the platform could have some knowledge about the distribution of the whole sequence of tasks, and this distribution (i.e., $p^t : \mathcal{M} \to [0, 1]$) could be estimated by observing the data logs (history of the assignments).

Definition 1 *(Cluster). We define a cluster as a set of locations in a circle region of the space $\mathcal{M}$, where tasks are frequently requested. Each cluster (i.e., CL_i) is characterized by a weight (i.e., α_i), a center (i.e., μ_i) and the radius of the circle (i.e., $radius_i$).*

Tasks are distributed in forms of clusters, which is often the case in real world applications. The weight of a cluster is proportional to how many tasks were observed inside the cluster.

Moreover, distributions are different for different time period according to some time slots (morning, lunch time, afternoon, etc.) or to the week days, i.e., $p^{t+1} \neq p^t$ for some t.

3.2   The Assignment Cost Minimization Problem

Our spatial task assignment problem consists in moving around the m workers to perform the tasks that arise over time at some locations of the metric space $\mathcal{M}$. Tasks arrived at time step t must be treated at this step, and the assigned workers have to move to the corresponding locations to perform their tasks.

Let $c(l_1, l_2)$ denote the cost of moving from location l_1 to l_2. At each time step $t \in \mathbb{Z}^+$, n^t tasks $\{r_1^t, r_2^t, ..., r_{n^t}^t\}$ are requested at some locations $\{l_1^t, l_2^t, ..., l_{n^t}^t\}$, and each task is assigned to exactly one of the m registered workers. Let $a^t = (a_1^t, a_2^t, ..., a_{n^t}^t)$ be the assignment at time step t, each a_j^t represents the worker assigned to task r_j^t. Then, the cost at time step t is $\sum_{j=1}^{n^t} c(l(a_j^t), l_j^t)$. The unassigned workers stay their locations unchanged, i.e. $l(w_i)^{t+1} = l(w_i)^t$ for all $w_i \notin a^t$.

The goal is to determine the assignment strategy a^t at each time step t such that the cost is minimized after the completion of all the tasks, i.e. to minimize

$$\sum_{t=1}^{\tau} \sum_{j=1}^{n^t} c(l(a_j^t)^t, l_j^t) \tag{1}$$

where τ is the total number of time steps.

4   Algorithms

Our online optimization problem is a Metrical Task System [4] and is a slight variation of the k-server problem [6]. While the k-server problem considers the assignment of single workers to a schedule of individual tasks, we consider that tasks arrive set-at-a-time. Several classic algorithms [2,3,9,13,14] exist for the k-server problem that can be adapted. Noticeably we should mention the Greedy Algorithm and the Work Function Algorithm. The Greedy Algorithm simply assigns a task to the nearest worker. The Work Function Algorithm assigns tasks to workers according to the work function [13]. The work function balances between the distance of a set of worker to the set of tasks and the hypothetic distance of the workers to the tasks if an optimal algorithm had been used. The Work Function Algorithm has competitive ratio $2 \cdot \binom{|\mathcal{M}|}{m} - 1$, where $|\mathcal{M}|$ is the number of locations in the finite metric space $\mathcal{M}$. Time complexity of the Work Function Algorithm is at least $O(\binom{m}{n^t} mn^2)$, which makes it perform rather poorly in practice. The Greedy Algorithm has an unbounded competitive ratio, but it performs very well in practice with low costs. Randomized versions of the Greedy Algorithm [14], with their statistical bounds $O(m2^m)$ on the competitive ratio [9], do not yield equally good results on practical problems. Time complexity of both the Greedy and the Randomized versions of the Greedy Algorithm is $O(nm)$.

The typical situation that causes the unbounded competitive ratio for the Greedy Algorithm in the k-server problem is that where there are two workers and two tasks on a one dimensional space, a line. The workers are on the left and the tasks on the right. The Greedy Algorithm sends the nearest worker to the first task arriving, then it sends the same worker to the second task arriving. If the schedule of task alternates between tasks at the same two positions, the first worker moves between the two tasks frantically while the second one remain idle. This is not only a situation where the Greedy Algorithm is inefficient, but also a situation where Greedy is socially unfair. A simple and natural fix is to randomize the Greedy Algorithm. The Randomized Greedy Algorithm that we consider chooses the worker to whom a task is assigned with a probability inversely proportional to the cost of the worker to the task. In the above example, randomization could eventually give a chance to the second worker to be assigned a task thus definitely unlocking the initial deadlock.

All these online algorithms are memory-less. Another design direction is, as usual with online algorithms, to add memory to the algorithm. Here we propose to devise algorithms that learn and use their knowledge of the distribution of tasks. For the sake of simplicity and because of space limitations, we do not discuss the orthogonal issue of incrementally learning the distributions. The reader can refer to the extensive literature on clustering algorithm and their incremental versions [8,17]. We consider that we are able to obtain a fairly good estimation of the distribution of tasks. Namely, we know the number of circular clusters of tasks, the coordinates of the centers, the radiuses, and the weights. This knowledge is updated as new tasks arrive. How the knowledge is updated depends on the incremental clustering or learning algorithm.

The idea that we propose in order to leverage the knowledge of the distribution is to assign workers to clusters of tasks according to the learned distribution. For each cluster, a quota of workers is calculated based on the distribution. In a first algorithm, we proactively distribute the workers to stand on the circles of the clusters. The workers are therefore on standby, waiting to be assigned a task. When a change in the distribution is observed, the workers are redeployed accordingly. However, it may not be necessary to proactively deploy the workers. In a second algorithm, we assign tasks inside a cluster to workers located outside all clusters on demand until the quota of the cluster is reached. These two algorithms can also mitigate the situation where some workers are overloaded while some others are starving, thus improve social fairness.

We present two Distribution Aware Algorithms, namely, the Proactive Distribution Aware Algorithm and the on Demand Distribution Aware Algorithm in Sect. 4.1.

4.1 The Distribution Aware Algorithm

The idea of the Distribution Aware Algorithm is to move a sufficient number of workers who are located outside the clusters to perform tasks requested inside the clusters (we call this movement DEPLOY), such that the tasks that are frequently requested inside the clusters can be served by a closer worker.

The algorithm consists of two phases: (i). Estimate this "sufficient number" (i.e., quota: quo_i) of every cluster CL_i. (ii). Real assignment based on the quota. Algorithm 1 gives the outline of the Distribution Aware Algorithm.

Algorithm 1. The Distribution Aware Algorithm

Input: Set of workers W, a sequence of set of tasks $R^1, ..., R^\tau$, distributions
        for each time period $p^1, p^2, ...$
Output: assignment $a^1, ..., a^\tau$

1 **for** *each time period* $tp = 1, 2, ...$ **do**
2   /* Acquire distribution information of tasks */
3   CL_i: $(\alpha_i, \mu_i, radius_i)$ $(i = 1, ..., k) \leftarrow p^{tp}$
4   /* Phase (i). Estimate the quota */
5   **for** *each* CL_i *(i = 1, ..., k)* **do**
6   calculate quo_i for each cluster CL_i;
7   /* Phase (ii). Assignment */
8   Subroutine 2 (or Subroutine 3)

Phase (i), moving a worker located outside the clusters to perform the tasks inside the clusters incurs a heavier cost than moving a nearest worker, thus quo_i should not be too large. We set a upper bound of quo_i to be the minimum of a value proportional to the weight of CL_i and the number of locations in CL_i, minusing the existing number of workers (i.e., $exist_i$) inside CL_i. That is,

$$quo_i \leq \min\{\alpha_i \cdot m, \ |CL_i|\} - exist_i \tag{2}$$

$|X|$ is the cardinality of the set X (e.g., $|CL_i|$ is the number of locations inside CL_i).

On the other hand, if quo_i is insufficiently large, DEPLOY will not be effective, either. Here is the **benefit analysis** for the DEPLOY. The benefit of moving quo_i workers to CL_i is the cost saved by these quo_i workers within CL_i minus the cost of the DEPLOY. We want to maximize the following benefit which is the summed benefit of all clusters.

$$\sum_{Dep_1, ..., Dep_k} saved\ cost\ by\ Dep_i \ - \ deploy\ cost\ of\ Dep_i; \tag{3}$$

$Dep_i \subseteq W$ $(i = 1, ..., k)$ is the set of DEPLOY workers to cluster CL_i, and $Dep_i \cap Dep_j = \emptyset$ if $i \neq j$. We want to determine the optimal set $\{Dep_1, ..., Dep_k\}$ that maximize Eq. 3. However, the "saved cost" and the "deploy cost" is unknown since we do not know the subsequent tasks. What is more, it is inefficient to traverse all possibilities of $\{Dep_1, ..., Dep_k\}$. Thus, we calculate the benefit of the DEPLOY of each cluster using the following heuristic and estimation.

We sort the clusters according to their weights α_i in descending order, and estimate the benefit for each cluster as in Eq. 4.

$$BNF_i = \frac{|Dep_i|}{|CL_i|} \cdot radius_i \cdot \alpha_i N^{tp} - \sum_{j=1}^{|Dep_i|} c(NN_j, \mu_i) \tag{4}$$

where $\frac{|Dep_i|}{|CL_i|}$ is the probability that a task in CL_i is served by the deployed worker. $radius_i$ is the estimated cost between two locations in the cluster.

With N^{tp} being the total number of tasks in this period, we could have the left part of BNF_i being the total estimated "saved cost by Dep_i". The right part of BNF_i is the estimated "deploy cost of Dep_i", i.e., the cost of moving the nearest $|Dep_i|$ workers (i.e., NN_j in Eq. 4) located outside all clusters to the center of CL_i (i.e., μ_i). We set quo_i to be the value of $|Dep_i|$ that maximizes BNF_i under the constraint in Eq. 2.

Subroutine 2. The Proactive Distribution Aware Algorithm

  Input: Set of workers W, a sequence of set of tasks $R^1, ..., R^\tau$
  Output: assignment $a^1, ..., a^\tau$
1 Sort the clusters in descending order according to the weight;
2 **for** *each cluster CL_i $(i = 1, ..., k)$* **do**
3   Find the nearest quo_i workers to μ_i located outside all clusters;
4   Move each of the quo_i workers to the nearest point on the circle of CL_i.
5 **for** *each time step t* **do**
6     $R^t \leftarrow \{r_1^t, ..., r_{n^t}^t\}$; $W = \{w_1, w_2, ..., w_m\}$;
7   **while** $R^t \neq \emptyset$ **do**
8         $r_j^t \leftarrow$ Randomly pick a task from R^t;
9         $a_j^t \leftarrow$ The nearest worker $\in W$ to r_j^t;
10        $R^t \leftarrow R^t \backslash \{r_j^t\}$; $W \leftarrow W \backslash \{a_j^t\}$
11   Output a^t;

Phase (ii). Assignment.

The Proactive Distribution Aware Algorithm (Subroutine 2). This algorithm firstly sorts the clusters in descending order according to the weight α_i (Line 1). Then, for each cluster CL_i, this algorithm proactively moves the nearest quo_i workers (nearest to the center μ_i) who are located outside all clusters to the nearest point on the circle of CL_i (Line 3,4). Workers are therefore on standby, waiting to be assigned a task. Finally, assign the tasks at one step in a random order, move the nearest available worker to perform each task (Line 8–10).

The on Demand Distribution Aware Algorithm (Subroutine 3). This algorithm deploys the workers into the clusters on demand. It assigns the tasks at one step in a random order. For each task r_j^t at time step t, if it is located inside cluster CL_j^t (that has the quota quo_j^t). If $quo_j^t > 0$, move the nearest worker who is located outside all clusters to serve r_j^t (Line 6). In other cases, assign r_j^t to the nearest worker (Line 9).

In phase (i), calculating the quota for each cluster requires sorting workers with respect to cost, which takes $O(mlog(m))$. Time complexity of phase (i) is $O(kmlog(m))$ for k clusters. In phase (ii), the assignment takes $O(nm)$. Thus, Time complexity of the Distribution Aware Algorithms is $O(kmlog(m) + nm)$.

Go back to the example of two workers on a line, the Distribution Aware Algorithms will send the two workers to the two locations of the tasks based on the distribution. So, the cost will be zero afterwards. We do not have theoretical bounds for the Distribution Aware Algorithms.

Subroutine 3. The on Demand Distribution Aware Algorithm

Input: Set of workers W, a sequence of set of tasks $R^1, ..., R^\tau$
Output: assignment $a^1, ..., a^\tau$

1 **for** *each time step* t **do**
2      $R^t \leftarrow \{r_1^t, ..., r_{n^t}^t\}$; $W = \{w_1, w_2, ..., w_m\}$;
3   **while** $R^t \neq \emptyset$ **do**
4          $r_j^t \leftarrow$ Randomly pick a task from R^t;
5   **if** $r_j^t \in CL_j^t$ *AND* $quo_j^t > 0$ **then**
6              $a_j^t \leftarrow$ The nearest worker $\in W$ to r_j^t who is not in any clusters;
7              $quo_j^t \leftarrow quo_j^t - 1$;
8   **else**
9              $a_j^t \leftarrow$ The nearest worker $\in W$ to r_j^t;
10         $R^t \leftarrow R^t \backslash \{r_j^t\}$; $W \leftarrow W \backslash \{a_j^t\}$
11   Output a^t;

---

5   Performance Evaluation

5.1   Experimental Methodology

We conduct experiments on both real world and synthetic datasets to evaluate the performance of the three baseline online algorithms, i.e., the Greedy Algorithm, the Randomized Greedy Algorithm and the Work Function Algorithm, and two Distribution Aware Algorithms, i.e., the Proactive Distribution Aware Algorithm and the on Demand Distribution Aware Algorithm. For simplicity, we will name these algorithms as Greedy, Random, WFA, Proactive and onDemand, respectively, in this section.

The real world dataset "Delivery" consists of records of the delivery failure packages in Singapore from April 2014 to June 2014. The distribution of those failed deliveries is presented in Fig. 1(a). The color, from white to red, represents the density of the failed deliveries on the corresponding coordinates. We identify 12 clusters according to this observed distribution. The centers of the clusters are denoted by black stars in Fig. 1(a). We set the radius of each cluster to be 2 Km and the weight of each cluster to be the density of tasks inside the cluster between April 2014 and May 2014.

We evaluate the five algorithms on the dataset of June based on the learned distributions and clusters. The number of tasks is 19 k (i.e., $n = 19$ k). Tasks are all in one time period. The workers are generated uniformly on the space. We vary the number of workers (i.e., m) and also the number of tasks in one time step in order to observe the behaviors of the algorithms. Euclidean distance (i.e., traveled distance) is used to measure the commuting cost between two locations for all datasets. We use the window version of WFA [15] (i.e., ω-WFA) in the experiments considering the running time. We evaluate the 50-WFA on the Delivery data where there is 1 task per step and $m = \{50, 100, 200\}$. In other evaluations, we only compare the performances of Greedy, Random, Proactive and onDemand.

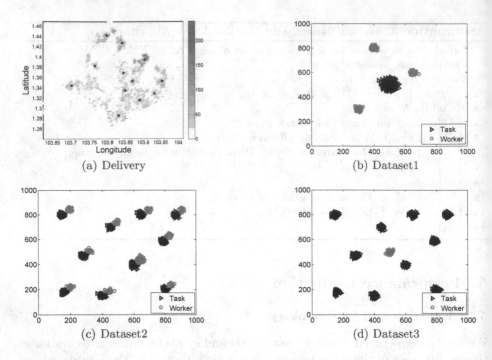

(a) Delivery   (b) Dataset1

(c) Dataset2   (d) Dataset3

Fig. 1. Distributions of the real and synthetic datasets (Color figure online)

In order to evaluate the scalability and watch the behaviors of the algorithms, we generate 3 synthetic datasets named "Dataset1" (Fig. 1(b)), "Dataset2" (Fig. 1(c)) and "Dataset3" (Fig. 1(d)) from Gaussian Mixture Distribution (i.e., GMM) on a 1000 × 1000 grid space. Figure 1(b), (c) and (d) show the distributions of the tasks and the workers in the three synthetic datasets where the red circles represent the distribution of the workers and the black triangles represent the distribution of the tasks. Specifically, we use diagonal covariance matrixes for the multivariate Gaussian Distributions, and the variances for the two dimensions are set to be equal. In this way, each component in the GMM is corresponding to a cluster with the center locating at the mean, the radius equaling to the standard deviation (σ) of the x-dimension (or y-dimension). In Dataset1, 100 k tasks are generated from a 1-component-GMM. All the tasks are in one time period. Workers are generated from a 3-component-GMM, the centers of which are uniformly distributed on the space. In Dataset2, there are 2 time periods. In the first time period, 50 k tasks are generated from a 10-component-GMM, the means of which are uniformly distributed on the space. One of the components has weight 0.5, the others have weights 0.056. In the second time period, 10 components with same means as in the first period evolve all their weights to 0.1. Workers are generated from a 10-component-GMM that has different means but the same weights with the clusters of tasks in the first period. In Dataset3, there are 10 time periods. In each time period, 20 k tasks are generated from a 1-component-GMM. The center of this component changes at every time period. Workers are generated from a 1-component-GMM.

Table 1. Summary of the synthetic datasets

	n	Variance	# periods	# clusters/period	Rationale
Dataset1	100 k	400	1	1	Worst case of greedy
Dataset2	100 k	100	2	10	Evolution of weights
Dataset3	100 k	100	10	1	Moving clusters

The summary of the datasets is in Table 1. We evaluate the performances of the algorithms by varying the number of workers and the number of tasks in one step. Each evaluation is averaged over 20 independent random cases.

5.2 Experimental Results on Cost Minimization

Experimental Results on Real Data. Figure 2 shows the experimental results on the Delivery dataset. Generally, as the number of worker m increases, the costs by all five algorithms decrease since there are more opportunities to assign the tasks to close workers when m is larger. Random performs worse than the other algorithms due to its randomness. The Distribution Aware Algorithms, Proactive and onDemand, perform better than the baseline memory-less algorithms, onDemand is better than Proactive.

Comparing Fig. 2(a), (b) and (c), we can see that all algorithms perform better when the number of tasks in one step is smaller. This is because the available worker set is larger for each task when there are fewer competitive tasks in the same time step.

When the number of workers (m) is sufficiently large, all algorithms tend to have similar performance since there are sufficient number of close workers to serve each task. As a result, the costs by all algorithms are low.

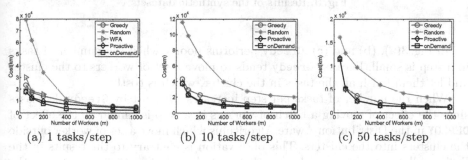

(a) 1 tasks/step   (b) 10 tasks/step   (c) 50 tasks/step

Fig. 2. Results of the delivery data

Experimental Results on Synthetic Data. Figure 3 shows the experimental results on Dateset1, Dataset2 and Dataset3. Generally, the performance of Random is poor, and the Distribution Aware Algorithms, Proactive and onDemand, perform better than the baseline memory-less algorithms, onDemand is better than Proactive.

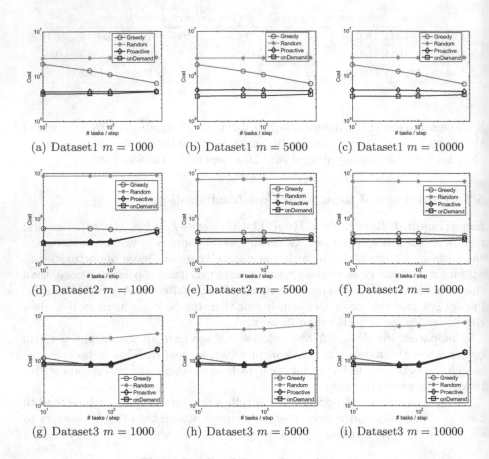

(a) Dataset1 $m = 1000$ (b) Dataset1 $m = 5000$ (c) Dataset1 $m = 10000$

(d) Dataset2 $m = 1000$ (e) Dataset2 $m = 5000$ (f) Dataset2 $m = 10000$

(g) Dataset3 $m = 1000$ (h) Dataset3 $m = 5000$ (i) Dataset3 $m = 10000$

Fig. 3. Results of the synthetic datasets

In Fig. 3(a), (b) and (c), Greedy performs poorly when the number of tasks in a step is small. Because Greedy tends to move a set of workers to the cluster and let them finish all the tasks in the cluster, which is costly.

When the number of tasks per step increases, the cost by Greedy decreases since the effect of moving a batch of workers in a step is similar to the effect of DEPLOY in the Distribution Aware Algorithms which move a set of worker outside the clusters into the clusters. This observation is contrary to the results in the Delivery dataset where we observe that costs by all algorithms increase as the number of tasks per step increases. This is because the benefit of the DEPLOY of

Greedy is counteracted by a heavier cost because of the unavailability of workers in one step, which results in increasing cost in the Delivery dataset.

In Fig. 3(d), (g), (h) and (i), costs by the Distribution Aware Algorithms increase as the number of tasks in a step increases because of the unavailability of workers in one step. In this case, Greedy and the Distribution Aware Algorithms tend to have similar performance.

The advantage of the Distribution Aware Algorithms is that it spends a heavier cost at the beginning of each time period to deploy a number of workers to serve the tasks that are frequently requested inside clusters. This deployment benefits the following sequence of tasks. When the distributions evolves frequently as in Dataset3, the performances of Distribution Aware Algorithms are similar to Greedy, as shown in Fig. 3(g), (h) and (i), since the benefit of deployment is counteracted by the heavier cost at the beginning of every time period.

In conclusion, the Distribution Aware Algorithms have a lower cost of assignment compared to the baseline algorithms when the tasks are distributed in clusters.

5.3   Experimental Results on Social Fairness

We analyse the social fairness of the four algorithms, Greedy, Random, Proactive and onDemand, based on the Delivery data with one task per step and a number of workers varying from 200 to 600. We quantify the workload in two ways: the number of assigned tasks and the traveled distance, for every worker in the four algorithms.

Figure 4 shows the number of assigned tasks per worker by the different algorithms. We can see that Greedy is generally unfair, few workers are assigned a large number of tasks while most of the other workers remain idle.

This is because Greedy always assigns the tasks to the nearest workers. Some workers who are initially located close to a cluster will have to finish all the tasks inside this cluster since they are always the nearest workers. This results in heavy workload of these workers.

The Distribution Aware Algorithms alleviate this problem by deploying an appropriate number of workers into each clusters. Thus, we can see from Fig. 4

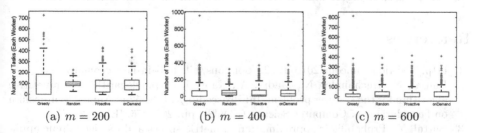

(a) $m = 200$ (b) $m = 400$ (c) $m = 600$

Fig. 4. Assigned number of tasks of every worker

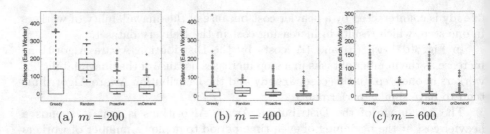

(a) $m = 200$ (b) $m = 400$ (c) $m = 600$

Fig. 5. Traveled distance of every worker

that Proactive and onDemand have smaller gaps between the max-workload and the min-workload than Greedy. Random is the fairest by construction. Indeed, every worker has a chance at every time step to be assigned to perform tasks.

Figure 5 shows the traveled distance per worker under the different algorithms. We can observe similar patterns as in Fig. 4. Few workers traveled long distances under Greedy, while most of the others remain idle. Random is fair, however, the median cost of Random is higher than that of the other three algorithms. The Distribution Aware Algorithms, Proactive and onDemand, are relatively fair compared to Greedy, having a more balanced workload.

Clearly, fairness was built-in the design of our two Distribution Aware algorithms. It is remarkable, however, that this can be done at the benefit of a lower cost as well.

6   Conclusion

In this paper, we study the problem of assigning spatial tasks to crowd workers in the spatial crowdsourcing scenario. We formalize the assignment cost minimization problem, which is to minimize the cost of moving a set of workers to complete a sequence of spatial tasks. Then, we present three baseline and two Distribution Aware algorithms to solve this problem. We analyse the competitiveness of these five algorithms in terms of competitive ratio and social fairness. In the experiments, we compare the cost and social fairness of the five algorithms. The results show that the Distribution Aware Algorithms outperform the three baseline algorithms in terms of cost and yield a balanced workload.

References

1. http://edition.cnn.com/2015/10/22/opinions/hill-jobs-in-new-economy/
2. Bansal, N., Buchbinder, N., Madry, A., Naor, J.: A polylogarithmic-competitive algorithm for the k-server problem. In: Proceedings of the 52nd Annual Symposium on Foundations of Computer Science (FOCS), pp. 267–276. IEEE (2011)
3. Bartal, Y.: Probabilistic approximation of metric spaces and its algorithmic applications. In: Proceedings of the 37th Annual Symposium on Foundations of Computer Science (FOCS), pp. 184–193. IEEE (1996)

4. Borodin, A., Linial, N., Saks, M.E.: An optimal on-line algorithm for metrical task system. J. ACM (JACM) **39**(4), 745–763 (1992)
5. Cheng, P., Lian, X., Chen, Z., Fu, R., Chen, L., Han, J., Zhao, J.: Reliable diversity-based spatial crowdsourcing by moving workers. Proc. VLDB Endow. **8**(10), 1022–1033 (2015)
6. Chrobak, M., Karloff, H., Payne, T., Vishwanathan, S.: New results on server problems. In: Proceedings of the First Annual ACM-SIAM Symposium on Discrete Algorithms, SODA 1990, pp. 291–300. Society for Industrial and Applied Mathematics, Philadelphia (1990)
7. Deng, D., Shahabi, C., Demiryurek, U.: Maximizing the number of worker's self-selected tasks in spatial crowdsourcing. In: Proceedings of the 21st ACM SIGSPATIAL International Conference on Advances in Geographic Information Systems (SIGSPATIAL), pp. 324–333 (2013)
8. Ester, M., Kriegel, H.P., Sander, J., Wimmer, M., Xu, X.: Incremental clustering for mining in a data warehousing environment. In: Proceedings of the VLDB Endowment, pp. 323–333 (1998)
9. Grove, E.F.: The harmonic online k-server algorithm is competitive. In: Proceedings of the 23rd Annual ACM Symposium on Theory of Computing (STOC), pp. 260–266. ACM (1991)
10. Kazemi, L., Shahabi, C.: A privacy-aware framework for participatory sensing. ACM SIGKDD Explor. Newsl. **13**(1), 43–51 (2011)
11. Kazemi, L., Shahabi, C.: Geocrowd: enabling query answering with spatial crowdsourcing. In: Proceedings of the 20th ACM SIGSPATIAL International Conference on Advances in Geographic Information Systems (SIGSPATIAL), pp. 189–198. ACM (2012)
12. Kazemi, L., Shahabi, C., Chen, L.: Geotrucrowd: trustworthy query answering with spatial crowdsourcing. In: Proceedings of the 21st ACM SIGSPATIAL International Conference on Advances in Geographic Information Systems (SIGSPATIAL), pp. 304–313 (2013)
13. Koutsoupias, E., Papadimitriou, C.H.: On the k-server conjecture. J. ACM (JACM) **42**(5), 971–983 (1995)
14. Raghavan, P., Snir, M.: Memory Versus Randomization in On-line Algorithms, vol. 372. Springer, Heidelberg (1989)
15. Rudec, T., Baumgartner, A., Manger, R.: A fast work function algorithm for solving the k-server problem. CEJOR **21**(1), 187–205 (2013)
16. To, H., Ghinita, G., Shahabi, C.: A framework for protecting worker location privacy in spatial crowdsourcing. Proc. VLDB Endow. **7**(10), 919–930 (2014)
17. Xu, R.: Survey of clustering algorithms. IEEE Trans. Neural Netw. **16**(3), 645–678 (2005)

Crowdsourced Query Processing on Microblogs

Weikeng Chen[1(✉)], Zhou Zhao[2], Xinyu Wang[1], and Wilfred Ng[1]

[1] The Hong Kong University of Science and Technology, Hong Kong, China
{wchenad,xwangau,wilfred}@cse.ust.hk
[2] Zhejiang University, Hangzhou, China
zhaozhou@zju.edu.cn

Abstract. Currently, crowdsourced query processing is done on reward-driven platforms such as *Amazon Mechanical Turk (AMT)* and *Crowd-Flower*. However, due to budget constraints for conducting a crowdsourcing task in practice, the scalability is inherently poor. In this paper, we exploit microblogs for supporting crowdsourced query processing. We leverage the social computation power and decentralize the evaluation of the crowdsourcing platforms queries towards social networks. We propose a new problem of minimizing the cost of processing crowdsourced queries on microblogs, given a specified accuracy threshold of users' votes. This problem is NP-hard and its computation is #P-hard. To tackle this problem, we develop a greedy algorithm with a quality guarantee. We demonstrate the performance on real datasets.

1 Introduction

Crowdsourcing techniques [10,15,27–36] have attracted considerable attention due to their effectiveness in many applications such as entity resolution and image detection. An essential property of crowdsourcing is that the technique relies on a human workforce to at least partially complete an evaluation of the queries. Typically, a crowdsourcing application publishes its queries assigning with a fixed reward to the workers on *Amazon Mechanical Turk*[1] and *Crowd-Flower*[2]. Each crowdsourced query is then assigned with a fixed reward to the workers.

However, humans are prone to error and may provide poor quality crowdsourcing results. To address the problem, crowdsourcing applications often enroll a number of workers to process the replicated queries. If the collected results are conflicting, the majority vote is adopted to determine which is correct. However, the replication strategy may not be able to fully handle the diversity of answers. Suppose the tasks involved are hard, we have to enroll more workers to reduce the diversity of answers. The cost of this *Human Intelligence Task* (HIT) could then be very high. Thus, a limitation of the existing crowdsourcing approach is, that we may not have a sufficient number of workers to process the query under the budget constraint, which results in poor answer quality.

---

[1] Amazon Mechanical Turk (or simply AMT) platform at https://www.mturk.com.
[2] CrowdFlower platform at https://www.crowdflower.com.

© Springer International Publishing Switzerland 2016
S.B. Navathe et al. (Eds.): DASFAA 2016, Part I, LNCS 9642, pp. 18–32, 2016.
DOI: 10.1007/978-3-319-32025-0_2

To tackle this problem, we formulate a new problem of processing crowd-sourced queries on microblog, which provides new incentives to encourage microblog users to process crowdsourced queries. By developing a new crowd-sourcing model, we aim to reduce both the diversity of the answers and the cost of processing. We focus on addressing the following main issues:

- **Answer Diversity.** If the number of replicated queries are too few, we may not have enough confidence to infer a reliable answer. On the other hand, if we replicate too many queries, we may have to suffer high cost.
- **Incentive Mechanisms.** Crowdsourcing workers may be reluctant to process any queries until they know they will receive a reward. Thus, we aim to design an incentive mechanism that is able to reduce the workforce cost and at the same time, meet a given specified accuracy threshold.
- **Query Sharing.** We utilize *the word of mouth* effect to model the worker behavior of query sharing on microblogs. Intuitively, microblog users are more willing to answer a crowdsourced query under a social influence and to send messages to an interested group. Thus, crowdsourced queries can be diffused more efficiently and effectively over microblogs.

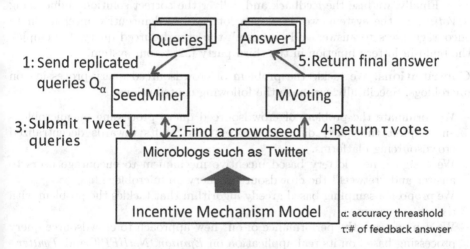

Fig. 1. A microblog-based crowdsourcing system

There are already some works studying the problem of answer diversity [12,18,20]. However, they all rely on centralized platforms such as *AMT* or *CrowdFlower*. In contrast, we study how to address the problem by exploiting the social influence of microblogs.

The general process of crowdsourced query processing on microblogs is given in Fig. 1. For instance take the crowdsourced query Q "Is Paris the City of Lights?" and the specified *accuracy threshold* α. Owing to the conflicting answers, we should have a sufficient number of replicated queries Q such that the accuracy of the final answer obtained from using the *majority voting rules* (i.e. the MVoting module) is greater than α. Let Q_α be the set of such replicated

$1 $1 $1 $1 $1 $1　　　　$3 $2 $1 $0 $0 $0

common crowdsoucing platforms　　Probablistic Incentive Mechanism

Fig. 2. The difference between common crowdsoucing platforms and the Probabilistic Incentive Mechanism Model

queries. The estimation of the number of queries in Q_α is based on the historic degree of the skill of the microblog users (Fig. 2).

In the first step, we take Q_α as the input to *SeedMiner*. The *SeedMiner* estimates the minimum crowdseed size needed to diffuse this query in microblogs such that we can have τ feedback. The *SeedMiner* algorithm is based on the *"word of mouth"* effect [13] of query diffusion on microblogs. In the second step, the *SeedMiner* returns the crowdseed and we issue the query to the crowdseed in the third step. We collect answers incrementally from microblogs in the fourth step. Finally, we fuse the feedback and deliver the correct solution online using *MVoting*. In the system, we develop a lottery based incentive mechanism to encourage users to answer and "retweet" the crowdsourced query. We employ the build-in lottery function in the third party [22] to our system.

Contributions. We tackle the problem of crowdsourced query processing on microblogs. Specifically, we make the following contributions.

- We formulate the problem of crowdsourced query processing on microblogs and exploit the query diffusion process that shifts towards decentralized crowdsourcing platforms.
- We design a new lottery based incentive mechanism to encourage users to answer and "retweet" the crowdsourced query on microblogs.
- We propose a sampling-based greedy algorithm that tackles the problem with a quality guarantee.
- We demonstrate the performance of our new approach to crowdsource query processing based on its real application on *Epinion*, *NetHEPT* and *Twitter*.

This paper is organized as follows. Section 2 introduces the crowdsourcing models. Section 3 formulates the problem of crowdsourced query processing on microblogs and analyze the complexity of the problem. Section 4 then surveys the related work while Sect. 5 presents the details of our algorithm. Section 6 presents the experimental results and we conclude the paper in Sect. 7.

2　Crowdsourcing Model

In this section, we first introduce the popular voting model, which is widely used in a crowdsourcing environment. Then we propose the incentive model and the diffusion model in our approach.

Voting Model. Given a query, the judgement among the crowd may be different. As a result, the human answers for the query posed on microblogs may easily conflict. To resolve this problem, we develop a voting model that consists of a *majority voting rule* given by

$$f(V) = \begin{cases} 1 \text{ if } \sum_{v_i \in V} v_i \geq \frac{\tau+1}{2} \\ 0 \text{ otherwise} \end{cases} \tag{1}$$

where the vote v_i is the answer of user u_i and V represents a collection of τ votes (or equivalently, τ feedback answers). Among the conflicting answers from the votes, we choose one that is supported by more than half of the votes. For ease of presentation, we consider the crowdsourced decision-making query in this paper. The vote v_i can be a binary random variable (i.e. either 0 or 1). It can also easily be extended to queries with K possible answers. In such a scenario, we can also assume the answer with more than half the votes is the correct one.

However, the output of the *majority voting rule* may not be reliable, if the number of votes is too few. On the other hand, the cost is high if we enroll too many votes. We thus propose a probabilistic model to estimate the number of replicated queries that are sufficient to make the *majority voting rule* reliable.

Suppose the accuracy of users' votes on microblogs that have processed the query are $\{a(u_1), \ldots, a(u_\tau)\}$, where $a(u_i)$ is the vote accuracy of user u_i (i.e. the probability of vote v_i being correct). We utilize the historic records of the degree of skill of the users to estimate the voting accuracy. In this paper, we do not focus on the estimation of the degree of skill of the users, as a variety of solutions have already been proposed in the literature [15, 20, 28, 36].

Given a collection of votes V of size τ, we consider the correctness probability of the *majority voting rule* as $p(f(V))$. We denote the random variable s_A consisting of all the accurate votes in V (i.e. $s_A \subseteq V$). The output result of the majority voting is correct only when at least half of the votes are accurate (i.e. $|s_A| \geq \frac{\tau+1}{2}$). Thus, the correctness probability of the *majority voting rule* is given by

$$p(f(V)) = Pr(|s_A| \geq \frac{\tau+1}{2}) = \sum_{k=\frac{\tau+1}{2}}^{\tau} Pr(|s_A| = k)$$

$$= \sum_{k=\frac{\tau+1}{2}}^{\tau} \sum_{s_A \in F_K} \prod_{u_i \in s_A} a(u_i) \prod_{u_j \notin s_A} (1 - a(u_j)) \tag{2}$$

where F_k comprises all possible combinations of users giving accurate votes for s_A with size k. We note that $1 - p(f(V))$ is a cumulative Poisson binomial distribution, since the accurate probability of each vote v_i is different.

We consider the accuracy of the result by the *majority voting rule* as the expectation of correctness probability (i.e. $E[p(f(V))]$). Then, the expected correctness of the *majority voting rule* is given by

$$E[p(f(V))] = \sum_{k=\frac{\tau+1}{2}}^{\tau} \binom{\tau}{k} \mu^k (1-\mu)^{\tau-k} \tag{3}$$

where μ denotes the average degree of skill of the users (i.e. the average accuracy of the votes). By using the Chernoff Bound, the lower bound of the expected correctness is given by

$$\sum_{k=\frac{\tau+1}{2}}^{\tau} \binom{\tau}{k} \mu^k (1-\mu)^{\tau-k} \geq 1 - e^{\frac{-2\tau(\mu-\frac{1}{2})^2}{4\mu}} \tag{4}$$

Then, we let $1 - e^{\frac{-2\tau(\mu-\frac{1}{2})^2}{4\mu}} \geq \alpha$, where α is the specified accuracy and the number of replicated queries is given by

$$\tau \geq \frac{-4\mu \ln(1-\alpha)}{2(\mu-\frac{1}{2})^2} \tag{5}$$

Thus, the next goal is to seek τ users to answer the crowdsourced query. In the next section, we present incentive mechanisms to tackle this problem.

Incentive Mechanism Model. To encourage users to answer an imposed crowdsourced query, the existing centralized crowdsoucing platforms commonly pay each user a fixed amount of money as a reward for completing the task.

However, we find that such fixed reward incentive mechanisms do not work well on microblogs. In our experiment, we firstly distribute several crowdsourced queries with a reward of $ 0.05 to the crowd on microblogs. We found that few users answer these queries. Our observations are as follows: unlike the workers from *AMT*, users on social networks are reluctant to answer any unexpected problem with too little reward (e.g. $ 0.05). On the other hand, microblog users do not primarily aim to earn money but rather gain social interactions.

In this paper, we devise and evaluate a new incentive mechanism for crowd-sourced query processing based on the platform *Sojump* [22]. *Sojump* is a well established Q&A platform which is able to process a reward payment. Similar to other common Q&A systems, we generate problem sheets with a URL under the homepage of *Sojump* and specify the payment conditions. In our model, the payment condition is that a user completes the problem sheet and shares the URL of the sheet in the microblog.

Given a crowdsourced query budget B, we devise a *probabilistic incentive mechanism* in order to encourage the crowd on the microblog to answer and "retweet" the problem sheet. For example, we specify the budget of 6 crowd-sourced queries to $ 6 in this experiment. First, we decompose the total budget into three rewards such as $ 3, $ 2 and $ 1, as illustrated in Fig. 1. Suppose that the number of replicated crowdsourced queries is τ, we set the probability of getting each reward to $\frac{1}{\tau}$.

Crowdseed. The basic idea for processing crowdsourced queries under our incentive mechanism model is to issue the queries to a large number of users on microblogs. However, this approach incurs two problems. First, the manager of the microblog may suspect our application to be a type of spammer and forbid its usage. Second, the users on microblogs may become annoyed when they receive unexpected problem sheets from unknown sources.

To tackle the above problems, we include the concept of *crowdseed* in the process such that, after we issue the problem sheets to the users in the crowd-seed, the problem sheets may be diffused through their microblog relations to τ other users within the probabilistic incentive mechanism model. Our system encourages users to subscribe to crowdseed where the users can set the expected reward to accept the posting of crowdsourced queries. We consider the subscription cost of user u_i as $c(u_i)$. Then, the initialization cost of issuing the crowd-sourced query to the microblog is given by $c(S) = \sum_{u_i \in S} c(u_i)$, where S represents the *crowdseed* (i.e. a set of users who are willing to accept our crowdsourced queries as the seed). The seed cost $c(S)$ depends on the specified accuracy of the crowdsourced query (i.e. α).

Diffusion Model. Under the word of mouth effect [14], users' behaviors are probabilistically influenced by their friends. For example, microblog users may follow their friends' actions by "retweeting" the messages being shared. Due to this observation, we propose a probabilistic model based on the well-known *word of mouth* effect to model the crowdsourced query diffusion.

We denote the event that user u_i successfully diffuses the crowdsourced query to his/her friend u_j as $I(u_i, u_j)$. In practice, we found that microblog users tend to share "tweets" from their close friends. Based on this observation, we explore the query diffusion probability model between two users based on their closeness in microblogs. In this work, we utilize the *Jaccard Distance* of two users' friends to measure their closeness. The closeness of two users u_i and u_j is given by

$$J(u_i, u_j) = \frac{|N(u_i) \bigcap N(u_j)|}{|N(u_i) \bigcup N(u_j)|} \tag{6}$$

where $N(u_i)$ denotes the set of users that u_i follows and $|N(u_i) \bigcap N(u_j)|$ is the number of common following users of u_i and u_j.

We propose to use the popular *Sigmoid* function to explore the relationship between query diffusion and the closeness of users. We denote the probability of query diffusion between two users as $p(I(u_i, u_j))$. Then, the formula of the crowdsourced query diffusion model from user u_i to user u_j is given by

$$p(I(u_i, u_j)) = \frac{1}{1 + e^{aJ(u_i, u_j)+b}} \quad (a < 0, b > 0) \tag{7}$$

where a and b are the parameters of the probabilistic query diffusion model.

To estimate the values of parameters a and b, we conduct user study by posing 700 queries on *Twitter*. We collect the data in the format of

$(u_i, u_j, N(u_i), N(u_j), I)$ where I is the indication of the query diffusion. Next, we aggregate the collected data in the format of (J, p) where J is the *Jaccard Distance* value and p is the diffusion probability. We notice that Eq. 7 implies that $\ln\left(\frac{1}{p} - 1\right) = aJ + b$. Next, we employ the least square method to estimate the parameters a and b, using a transformed set of pairs such as $\{\ln\left(\frac{1}{p} - 1\right), J\}$.

Using Eq. 7, we transform the microblog into a probabilistic graph given by $G = (V, E, P)$, where V consists of all users, E is composed of all relations and P records the pairwise diffusion probability of two users (i.e. $p(I(u_i, u_j))$).

We consider the event in which the crowdsourced query diffuses from the crowdseed S to any user u_k as a graph reachability problem. However, the query diffusion between two users is uncertain. For a probabilistic graph with E edges, we have 2^E possible cases of query diffusion which are denoted as $\{G_1 = (V, E_1), \ldots, G_{2^{|E|}} = (V, E_{2^{|E|}})\}$. Then, the query diffusion from the crowdseed S to any user u_k is given by

$$p(I(u_k)|S) = \sum_{i=1}^{2^{|E|}} p(G_i)R_{G_i}(u_k, S) \qquad (8)$$

where $p(G_i)$ is the probability of the ith query diffusion case. The value of $p(G_i)$ can be computed by the product of the edge probabilities. We denote $R_{G_i}(u_k, S)$ as the indication of the reachability from the crowdseed S to the user u_k in case G_i (i.e. either zero or one). In other words, the query diffusion probability is the sum of the probability of cases that the query from crowdseed S can diffuse to the user u_k. We consider the expected diffusion size of the crowdseed S in the probabilistic graph G as $\delta(G|S)$. Then, the expected size $\delta(G|S)$ is given by

$$\delta(G|S) = \sum_{i=1}^{|V|} p(I(u_i)|S) \qquad (9)$$

where $|V|$ denotes the number of users in G.

3   Problem Statement

We formulate the problem of crowdsourced query processing on a microblog as *Crowdseed Selection* as follows. In a nutshell, we aim to seek a crowdseed set S from the microblogs such that: (1) we could have at least a collection of τ feedback answers, and (2) the cost of the crowdsourced query Q can be reduced as much as possible.

Problem 1 (Crowdseed Selection). Given a probabilistic graph $G(V, E, P)$ of the microblog and a crowdsourced query Q, we aim to find a crowdseed set such that Q can be processed with an expected accuracy larger than α.

However, Theorems 1 and 2 show that the complexity of this problem is NP-hard and even the computation complexity of the expected query diffusion of a crowdseed set S is #P-hard. The detailed proof can be found in Appendix [1].

Theorem 1. *The problem of* Crowdseed Selection *is NP-hard.*

Theorem 2. *The computation complexity of expected query diffusion of a crowdseed set S is #P-hard.*

4   Related Work

In this section, we survey some proposed crowdsourcing systems and introduce some work about processing various kinds of crowdsourcing queries.

Crowdsourcing Systems. Many crowdsourcing database systems like *Qurk* [17], *Deco* [19] and *Hog* [4] have recently been proposed as the plug-in components for traditional database systems. These systems integrate existing crowdsourcing platforms such as *Amazon MTurk* and *CrowdFlower* as an external data source. The crowdsearcher [2], a novel search paradigm, embodies crowds as first-class sources for the information seeking process. The Crowdturfing system [25] aims to study and understand the Crowdturfing campaigns in today's Internet. The work ZenCrowd [8] systematizes and automatizes manual matching techniques by dynamically creating micro matching tasks and by publishing them on a popular crowdsourcing platform. The work Cogos [9] leverages Twitter Lists to find topic experts in Twitter.

Queries with the Crowd. The unreliability of workers is a significant challenge for query processing using the crowdsourcing strategy, thus, different approaches have been proposed to tackle the problem of conflicting answers. The work in [6, 12,24] resolves the ordinal query problem of conflicting rankings. The work in [3,16,18] studies the screen query problem of conflicting answers. The work in [23] studies the crowdsourced enumeration query. The work in [21] studies the query-driven schema expansion. The work in [26] studies the crowdsourced join query to find all pairs of matching objects from two collections. The work in [7] studies the top-k and group-by queries with the crowd. The work in [13] studies how to choose the right question to answer the planning query with the crowd.

Some recent works [3,15] have studied the crowd selection problem on microblog and built a probabilistic voting model to estimate the reliability of the enrolled crowd. The dynamic programming based and greedy based algorithms are proposed to select the crowd. The selection of the crowd is only based on the reliability of the workers.

However, none of the above-mentioned works utilize the power of social influence for processing crowdsourced queries on microblogs. Our work shows that Twitter users send crowdsourced queries to their friends and that people also answer the Twitter queries based on their friendships. By taking the social influence into consideration, we further study how to mine crowdseed for crowdsourced query processing on microblogs.

5   Algorithms

In this section, we show how to tackle the complexity of the problem of *Crowdseed Selection*. We first define an objective function based on the selected crowdseed set S and then propose a greedy algorithm in order to maximize the function.

We aim to select a crowdseed set S such that τ feedback answers are obtained from the users. Thus, we set the constraint of the crowdseed set S at $\delta(G|S) \geq \tau$ and formulate the objective function as: $\min_S \sum_{u_i \in S} c(u_i)$ such that $\delta(G|S) \geq \tau$, where τ is the expected query diffusion size. The value of τ can be computed by Eq. 5.

We now present a greedy algorithm that iteratively selects the subscribed users to the crowdseed set S in order to satisfy the constraint. We denote the ratio of the expected query diffusion (i.e. $\delta(G|S)$) to the cost of S (i.e. $c(S)$) as Δ_S (i.e. $\Delta_S = \frac{\delta(G|S)}{c(S)}$) In each iteration, we aim to select a subscribed user u_i to the crowdseed S that is able to maximize this ratio, given by

$$u_i = \text{argmax}_{u_i}\left(\frac{\delta(G|S \cup \{u_i\})}{c(S \cup \{u_i\})} - \frac{\delta(G|S)}{c(S)}\right) = \text{argmax}_{u_i}(\Delta_{S \cup \{u_i\}} - \Delta_S) \quad (10)$$

The greedy algorithm terminates until the expected query diffusion is larger than τ (i.e. $\delta(G|S) \geq \tau$).

However, given a crowdseed set S, the computation of its expected query diffusion is #P-hard. Thus, we propose a sampling algorithm to estimate the expected query diffusion efficiently and effectively. As $G = (V, E, P)$ is a probabilistic graph, the samples can be obtained by flipping the edges according to the probabilities in P. For example, we can have k sample graphs, $G_1 = (V, E_1), ..., G_k = (V, E_k)$. Thus, the expected query diffusion can be obtained by taking the average of these sample graphs and is given by

$$\overline{\delta(G|S)} = \frac{\sum_{i=1}^k \delta(G_i|S)}{k} \quad (11)$$

where the sample G_i is a deterministic graph. The $\delta_B(G_i|S)$ is the size of the connected nodes from the set S which can be computed using the *BFS* algorithm.

We also show that the sampling algorithm can achieve (ε, η) approximation of estimating the expected query diffusion (i.e. $\delta(G|S)$). Using *Hoeffding's Inequality*, we have

$$Pr(|\overline{\delta(G|S)} - \delta(G|S)| \geq \epsilon) \leq 2\exp\left(-\frac{2\epsilon^2 k^2}{\sum_{i=1}^k (|V| - 1)^2}\right) \leq \eta \quad (12)$$

where ϵ is the error rate and η is the confidence of the estimation. Then it follows that we achieve (ϵ, η) approximation if the number of samples

$$k \geq \frac{(|V| - 1)^2 \ln \frac{2}{\eta}}{2\epsilon^2 |S|^2} \quad (13)$$

where $|S|$ is the size of the crowdseed S. Details are shown in Appendix [1]. The details of the greedy and sampling algorithms are presented in Algorithms 1 and 2. First, in Algorithm 1 we transform the obtained microblog into a probabilistic graph $G = (V, E, P)$ based on the similarity between two users on Line 1. The diffusion process of the crowdsourced query is based on G. Next, we compute the expected accuracy (i.e. μ) based on the average value of the historical accuracy record of the users on Line 2. Then, we estimate the lower bound of the number of replicated queries by Eq. 5 on Line 3. When the crowdseed S is empty, the expected query diffusion is assumed to be zero (i.e. $\delta(S|\phi) = 0$). At each iteration, we select a subscribed user u_i that is able to maximize the objective function on Line 6 and add it to the crowdseed S on Line 7. Algorithm 1 terminates when the expected query diffusion of the current crowdseed (i.e. $\delta(G|S)$) is larger than the number of required replicated queries (i.e. $\delta(G|S)) \geq \tau$) on Line 5.

Algorithm 1. $Crowdseed(Q_\alpha)$

Input: Q_α : a crowdsourced query with specified accuracy α; $a(u_1, \ldots, a(u_{|V|}))$: a historic accuracy record
Output: S : a crowdseed
1: Build a probabilistic graph $G = (V, E, P)$ by Equation 7
2: Expected accuracy $\mu \leftarrow \frac{\sum_{i=1}^{|V|} a(u_i)}{|V|}$
3: Set # of replicated queries τ by Equation 5
4: Set crowdseed $S \leftarrow \emptyset$
5: **while** $\delta(G|S) < \tau$ **do**
6:     $u_i = \text{argmax}_{u_i}(\Delta_{S \cup \{u_i\}} - \Delta_S)$
7:     $S \leftarrow S \cup \{u_i\}$
8: **end while**
9: **return** S

We illustrate the sampling process of computing the expected query diffusion $\delta(G|S))$ in Algorithm 2. In order to achieve (ϵ, η) approximation of the expected query diffusion, we choose K samples for estimation by Eq. 5 on Line 1. At each time, we sample K certain graphs from the probabilistic graph $G = (V, E, P)$ and take the average of the query diffusion of the current crowdseed S on the

---

Algorithm 2. $GetDiffusion(G, S)$

Input: G : a probabilistic graph; S : a candidate crowdseed
Output: $\delta(G|S)$: an expected query diffusion
1: Set # of samples K by Equation 13
2: Set expected query diffusion $\delta(G|S) \leftarrow 0$
3: **for** $i = 1 \rightarrow K$ **do**
4:     Sample G_i form $G = V, E, P$
5:     $\delta(G_i|S) \leftarrow BFS(G_i, S)$
6:     $\delta(G|S) \leftarrow \delta(G|S) + \delta(G_i|S)/K$
7: **end for**
8: **return** $\delta(G|S)$

---

samples from Lines 3 to 8. We employ the *BFS* algorithm to compute the query diffusion on certain graphs on Line 5. In the implementation, we find that the time duration of generating K sample graphs is very lengthy. To tackle this problem, we sample K certain graphs offline and store them in the memory. Thus, Algorithm 2 is able to compute query diffusion on these stored graphs without generating new samples.

6   Experimental Studies

The studies are performed in a Linux box with an 8-core Intel(R) Xeon(R) CPU X5450 3.00 GHz and 16 GB memory.

We investigate the robustness of our algorithms by varying the size of the crowdseed and the query cost.

Datasets. In order to generate the synthetic datasets, we simulate the crowd-sourced query diffusion process using three real datasets. The *Epinion* is a Who-trusts-whom network where the vertices represent the sites and the edges represent the trust relations between two sites. The *NetHEPT* is a large academic collaboration network where the vertices represent authors and the edges represent coauthorship relations. *Twitter* is microblog dataset where the vertices represent the users and the edges represent the following relationships. Thus, we build the probabilistic graph $G = (V, E, P)$ based on these three datasets in order to model the query diffusion. The statistics of these three real social graphs are given in Table 1.

Table 1. Statistics of datasets

Dataset	Epin	NetHEPT	Twitter
No. of nodes	75888	15233	11555
No. of edges	508837	58891	500000

Baseline Algorithms. We evaluate the effectiveness and robustness of the proposed algorithms using the three real social graphs aforementioned. We also propose four baseline algorithms such as *Degree-based* algorithm, *Centrality-based* algorithm, the CELF++ [11] Algorithm and the LDAG [5] The *Degree-based* and *Centrality-based* algorithms rank all the vertices based on vertices degree and centrality first. The CELF++ Algorithm and LDAG algorithm picks the most influential seeds by their definition. Then, these four algorithms select the vertices and add them to the crowdseed S according to their order.

Measurement. We demonstrate the effectiveness of the proposed algorithms by generating crowdsourced queries for each dataset with different error rate ϵ (i.e. $1 - \alpha$). We propose two quality measurements: (1) size of Crowdseed (# of seeds) and (2) cost of crowdsourced query. In the first phase, we assume that the cost of each user joining the crowdseed S is the same. Thus, the algorithm which outputs the smallest crowdseed works the best. In the second phase, we generate the cost of each user to join crowdseed S from a uniform distribution (i.e. $Uni(1, 100)$). Then, we compare the cost of the crowdsourced query for different algorithms, where the one with the least cost is the most effective.

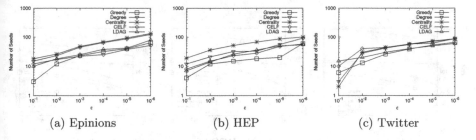

(a) Epinions   (b) HEP   (c) Twitter

Fig. 3. Crowdseed size v.s. error rate

Crowdseed Size. Figure 3(a), (b) and (c) illustrate the size of the crowdseed of the proposed algorithms by generating six types of crowdsourced queries for each dataset with different error rate ϵ (i.e. $10^{-1}, 10^{-2}, \ldots, 10^{-6}$). For example, the size two crowdseed means that we need to issue the crowdsourced query to two users as the seeds for later diffusion.

To mitigate the unreliability of the *majority voting rule*, we need to enroll many users to improve its accuracy. In the microblog, the expected query diffusion is proportional to the size of the crowdseed. Thus, as the error rate decreases, the size of the crowdseed increases. However, the size of crowdseed by the greedy algorithm is smaller than other algorithms.

The baseline algorithms select the seeds based on their independent query diffusion. However, selecting a vertex with a high diffusion value may not always increase the query diffusion of the crowdseed by much. For example, selecting a vertex with a lot of neighbors in the crowdseed may not increase the total query diffusion by much. Thus, our proposed algorithm considers the joint expected query diffusion of the selected seeds such that this problem can be avoided.

Query Cost. We study the cost of processing a crowdsourced query using our algorithms. Figure 4(a), (b) and (c) show the query cost of our proposed algorithms on different error rates. To make a fair comparison with the baseline algorithms, we penalize the users with high costs in both *Degree-based* and *Centrality-based* algorithms. For example, the *Degree-based* algorithm first sorts the users based on their degree (i.e. $deg(u_i)$) and then selects the users according to that order. In the new *Degree-based* algorithm, we set the score of each user to be the degree divided by its cost (i.e. $deg(u_i)/c(u_i)$). Similarly, we build a new *Centrality-based* algorithm for comparison.

Figure 4(a), (b) and (c) show that the cost of the crowdsourced query increases when reducing the error rate threshold. However, our algorithm also outperforms four other baseline algorithms in terms of the query cost. This nice result is attributed to the fact that we model the joint query diffusion in the proposed objective function while others treat the query diffusion of each vertex independently. The results show that our algorithm not only mines the crowd-seed effectively, but is also very efficient compared with the other three baseline algorithms (Fig. 5).

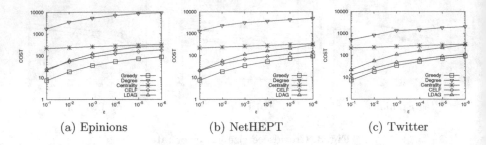

(a) Epinions   (b) NetHEPT   (c) Twitter

Fig. 4. Query cost v.s. error rate

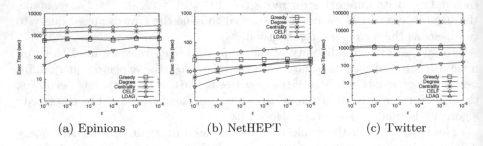

(a) Epinions   (b) NetHEPT   (c) Twitter

Fig. 5. Execution time v.s. error rate

7 Conclusions

We explore a new approach to processing crowdsourced queries on microblogs. Our goal is to minimize the cost of the crowdsourced query processing while the aggregated answer satisfies a specified accuracy threshold. We develop a new query diffusion model and formulate the problem of *Crowdseed Selection*. However, we prove that this problem is NP-hard and, given a crowdseed set S, the computation of query diffusion is #P-hard. We then develop a greedy algorithm to tackle the problem and a sampling algorithm to compute the query diffusion of the selected crowdseed set. We also derive an error bound for the proposed sampling algorithm. We validate the performance of our algorithm using three real datasets. The experimental results clearly demonstrate that our algorithms are able to reduce the cost of the crowdsourced query effectively. We also show that our new incentive mechanism helps the queries to be more efficiently evaluated. As the number of smart phone users and microblog subscribers are increasing every year, our proposed approach is promising to gain further superiority over existing approaches relying mainly on a centralized platform.

References

1. Appendix. http://www.cse.ust.hk/~wilfred/CQP.html
2. Bozzon, A., Brambilla, M., Ceri, S.: Answering search queries with crowdsearcher. In: WWW, pp. 1009–1018 (2012)
3. Cao, C.C., She, J., Tong, Y., Chen, L.: Whom to ask?: jury selection for decision making tasks on micro-blog services. VLDB 5(11), 1495–1506 (2012)
4. Chai, X., Vuong, B.Q., Doan, A., Naughton, J.F.: Efficiently incorporating user feedback into information extraction and integration programs. In: SIGMOD, pp. 87–100 (2009)
5. Chen, W., Yuan, Y., Zhang, L.: Scalable influence maximization in social networks under the linear threshold model. In: ICDM (2010)
6. Chen, X., Bennett, P.N., Collins-Thompson, K., Horvitz, E.: Pairwise ranking aggregation in a crowdsourced setting. In: WSDM, pp. 193–202 (2013)
7. Davidson, S.B., Khanna, S., Milo, T., Roy, S.: Using the crowd for top-k and group-by queries. In: ICDT, pp. 225–236 (2013)
8. Demartini, G., Difallah, D.E., Cudré-Mauroux, P.: Zencrowd: leveraging probabilistic reasoning and crowdsourcing techniques for large-scale entity linking. In: WWW, pp. 469–478 (2012)
9. Ghosh, S., Sharma, N., Benevenuto, F., Ganguly, N., Gummadi, K.: Cognos: crowdsourcing search for topic experts in microblogs. In: SIGIR, pp. 575–590 (2012)
10. Gomes, R.G., Welinder, P., Krause, A., Perona, P.: Crowdclustering. In: NIPS, pp. 558–566 (2011)
11. Goyal, A., Lu, W., Lakshmanan, L.V.: Celf++: optimizing the greedy algorithm for influence maximization in social networks. In: WWW, pp. 47–48 (2011)
12. Guo, S., Parameswaran, A., Garcia-Molina, H.: So who won?: dynamic max discovery with the crowd. In: SIGMO, pp. 385–396 (2012)
13. Kaplan, H., Lotosh, I., Milo, T., Novgorodov, S.: Answering planning queries with the crowd. VLDB 6(9), 697–708 (2013)
14. Kempe, D., Kleinberg, J., Tardos, É.: Maximizing the spread of influence through a social network. In: KDD, pp. 137–146 (2003)
15. Liu, Q., Peng, J., Ihler, A.T.: Variational inference for crowdsourcing. In: NIPS, pp. 692–700 (2012)
16. Liu, X., Lu, M., Ooi, B.C., Shen, Y., Wu, S., Zhang, M.: Cdas: a crowdsourcing data analytics system. VLDB 5(10), 1040–1051 (2012)
17. Marcus, A., Wu, E., Karger, D., Madden, S., Miller, R.: Human-powered sorts and joins. VLDB 5(1), 13–24 (2011)
18. Parameswaran, A.G., Garcia-Molina, H., Park, H., Polyzotis, N., Ramesh, A., Widom, J.: Crowdscreen: algorithms for filtering data with humans. In: SIGMOD, pp. 361–372 (2012)
19. Parameswaran, A.G., Park, H., Garcia-Molina, H., Polyzotis, N., Widom, J.: Deco: declarative crowdsourcing. In: CIKM, pp. 1203–1212 (2012)

20. Raykar, V.C., Yu, S., Zhao, L.H., Valadez, G.H., Florin, C., Bogoni, L., Moy, L.: Learning from crowds. JMLR **11**, 1297–1322 (2010)
21. Selke, J., Lofi, C., Balke, W.-T.: Pushing the boundaries of crowd-enabled databases with query-driven schema expansion. VLDB **5**(6), 538–549 (2012)
22. Sojump. http://www.sojump.com
23. Trushkowsky, B., Kraska, T., Franklin, M.J., Sarkar, P.: Crowdsourced enumeration queries. In: ICDE, pp. 673–684 (2013)
24. Venetis, P., Garcia-Molina, H., Huang, K., Polyzotis, N.: Max algorithms in crowdsourcing environments. In: WWW, pp. 989–998 (2012)
25. Wang, G., Wilson, C., Zhao, X., Zhu, Y., Mohanlal, M., Zheng, H., Zhao, B.Y.: Serf and turf: crowdturfing for fun and profit. In: WWW, pp. 679–688 (2012)
26. Wang, J., Li, G., Kraska, T., Franklin, M.J., Feng, J.: Leveraging transitive relations for crowdsourced joins. In: SIGMOD, pp. 229–240 (2013)
27. Wang, X., Zhao, Z., Ng, W.: A comparative study of team formation in social networks. In: Renz, M., Shahabi, C., Zhou, X., Cheema, M.A. (eds.) DASFAA 2015. LNCS, vol. 9049, pp. 389–404. Springer, Heidelberg (2015)
28. Welinder, P., Branson, S., Perona, P., Belongie, S.J.: The multidimensional wisdom of crowds. In: NIPS, pp. 2424–2432 (2010)
29. Yi, J., Jin, R., Jain, S., Yang, T., Jain, A.K.: Semi-crowdsourced clustering: generalizing crowd labeling by robust distance metric learning. In: NIPS, pp. 1772–1780 (2012)
30. Zhao, Z., Cheng, J., Wei, F., Zhou, M., Ng, W., Wu, Y.: Socialtransfer: transferring social knowledge for cold-start cowdsourcing. In CIKM, pp. 779–788 (2014)
31. Zhao, Z., Ng, W., Zhang, Z.: Crowdseed: query processing on microblogs. In: EDBT, pp. 729–732 (2013)
32. Zhao, Z., Wei, F., Zhou, M., Chen, W., Ng, W.: Crowd-selection query processing in crowdsourcing databases: a task-driven approach. In: EDBT (2015)
33. Zhao, Z., Wei, F., Zhou, M., Ng, W.: Cold-start expert finding in community question answering via graph regularization. In: Renz, M., Shahabi, C., Zhou, X., Cheema, M.A. (eds.) DASFAA 2015. LNCS, vol. 9049, pp. 21–38. Springer, Heidelberg (2015)
34. Zhao, Z., Yan, D., Ng, W., Gao, S.: A transfer learning based framework of crowd-selection on twitter. In: KDD, pp. 1514–1517 (2013)
35. Zhao, Z., Zhang, L., He, X., Ng, W.: Expert finding for question answering via graph regularized matrix completion. IEEE Trans. Knowl. Data Eng. **27**, 993–1004 (2015)
36. Zhou, D., Basu, S., Mao, Y., Platt, J.C.: Learning from the wisdom of crowds by minimax entropy. In: NIPS, pp. 2195–2203 (2012)

Effective Result Inference for Context-Sensitive Tasks in Crowdsourcing

Yili Fang[1], Hailong Sun[1(✉)], Guoliang Li[2], Richong Zhang[1], and Jinpeng Huai[1]

[1] School of Computer Science and Engineering, Beihang University, Beijing, China
{fangyili,sunhl,zhangrc}@act.buaa.edu.cn, huaijp@buaa.edu.cn
[2] Department of Computer Science, Tsinghua University, Beijing, China
liguoliang@tsinghua.edu.cn

Abstract. Effective result inference is an important crowdsourcing topic as workers may return incorrect results. Existing inference methods assign each task to multiple workers and aggregate the results from these workers to infer the final answer. However, these methods are rather ineffective for context-sensitive tasks (CSTs), e.g., handwriting recognition, due to the following reasons. First, each CST is rather hard and workers usually cannot correctly answer a whole CST. Thus a task-level inference strategy cannot achieve high-quality results. Second, a CST should not be divided into multiple subtasks because the subtasks are correlated with each other under certain contexts. So a subtask-level inference strategy cannot achieve high-quality results as it neglects the correlation between subtasks. Thus it calls for an effective result inference method for CSTs. To address this challenge, this paper proposes a smart assembly model (SAM), which can assemble workers' complementary answers in the granularity of subtasks without losing the context information. Furthermore, we devise an iterative decision model based on the partially observable Markov decision process, which can decide whether we need to ask more workers to get better results. Experimental results show that our method outperforms state-of-the-art approaches.

Keywords: Crowdsourcing · Result inference · Context-sensitive tasks · Smart assembly model

1 Introduction

Crowdsourcing is a fast-growing field that seeks to harness the cognition superiorities of humans to solve the problems that are hard to solve by computers [7,18], such as sentiment analysis, handwriting recognition, and image labelling. As workers can earn rewards for answering questions, workers may return incorrect results to cheat for more rewards. To infer the best answer from noisy results, effective result inference in crowdsourcing is an important problem [17].

To obtain high-quality results, an iterative framework is widely used in existing crowdsourcing systems as shown in Fig. 1. After a requester submits a crowdsourcing task, the *Task Assignment* component assigns the task to n workers.

© Springer International Publishing Switzerland 2016
S.B. Navathe et al. (Eds.): DASFAA 2016, Part I, LNCS 9642, pp. 33–48, 2016.
DOI: 10.1007/978-3-319-32025-0_3

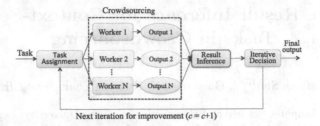

Fig. 1. A general framework of iterative crowdsourcing result inference.

Once workers complete their tasks, the *Result Inference* component aggregates the results from these n workers and generates the inference result. Then, the *Iterative Decision* component checks whether the inference result can be further improved: if the result is good enough, the iterative process will be terminated; otherwise it will ask another n workers and repeat the above three steps. In other words, the *Iterative Decision* component decides whether the iterative improvement process should be terminated or not. In this paper, we study the two core issues in this framework: (1) how to efficiently infer high-quality results based on the outputs of different workers in the *Result Inference* component, and (2) how to decide whether to terminate the iterative process in the *Iterative Decision* component.

There are two categories of literatures on the *Result Inference* and the *Task Assignment*: task-level inference (`Task-Inf`) and subtask-level inference (`Subtask-Inf`). `Task-Inf` first assigns each task to multiple workers and then utilizes machine-learning techniques [13,15] or crowd-voting techniques [1,3,5,6,10] to infer result. `Subtask-Inf` splits a complex task into some subtasks, crowd-sources the subtasks, utilizes `Task-Inf` to infer the result of every subtask [7,18], and assembles the best results of each subtask to generate the final answer of the complex task. These inference methods can improve the result quality for simple tasks, such as image annotation [4,12], named entity recognition [3], and search tasks [8].

However, existing methods are ineffective for context-sensitive tasks (`CST`s), e.g., handwriting recognition, speech recognition and route recommendation, due to the following reasons. First, `CST`s are rather hard and workers usually cannot correctly answer a whole task. For example, for the handwriting recognition task in Fig. 2, all of the three workers cannot correctly answer the task (see Fig. 3). Thus, `Task-Inf` cannot obtain high-quality results. Second, a `CST` should not be divided into multiple subtasks because the subtasks are correlated with each other under certain contexts. For example, the second worker recognizes the 9th word as "`next`" but is not sure about the 10th word ("`tip`" or "`time`"). Based on the context, she can label it as "`time`". So `Subtask-Inf` cannot achieve high-quality results as it neglects the correlation between subtasks. Thus, it calls for an effective result inference method for `CST`s in crowdsourcing.

In this paper, we propose a smart assembly model (`SAM`) to aggregate the results from different workers. Each `CST` is assigned to multiple workers, and then we assemble workers' answers by selecting the good results and discarding the bad results from these workers through the SAM model. Next, we devise an iterative decision model based on the partially observable Markov decision process to improve the results.

To summarize, we make the following contributions.

- We identify a common category of tasks, namely context-sensitive task (CSTs). Each CST consists of a certain number of subtasks under a certain context, but which is not suitable to split for crowdsourcing subtasks.
- We propose a smart assembly model (SAM) to effectively assemble workers' answers, which selects the good results and discards the bad results to generate the inference result.
- We devise an iterative decision model based on the partially observable Markov decision process, which decides whether to terminate the iterative process. This method significantly reduces the decision complexity.
- We conduct both simulated and real experiments for our proposed model on real crowdsourcing platforms and compare with existing models in a variety of complex scenarios. The experimental results show that our method outperforms existing approaches.

The remainder of this paper is organized as follows. In Sect. 2, we discuss related work on result inference and quality control in crowdsourcing. We then formalize the problem in Sect. 3. Our proposed result inference model and iterative decision method are respectively presented in Sects. 4 and 5. Experimental setup and results are described in Sect. 6. Finally, we conclude this work in Sect. 7.

2   Related Work

In crowdsourcing, one of the challenging issues with quality control is to effectively infer results from noisy results provided by workers, which is also closely related to the type of crowdsourcing tasks. In the following we describe the related work in two aspects as follows.

Majority voting [19] was a well-known method to infer the results from the outputs of multiple workers. Early works with majority voting do not consider the difference of worker ability, which is effective for simple tasks that most workers can process correctly. But this method can lead to low quality results in the case of difficult tasks or less capable workers. By considering workers' ability, weighted majority voting was proposed [15]. In this regard, various solutions like EM and Bayesian methods [15] were proposed to estimate workers' ability.

To further improve the majority voting strategy, there were two lines of studies related to specific types of crowdsourcing tasks. For isolated tasks, some literature [5–7,10,18] proposed result inference methods based on Task-Inf (i.e., different workers are asked to perform the same task until a consensus is achieved on the outcome) to deal with inaccuracies and minimize the cost; while for complex tasks, some works proposed other inference methods based on Subtask-Inf that utilize a multiple phrase strategy to process the tasks [3,17]. CDAS [11] used a quality-sensitive answering model. Askit [2] utilized entropy-like techniques. QASCA studied quality-aware inference method [20]. Compared with these systems, our method considered context-sensitive tasks and achieved better result quality. Find-Fix-Verify (FFV) [17] was proposed to correct and shorten text in a three-step strategy. The 'Find' step identified some mistakes in sentences by a set of workers and the 'Fix' step fixed these mistakes by another set of workers. The 'verify' step verified these mistakes and applied Task-Inf to aggregate the results. Our CSTs were different from FFV tasks. A FFV-task involved an ordered three-step processing workflow of find, fix and verify while a CST should not be split into a set of subtasks no matter how many processing steps are involved. Our method is orthogonal to FFV. In practice, for simplifying crowdsourcing processing, many tasks are split into subtasks. For example, in CAPTCHA/reCAPTCHA, a single word is considered as a task, while a CST (e.g. recognizing a handwritten sentence) is not fit for crowdsourcing in the granularity of single words. Other CSTs include translation, transcription, text recognition and etc.

In addition, to obtain workers' ability or eliminate workers with low ability, there were also many algorithms to control workers' quality by qualification test and golden test in order to obtain the ability of the anonymity worker or eliminate the bad workers through the test [6,10].

Different from these studies, we focus on inferring high-quality results for context-sensitive tasks. Since Task-Inf may overestimate workers' ability and Subtask-Inf neglects the context of CSTs, they cannot obtain high-quality results for complexed crowdsourcing tasks like CSTs. Thus, we propose a smart assembly model that assembles workers's answers by selecting the good results and discarding the bad results without losing task contexts. Different from traditional Markov models, our POMDP significantly reduced the decision complexity by reducing the number of states. Experimental results also validated the superiority of our techniques.

3   Problem Description

In this paper, we focus on context-sensitive tasks (CSTs). A CST contains multiple subtasks, which are correlated within a certain context. If these subtasks are split and crowdsourced separately, the context will be lost. For instance, Fig. 2 shows a context-sensitive task, handwriting recognition. If the handwritten sentence is split into 10 separated subtasks (e.g., words) to crowdsource, the context correlation among these subtasks will be lost and increase the difficulty of handwriting recognition.

Given a CST $T = \{t_1, t_2, \cdots, t_m\}$ with m subtasks, which is answered by n workers $W = \{w_1, w_2, \cdots, w_n\}$, we crowdsource the CST as a whole task and use an output matrix to denote the answer of these workers on T, which is formally defined as below.

Definition 1 (Output Matrix). *The answer of n workers on T is denoted by an $n \times m$ matrix, $O = \{o_{ij}\}$, where o_{ij} is the output produced by worker w_i for*

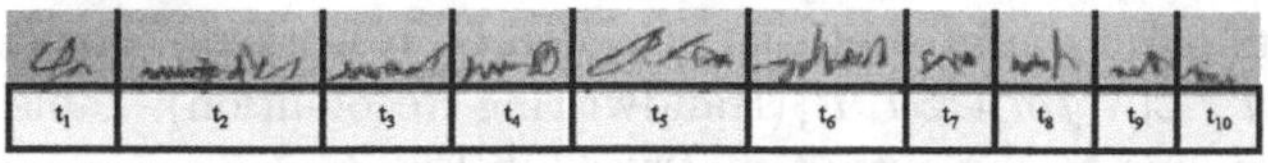

Fig. 2. Handwriting recognition with 10 words.

$$
\begin{bmatrix}
\text{You misspelled} & \perp & \text{words, Plan spellcheck} & \perp & \text{work} & \perp & \text{tip.} \\
\text{You misspelled several words, Please} & & \perp & \text{your} & \perp & \text{next time.} \\
\perp & \text{misspelled several works, Play} & & \perp & \text{your work} & \perp & \text{time.}
\end{bmatrix}
$$

Fig. 3. The output matrix of handwriting recognition.

Table 1. The COVs of handwriting recognition.

I_T^1	You misspelled several work, Plan spellcheck your work next tip
I_T^2	You misspelled several work, Plan spellcheck your work next time
I_T^3	You misspelled several work, Please spellcheck your work next tip
I_T^4	You misspelled several work, Please spellcheck your work next time
I_T^5	You misspelled several work, Play spellcheck your work next tip
I_T^6	You misspelled several work, Play spellcheck your work next time
I_T^7	You misspelled several words, Plan spellcheck your work next tip
I_T^8	You misspelled several words, Plan spellcheck your work next time
I_T^9	You misspelled several words, Please spellcheck your work next tip
I_T^{10}	You misspelled several words, Please spellcheck your work next time
I_T^{11}	You misspelled several words, Play spellcheck your work next tip
I_T^{12}	You misspelled several words, Play spellcheck your work next time

subtask t_j. If worker w_i does not provide an output for subtask t_j, $o_{ij} = \perp$. We use O_{*j} to denote the output vector of subtask t_j and O_{i*} to denote the output vector of worker w_i.

Note that o_{ij} is $\perp$ if and only if worker w_i cannot process subtask t_j. For example, Fig. 3 illustrates the output matrix of the handwriting recognition CST in Fig. 2, which is completed by three workers. $o_{31} = \perp$, because the third worker cannot recognize the first word.

Given a CST T, we aim at inferring a high-quality result of T through aggregating all subtask outputs from different workers. To generate the inference result of T, we define the candidate output vector to represent the inference result.

Definition 2 (Candidate Output Vector − COV). *Let* T *be a* CST *with* m *subtasks, a candidate output vector is defined as a vector of subtask outputs* $C_T = \langle o_{i_1 1}, ... o_{i_j j}, ..., o_{i_m m} \rangle$, *where* $o_{i_j j} \in O_{*j}$ *(i.e.,* $i_j \in [1, n]$).

If a CST T with m subtasks is processed by n workers, there may be large numbers of possible COVs (in the worst case, there are n^m COVs by combining all the outputs of all the m subtasks provided by n workers). For example, $\mathcal{I} = \{I_T^l | I_T^l$ is the lth COV for CST $T\}$(Handwriting Recognition), Table 1 illustrates set $\mathcal{I}$ corresponding to the output matrix O in Fig. 3.

With Definition 2, we can transform the problem of inferring the answer of a CST into the best COV identification problem, which finds the best COV from all possible COVs.

Definition 3 (Best COV Identification). *Let* T *be a* CST, O *be the output matrix of* T, *and* Z *be the set of all possible* COV*s, the problem is to find the best* COV *from* Z *so as to obtain the best result of* T.

For instance, assembling the best result based on O in Fig. 3 is transformed into choosing the best COV from its COV set in Table 1. We will discuss how to quantify a COV and how to identify the best COV in Sect. 4.

To improve the inference quality, we usually adopt an iterative method. To determine whether to go to the next iteration, we need to address the iterative decision problem.

Definition 4 (Iterative Decision Problem – IDP). *Let* T *be a* CST, O *be the output matrix of* T, I_T *and* I_T' *be the inference result of the current iteration and the previous iteration respectively. The* IDP *problem is to determine whether to terminate the iterative process.*

Definition 4 defines a decision problem of iteratively obtaining a satisfactory inference result for CSTs. We propose an effective iterative model to make a decision in Sect. 5.

According to Definition 2, the size of the COV set grows exponentially with the number of subtasks. Specifically, the number of COVs can be n^m. Thus in theory the complexity of *Best COV Identification* defined in Definition 3 should be an NP hard problem. In practice, as mentioned in [3], the state space of COVs is not very large due to the limited size of a CST and the repetition of subtask results provided by workers. For instance, as shown in Table 1, the size of COVs is only 12 instead of 3^{10}.

4   The SAM Model for Result Inference

This section studies the best COV identification problem. We propose a smart assembly model (SAM) to assemble workers' results in the granularity of subtasks instead of the whole task. We first use the partially ordered set theory to compute a set of greatest COVs and then select the best COV from the greatest COVs as the inference result by considering workers' ability on the CSTs.

Given a CST T, n workers generate an output matrix O for T. If $o_{ij} \neq \perp$, we use $p_{ij} = k_{ij}/n$ to measure the quality of the subtask output o_{ij}, where k_{ij} denotes that k_{ij} workers have the same output with $o_{ij} \in O_{*j}$ for the subtask t_j; otherwise ($o_{ij} = \perp$), $p_{ij} = 0$. With all the outputs of subtask t_j, we introduce a subtask probability vector.

Definition 5 (Subtask Probability Vector – SPV). *Given a CST T with m subtasks and the output matrix O of T, a subtask t_j's SPV consists of the probabilities of the outputs of n workers for t_j, which is denoted by*

$$P_{t_j} = \langle p_{1j}, p_{2j}, \cdots, p_{nj} \rangle \tag{1}$$

Basing on Definition 5, we can figure out that the larger p_{ij} is, the more likely the output o_{ij} is the correct answer to subtask t_j. To determine the quality of a COV, we define a relation $\preceq$ to compare two COVs by comparing the corresponding SPVs. Let I_T and I'_T denote two COVs, we show the partial order as follows:

$$I_T \preceq I'_T \text{ iff. } \forall t_j \in T, o_{ij} \in I_T, \ o_{kj} \in I'_T, p_{ij} \leq p_{kj}.$$
$$I_T \prec I'_T \text{ iff. } I_T \preceq I'_T \ \& \ \exists o_{ij} \in I_T, o_{kj} \in I'_T, p_{ij} < p_{kj}. \tag{2}$$

Similarly, we can define $\succ$ and $\succeq$. If $I_T \preceq I'_T (I_T \succeq I'_T)$, we say I_T is worse (better) than or equal to I'_T. If $I_T \prec I'_T (I_T \succ I'_T)$, we say I_T is worse (better) than I'_T. Let $\mathcal{I}$ denote the set of all COVs of T, $\mathcal{I} = \{I^l_T | I^l_T$ is the lth COV for CST $T\}$. Then, we can deduce that $\langle \mathcal{I}, \prec \rangle$ is a partially ordered set, and thus we can sort COVs based on the partially ordered set. We can deduce the greatest (least) COVs which has no COV that is better (worse) than them.

Based on the partially ordered set, we can construct a graph for the COVs, where nodes are COVs and edges are the partial orders between two COVs. For example, for all the COVs in Table 1, if we add $I^0_T = \emptyset$ to $\langle \mathcal{I}, \prec \rangle$, the graph is illustrated in Fig. 4(a), which has three greatest COVs, I^8_T, I^{10}_T, and I^{12}_T.

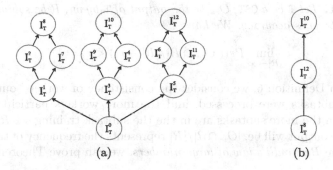

Fig. 4. (a) Partially ordered set of handwriting recognition, and (b) The relationship of the maximal COVs.

When $\langle \mathcal{I}, \prec \rangle$ has more than one greatest COVs, we need to break the tie and select the best one. To address this issue, we consider the workers' abilities and utilize them to select the best COV. To quantify the worker's ability, we need to estimate workers' accuracy on the CST. Then we identify a subtask training set where the subtasks in the set have enough support (that is many workers return the same result for the task), and take them as ground truth so as to evaluate workers' accuracy. Next, we formally define the subtask training set.

Definition 6 (Subtask Training Set). *Given a* CST *T with m subtasks,* T^* *is a subset of* T*, where each subtask* $t_j \in T^*$ *satisfies that there exist more than one workers that return the same result* o_{ij} *with the largest probability* p_{ij}. *Let R denote the set of all such results, i.e.,* $R = \{o_{ij}\}$, *where* o_{ij} *is the majority output for* $t_j \in T^*$ *with the largest probability* p_{ij}.

For example, for the handwriting recognition task in Fig. 1, since the first worker and the second worker return the same answer "You" with the largest probability, t_1 is in T^* and "You" is in R. We can get $T^* = \{t_1, t_2, t_3, t_4, t_7, t_8, t_{10}\}$ and $R = \{You, misspelled, several, words, your, work, time\}$.

Based on the subtask training set, we can measure each worker's accuracy r_i as below

$$r_i \sim |O_{i*} \cap R|/|R| \tag{3}$$

For example, for the handwriting recognition task in Fig. 1, the accuracy of the three workers are $4/7 = 0.57$, $6/7 = 0.86$ and $5/7 = 0.71$ respectively. Based on the workers' accuracy, w_2 is better than w_3, which in turn is better than w_1. Except for the 5th subtask, the three maximal COVs (I_T^8, I_T^{10}, I_T^{12}) have the same output on other subtasks. Thus I_T^{10} (the 5th word is answered by w_2) is better than I_T^{12} (by w_3), which in turn is better than I_T^8 (by w_1). Hence the most likely COV of T is $I_T^{10} = \langle T,$ "You misspelled several words, Please spellcheck your work next time."$\rangle$. Note that our estimation method on the worker's accuracy in Eq. 3 has theoretical guarantee based on the *Bernoulli's law of large number*, as proved in Theorem 1.

Theorem 1. *Let* T *be a* CST, O_{i*} *be the output of* T *by* w_i, *R be subtask training set, and* r_i *be* w_i's *accuracy. We have*

$$\lim_{|R| \to \infty} Pr((|O_{i*} \cap R|/|R| - r_i) < \epsilon) = 1 \tag{4}$$

Proof. With Definition 6, we consider the consistency of workers' outputs. The more the subtasks were processed, and the more workers participate in this process, then the more subtasks are in the the subtask training set R, the more accurate the output will be. $|O_{i*} \cap R|/|R|$ represents the frequency of true results. Based on the *Bernoulli's law of large numbers*, we can prove Theorem 1.

We then can utilize the subtask training set to measure each worker's accuracy r_i. With this theorem we can conclude the more subtasks are in the the subtask training set R, the more accurate of the estimation of r_i is. For each greatest COV, we compute its overall score based on Definition 7 and select the COV with the largest score as the best COV.

Definition 7 (COV Score). *Given a* COV $C_T = \langle o_{i_1 1}, ... o_{i_j j}, ..., o_{i_m m} \rangle$, *its score is computed as below*

$$\text{SCORE}(C_T) = \sum_{j=1}^{m} p_{i_j j} \cdot r_{i_j} \tag{5}$$

A question is that given workers with diverse ability, how can we compare the inference result of our SAM model with that of existing Task-Inf model. If all subtasks are in the same difficulty level, and all workers' abilities are averagely consistent, our SAM model is reliably better than Task-Inf as shown in Theorem 2.

Theorem 2. *Let* T *be a* CST *and* r_i *denote* w_i*'s accuracy,* I_T^a *denote the best* COV *of our* SAM *model and* I_T^c *denote the best* COV *of* Task-Inf. *There exists a* $\tilde{r}$, *if* $\forall r_i \geq \tilde{r}$, *then* $Pr(I_T^c \prec I_T^a) \geq Pr(I_T^a \prec I_T^c)$.

Proof. For every subtask $t_j \in T$ and its SPV P_{t_j}, there are two cases for the output of subtask t_j as follows.

- If P_{t_j} has only one maximal element, according to our model, the best subtask output is chosen from O_{*j} based on SPV. For every two outputs of subtasks, it is a Condorcet model [14,16,19]. If $\tilde{r} = 0.5$, $o_{ij} \in O_{*j}$ if p_{ij} is a maximum in the P_{t_j}, then o_{ij} is most likely the true answer. For all $o_{kj} \in O_{*j}$, we have

$$Pr(p_{ij} \geq p_{kj}) \geq Pr(p_{ij} < p_{kj}) \tag{6}$$

- If P_{t_j} has more than one maximal elements, since two subtasks' outputs generated by the two models have the same possibility to be the correct output of t_j. And we use workers' accuracy to estimate the possibilities. If worker w_i has the best ability, then her answer is chosen.

Hence, for every subtask t_j, $\exists \tilde{r}$, if every worker's ability $r \geq \tilde{r}$, and $o_{ij} \in I_T^a$, $o_{kj} \in I_T^c$, we have

$$Pr(p_{ij} \geq p_{kj}) \geq Pr(p_{ij} \leq p_{kj}) \tag{7}$$

Note that I_T^a and I_T^c consist of the same number of o_{ij}, and each of two subtasks does not have intersection. Considering all elements of I_T^c and I_T^a and Eq. 7, we have

$$Pr(I_T^c \prec I_T^a) \geq Pr(I_T^a \prec I_T^c) \tag{8}$$

We conclude that when worker's accuracy is larger than 0.5, our model is reliably better than Task-Inf.

Another question is that given workers with diverse ability, how can we compare the inference result of our SAM model with those of existing Subtask-Inf models, which neglect the context of CSTs. Since the contexts reflect the semantic relations between subtasks, crowdsourcing a CST without splitting it into a group of subtasks can help workers complete the subtasks better and more easily. Therefore we can conclude that SAM can generate better aggregated results than Subtask-Inf even with the same aggregation methods.

5   Iterative Decision with POMDP

This section studies the IDP problem to determine when to terminate the iterative process. Formally, let $q_c \in [0,1]$ and $q_{c+1} \in [0,1]$ denote the quality of

I'_T and I_T respectively, we know that a worker has the probability of $1 - q_{c+1}$ to choose I'_T. Since (q_c, q_{c+1}) is only partially observable and the result of the current iteration only depends on the previous iteration, we can formulate this problem as a Partially Observable Markov Decision Problem (POMDP).

Definition 8 (POMDP for CSTs). *The* POMDP *for* CSTs *is a six-tuple* $\langle S, A, R, T, O, P \rangle$, *where*

- $S = \{\langle q_{c+1}, q_c \rangle\}$ *is a finite set of discrete states, where q_{c+1} and q_c are the quality of I'_T and I_T respectively;*
- $A = \{$ *Create new crowdsourcing tasks, Submit the best inference output* $\}$ *is the action set;*
- $R = R_0 + R_s$ *is the rewarding function, where R_0 is a constant amount of money paid to a worker for her participation, and R_s is a dynamic reward based on improved quality q of a worker's contribution;*
- $T: S \times O \times S \to [0,1]$ *is the transition function and is specified below;*
- $O = \{I'_T, I_T\}$ *is a finite set of observations, where I'_T and I_T are defined in our SAM model;*
- $P: S \times O \to [0,1]$ *is the observation function which will be discussed later.*

Transition Function. The state transition in our problem is invoked when a new COV is generated and the current state (q_c, q_{c+1}) is not known. The quality of I_T depends on all workers $(w_1, w_2, \ldots w_n)$. Since each worker is independent of others and every worker can improve the COV and the inference result follows a conditional distribution $f(q_{c+1}|q_c, w_i)$, we can compute the transition function that is the mixture of conditional distributions of $f(q_{c+1}|q_c, w_1, w_2, \ldots, w_n)$.

$$f(q_{c+1}|q_c, w_1, w_2, \ldots, w_n) = \sum_{i=1}^{n} \lambda_i \cdot f(q_{c+1}|q_c, w_i) \tag{9}$$

where λ_i is the parameter of the mixture conditional distribution, and $\sum_{i=0}^{n} \lambda_i = 1$. We use $\lambda'_i = \frac{|O_{i*} \cap I_T|}{|I_T|}$ to normalize λ_i. Thus we can get $\lambda_i = \frac{\lambda'_i}{\sum_{i=1}^{n}(\lambda'_i)}$.

Belief Update. In comparison with the Task-Inf model, we evaluate quality of each COV basing on the SPV. The global evaluation metric of a COV is defined as below:

$$Pr(I_T) = \prod_{j=0}^{m} p_{ij}, o_{ij} \in I_T \tag{10}$$

Basing on Eq. 10, for the COV I_T of the current iteration and COV I'_T of the previous iteration, we can compute the quality distribution as follows:

$$Pr(q_{c+1} > q_c | Pr(I_T), Pr(I'_T)) = \frac{Pr(I_T)}{Pr(I'_T) + Pr(I_T)}. \tag{11}$$

Observation Function. Our observation function is only related to SPVs. If the current (hidden) state is (q_c, q_{c+1}) with corresponding outputs I'_T and I_T. This problem can be translated into the problem of majority voting [16]. Let $o_{ij} \in I_T$, $o_{kj} \in I_T'$, v_j denote the probability of $p_{ij} \geq p_{kj}$, and r denote the probability that we obtain the true result from workers. According to the previous study [3], we compute $r = (1 + (1 - d)^{r_i})/2$, where d is the difficulty of the task, and r_i is the ability of worker w_i. If all workers have roughly the same ability, we have

$$Pr(v_j = 1|q_c, q_{c+1}) = \sum_{k=o}^{n} \sum_{l=o}^{\lfloor k_{ij}/2 \rfloor} \binom{l}{k} r^{k-l}(1-r)^l \tag{12}$$

Since T has m subtasks, we have

$$Pr(I_T' \prec I_T|q_c, q_{c+1}) = Pr(v_j = 1|q_c, q_{c+1})^m \tag{13}$$

Similarly, we have

$$Pr(I_T \prec I_T'|q_c, q_{c+1}) = \left(1 - Pr(v_j = 1|q_c, q_{c+1})\right)^m \tag{14}$$

With Eqs. 13 and 14, we have

$$Pr(I_T' \prec I_T|q_c, q_{c+1}) = \frac{Pr(I_T' \prec I_T|q_c, q_{c+1})}{Pr(I_T \prec I_T'|q_c, q_{c+1}) + Pr(I_T' \prec I_T|q_c, q_{c+1})} \tag{15}$$

Definition 8 mainly describes a control model of processing CSTs with crowd-sourcing, and our goal is to receive a better utility, so we define Utility Estimations as follow.

Utility Estimations. We then discuss how to estimate the utility of an improvement of the crowdsourced task. At this point, we have already received n outputs from k workers. Based on our SAM model, we can get the assembly result and its quality q_c. Let $\triangle(q) = q_c - q_{c-1}$, $\mathcal{R}_s = \mu_S(q_c) = \frac{e^{\triangle q} - 1}{e - 1}$ represent the utility of current iteration. If $\triangle q = 0$, $\mu_S(\triangle q) = 0$; otherwise (if $\triangle q = 1$), $\mu_S(\triangle q) = \mathcal{R}_s$, which represents the probability that workers completely and accurately finish the task in the first iteration. Then our utility estimation equation is as follows:

$$\mathcal{V}(\mathcal{S}) = -\mathcal{R}_0 * k + \mu_S(\triangle q) \tag{16}$$

Given the current state $\mathcal{S}$ and next sate $\mathcal{S}'$, if $\mathcal{V}(\mathcal{S}') > \mathcal{V}(\mathcal{S})$, we go to state $\mathcal{S}'$; otherwise we keep the current state $\mathcal{S}$.

With Definition 8 and the defined utility estimation function (Eq. 16), we can improve the quality of a CST processing with crowdsourcing iteratively. The iterative process will be terminated when either the cost reaches the reward budget provided by requesters or the quality cannot be further improved through more iterations. Note that different from existing methods, our model has smaller numbers of states and thus significantly reduces the decision complexity.

6   Experimental Evaluation

Our experimental goal is to evaluate the superiority of SAM and POMDP. We compared with two state-of-the-art inference methods: Task-Inf [3] and Subtask-Inf [6,18], where we implemented the best inference methods. Task-Inf crowdsourced the whole task and aggregated the results from multiple workers. Subtask-Inf spilt CSTs into subtasks and each subtask's result was aggregated by Task-Inf. To ensure the reliability of the experimental results, we conducted both simulations and real-world experiments.

6.1   Real-World Experiments

Experiments Setups. In this set of experiments, we chose two tasks(i.e. handwritten text recognition and speech recognition) that could not be well processed with computer algorithms and were very suitable for crowdsourcing. Especially the chosen two tasks fell well into the CST tasks targeted by this work because either of the two tasks was not fit for being split into subtasks. The handwritten text recognition task was to recognize a uncommon ancient Chinese poem written with traditional Chinese cursive calligraphy. And the speech recognition task was to recognize the talking content given a clip of audio recorded in a noisy KTV in Beijing. Since both of the two tasks are closely related to Chinese language, Thus, we crowdsourced our tasks to a real Chinese crowdsourcing platform **Zhubajie**[1], which has been widely used in existing studies [9]. Zhubajie had more than 10 million workers. We utilized golden tests to eliminate workers with low ability. The Chinese cursive calligraphy recognition task had 118 words and speech recognition contained 130 words. For handwriting recognition, we asked 5 workers to finish the task in each iteration; for speech recognition, we asked 7 workers to finish the task in every iteration. For SAM and Task-Inf, the cost for each task was 1 RMB. For Subtask-Inf, we split each task to 118 subtasks (130 subtasks) for Chinese cursive calligraphy recognition (speech recognition). The price for each subtask was 0.1 RMB.

Chinese Cursive Calligraphy Recognition. The experimental results are shown in Figs. 5(b) and 6(a), where characters with the gray background were not recognized or not correctly identified. It can be seen that SAM outperformed Task-Inf by a big margin. Figure 5(a) plots the obtained recognition accuracy under different number of participated workers for Task-Inf, Subtask-Inf and SAM, in which SAM achieves the best accuracy among the three methods in all the cases. Especially when the number of workers reaches 5, SAM performs the best, which means it is unnecessary to ask too many workers to participate with SAM. As shown in Fig. 6(a), SAM took for 7 iterations and spent 54 RMB. 111 out of 118 words were correctly identified and the accuracy was 0.94. Task-Inf took for 8 iterations and spent 64 RMB. 94 out of 118 words were correctly identified and the accuracy was 0.79. Subtask-Inf ran only once and spent 59 RMB. 86 out of 118 words were correctly identified and the accuracy was 0.73. Thus SAM

---

[1]   http://www.zhubajie.com.

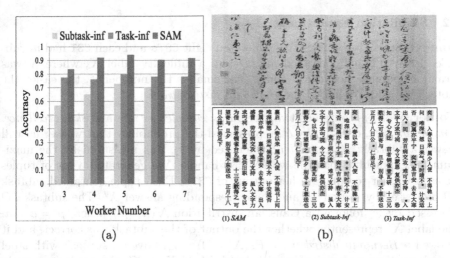

(a)   (b)

Fig. 5. (a) The inference result of three methods with different worker number, (b) Result of chinese cursive calligraphy recognition.

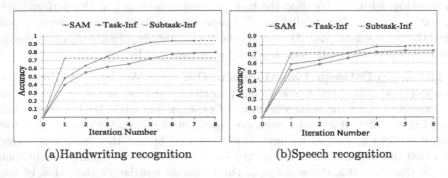

(a)Handwriting recognition   (b)Speech recognition

Fig. 6. Comparisons on a real crowdsourcing platform.

outperformed the state-of-the-art methods by 15 %, because SAM considered the context between subtasks. The result verified that SAM had higher accuracy than Task-Inf, which in turn was better than Subtask-Inf. SAM took less money than Subtask-Inf as Subtask-Inf generated many subtasks.

Speech Recognition. The experimental result is shown in Fig. 6(b). SAM took 5 iterations and took 42.5 RMB. 102 out of 130 words were correctly identified and the accuracy was 0.78. Task-Inf finished in 6 iterations and spent 54 RMB. 96 out of 130 words were correctly identified and the accuracy was 0.74. Subtask-Inf took for once and spent 65 RMB. 93 out of 130 words were correctly identified and the accuracy was 0.71. The result verified the superiority of SAM, which not only outperformed Task-Inf and Subtask-Inf in terms of accuracy, but took less money than Subtask-Inf.

6.2   Simulation Experiment

Context-Sensitive Tasks. We generated eight CSTs and each CST had m subtasks. Each subtask contained an ID and a difficulty label X, where X was a random number between 0 and 1 (the smaller the number is, the easier the subtask is), which followed the *Bernoulli distribution*.

Worker. Workers contained two attributes: *completion rate* and *skill excellence*. *Completion rate*, denoted by c_i, represented the probability that worker w_i will answer a subtask; *skill excellence*, denoted by r_i, was measured by the maximum task difficulty that the worker can handle. We generated a set of tuples for each worker to simulate her output of CSTs. Each tuple contained subtask ID processed by worker w_i and the corresponding answer X'. The subtask ID processed by w_i followed a Gaussian distribution $\mathcal{N}(\mu, \sigma_i^2)$, where $\mu = c_i \cdot m$. The label X' represented whether the output of the subtask was correct, and it followed a *Bernoulli distribution*, $Pr(X' = 0) = r_i$. Given a subtask with label X and an answer of a worker with label X', if $X = X'$, the worker correctly answered this subtask.

Evaluation Metric. We used the task processing accuracy as the performance evaluation metric. Specifically, let Z denote the number of subtasks processed correctly by workers, and M denote the total number of subtasks processed. Then the task processing accuracy can be represented by $\frac{Z}{M}$.

Relationship Between Two Models in Different Workers' Ability. Suppose workers' skill excellence is roughly in the same level and $r_i \geq \tilde{r}$. To show the impact of $\tilde{r}$, we ran a large number of experiments with the SAM model and Task-Inf in different $\tilde{r}$, and results are shown in Fig. 7(a). We can see that the quality of COVs processed by our model (denoted by the rhombus) was better than that of Task-Inf (denoted by the triangle). When the accuracy of output was greater than 0.5, the number of the rhombus was larger than that of the triangles. The assembly COVs from the output matrix given by all workers was more reliable than that by any single worker. Since the quality of task processing was often used to measure the skill excellence, we can conclude that while all workers' skill excellence was above 0.5, and the quality of output produced by all workers was often higher than a single worker.

Comparison with Subtask-Inf and Task-Inf. We used the ZMDP package[2] to implement POMDP. We set $d \in [0.7, 0.91]$, $\tilde{r} \geq 0.7$, and $c_i \geq 0.6$. Let N denote number of subtasks. We simulated 8 different CSTs and each CST had different difficulties and number of subtasks. Each CST had been repetitively executed by 100 times. Our simulation showed the quality of the results of the SAM and Task-Inf at the finally and Subtask-Inf at the first iteration. The expectation of the quality of results and their variances was illustrated in Fig. 7(b). The result showed that the SAM model consistently achieved the best results and significantly outperformed existing studies, whatever the difficulty of tasks (hard or easy) and the number of subtasks (more or less). For example, our SAM model

---

[2] http://www.cs.cmu.edu/trey/zmdp/.

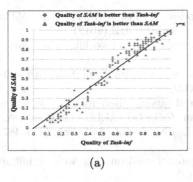
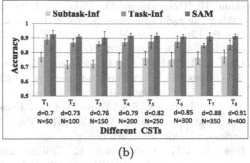

(a)   (b)

Fig. 7. (a) Comparison between SAM and Task-Inf with different $\bar{r}$. (b) The final result of simulation comparison

outperformed Subtask-Inf by 7 % and Task-Inf by 18 %. Although the context in the simulation was weak, the Task-Inf model was still better than the Subtask-Inf model, because the context will reduce the difficulty of the CST, and improve the accuracy of the worker to finish the CST.

7 Conclusions

In this paper, we have studied the result inference problem in crowdsourcing. We identified a category of tasks, context-sensitive tasks, which should not be split into a group of micro tasks to process with crowdsourcing. To best aggregate the results provided by workers, we proposed a novel smart assembly model (SAM) to aggregate contributions of different workers for CSTs. With SAM, a CST was allocated to multiple workers as a whole while the result was obtained by subtask level aggregation with partially ordered set theory. Since CSTs are usually very complex and cannot be well processed with a single iteration of crowdsourcing, we also presented a POMDP model to control the iterative quality. Both simulation and real experiments on a popular crowdsourcing site in China illustrated that our SAM model achieved much higher quality than state-of-the-art Task-Inf and Subtask-Inf under various applications.

Acknowledgment. This work was supported partly by China 973 program (2015CB358700, 2014CB340304), and National Natural Science Foundation of China (61370057). We thank Prof. Yongyi Mao from University of Ottawa for his valuable suggestions.

References

1. Bernstein, M.S., Little, G., Miller, R.C., Hartmann, B., Ackerman, M.S., Karger, D.R., Crowell, D., Panovich, K.: Soylent: a word processor with a crowd inside. In: UIST, pp. 313–322. ACM (2010)

2. Boim, R., Greenshpan, O., Milo, T., Novgorodov, S., Polyzotis, N., Tan, W.C.: Asking the right questions in crowd data sourcing. In: ICDE (2012)
3. Dai, P., Lin, C.H., Weld, D.S., et al.: Pomdp-based control of workflows for crowdsourcing. Artif. Intell. **202**, 52–85 (2013)
4. Dredze, M., Talukdar, P.P., Crammer, K.: Sequence learning from data with multiple labels. In: Workshop Co-Chairs, p. 39 (2009)
5. Feng, J., Li, G., Wang, H., Feng, J.: Incremental quality inference in crowdsourcing. In: Bhowmick, S.S., Dyreson, C.E., Jensen, C.S., Lee, M.L., Muliantara, A., Thalheim, B. (eds.) DASFAA 2014, Part II. LNCS, vol. 8422, pp. 453–467. Springer, Heidelberg (2014)
6. Ipeirotis, P.G., Gabrilovich, E.: Quizz: targeted crowdsourcing with a billion (potential) users. In: WWW, pp. 143–154 (2014)
7. Law, E., Ahn, L.V.: Human computation. Synth. Lect. Artif. Intell. Mach. Learn. **5**(3), 21–22 (2011)
8. Law, E., Zhang, H.: Towards large-scale collaborative planning: answering high-level search queries using human computation. In: AAAI (2011)
9. Lee, K., Tamilarasan, P., Caverlee, J.: Crowdturfers, campaigns, and social media: tracking and revealing crowdsourced manipulation of social media. In: ICWSM (2013)
10. Liu, Q., Peng, J., Ihler, A.T.: Variational inference for crowdsourcing. In: NIPS, pp. 692–700. Curran Associates, Inc. (2012)
11. Liu, X., Lu, M., Ooi, B.C., Shen, Y., Wu, S., Zhang, M.: Cdas: a crowdsourcing data analytics system. PVLDB **5**(10), 1040–1051 (2012)
12. Rodrigues, F., Pereira, F., Ribeiro, B.: Sequence labeling with multiple annotators. Mach. Learn. **95**(2), 165–181 (2014)
13. Salek, M., Bachrach, Y., Key, P.: Hotspotting-a probabilistic graphical model for image object localization through crowdsourcing. In: AAAI (2013)
14. Sheng, V.S., Provost, F., Ipeirotis, P.G.: Get another label? improving data quality and data mining using multiple, noisy labelers. In: SIGKDD. ACM (2008)
15. Sheshadri, A., Lease, M.: Square: a benchmark for research on computing crowd consensus. In: HCOMP (2013)
16. Snow, R., O'Connor, B., Jurafsky, D., Ng, A.Y.: Cheap and fast-but is it good?: evaluating non-expert annotations for natural language tasks. In: Proceedings of the Conference on Empirical Methods in Natural Language Processing (2008)
17. Tran-Thanh, L., Huynh, T.D., Rosenfeld, A., Ramchurn, S.D., Jennings, N.R.: Crowdsourcing complex workflows under budget constraints. In: AAAI (2015)
18. Von Ahn, L., Maurer, B., McMillen, C., Abraham, D., Blum, M.: reCAPTCHA: human-based character recognition via web security measures. Science **321**(5895), 1465–1468 (2008)
19. Young, H.P.: Condorcet's theory of voting. Mathématiques et Sci. Humaines **111**, 45–59 (1990)
20. Zheng, Y., Wang, J., Li, G., Cheng, R., Feng, J.: Qasca: a quality-aware task assignment system for crowdsourcing applications. In: SIGMOD (2015)

Data Quality

CrowdAidRepair: A Crowd-Aided Interactive Data Repairing Method

Jian Zhou[1], Zhixu Li[1(✉)], Binbin Gu[1], Qing Xie[3], Jia Zhu[2],
Xiangliang Zhang[3], and Guoliang Li[4]

[1] School of Computer Science and Technology, Soochow University, Suzhou, China
jzhou_jz@hotmail.com, gu.binbin@hotmail.com, zhixuli@suda.edu.cn
[2] School of Computer Science, South China Normal University, Guangzhou, China
jzhu@m.scnu.edu.cn
[3] The King Abdullah University of Science and Technology, Jeddah, Saudi Arabia
{qing.xie,xiangliang.zhang}@kaust.edu.sa
[4] Department of Computer Science and Technology,
Tsinghua University, Beijing, China
liguoliang@tsinghua.edu.cn

Abstract. Data repairing aims at discovering and correcting erroneous data in databases. Traditional methods relying on predefined quality rules to detect the conflict between data may fail to choose the right way to fix the detected conflict. Recent efforts turn to use the power of crowd in data repairing, but the crowd power has its own drawbacks such as high human intervention cost and inevitable low efficiency. In this paper, we propose a crowd-aided interactive data repairing method which takes the advantages of both rule-based method and crowd-based method. Particularly, we investigate the interaction between crowd-based repairing and rule-based repairing, and show that by doing crowd-based repairing to a small portion of values, we can greatly improve the repairing quality of the rule-based repairing method. Although we prove that the optimal interaction scheme using the least number of values for crowd-based repairing to maximize the imputation recall is not feasible to be achieved, still, our proposed solution identifies an efficient scheme through investigating the inconsistencies and the dependencies between values in the repairing process. Our empirical study on three data collections demonstrates the high repairing quality of CrowdAidRepair, as well as the efficiency of the generated interaction scheme over baselines.

Keywords: Data repairing · Interaction · Rule-based repairing · Crowd-based repairing

1 Introduction

Data repairing aims at discovering and correcting erroneous data in databases. So far, various data repairing solutions have been developed to automatically detect and repair erroneous data in databases [12]. The main stream of rule-based

S.B. Navathe et al. (Eds.): DASFAA 2016, Part I, LNCS 9642, pp. 51–66, 2016.
DOI: 10.1007/978-3-319-32025-0_4

solutions [2,8,9] rely on a variety of quality rules such as FD/CFDs [1,4,11] to detect violations and conflicts between data. By resolving these violations and conflicts, they expect to fix the erroneous data. However, without having the background knowledge, the existing rule-based method just follows some simple modification-strategy (such as minimum-modification) to make modifications [2,9], which as a result, may produce more errors.

Recent efforts use the power of Crowd for data repairing, which let the crowd to help make right modification decisions according to predefined quality rules. Basically, these crowd-based methods can effectively improve the quality of the data after the repairing. For instance, Yakout et al. [13] use user's feedback to repair a database and to adaptively refine the training set for a repairing model. However, dislike using rules, no repairing method can solve all the conflicts with one single repairing model. On the other hand, the NADEEF system [3] allows the users to specify data quality rules and how to repair it through writing code that implements predefined classes. However, although some efforts are made, it still requires high labor cost for data repairing. In addition, any methods relying on humans can not be very efficient since humans need to take rest anyway.

In this paper, we propose a novel combined repairing method, CrowdAidRepair, which performs crowd-based repairing and rule-based repairing alternatively for achieving a high repairing quality at the minimum crowd cost. Specifically, we still rely on FD/CFDs to identify conflicts between values, but we do rule-based repairing to a conflict only when this repairing operation can satisfy a predefined quality constraint. When no more conflict can be repaired by the rules, we select some values for crowd-based repairing to let more values be repairable to rule-based method. We continue with this interactive repairing process iteratively until no more values can be modified. To this end, CrowdAidRepair faces a challenge of selecting the least number of values for crowd-based repairing to maximize the number of values for rule-based repairing. Ideally, an optimal interaction scheme minimizes the number of issued crowd-based repairing operations for resolving the detected conflicts correctly.

This scheduling problem for the interaction is nontrivial: Primarily, to reach the minimum crowd cost, we hope to do crowd-based repairing only to those conflicts that can never be resolvable to rule-based repairing. However, we do not know a priori which conflicts can never be resolved correctly by rule-based repairing until all the other conflicts are resolved. Furthermore, the whole interaction issue is considered in a dynamic setting. As more and more conflicts are resolved, the rule-based repairing result to every unresolved conflict might be changed, and the set of unresolved conflicts will also be changed as some new conflicts will be generated while some old ones will be solved/dismissed.

We analyze in theory that the optimal interaction scheme is not feasible to be achieved, and thus we propose our alternative algorithm that can generate an efficient scheme for the interaction between rule-based repairing and crowd-based repairing. In particular, we investigate the inconsistency that each value brings to the database, according to which we estimate a *disharmonious score* for each value. We will justify that a value with a higher disharmonious score should

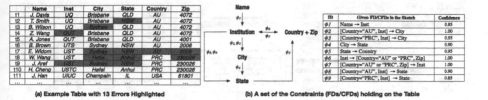

	Name	Inst	City	State	Country	Zip
t1	J. Davis	UQ	Brisbane	QLD	AU	4072
t2	T. Smith	UQ	Brisbane	NSW	AU	4072
t3	B. Wilson	UQ	Sydney	QLD	AU	4072
t4	Z. Wang	QUT	Brisbane	QLD	AU	4072
t5	A. Jones	QUT	Brisbane	QLD	AU	4001
t6	B. Brown	UTS	Sydney	NSW	AU	2006
t7	E. Wldom	UST	Sydney	NSW	AU	2006
t8	W. Wang	UST	Hefei	Anhul	PRC	230026
t9	J. Aref	UST	Sydney	NSW	PRC	230026
t10	H. Cheng	USTC	Hefei	Anhui	PRC	230026
t11	J. Han	UIUC	Champaln	IL	USA	61801
...	...	...	...	...	...	...

(a) Example Table with 13 Errors Highlighted

ID	Given FDs/CFDs in the Sketch	Confidence
$\phi 1$	Name → Inst	0.85
$\phi 2$	[Country="AU", Inst] → City	1.00
$\phi 3$	[Country="PRC", Inst] → City	0.95
$\phi 4$	City → State	0.90
$\phi 5$	State → Country	0.95
$\phi 6$	Inst → [Country="AU" or "PRC", Zip]	1.00
$\phi 7$	[Country="AU" or "PRC", Zip] → Inst	1.00
$\phi 8$	[Country="AU", Inst] → State	0.90
$\phi 9$	[Country="PRC", Inst] → State	0.85

(b) A set of the Constraints (FDs/CFDs) holding on the Table

Fig. 1. A running example for illustration

have a high priority to be checked with the crowd. Besides, although we are in a dynamic setting to schedule conflicts and their covered values for repairing, we can still fix something based on the dependency relations between conflicts and then make decisions accordingly. A challenge is to solve the dependency loop between conflicts and some greedy heuristic algorithm will be proposed to tackle with this NP-hard problem.

Contributions. We develop CrowdAidRepair, a novel Crowd-Aided Data Repairing approach, which performs crowd-based and rule-based repairing alternatively for achieving a high repairing quality at the minimum crowd cost. We identify and study the quality-constrained interaction problem between crowd-based and rule-based repairing, targeting at a balance between repairing quality and repairing cost. After proving in theory that the optimal interaction scheme is unlikely to be identified, we propose our algorithm to generate efficient interaction schemes.

Roadmap. The rest of the paper is organized as follows: We define the problem in Sect. 2, and then present our algorithm in Sect. 3. The experiments are reported in Sect. 4, followed with related work in Sect. 5. We conclude in Sect. 6.

2 Preliminary and Problem Statement

2.1 Preliminary on Rule-Based Repairing

Definition 1. *We say a set of values are correct if all values in this set are correct. We say a* **Conflict** *happens between two sets of values if the two sets of values cannot be both correct.*

The rule-based repairing method relies on a set of predefined quality rules to detect conflicts between data, and then work to resolve these conflicts with expecting to clean relevant errors that have aroused these conflicts. Particularly in this paper, we take FD/CFD as an example of quality rules to show how our method works. For easier understanding, we present the preliminary with a running example.

Example 1. Given a personal contact data depicted in Fig. 1(a), where each tuple contains the Name, Email and Inst (Institution) of a person, in addition to one's

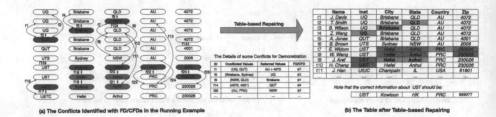

(a) The Conflicts Identified with FD/CFDs in the Running Example   (b) The Table after Table-based Repairing

Fig. 2. The performance of rule-based repairing method

address information: City, State, Country and Zip. We highlight errors waiting to be unveiled and corrected in the table. A set of FD/CFDs holding on the table are listed in Fig. 1(b).

(1) Conflicts Detection. According to the given FD/CFDs, a number of conflicts between data can be detected from the table. For example, according to ϕ_2, t_1[City] ("Brisbane") and t_3[City] ("Sydney") are conflicted with each other as they both correspond to the same Inst ("UQ"). Figure 2(a) shows that 26 conflicts can be identified according to the constraints in Fig. 1(a), where each node denotes an attribute value in the table (erroneous values are highlighted), and each line between two nodes denotes a conflict between the two nodes.

(2) Conflicts Resolution. When a conflict happens between values, some values should be modified in order to resolve the conflict. In order to resolve all the conflicts in a database, some works tend to make the least changes to the data set [2,9], while others prefer to make the most likely correct changes based on some simple prediction model [8,12]. For example in Fig. 2(a), since t_4[Inst] ("QUT") is conflicted with three other values ("UQ"). To resolve the three conflicts, we either change t_4[Inst] ("QUT") into "UQ" (cost is 1), or change the three "UQ" into "QUT" (cost is 3). The first modification way is preferred according to either of the two criterions. Fortunately, this is also the correct modification way.

However, the criterions will make wrong decisions in three situations below:

(1) It is very likely to make wrong decisions based on a simple criterion. For example, both t_8[Inst] ("UST") and t_9[Inst] ("UST") are conflicted with t_{10}[Inst] ("USTC"), to make the least change, t_{10}[Inst] ("USTC") will be changed into "UST". As a result, one more error is produced.
(2) Some conflict contains no errors, such as the conflict f_{10} between t_8[Inst] ("UST") and t_{10}[Inst] ("USTC") in Fig. 2, but the method tends to make corrections to every conflict;
(3) The method can not make right corrections when there is no correct correction value to a position within the data set. For instance, in the conflict f_{13} and f_{14}, both the two values ("Sydney" and "Hefei") are incorrect City that "UST" locates at (which should be "Kowloon (HK)"), but the method will still pick one from these erroneous values as the correction value.

(3) Correction Confidence Estimation. A correction by rule-based repairing is decided jointly by the FD/CFD and all relevant values that used in deducing this correction value. Therefore, the quality of a correction is also determined by the quality of these referred values that are used to deduce this correction value, and the confidence of the referred FD/CFD, that is,

$$c(v_c) = c(\phi) \times \prod_{v_i \in V_R} c(v_i), \tag{1}$$

where V_R contains a set of referred values that used to deduce v_c for the position, $c(v)$ denotes the confidence of a value v, and $c(\phi)$ denotes the confidence of ϕ.

2.2   Problem Statement in the Interaction

We still rely on FD/CFDs to identify conflicts between data, but to identify and correct erroneous values in these conflicts, we consider to involve the crowd into the repairing process to help improve the repairing quality in an efficient interactive way. Particularly, we temporarily neglect the wrong modifications that might be made by the crowd in this paper, and will take it as future work.

The basic interaction can be described as: We set a quality constraint and do rule-based repairing to those conflicts that can satisfy the quality constraint. Then we select some values for crowd-based repairing to let more values be repairable (or so-called *deducible*) to rule-based method. We continue with this interactive repairing process iteratively until no more values can be modified.

The interaction between crowd-based and rule-based repairing can be represented by a sequence of value sets, denoted as $S = \langle T_0, W_1, T_1, W_2, T_2, \cdots, W_n, T_n \rangle$, where W_i is a set of values for repairing at the i-th crowd-based repairing step and T_i is a set of values for repairing at the i-th rule-based repairing step, $\forall i \neq j, W_i \cap T_i = W_i \cap T_j = W_i \cap W_j = T_i \cap T_j = \emptyset$, and $\forall i, W_i \subseteq \mathbb{V}, T_i \subseteq \mathbb{V}$, where $\mathbb{V}$ denotes the domain of all values in the data set. Note that there is no fix number of values for repairing. An interaction scheme is a qualified one as long as it resolves all the detectable conflicts in the data set.

Since the cost of a crowd-based repairing operation is much more expensive than a rule-based repairing operation or any other computational process, the cost of CrowdAidRepair following an interaction scheme S can be roughly represented by the number of values for crowd-based repairing in S, i.e., $cost(S) = \sum_{1 \leq i \leq n} |W_i|$, where $|\cdot|$ is the size of a set.

Definition 2. (Quality-Constrained Interaction Problem). *Given a relational table T for repairing, a set of predefined FD/CFDs Φ holding on T, a quality measuring scheme $c(\cdot)$ and a quality threshold τ $(0 \leq \tau \leq 1)$, the object is to identify an optimal interaction scheme S_{op} for repairing values in T, which satisfies: (1) resolving all the conflicts in T w.r.t. Φ; (2) $\forall v_c, c(v_c) \geq \tau$, where v_c denotes a correction value; (3) $\forall S'$ satisfying the above two conditions, we have $cost(S_{op}) \leq cost(S')$.*

See the situation in the running example, an optimal interaction scheme constructed manually can be:

$<\{t_4[\text{Inst}]\}_w, \{t_3[\text{City}], t_2[\text{State}]\}_t, \{t_7[\text{Country}], t_7[\text{Zip}]], t_8[\text{Zip}]], t_9[\text{Inst}]\}_w,$ $\{t_8[\text{City}]], t_8[\text{State}]]\}_t, \{t_7[\text{City}], t_7[\text{State}]]\}_w, \{t_8[\text{City}]], t_8[\text{State}]]\}_t>$, which has 7 values for crowd-based repairing, and the left 6 values for rule-based repairing.

However, the optimal interaction scheme is not feasible to be constructed automatically as described below. Proof to Theorem 1 can be found with the link: http://ada.suda.edu.cn/Uploads/File/201512/04/1449212591654/proof-dasfaa.pdf.

Theorem 1. *The optimal interaction scheme to the Quality-Constrained Interaction problem is not feasible to be achieved.*

3   A Quality-Constrained Interaction Algorithm

We present our algorithm for generating an efficient interaction scheme. The key problem here lies on how to select values for crowd-based repairing at each crowd-based repairing step.

Initially, we tend to choose the value that has aroused the most conflicts between data for crowd repair, such that the most values will become deducible in the next rule-based repairing step. In order to find out the value that has aroused the most conflicts between data, we estimate a so-called "disharmonious degree" (or *dScore* for short) for each value, to denote the "disharmony" between this value and all the other values in the data set. We will introduce how we estimate the dScore of each value in Sect. 3.1.

In addition to dScore, the dependency relations between conflicts should also be taken into account. We say a conflict f_a depending on another conflict f_b, if some values in f_b are the reasons (or part of the reasons) that have aroused the conflict in f_a according to some FD/CFD. Let a conflict f_a depends on another conflict f_b, we normally should process f_b prior to processing f_a for three reasons below: (1) Initially, it is possible that after the conflict in f_b is resolved, the conflict in f_a is dismissed automatically without any repairing operations. (2) To say the least, even if f_a is a true conflict and we need to do crowd-based operations to check the values inside it, sometimes we have no other choices but to rely on those values in the conflicts that f_a depends on to formulate crowd-based repairing queries. (3) Lastly, after we process all conflicts it depends on, we can update the *dScores* for the values in f_a for better judging which value is more likely an error.

Although we are in a dynamic setting to schedule conflicts and their covered values for repairing, we can still fix something based on the dependency relations between conflicts and then make decisions accordingly. A challenge is to solve the dependency loop between conflicts. We will discuss this in Sect. 3.2.

3.1 dScore: Estimating the Incorrectness of Values

The dScore of a value can be roughly reflected by the number of conflicts it brings to the data set. We first introduce how to calculate the *dScore* for each value in a simplified case. To begin with, we assume that the data set is consistent without the value at a position, that is, all the other values in the data set appear to be in harmony. Then the value at this position comes, which may bring conflicts in two ways: (1) itself conflicts with some values; (2) it may let some values involved in a conflict. Usually, the more conflicts it brings to the data set, the higher probability it is an erroneous value. In other words, the *dScore* of a value can be manifested as the number of conflicts it caused in this simple setting.

We now consider the situation in real case, where there are already erroneous values and conflicts in the data set. When a new value at a position comes, either an erroneous one or not, it brings some changes anyway, such as producing new conflicts, or voting for existing conflicts. In this case, the *dScore* of a value can be manifested by two things: (1) the new conflicts produced, and the "credibility", or what we call the *cScore* of these conflicts, which will be discussed in Eq. (3); (2) the changes on the *cScore* of existing conflicts. Specifically, *dScore(v)* of a value v can be calculated by:

$$dScore(v) = \alpha \times \sum_{f \in F(v)} \Delta(cScore(f)) \qquad (2)$$

where α is a normalization factor to scale $dScore(v)$ between 0 and 1, $F(v)$ contains all conflicts that are influenced by putting v into the data set, and $\Delta(cScore(f))$ is the change on the *cScore* of a conflict f.

Fig. 3. An example conflict with relevant values and CFD

In particular, the *cScore* of a conflict f is decided by four relevant values as given in Fig. 3. Previous work considers that a conflict is consisted of two values such as v_1 and v_2 in the figure, but a conflict is also closely related to another two values which are referenced to identify the conflict according to a certain CFD, such as the two v_3 in the figure. Thus, the correctness of the four values jointly decide the *cScore* of a conflict f. Furthermore, when a conflict is voted as a conflict by several groups of values w.r.t. different CFDs, we only pick the one with the highest *cScore* as the final *cScore* of the conflict. More specifically,

$$cScore(f) = \underset{\phi \subseteq \Phi(f)}{ArgMax}[c(\phi) \times \prod_{v_i' \in V(f,\phi)} (1 - dScore(v_i'))] \qquad (3)$$

where $\Phi(f)$ is the set of CFDs that voted f as a conflict, and $V(f, \phi)$ contains all values related to the conflict f w.r.t. ϕ.

3.2 Employing Dependencies Between Conflicts

We consider the dependencies between conflicts in scheduling conflicts for repairing. We first get the dependency relations among all conflicts, and then build a conflicts dependency graph based on these relations.

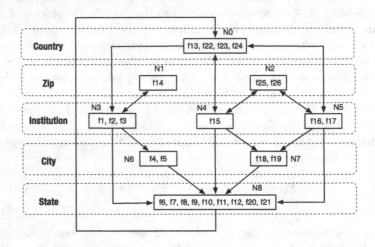

Fig. 4. The dependency graph of the conflicts in Fig. 1(a)

(1) Relations Between Conflicts. Basically, there are three kinds of relationships between each pair of conflicts. The first is the *Dependency Relation* as we introduced above. Note that the dependency relation is transitive, that is, if f_a depends on f_b, and f_b depends on f_c, then f_a also depends on f_c. Secondly, we say two conflicts are in a *Overlapped Relation* if they share some positions, such as f_1 and f_2 sharing $t_4[\mathtt{Inst}]$("QUT") in Fig. 2(a). Finally, If two conflicts are in neither of the two relations above, they are *Independent* from each other.

(2) Building Conflicts Dependency Graph. With the relations between all conflicts, we can built a conflicts dependency graph as in Fig. 4 (which is built on Fig. 1(a)) through the following steps:

(1) Initially, we take each conflict as a node in the dependency graph.
(2) We then put a directed edge pointing from every conflict f_a to every other conflict f_b if f_b depends on f_a. Note that we only need to put an edge between two conflicts if one directly depends on the other.
(3) Finally, to make the graph easier to process, we merge nodes sharing at least one value into one node (i.e., we put overlapped conflicts into one node), and the directed edges of the same direction between the two nodes are merged into one directed edge.

As introduced above, a conflict should be processed after all the conflicts it depends on are processed. But for those overlapped conflicts in the same node, we need to consider the priority of each value that involved in the conflicts for checking. Here we can still rely on the *dScores* of these values. A value with a highest *dScore* in a node can be checked firstly. Each time a value is modified, the graph needs to be updated accordingly.

3.3 Tackling Dependency Loops

The main challenge here is how to schedule those conflicts in dependency loops for processing. We say a number of conflicts are in dependency loop if they depend on each other such as f_1, f_2, f_3 and f_{14}. In this situation, the dependency-based interaction principle mentioned above dose not work at all. Things become more intractable when there are several loops overlapped with each other at different nodes. As in Fig. 4, there are 19 loops in total and almost every loop is overlapped with some other loops at some nodes. Basically, we have to choose one (or more than one) node in a loop to process to "break up" the loop. In order to minimize the cost, we have to be very careful in selecting the *break-up node* for a loop as different break-up nodes will bring different costs.

Theorem 2. *It is an NP-hard problem to break up loops in a dependency graph with the minimum crowd-based repairing cost.*

We put the proof to Theorem 2 online. In the following, we give our greedy algorithm to break up loops in a dependency graph.

(1) Breaking up a Single Loop. We basically consider two factors in selecting the break-up node for a loop: (1) factor 1: the number of values that must be verified in a node for breaking up the loop (for easier presentation, we call these values as break-up values); (2) factor 2: the *dScores* of these break-up values in a node. Usually, we tend to select the node with the least number of break-up values holding the highest *dScores* as the break-up node for the loop. More specifically, we calculate a break-up score, or *bScore* for short, for each node in a loop as given in Eq. 4 below. Among all nodes in a loop, the node with the highest *bScore* will be selected as the break-up node in priority.

$$bScore(\mathcal{N}, \mathcal{L}) = \prod_{v \in V_b(node, loop)} dScore(v) \tag{4}$$

where $V_b(\mathcal{N}, \mathcal{L})$ is the set of break-up values in *node* for breaking up *loop*.

(2) Breaking up Multiple Loops. For a number of loops overlapped with each other, we can not simply decide the break-up nodes for a single loop. Otherwise, we may not be able to reach the best performance in minimizing the number of crowd-based operations. For each node, we consider a global *bScore*, or *gbScore* for short, to denote its break-up score for all loops in the graph, and the one with the highest *gbScore* will be selected as the break-up node in priority. The *gbScore* of a node is decided by two factors: (1) the local *bScore* of the node in

---

Algorithm 1. Dependency-Aware Interaction

Input : A table with a set of conflicts $\mathbb{F}$
Output: A repairing scheme $\mathcal{S} = \langle \mathcal{T}_0, \mathcal{W}_1, \mathcal{T}_1, \cdots, \mathcal{W}_n, \mathcal{T}_n \rangle$

Set $i = 0$;
while $\mathbb{F} \neq \emptyset$ **do**
    1. $\mathcal{T}_i \leftarrow$ All deducible values at the moment;
    2. Deducing all values in $\mathcal{T}_i$;
    3. Updating $\mathbb{F}$;
    4. $i++$;
    5. Calculating $dScores$ for all values in $\mathbb{F}$ with Eq. 2;
    6. Building the Dependencies Graph on $\mathbb{F}$;
    7. **while** *no new deducible values* **do**
        $V \leftarrow$ Values in conflicts depending on nothing;
        if $V \neq \emptyset$ **then** $\mathcal{W}_i \leftarrow \mathcal{W}_i \cup V$;
        else
            Calculating $gbScores$ for all conflicts in $\mathbb{F}$ with Eq. 5;
            $V \leftarrow$ Values with the highest $dScore$ in conflicts with the highest $gbScore$;
            $\mathcal{W}_i \leftarrow \mathcal{W}_i \cup V$
        end
        Checking/Repairing V with the Crowd;
        if V *is updated with correction values* **then**
            Updating $\mathbb{F}$ and $dScores$;
        end
    end
end
return $\langle \mathcal{T}_0, \mathcal{W}_1, \mathcal{T}_1, \cdots, \mathcal{W}_n, \mathcal{T}_n \rangle$;

---

each loop; and (2) the benefit of solving each loop, which is actually the number of values that can be moved out from the loops. More specifically,

$$gbScore(\mathcal{N}) = \sum_{\mathcal{L} \in L(\mathcal{N})} [bScore(\mathcal{N}, \mathcal{L}) \times benefit(\mathcal{N}, \mathcal{L})] \qquad (5)$$

where $L(\mathcal{N})$ is the set of loops having $\mathcal{N}$ as its node in the graph, and the $benefit(\mathcal{N}, \mathcal{L})$ is the benefit of breaking up $\mathcal{L}$ by solving $\mathcal{N}$, which is mainly decided by the number of values in $\mathcal{L}$.

3.4   Dependency-Aware Interaction Algorithm

A formal description of this algorithm is given in Algorithm 1. Initially, we build the conflicts dependency graph for a data set. For those nodes depending on nothing, we keep on choosing the value with the highest $dScore$ within each node for crowd-based repairing until all the conflicts in the node are resolved. When there is no node of this kind but only loops, we calculate the $gbScores$ for all nodes in these loops, and choose the one with the highest $gbScore$ to process to break up the loops. Each time a value is modified, we need to update the graph and all $bScores$ and $gbScores$. The algorithm stops when the graph is empty. Basically, the computation complexity of Algorithm 1 is $O(m\log m + n)$, where m is the number of nodes which are in the loops of the graph and n is the number of nodes which are not in the loops of the graph.

Example 2. We apply the algorithm to the running example and the interaction scheme generated is depicted in Table 1. Overall, we issue crowd-based operations for 10 values, among which 8 values are true erroneous values while the other 2 values are correct values. Meanwhile, 5 erroneous values are corrected by rule-based repairing.

Theorem 3. *The crowd cost of the scheme generated by Algorithm 1 is not larger than* $\frac{e-1}{e}$ *times the cost of the optimal interaction scheme.*

We also put the proof to Theorem 3 online.

4 Experiments

We perform our experiments on two real and one synthetic data sets. We also employ 20 users in our research group to act as crowd, all of whom have known part of the ground-truth knowledge a priori.

(1) Personal Information Table (**PersonInfo**): This is a 50k-tuples, 9-attributes table, which contains contact information for academics including name, email, title, university, street, city, state, country and zip code.
(2) DBLP Publication Table (**DBLP**): This is a 100k-tuples, 5-attributes table. Each tuple contains information about a published paper, including its title, first author and his/her affiliation, conference name, year and venue.
(3) Synthetic Table (**Syn**): We also generate a 1million-tuples, 100-attributes table following a scheme containing 100 randomly generated approximate attribute dependencies with confidences near-uniformly distributed between 0.7 and 1, where the first attribute is the key attribute.

All the three data sets are relational tables without erroneous data. To generate tables with errors for the experiments, we keep the key attribute value in each tuple and replace non-key attribute values at random positions with attribute values selected from random picked tuples of the table.

4.1 Repairing Quality Evaluation

In the following experiments, we compare the repairing quality of PureCrowdRepair (Pure crowd-based Repairing) and CrowdAidRepair with four state-of-the-art general textual data repairing approaches on the three data sets.

Table 1. The interaction scheme generated by Algorithm 1

$\mathcal{T}_0$	$\emptyset$		
$\mathcal{W}_1$	$t_7[\texttt{Inst}]$("UST") is correct, not changed; $t_4[\texttt{Inst}]$("QUT") is incorrect, modified; $t_3[\texttt{City}]$("Sydney") is incorrect, modified; $t_2[\texttt{State}]$("NSW") is incorrect, modified; $t_8[\texttt{Inst}]$("UST") is correct, not changed; $t_9[\texttt{Inst}]$("UST") is incorrect, modified;	$\mathcal{W}_2$	$t_8[\texttt{Zip}]$("230026") is incorrect, modified;
		$\mathcal{T}_2$	$t_7[\texttt{Zip}]$("2006") is incorrect, modified;
		$\mathcal{W}_3$	$t_8[\texttt{City}]$("Hefei") is incorrect, modified;
		$\mathcal{T}_3$	$t_7[\texttt{City}]$("Sydney") is incorrect, modified.
		$\mathcal{W}_4$	$t_8[\texttt{State}]$("Anhui") is incorrect, modified;
		$\mathcal{T}_4$	$t_7[\texttt{State}]$("NSW") is incorrect, modified;
$\mathcal{T}_1$	$t_9[\texttt{City}]$("Sydney") is incorrect, modified; $t_9[\texttt{State}]$("NSW") is incorrect, modified; $t_9[\texttt{Country}]$("PRC") is correct, not changed;	$\mathcal{W}_5$	$t_7[\texttt{Country}]$("AU") is incorrect, modified;
		$\mathcal{T}_4$	$t_8[\texttt{Country}]$("PRC") is correct, not changed

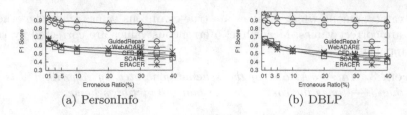

(a) PersonInfo   (b) DBLP

Fig. 5. Comparing the F1 scores of all methods

Table 2. Comparing the repairing quality with previous methods

	PersonInfo			DBLP		
	Prec	Recall	F1	Prec	Recall	F1
CFD-ML	0.535	0.521	0.528	0.645	0.451	0.531
ERACER	0.638	0.489	0.554	0.698	0.421	0.525
SCARE	0.655	0.512	0.574	0.743	0.453	0.563
GuidedRepair	**0.825**	**0.804**	**0.814!!**	**0.871**	**0.837**	**0.8535!!**
PureCrowdRepair	**0.932**	**0.891**	**0.911**	**0.975**	**0.911**	**0.942**
CrowdAidRepair	**0.903**	**0.876**	**0.889**	**0.958**	**0.895**	**0.925**

(1) Rule-based Most-Likely (**CFD-ML**): This approach relies on FD/CFDs to detect and correct erroneous data [2], and follows the most-likely correct modification criterion as introduced in Sect. 2.1.

(2) Model-based 1 (**ERACER**): This is a model-based repairing based on belief propagation and relational dependency networks [10]. In contrast to prior work that cleans tuples in isolation, this approach exploits the graphical structure of the data to propagate inferences throughout the database [10].

(3) Model-based 2 (**SCARE**): This is another model-based repairing approach based on maximizing the correctness likelihood of replacement data given the data distribution, which is modelled using statistical machine learning techniques [12].

(4) Crowd-based (**GuidedRepair**): We implement the state-of-the-art crowd-based repairing method proposed in [13], which collects feedbacks from users to adaptively refine the training set for a repairing model.

We first make a comprehensive comparison on the Precision, Recall and F1 of all the methods at an erroneous ratio of 10 % on the two real data sets. The parameter setting for each method lets the method reaches the best repairing quality (w.r.t. F1). As shown in Table 2, the precision and recall of the rule-based method (CFD-ML) are not high, as it can only make correct modifications to about half of the erroneous values in the data sets, and in 40–60% chances it makes wrong corrections. Comparatively, the precision of the two model-based methods (ERACER and SCARE) is a bit (5–10%) higher than the rule-based methods since the models they build can understand the correlation between data and thus make better judgements. On the other hand, their recall is as low as that of the rule-based method, since there are some non-quantitative attributes like email, street, author, and venue which can not be handled well by models.

Apparently, the precision and recall of GuidedRepair, PureCrowdRepair and CrowdAidRepair are much higher (85+% precision and 85+% recall) than the rule-based and model-based methods. In particular, PureCrowdRepair reaches the highest precision and recall as it lets the crowd to do every correction. The precision and recall of our method (CrowdAidRepair) is a bit less than the PureCrowdRepair method, but higher than the GuidedRepair method. This is because our method takes the advantages of both rules and crowd. which can work better than a model, even if the model is refined by the crowd.

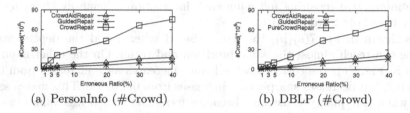

(a) PersonInfo (#Crowd) (b) DBLP (#Crowd)

Fig. 6. Comparing #Crowd on the real data sets ($\tau = 0.7$, erroneous ratio $= 10\%$)

We then compare the F1 scores of all methods at various erroneous ratios ($1\%, 3\%, 5\%, 10\%, 20\%, 30\%, 40\%$) by setting $\tau = 0.7$ over the two real-world data sets. As demonstrated in Fig. 5, CrowdAidRepair always reaches higher F1 scores than all the other four methods including the GuidedRepair method at various erroneous ratio, which also proves the advantage of CrowdAidRepair over the four other methods.

4.2 Repairing Cost Evaluation for Methods Using the Crowd

In this paper, we denote the number of values for crowd repair (**#Crowd**) as the crowd cost of a repairing method. We now compare the crowd cost of CrowdAidRepair with PureCrowdRepair and GuidedRepair on **#Crowd**.

As demonstrated in Fig. 6(c)(d), CrowdAidRepair only uses about 20 % crowd cost of that used by PureCrowdRepair and thus greatly reduces the overhead of the repairing process. However, compared with GuidedRepair, we use a bit higher overhead for reaching a higher repairing quality.

4.3 Interaction Schemes Evaluation

To further evaluate the effectiveness of CrowdAidRepair, we compare the effectiveness of two interaction schemes generated by (1) an algorithm relying on dScores only to find the interaction scheme (so called dScore-based scheme);

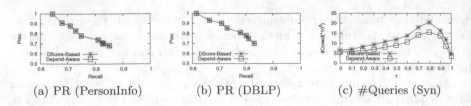

(a) PR (PersonInfo)   (b) PR (DBLP)   (c) #Queries (Syn)

Fig. 7. Comparing the schemes: precision and recall in subfigures (a)(b) (on the two real data set), and #Queries in Subfigures (c) (Syn data set)

and (2) the CrowdAidRepair algorithm that considers both dScores and conflict dependencies (so called depend-aware scheme). We set the erroneous ratio to 10 %, and then compare the repairing quality (precision and recall) and cost of the interactive repairing following each interaction scheme by changing the quality threshold τ from 0 to 1.

As shown in Fig. 7(a)(b), the dScore-based scheme and the depend-aware scheme can reach almost the same precision and recall. On the other hand, as shown in Fig. 7(c), the cost of both schemes increases as τ increases from 0 to about 0.8, but decreases sharply as τ increases from 0.8 to 1.0. This makes sense since when the quality constraint becomes too strict, much less values can be repaired to satisfy the constraint. Nonetheless, the cost of the depend-aware interaction scheme is always about 40 % less than the dScore-based scheme, which proves the advantage of the depend-aware scheme over the other scheme.

5   Related Work

Data repairing aims at discovering and correcting erroneous data in databases. So far, various data repairing solutions have been developed to automatically detect and repair erroneous data in databases [12]. All existing solutions can be roughly put into three categories below.

The traditional category of methods relies on a variety of constraints including FDs [1,11], CFDs [4], Integrity Constraints [9] and Inclusion Dependencies (INCs) [1] to detect inconsistency (or conflicts) between data aroused by erroneous data, and then work on resolving all the conflicts with expecting to fix all erroneous data in this way [2,8,9]. For general textual databases, most work in this category use FD/CFDs for repairing as they are the constraints within a single relational table, while some other work uses INCs for repairing between multiple relational tables. Usually, this category of methods can effectively detect a large percent of erroneous data involved in the identified conflicts in a wide range of databases, but to repair these errors and resolve the conflicts, some work tends to make the least changes to the data set [2,9], while others prefer to make the most likely correct changes based on some simple prediction model [8,12]. However, neither criterion can have all errors modified correctly.

The second category of solutions are model-based repairing, which usually build some prediction models for detecting and correcting erroneous values in

a data set [6,7,10,12,14,15]. The construction of the model employs statistical Machine Learning (ML) techniques for data cleaning, which can effectively
capture the dependencies and correlations between data in the dataset based on
various analytic, predictive or computational models [12,15]. However, not every
erroneous data can be identified and corrected in the right way since there are
always outliers that do not obey the captured constraints.

The third category of solutions are external source based repairing
approaches, which leverage the information in reference master data set [5] or
user's interaction data such as GuidedRepair [13] and NADEEF [3] for better
data cleaning performance. However, the required external information is not
always available and thus the methods can not be applied in general scenarios.
In this paper, we propose CrowdAidRepair, which is a hybrid repairing approach
using the crowd.

6 Conclusions and Future Work

We propose CrowdAidRepair, a novel crowd-aided data repairing approach that
can greatly enhance the repairing quality of the existing rule-based repairing
method with the Crowd help. Extensive experimental results based on several
data collections demonstrate that the generated interaction scheme decreases
on average 60 % cost of a baseline, and reaches almost the same high repairing
quality that was reached by a pure crowd-based retrieving approach. Future
work may consider combining CrowdAidRepair with state-of-the-art model-
based methods, and apply CrowdAidRepair to databases with both incorrect
values and missing values.

Acknowledgements. This research is partially supported by Natural Science Foundation of China (Grant No. 61303019, 61402313, 61472263, 61572336), Postdoctoral
scientific research funding of Jiangsu Province (No. 1501090B) National 58 batch of
postdoctoral funding (No. 2015M581859) and Collaborative Innovation Center of Novel
Software Technology and Industrialization, Jiangsu, China.

References

1. Bohannon, P., Fan, W., Flaster, M., Rastogi, R.: A cost-based model and effective
   heuristic for repairing constraints by value modification. In: SIGMOD, pp. 143–154
   (2005)
2. Cong, G., Fan, W., Geerts, F., Jia, X., Ma, S.: Improving data quality: consistency
   and accuracy. PVLDB, 315 326 (2007)
3. Dallachiesa, M., Ebaid, A., Eldawy, A., Elmagarmid, A., Ilyas, I.F., Ouzzani, M.,
   Tang, N.: Nadeef: a commodity data cleaning system. In: SIGMOD, pp. 541–552
   (2013)
4. Fan, W., Geerts, F., Jia, X., Kementsietsidis, A.: Conditional functional dependencies for capturing data inconsistencies. ACM Trans. Database Syst. (TODS)
   33(2), 6 (2008)

5. Fan, W., Li, J., Ma, S., Tang, N., Yu, W.: Towards certain fixes with editing rules and master data. PVLDB **3**(1–2), 173–184 (2010)
6. Hua, W., Wang, Z., Wang, H., Zheng, K., Zhou, X.: Short text understanding through lexical-semantic analysis. In: International Conference on Data Engineering (ICDE) (2015)
7. Koh, J.L.Y., Li Lee, M., Hsu, W., Lam, K.T.: Correlation-based detection of attribute outliers. In: Kotagiri, R., Radha Krishna, P., Mohania, M., Nantajeewarawat, E. (eds.) DASFAA 2007. LNCS, vol. 4443, pp. 164–175. Springer, Heidelberg (2007)
8. Kolahi, S., Lakshmanan, L.V.: On approximating optimum repairs for functional dependency violations. In: ICDT, pp. 53–62 (2009)
9. Lopatenko, A., Bravo, L.: Efficient approximation algorithms for repairing inconsistent databases. In: ICDE, pp. 216–225 (2007)
10. Mayfield, C., Neville, J., Prabhakar, S.: Eracer: a database approach for statistical inference and data cleaning. In: SIGMOD, pp. 75–86 (2010)
11. Wijsen, J.: Database repairing using updates. ACM Trans. Database Syst. (TODS) **30**(3), 722–768 (2005)
12. Yakout, M., Berti-Équille, L., Elmagarmid, A.K.: Don't be scared: use scalable automatic repairing with maximal likelihood and bounded changes. In: SIGMOD, pp. 553–564 (2013)
13. Yakout, M., Elmagarmid, A.K., Neville, J., Ouzzani, M., Ilyas, I.F.: Guided data repair. PVLDB **4**(5), 279–289 (2011)
14. Zheng, B., Yuan, N.J., Zheng, K., Xie, X., Sadiq, S., Zhou, X.: Approximate keyword search in semantic trajectory database. In: 2015 IEEE 31st International Conference on Data Engineering (ICDE), pp. 975–986. IEEE (2015)
15. Zhu, X., Wu, X.: Class noise vs. attribute noise: a quantitative study. Artif. Intell. Rev. **22**(3), 177–210 (2004)

Crowdsourcing-Enhanced Missing Values Imputation Based on Bayesian Network

Chen Ye[✉], Hongzhi Wang, Jianzhong Li, Hong Gao, and Siyao Cheng

Department of Computer Science and Technology,
Harbin Institute of Technology, Harbin, China
{yech,wangzhi,lijzh,honggao,csy}@hit.edu.cn

Abstract. Due to development of the Internet, the size of data continue to be large and rough. During the process of data collection, different kinds of data problems occurred, among where incompleteness is one of the most serious problems to deal with. The existing methods for missing values imputation have mostly relied on using statistics and machine learning. These methods are known to be limited in efficiency and accuracy, which are caused by high dimensional calculation and low quality of initial data. In this paper, we propose a new method combining Bayesian network and crowdsourcing to deal with missing values together. We use Bayesian network to inference missing values to improve efficiency while use crowdsourcing to obtain additional information in need to improve accuracy. Experiments on real datasets show that our methods achieve better performance compared to other imputation methods.

Keywords: Missing values · Bayesian network · Crowdsourcing

1 Introduction

Missing values have negative effects on data quality. Missing values may lead to the loss of important information. For example, missing values always exist in the medical area, and the incomplete information can lead to biased result which has a serious influence to our health [1].

Due to its importance, many efforts have been done to deal with missing values [2–9,17,23–25]. The job of filling missing values is also called imputation. Traditional imputation methods can mainly be divided into two categories, statistics-based methods [2–5] and machine-learning-based methods [6–9,17]. Even though these methods can solve incomplete problems in a specific situation, there still exist efficiency and accuracy issues.

Statistics-based methods such as EM algorithm [2], regression [3,4] and sampling methods [5] always estimate a large number of parameters, thus calculation is complex and cost is large.

Machine-learning-based methods such as naive Bayesian methods [6,7], clustering methods [8], neural network methods [9], and decision tree methods [17] use initial data to train the classifier to choose the most possible value as the

© Springer International Publishing Switzerland 2016
S.B. Navathe et al. (Eds.): DASFAA 2016, Part I, LNCS 9642, pp. 67–81, 2016.
DOI: 10.1007/978-3-319-32025-0_5

missing value. As these methods can only deal with one variable at one time, it is of low efficiency to fill the missing values of all the variables.

What's more, accuracy of both statistics-based methods and machine-learning-based methods is limited to the quality of initial data. When the proportion of missing values is too large to provide enough information for missing values imputation, the result tends to low accuracy.

To fill missing values efficiently and accuracy, we propose novel imputation algorithms based on Bayesian network and crowdsourcing.

As Bayesian network can overcome the high dimensional calculation of statistics methods and make a connection between variables based on causal relationship to fill the missing values of all the variables at one time, it can overcome the limitations of the above methods in terms of efficiency. Thus we choose Bayesian network as our basic imputation structure to deal with missing values.

However, we also notice that when the proportion of missing values is too large, the evidence maybe not enough for Bayesian network to estimate the missing value. We need additional evidence for Bayesian network to improve the accuracy. As calling for experts is limited to cost, we use crowdsourcing to reduce the cost and gather the wisdom of people. In recent years, crowdsourcing has been successfully used in many areas such as image processing [10,11], entity recognition [12,13], schema matching [14]. With the help of crowdsourcing, we can overcome the limitations of information shortage. To our best knowledge, we are the first one to bring crowdsourcing as human involvement combining with Bayesian network into the area of missing values imputation.

Challenges. Although combining Bayesian network with crowdsourcing for missing values imputation has many benefits, it also brings two new problems.

1. As the quality of Bayesian network strongly depends on precision of probability distribution, Bayesian network construction based on incomplete data tends to be hard. In the case of lacking a amount of missing values, we cannot achieve accurate parameters which may leads to ambiguity and mistakes. Then how to achieve a good Bayesian network structure becomes a problem.
2. As we use crowdsourcing to provide additional information to improve accuracy of Bayesian inference, cost of crowdsourcing should be minimized. Thus number of crowd questions is also limited. Therefore, we need to present a crowd question selection strategy to maximize accuracy and reduce cost.

We address two main research challenges that arise in Bayesian network with crowdsourcing. To address the first challenge, we develop a Bayesian network construction algorithm which is specific to incomplete data based on relevance relationship. The second challenge is how to design a crowd question selection strategy. We prove that selecting a certain number of questions that maximize accuracy is NP-hard and present a greedy crowd questions selection algorithm.

We summarize our contributions as follows:

1. *Bayesian Networks Construction.* We first define the reliability of variables and propose a reliability-based algorithm to construct Bayesian network for incomplete data.

2. *Missing Values Imputation.* Given a Bayesian network, we fill missing values with different categories: {a} Filling missing values based on Bayesian inference; {b} Filling missing values based on the answers from crowd. We present a missing value imputation algorithm based on Bayesian inference and improve it with the join of crowdsourcing.
3. We conducted experiments on real datasets to show that performance of our methods outperformed other imputation methods in terms of accuracy and efficiency.

In this paper, we introduce basic concepts about Bayesian network and problem definition in Sect. 2. We propose the method of Bayesian network construction as well as the missing values imputation algorithm based on Bayesian Inference in Sect. 3. In Sect. 4 we develop our missing values imputation method with the join of crowdsourcing. We evaluate all of our methods with other methods for comparison on real datasets and analyze experimental results in Sect. 5 and draw a conclusion in Sect. 6.

2 Bayesian Network and Problem Definition

As Bayesian network is a probability network based on probability inference, we first introduce some basic concepts about probability theory.

2.1 Probability Theory

Definition 1 *(Conditional Probability).* *For two events X and Y, x_i represents the finite state of X while y_j represents the finite state of Y. The conditional probability of x_i given y_j is defined as:*

$$P\left(x_i|y_j\right) = \frac{P(x_i, y_j)}{P(y_j)} \tag{1}$$

where $P(x_i, y_j)$ represents the joint probability that x_i and y_j occurred at the same time. If X is independent with Y, then Eq. 1 turns to:

$$P\left(x_i|y_j\right) = P(x_i). \tag{2}$$

Definition 2 *(Conditional Independence).* *For three events X, Y and Z, x_i, y_j, z_k represent the finite state of X, Y, Z respectively. If X is independent with Y given Z, the conditional probability is defined as:*

$$P\left(x_i|y_j, z_k\right) = P\left(x_i|z_k\right). \tag{3}$$

2.2 Bayesian Network

Bayesian network is a directed acyclic graph, in which each node represents a variable and each directed edge represents causal relationship between variables. For example, a possible Bayesian network structure with seven nodes is shown in Fig. 1.

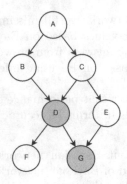

Attribute	Reliability
A	0.9753
B	0.9383
C	0.9136
D	0.8889
E	0.8025
F	0.7834
G	0.7525

Fig. 1. A Bayesian network example   **Table 1.** Reliability of attributes

Definition 3 *(Bayesian Network Rule). Due to Pearl's conclusion [15], given a Bayesian network G, each variable V is conditionally independent of all its non-descendants given its parents Pa(V), where Pa(V) refers to variables directly preceding variable V.*

According to Definition 3, conditioning on variable D in Fig. 1 makes variable F independent to other variables like B and C. It can be defined as $P(F|B, C, D) = P(F|D)$.

2.3   Problem Definition

In this section, we define our problem based on Bayesian network. Since Bayesian network can estimate the conditional probability distribution in the presence of missing value [21], we can use it to provide us probabilistic knowledge to inference missing values. With Bayesian network, we can find the most possible value by maximum conditional probability. Thus, we convert missing value imputation problem into maximum a posteriori hypothesis (MAP) problem on Bayesian network and define it as follows.

Definition 4 *(Problem Definition).   Given incomplete data set D and a Bayesian network G = (V, E) with V as variables, E as relationship between variables, for each missing value of variable V, the goal is to find the most possible value x conditioning with evidences E according to G. The most possible value is defined as follows.*

$$x^* = argmax_x P(V = x | E = e). \qquad (4)$$

3   Missing Values Imputation

In this section, we propose our method of missing values imputation in two steps. Firstly, we describe process of Bayesian network construction specific to incomplete data. Then we fill missing values with probability inference according to Bayesian network.

3.1 Bayesian Network Construction

As traditional methods of building Bayesian network like K2 [18] and Hill-climbing [19] are mainly suitable for complete data which can get causal relationship between variables according to strict score function. However, several problems arise when apply these methods to incomplete data.

1. Lack of data will influence computational accuracy of score function, thus influence judgement of causal relationship between variables.
2. Causal relationship may not always exists between variables in incomplete data. Absence of causal relationship may lead to low accuracy of Bayesian network construction.

Thus, we give up building Bayesian network according to causal relationship, and turn to find relevance relationship between variables. We build nodes according to their reliability instead of randomly selection, and add edges towards relevance score between variables. In the following parts of this section, we give definition of relevance score between variables and reliability of variables in turn, then propose Bayesian network construction algorithm.

As Pearson Correlation Coefficient [16] can accurately measure correlation between incomplete data, we use it as score function to measure relevance relationship between variables.

Definition 5 *(Relevance Score). For each pair of variables (V_i, V_j), relevance score of (V_i, V_j) is defined as:*

$$\rho_{ij} = |Corrcoef(V_i, V_j)| = \frac{|Cov(V_i, V_j)|}{\sqrt{Var(V_i)Var(V_j)}} \tag{5}$$

where $Cov(V_i, V_j)$ and $Var(V_i), Var(V_j)$ represent covariances and variances between V_i and V_j, respectively.

Before we use such relevance relationship to build Bayesian network, we need an order of variables $V = \{V_1, V_2, \ldots, V_n\}$ to build Bayesian network in turn. When we consider variable V_i, we measure relationship between V_i and variables $\{V_1, V_2, \ldots, V_{i-1}\}$, which appear ahead of V_i and select its parents among $\{V_1, V_2, \ldots, V_{i-1}\}$. As the order mainly determines scope of parents of variable V_i, a random order does not seem to be a good choice.

Since in most cases, missing values appear randomly. The distribution of missing values of each attribute is usually not the same. It can be easily inferred that variables with fewer missing values more reliable. Thus we prefer to order variables according to their reliability, which can make sure that during construction of Bayesian network, edges representing relevance relationships between variables are generated from the more reliable one to the less reliable one. Thus, we can infer the missing values of variable V_i from its more reliable parents.

We first define missing value set of each variable, and then give the definition of reliability of each variable.

Definition 6 *(Missing Value Set).* *Given incomplete data set D containing tuples $T = \{t_1, t_2, \ldots, t_n\}$, for each variable V, missing value set is defined as:*

$$M = \{t | t \in T, t_i = \text{``?''}\} \tag{6}$$

Definition 7 *(Reliability).* *For each variable V_i, given missing values set M_i, the reliability of V_i is defined as:*

$$Rel(V_i) = 1 - \frac{|M_i|}{n} \tag{7}$$

where $|M_i|$ and n represent the number of M_i and total number of tuples, respectively.

Algorithm 1 shows the main process of Bayesian network construction. We first sort attributes according to their reliability. A possible order according to the reliablility of variables is shown in Table 1. For each attribute we create nodes in order and find parents for each node according to relevance score. Finally we create edges from parents to each node. Node that in Line 8–13, we compare relevance score iteratively and select the nodes with higher score constantly. After Algorithm 1 is finished, we obtain a Bayesian network G with the most reliable node on the top and relative edges between each node.

Algorithm 1. BN_CONSTRUCTION(D)

Intput: Incomplete data set $D = \{V_1, V_2, \ldots, V_m\}$ with V_i as the $i - th$ variable.
Output: A Bayesian network G
 1: **for** $i = 1$ to m **do**
 2:     calculate $Rel(V_i)$
 3:     $X_1, X_2, \ldots, X_m$ = sort $V_1, V_2, \ldots, V_m$ according to $Rel(V_i)$ in descending order
 4: **for** $i = 1$ to m **do**
 5:     create a node for X_i
 6:     old-score=0
 7:     parents$(X_i) = \phi$
 8:     **for** $j = 1$ to i-1 **do**
 9:         add-node=ϕ
10:         new-score=ρ_{ij}
11:         **if** new-score $>$ old-score **then**
12:             old-score=new-score
13:             add-node=add-node $\cup X_j$
14:     **if** add-node $\neq \phi$ **then**
15:         add add-node to parents(X_i)
16:         add edges from parents(X_i) to X_i

Time Complexity Analysis. The complexity of calculating reliability in line 1–3 is $\mathcal{O}(m)$ which depends on the number of variables. Line 4–16 contain m iterations, and in each iteration, it seek for parents in $(i - 1)$ probable variables for each node i, whose cost is $\mathcal{O}(i * (i - 1)) = \mathcal{O}(m^2)$. Overall, complexity of Algorithm 1 is $\mathcal{O}(m^2)$.

3.2 Bayesian Inference

After construction phase, we obtain a Bayesian network G, from which we can infer relevance relationship between variables. In order to maximize the influence of variables with highly reliability and low missing percentage, we present an algorithm to fill missing values by variables from top to bottom, which means that we fill variables from the most reliable one to the least reliable one.

Filling missing values according to the reliability of variables make us impute a missing value only conditioning on more reliable variables that are directly linked to it. Specially, when we deal with the top variables which do not have parents, they are filled by randomly generating a values from the marginal distribution. When dealing with other variables, we fill them by calculating their conditional probability distribution. Then, we propose the inference approach formally.

As from Definition 3, we have concluded that missing values of variables only on condition of their parents, which can be defined as follows.

Definition 8 (*Bayesian Inference Based on Bayesian Network*).

$$x^* = argmax_{x_i} P\left(X = x_i | Pa(X)\right) \tag{8}$$

Notice that when $Pa(X) = \phi, x^* = argmax_{x_i} P(X = x_i)$.

As we attempt to find the optimal imputation method towards to each variable with the help of crowdsourcing below, we change our goal in a new format according to each missing value set. According to Definition 6, the whole missing value set can be defined as $M = \{M_1, M_2, \ldots, M_m\}$ in accordance with $X_1, X_2, \ldots, X_m$, where m is the number of variables. Thus our new problem is defined as follows.

Definition 9 (*Problem Definition based on Missing Values Sets*). *For each missing value set M, find the most possible value set M^* with the evidence obtained from Bayesian network G.*

Theorem 1 shows that the proposed algorithm could obtain optimal solution for the problem we define in Definition 9.

Theorem 1. *Give a Bayesian network G and the whole missing values set $M = \{M_1, M_2, \ldots, M_m\}$, for each missing value x_i, the most possible value x^* of x_i can be calculated according to Eq. 8.*

Proof.

(1) When i = 1, the missing values set M_1 refer to the missing values of variable X_1, which is at the top of G. As $Pa(X_1) = \phi$, the most possible value x^* for each missing value of X_1 is calculated by $x^* = argmax_{x_i} P(X = x_i)$.

(2) Assume that when i = m−1, the missing values in set M_{m-1} can be calculated by $x^* = argmax_{x_i} P(X = x_i | Pa(X_{m-1}))$. Then when i = m, the missing values of variables before X_{m-1} have been imputed before the above

step, and the missing values of variable X_{m-1} is imputed in the above step, thus variables $X = \{X_1, X_2, \ldots, X_{m-1}\}$ before X_m are all completed. As $Pa(X_m)$ is among the variables $X = \{X_1, X_2, \ldots, X_{m-1}\}$, thus values of $Pa(X_m)$ are all known. Missing values in set M_m can be calculated by $x^* = argmax_{x_i} P(X = x_i | Pa(X_m))$. Thus, we prove the theorem. $\qquad\square$

Algorithm 2 shows the whole process of probability inference based on Bayesian network. Line 2–4 finds the most possible value for each missing value in the missing values set.

Algorithm 2. Bayesian Inference(D,M,G)

Intput: Incomplete data set D, Missing value sets $M = \{M_1, M_2, \ldots, M_m\}$, Bayesian network G
Output: complete data set D'
1: **for** i=1 to m **do**
2:    **for** each $t_j \in M_i$ **do**
3:        $x_j^* = argmax_{x_j} P(X = x_j | Pa(X_i))$
4:        $t_j = x_j^*$

Time Complexity Analysis. The complexity of the first step is $\mathcal{O}(m)$ as it depends on the number of variables. And for each variable, the algorithm fills the missing value one by one in the second step. Its complexity is $\mathcal{O}(|M_i|)$ which can be treated as a constant. The total complexity of Algorithm 2 is $\mathcal{O}(m \cdot |M_i|)$.

Although we have proved that the most possible value can be calculated in order, we cannot ensure the most possible value is the correct value, especially in the situation of evidence is not enough for probability inference. For example, when calculating the conditional probability of D in Fig. 1, if $P(D = d_1 | B, C) = 0.3$, $P(D = d_2 | B, C) = 0.4$ and $P(D = d_3 | B, C) = 0.3$ when variable D only has three possible values. Bayesian inference will doubt about which one is the most possible value and may make mistakes when selecting the most possible value among similar probability of values. Thus we bring crowdsourcing to help improving accuracy of Bayesian inference.

4 Bayesian Inference with Crowdsourcing

In order to improve accuracy, we bring crowdsourcing to provide external knowledge. However, considering the cost of crowdsourcing, it is not realistic to send all the uncertain tuples which have low probability to the crowd. Especially when the data set is very large, uncertain tuples will be relatively large too. To increase accuracy of Bayesian inference, we prefer to select uncertain tuples with the maximum influence to other uncertain tuples as representatives to the crowd as crowd questions. We first define the uncertain set and then formalize the problem of crowd questions selection from uncertain set.

Definition 10 *(Uncertain Set). Given a Bayesian network G, the whole missing value sets $M = \{M_1, M_2, \ldots, M_m\}$, and a lower threshold θ, for each missing value set M_i, the uncertain set is defined as:*

$$M_i^u = \{t|p_{t^*} < \theta, t \in M_i\} \tag{9}$$

where p_{t^*} is the probability of the most possible value t^*.

The influence of M_i^q can be calculated by $INF(M_i^q) = \sum_{t_j \in M_i^u} I(|p'_{t_j} - p_{t_j^*}| \neq 0)$ where $I(\cdot)$ is an indicator function $I(|p'_{t_j} - p_{t_j^*}| \neq 0) = 1$ when $(|p'_{t_j} - p_{t_j^*}| \neq 0)$ and $I(|p'_{t_j} - p_{t_j^*}| \neq 0) = 0$ when $(|p'_{t_j} - p_{t_j^*}| = 0)$, p'_{t_j} is the probability of the most possible value of t_j when the value of $t^u \subseteq M_i^q$ is known from crowd.

Definition 11 *(Crowd Questions Selection). For each missing values set M_i, give the uncertain set M_i^u and a number q, it selects a subset of uncertain questions $M_i^q \subseteq M_i^u$ satisfying: (1) the size $|M_i^q| \leq q$. (2) the influence of M_i^q is maximized.*

Theorem 2 shows the difficulty of this problem.

Theorem 2. *Crowd questions selection is NP-hard.*

Proof. We prove the theorem by showing that the crowd questions selection problem and the maximum coverage problem are equivalent under L-reduction.

Recall that an instance of maximum coverage problem (U, S, k) consists of a set of elements $U = \{u_1, u_2, \ldots, u_{|U|}\}$, a collection of subsets $S = \{S_1, S_2, \ldots, S_{|S|}\}$ where $S_i \subseteq U$, and a number k. The problem aims to select k subsets $S^* \subseteq S$ to maximize the number of covered element $|\bigcup_{S \subseteq S^*} S|$.

Let $F = $ Maximum Coverage Problem(U, S, k) and $G = $ Crowd Question Selection Problem(M^u, C, q), where $M^u = \{t_1, t_2, \ldots, t_{|t|}\}$ is uncertain set, $C = \{C_1, C_2, \ldots, C_{|C|}\}$ is a collection of influence set of each uncertain tuple $C(t_i) = \{t_j|p'_{t_j} - p_{t_j^*}| \neq 0, t_j \in M^u\}$ and q is the number of crowd questions.

Define a transformation f from G to F by $M^u = U$, $C = S$ and $q = k$. Given the optimal solution $S^* = \{S_1, S_2, \ldots, S_k\}$ of F, the optimal solution of G is $M^q = \{t_i|C(t_i) = S_i\}$. A crowd question set now corresponds to a maximum coverage set of the same size. Thus f is an L-reduction with $\alpha = \beta = 1$.

Define a transformation g from F to G by $U = M^u$, $S = C$ and $k = q$. Given the optimal solution $M^q = \{t_1, t_2, \ldots, t_k\}$ of G, the optimal solution of F is $S^* = \{S_i|S_i = C(t_i)\}$. Since a maximum coverage set correspond to a crowd question set of the same size, g is an L-reduction with $\alpha = \beta = 1$.

As it is well-known that the maximum coverage problem is NP-hard, thus the crowd questions selection is also NP-hard. □

Due to the difficulty of crowd question selection problem, we present a greedy algorithm. It selects uncertain tuple which can influence the most unselected tuples at each stage as the crowd question. Algorithm 3 shows our greedy crowd question selection algorithm. Line 5–8 set a lower threshold θ for probability

Algorithm 3. Crowd Question Selection(D, M, G, q)

Intput: Incomplete data set D, missing value sets $M = \{M_1, M_2, \ldots, M_m\}$, Bayesian
network G, number of crowd questions q.
Output: Selected crowd questions set $M^q = \{M_1^q, M_2^q, \ldots, M_m^q\}$

1: **for** $i = 1$ to m **do**
2:      $M_i^u = \phi$
3:      **for** each $t_j \in M_i$ **do**
4:          $x_j^* = argmax_{x_j} P(X = x_j | Pa(X_i))$
5:          **if** $p_{x_j^*} >= \theta$ **then**
6:              $t_j^* = x_j^*$
7:          **else**
8:              add t_j to M_i^u ;
9:      **for** $k = 1$ to q **do**
10:         $t = argmax_{t \in M_i^u - M_i^q} INF(M_i^q \cup \{t\}) - INF(M_i^q)$
11:         where $INF(M_i^q) = \sum_{t_k \in M_i^u} I(|p_{t_k}' - p_{t_k^*}| \neq 0)$
12:         $M_i^q \leftarrow M_i^q \cup \{t\}$

of the most possible value. We consider that when the probability of the most
possible value has an absolute advantage compared to other candidate, we set
it as the correct value. And when the probability of the most possible value is
similar to other candidate, we treat the tuple as uncertain tuple and add it to
uncertain set. Line 9–12 we select a tuple that maximizes the marginal influence
in each iterations.

Theorem 3. *The approximation ratio of our greedy crowd question selection
algorithm is* $1 - \frac{1}{e}$.

Proof. As we have proved that the crowd questions selection problem and the
maximum coverage problem are equivalent under L-reduction in Theorem 2, in
both the reductions $|k| = |q|$. The reductions are S-reductions with size amplifi-
cation n in the number of subsets and questions. Thus, an approximation algorithm for one of the problems gives us an equally approximation algorithm for
the other problem.

Since the greedy algorithm for maximum coverage chooses set according to
the rule to choose the set containing the largest number of uncovered elements
at each stage, and it can achieve an approximation ratio of $1 - \frac{1}{e}$, where e is the
base of natural logarithm [22]. Therefore, our greedy algorithm which selects the
uncertain tuples that can affect the most unselected tuples at each stage as the
crowd question has the same approximation ratio of $1 - \frac{1}{e}$.   □

Time Complexity Analysis. The complexity of the first step is $\mathcal{O}(m)$ as it
depends on the number of variables. And for each variable, the algorithm fill the
missing value one by one in the second step, its complexity is $\mathcal{O}(|M_i|)$, and if the
maximum probability is lower than θ, we add it into the uncertain set which can
be treated as a constant. In the third step, it computes $INF(M_i^q \cup \{t\})$ for each
uncertain tuple $t \in M_i^u$ in k iterations, which needs to scan all the uncertain
tuples in M_i^u. Overall, the complexity of Algorithm 3 is $\mathcal{O}(m \cdot (|M_i| + k \cdot |M_i^u|))$.

5 Experiments

5.1 Experiment Setup

In this section, in order to show the performance of our method, we test our proposed algorithms on two real datasets from UCI Irvine Machine Learning Repository. We used categorical attributes in the data set to test our approaches. Basic information is shown in Table 2. All algorithms were implemented with C++. The algorithms were run on a Windows 7 machine with 2.2 GHZ processor, 4 GB memory.

Table 2. The experimental datasets

Dataset	Attribute number	Instance number
Zoo	18	101
US_Congressional_Voting	17	435

The datasets are described as follows.

(1) *Zoo* dataset: The dataset described various characteristics of different animals. It was complete with no missing value initially. We randomly eliminated different percent of values as missing values to test our algorithms.
(2) *US_Congression_Voting* dataset: The dataset which included votes for each of U.S. House of Representatives Congressmen was identified by different types of votes. It had approximately 5 % of missing values in total and the maximum percentage of missing values according to one attribute reached 23 %. As the value of class attribute was complete, we randomly eliminated different percent of values in the class attribute as missing values. We filled the missing values of class attribute in the presence of other incomplete attributes.

We evaluated the performance of our methods on accuracy and efficiency. Accuracy was computed as the ratio of tuples with correctly imputation results and efficiency was measured by processing time of missing values imputation.

5.2 Effectiveness of Bayesian Network Construction

In this section, we evaluated the performance of our method of Bayesian network construction based on relevance relationship named Rel_BN. We used the most common Bayesian network construction algorithm K2 for comparison. We calculated the accuracy of Rel_BN and K2 in the case of different mis_rate, and the results are shown in Fig. 2. From the figure we observe that the performance of our method outperformed K2 in both two datasets. With the increase of mis_rate, the accuracy of our method Rel_BN had a slower decline than K2, which shows that our method can make full use of the information of relevance attributes and achieve a better performance on incomplete data.

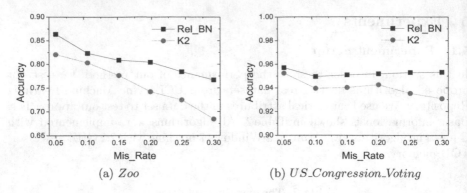

(a) *Zoo*   (b) *US_Congression_Voting*

Fig. 2. Effectiveness of Bayesian network construction

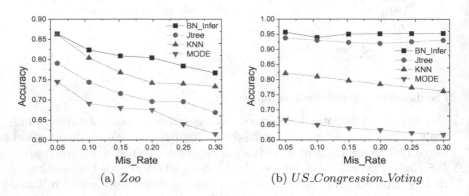

(a) *Zoo*   (b) *US_Congression_Voting*

Fig. 3. Effectiveness of missing values imputation

5.3   Effectiveness of Bayesiain Inference

In this section, we evaluated the performance of our algorithm based on Bayesian inference named BN_Infer and used the most common imputation methods Mode and KNN for comparison. In mode imputation, we took the most frequent value as the correct value. In KNN, we choosed candidate with the maximum probability as substitution and set the cluster number to 5. To show the advantage of our Bayesian inference method, we also used junction tree algorithm [20] named JTree for short as the competitor. In JTree, we used our method Rel_BN to build the Bayesian network. We evaluated the performance of these methods on both datasets, and results are shown in Fig. 3.

We observe that Mode achieved the worse performance in most of the cases. This is due to the lack of data analysis. KNN and Jtree performed better than Mode, and BN_Infer achieved the best accuracy in both datasets. The improvement was attributed to not only our Bayesian network construction based on relationships but also the Bayesian inference based on relevance attributes.

To evaluate the efficiency of our method, we also compared the processing time of our method with K2 and joint tree algorithm. Figure 4 shows the experimental results. We observe that our method BN_Infer also had a slight advantage

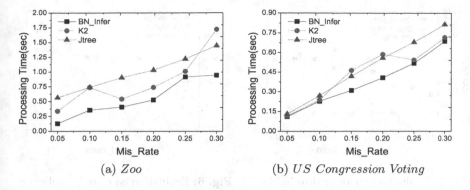

(a) *Zoo*   (b) *US Congression Voting*

Fig. 4. Evaluating processing time

over K2 and Jtree. Thus, we can conclude that our method achieves a better performance on both accuracy and efficiency.

5.4 Effectiveness of Crowd Questions Selection

In this section, we evaluated the effectiveness of our crowd questions selection strategy Crowd_inf. We used Crowd_ran which randomly selected a set of q tuples as crowd questions for comparison. To show the improvement of accuracy on Bayesian inference enhanced by crowdsourcing, we set the baseline method as BN_Infer.

Firstly we evaluated the impact of lower threshold θ. In the experiments, we used *Zoo* dataset, and set the number q and mis_rate as 10 and 30 %, respectively. The result is shown in Fig. 5. We observe that the accuracy of both Crowd_ran and Crowd_inf increased with the increase of lower threshold θ, and then tend to be stable. And our method Crowd_inf had a better performance than Crowd_ran over all different lower thresholds. This is due to our crowd question selection strategy which selects the most influential tuple each time. When we set the lower threshold θ as 0.7, Crowd_inf can achieve 90 % of accuracy. Crowd_inf improved the accuracy of BN_Infer to 5 %–17 %. The improvement is attributed to the extra knowledge from crowd.

Then we evaluated the influence of crowd question number q. In the experiments, we used *Zoo* dataset, and set lower threshold θ as 0.8 and mis_rate as 30 %. The result is shown in Fig. 6. We observe that the accuracy of both Crowd_ran and Crowd_inf increased with the increase of crowd questions number q, and then tend to be stable. When we set crowd questions number as 10, the accuracy of Crowd_inf reached almost 95 %. It had improved the accuracy of BN_Infer to 16 %–17 %. We can also infer that the accuracy did not always increase with the increase of the crowd questions number q. This is because when the accuracy of Bayesian inference was very high, the crowd answers was consistent with the inference results. Thus, the final result did not improve significantly.

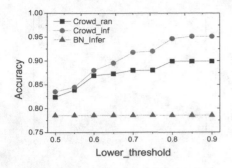
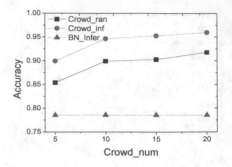

Fig. 5. Evaluation on lower threshold θ **Fig. 6.** Evaluation on crowd numbers q

6 Conclusion

In this paper, we proposed a new method for missing values imputation combining Bayesian network and crowdsourcing. We first build Bayesian network based on reliability between variables specific to incomplete data and then present our Bayesian inference algorithm. To obtain enough evidence, we bring crowdsourcing to improve the accuracy of Bayesian inference. We deployed our methods on real datasets. Experimental results on two real datasets show that our methods achieved much higher quality than the existing methods. Our future work includes taking mixture attributes into consideration and evaluating the quality of workers on crowdsourcing platform.

Acknowledgement. This paper was supported by NGFR 973 grant 2012CB316200, NSFC grant U1509216, 61472099, 61133002 and National Sci-Tech Support Plan 2015BAH10F01.

References

1. Janssen, K.J.M., Donders, A.R.T., Harrell, F.E., et al.: Missing covariate data in medical research: to impute is better than to ignore. J. Clin. Epidemiol. **63**(7), 721–727 (2010)
2. Dempster, A.P., Laird, N.M., Rubin, D.B.: Maximum likelihood from incomplete data via the EM algorithm. J. R. Stat. Soc. Ser. B (Methodological) **39**, 1–38 (2011)
3. Shan, Y., Kernel, D.G.: PCA regression for missing data estimation in DNA microarray analysis. In: IEEE International Symposium on Circuits and Systems, ISCAS 2009, pp. 1477–1480. IEEE (2009)
4. Lakshminarayan, K., Harp, S.A., Goldman, R.P., et al.: Imputation of missing data using machine learning techniques. In: KDD, pp. 140–145 (1996)
5. Yang, K., Li, J., Wang, C.: Missing values estimation in microarray data with partial least squares regression. In: Alexandrov, V.N., van Albada, G.D., Sloot, P.M.A., Dongarra, J. (eds.) ICCS 2006. LNCS, vol. 3992, pp. 662–669. Springer, Heidelberg (2006)

6. Li, X.B.: A Bayesian approach for estimating and replacing missing categorical data. J. Data Inf. Qual. (JDIQ) 1(1), 3 (2009)
7. Di Zio, M., Scanu, M., Coppola, L., et al.: Bayesian networks for imputation. J. R. Stat. Soc. Ser. A (Statistics in Society) 167(2), 309–322 (2004)
8. Zhang, S.: Shell-neighbor method and its application in missing data imputation. Appl. Intell. 35(1), 123–133 (2011)
9. Setiawan, N.A., Venkatachalam, P.A., Hani, A.F.M.: Missing attribute value prediction based on artificial neural network and rough set theory. In: International Conference on BioMedical Engineering and Informatics, BMEI 2008, vol. 1, pp. 306–310. IEEE (2008)
10. Nowak, S., Rger, S.: How reliable are annotations via crowdsourcing: a study about inter-annotator agreement for multi-label image annotation. In: Proceedings of the International Conference on Multimedia Information Retrieval, pp. 557–566. ACM (2010)
11. Noronha, J., Hysen, E., Zhang, H., et al.: Platemate: crowdsourcing nutritional analysis from food photographs. In: Proceedings of the 24th Annual ACM Symposium on User Interface Software and Technology, pp. 1–12. ACM (2011)
12. Whang, S.E., Lofgren, P., Garcia-Molina, H.: Question selection for crowd entity resolution. Proc. VLDB Endowment 6(6), 349–360 (2013)
13. Wang, J., Kraska, T., Franklin, M.J., et al.: Crowder: crowdsourcing entity resolution. Proc. VLDB Endowment 5(11), 1483–1494 (2012)
14. Zhang, C.J., Chen, L., Jagadish, H.V., et al.: Reducing uncertainty of schema matching via crowdsourcing. Proc. VLDB Endowment 6(9), 757–768 (2013)
15. Pearl, J.: Probabilistic Reasoning in Intelligent Systems: Networks of Plausible Inference. Morgan Kaufmann, SanMateo (1988)
16. Lawrence, I., Lin, K.: A concordance correlation coefficient to evaluate reproducibility. Biometrics 45, 255–268 (1989)
17. Stekhoven, D.J., Bhlmann, P.: MissForestnon-parametric missing value imputation for mixed-type data. Bioinformatics 28(1), 112–118 (2012)
18. Cooper, G.F., Herskovits, E.: A Bayesian method for the induction of probabilistic networks from data. Mach. Learn. 9(4), 309–347 (1992)
19. Tsamardinos, I., Brown, L.E., Aliferis, C.F.: The max-min hill-climbing Bayesian network structure learning algorithm. Mach. Learn. 65(1), 31–78 (2006)
20. Huang, C., Darwiche, A.: Inference in belief networks: a procedural guide. Int. J. Approximate Reasoning 15(3), 225–263 (1996)
21. Lauritzen, S.L.: The EM algorithm for graphical association models with missing data. Comput. Stat. Data Anal. 19(2), 191–201 (1995)
22. Hochbaum, D.S.: Approximating covering and packing problems: set cover, vertex cover, independent set, and related problems. In: Approximation Algorithms for NP-Hard Problems, pp. 94–143. PWS Publishing Co. (1996)
23. Li, J., Cai, Z., Yan, M., Li, Y.: Using crowdsourced data in location-based social networks to explore influence maximization. In: The 35th Annual IEEE International Conference on Computer Communications (INFOCOM 2016) (2016)
24. Wang, Y., Cai, Z., Stothard, P., et al.: Fast accurate missing SNP genotype local imputation. BMC Res. Notes 5(1), 404 (2012)
25. Cai, Z., Heydari, M., Lin, G.: Iterated local least squares imputation for microarray missing values. J. Bioinform. Comput. Biol. 4(5), 935–957 (2006)

One-Pass Inconsistency Detection
Algorithms for Big Data

Meifan Zhang, Hongzhi Wang$^{(\boxtimes)}$, Jianzhong Li, and Hong Gao

Department of Computer Science and Technology,
Harbin Institute of Technology, Harbin, China
`miffy_zhang@126.com`,
`{wangzh,lijzh,honggao}@hit.edu.cn`

Abstract. Data in the real world is often dirty. Inconsistency is an important kind of dirty data. Before repairing inconsistency, we need to detect them first. The time complexities of current inconsistency detection algorithms are super-linear to the size of data and not suitable for big data. For inconsistency detection for big data, we develop an algorithm that detects inconsistency within one-pass scan of the data according to both the functional dependency (FD) and the conditional functional dependency (CFD). We compare our detection algorithm with existing approaches experimentally. Experimental results on real datasets show that our approach could detect inconsistency effectively and efficiently.

Keywords: Inconsistency detection · Big data · One-pass algorithm · Data quality

1 Introduction

Data quality problems in real world may cost not only billions of dollars in businesses, but also precious lives when they exist in medical data [1]. With the consideration of the serious consequences caused by data quality problems, techniques for detecting and fixing errors are in great demand. For big data, due to the volume feature, it has higher possibility to have data quality problems.

Inconsistency is an important aspect of data quality problems. Inconsistency means that some tuples violate given rules. For effective inconsistency detection, some forms of rules are proposed such as functional dependency (FD) and conditional functional dependency (CFD) [8].

In [2, 3], researchers proposed a SQL-based automatically detection method to identify the tuples violating the CFDs. In [2], they get a tableaux merged by multiple CFDs and translate that into a single pair of SQL queries. With this pair of SQL queries, they only need two passes of database. In the queries, they get a Macro by joining the tableaux with the whole dataset. However, join operation for big data still costs much. In [20], a method is proposed to detect and repair data structure inconsistencies automatically. However, they are not suitable for inconsistency detection on big data due to efficiency issues. As a result, they are difficult to scale to big data. We use an example to illustrate this point.

© Springer International Publishing Switzerland 2016
S.B. Navathe et al. (Eds.): DASFAA 2016, Part I, LNCS 9642, pp. 82–98, 2016.
DOI: 10.1007/978-3-319-32025-0_6

Example 1: Consider the schema in Table 1 and two rules: FD $\phi1(R : A, C \rightarrow D,$ $(\_, \_||\_))$, and CFD $\phi2(R : A, C \rightarrow D, (a1, c2||d2))$.

Table 1. An instance relation

	A	B	C	D	E	F
t1	a1	b1	c1	d1	e1	f1
t2	a1	b2	c1	d1	e1	f1
t3	a1	b1	c2	d1	e2	f2
t4	a1	b1	c3	d2	e2	f2
t5	a1	b2	c2	d2	e2	f3
t6	a1	b1	c2	d2	e3	f2

Table 2. The result of Q_C

	A	B	C	D	E	F
t3	a1	b1	c2	d1	e2	f2

Table 3. The result of Q_V

A	C
a1	c2

Taking the SQL-based detection method, we need following two queries to detect the inconsistencies with these two rules:

Q_C: **select** t from R
   where $t[A] = tp[A]$ AND $t[C] = tp[C]$ AND $t[D] \neq tp[D]$
Q_V: **select distinct** $t[A]$, $t[C]$ from R
   where $t[A] \times tp[A]$ AND $t[C] \times tp[C]$ AND $t[D] \neq tp[D]$
   group by $t[A]$, $t[C]$
   having count(distinct $t[A]$, $t[C])> 1$

In the queries above, t is a tuple in Table 1 and tp is a single pattern tuple of a rule. The rule $\eta1 \times \eta2$ means that if $\eta2$ is not "_", $1 = \eta2$; else $\eta2$ is "_".

The first query returns the tuples violating $\phi2$, which is shown in Table 2. And the second query returns value A and C of inconsistent tuples, the tuples matching the records in Table 3 are those involved in the consistencies.

It is obvious that these two queries need two passes scanning of the data. And the results of FD detection only return the distinct values of the left hand side (LHS) of inconsistent tuples. We still need a list of queries for the inconsistent tuples involved in the inconsistency. In this example, with the result of Q_v in Table 3, we still need one more query Q_R.

Q_R: **select** t from R
          where $t[A] = a_1$ AND $t[C] = c_2$

This query returns the violations of $\phi 1$, and the result is $\{t3, t5, t6\}$.

One query requires at least one pass of database scanning, and the efficiency of a query is affected by the database size. This method may cost too much time for detecting inconsistencies in big data.

The inconsistency detection for big data motives us to design a new algorithm that could accomplish the detection within one-pass of scanning data. Considering reducing the times of accessing database, we group each attribute by its value. And for the tuples with the same value on same attribute, we group their IDs together in a tuple set. Then we get the tuples matching the LHS of a rule by calculating a list of intersections of the grouped tuple sets. At last, we determine the inconsistencies by checking that whether the tuple set matching the LHS is the subset of a tuple set of the right hand side (RHS). As a result, the whole process scans the data set only once.

We make following contributions in this paper.

Our first contribution is that we present a one-pass algorithm for detecting all violations with both FDs and CFDs. Our method is suitable for big data. It is a linear algorithm for inconsistency detection according to FDs and CFDs.

Our second contribution is that we consider the cells as the units of the inconsistencies instead of tuples, such that the inconsistent data can be narrow down to some certain cells.

Our third contribution is the extensive experiments on real data. The experimental results demonstrate that the proposed method is efficient and effective.

2   Related Work

As an important problem in data quality management, the inconsistency problem consists of both the detecting and repairing problem has been well studied [2–9]. Some researchers use statistical model and raise some threshold to identify the inconsistencies [10, 11]. Other researchers use constraints [2–8, 12, 13] to detect inconsistencies.

It is necessary to obtain rules before the rule-based detection method. Some methods have been proposed many methods to discover rules automatically [13, 14], which are mainly extensions to traditional FD. [14] raised a search algorithm with pruning strategy which can discover rules effectively. In [15], a method is proposed to automatically discover CFD. The rules used for detection may have conflicts. In order to solve this problem, a data cleaning framework is proposed to resolve conflicts in rules [16].

Recently researchers have been looking into automatic inconsistency detecting method [20], which regards FDs as the main rules. Detecting methods based on CFDs are also proposed after the definition of CFD [2, 18]. Many previous method only based on one kind of rules [6–8, 19, 20], few works combine different kind of rules together to detect inconsistencies. The previous work on inconsistency detection mostly needs redundant data inquiries [2]. There have been no efforts on detecting tuples involved in the consistencies with only one pass database scan.

3 Preliminaries

In this section, we introduce some background of this paper including the definition of inconsistencies and the rules used to detect inconsistencies.

3.1 Rules

For a relational schema R, we use the rules containing both normal form FDs and normal form CFDs in this paper.

Definition 1 (Normal Form): A FD or a CFD $\phi(R : X \rightarrow Y, T\phi)$ is in normal form if (1) $T\phi$ consists of a single pattern tuple tp. If ϕ is a FD, $tp = (\_, \_, \ldots, \_||\_)$, which means tp consists "_" only. (2) Y consists of a single attribute A.

After the definition of normal form, we write ϕ simply as $\phi(R : X \rightarrow A, tp)$ Each CFD (FD) not in normal form can be written as a set of CFDs (FDs) in normal form. For example, a CFD $\phi(R : A, B, D \rightarrow C, E, (a1, b1, \_||c1, \_))$ not in normal form can be written into two rules in normal form: (1) $\phi1(R : A, B \rightarrow C, (a1, b1||c1))$, (2) $\phi3(R : A, B, D \rightarrow E, (a1, b1, \_||\_))$. With the help of rules in normal form we can easily know the cells involved in the inconsistencies, since that the RHS of the rules only contain one attribute. If a tuple t violates $\phi1(R : A, B \rightarrow C, (a1, b1||c1))$, it is clear that $t[A] = a1$, $t[B] = a1$ and $t[C] \neq c1$.

The normal form rules can be classified according to whether their RHS is a constant or not.

Definition 2 (Constant RHS Rules): The constant RHS rules contain the constant CFDs only. It is called a constant CFD [9] if its pattern tuple tp consists of constants only. That is, $tp[A]$ is a constant and for all attributes $B \in X$, $tp[B]$ is a constant.

Definition 3 (Variable RHS Rules): The variable RHS rules contain both FDs and variable CFDs. It is called a variable CFD [9] if $tp[A] = $ "_". That is, the right hand side (RHS) of its pattern tuple is the unnamed variable "_". The RHS of FDs is also obvious unnamed variable "_".

3.2 Inconsistency Detection

Inconsistency detection problem is to find the data violating given rules. An instance I of a schema R satisfies a rule $\phi(R : X \rightarrow A, tp)$, denoted by $I \vDash \phi$, (1) if ϕ is a constant RHS rule, then for each tuple t in I, $t[X] = tp[X]$ implies $t[A] = tp[A]$, (2) if ϕ is a variable RHS rule, then for each pair of tuples $t1$ and $t2$ in I, $t1[X] = t2[X] \asymp tp[X]$ implies $t1[A] = t2[A] \asymp tp[A]$. Here $\eta1 = \eta2 \asymp \eta$ denotes that if η is not "_", $\eta1 = \eta2 = \eta$ and otherwise, if η is "_", $\eta1 = \eta2$.

The violation can be separated into two kinds according to the kind of rules.

Definition 4 (Inconsistent Tuples Set): The inconsistent tuple set S is the set of tuples that violate the constant RHS rule.

Definition 5 (Inconsistent Tuples Sets Group): A group of inconsistent tuples sets G is a group of tuple sets violating the variable RHS rule. Tuples in the same group have the same LHS. Tuples in different sets of the same group have different RHS values.

The inconsistent tuples detected by a constant RHS rule $\phi(R : X \rightarrow A, tp)$ can be grouped into one inconsistent tuple set S. $\forall t \in S$, $t[X] = tp[X]$ and $t[A] \neq tp[A]$.

Example 2: In Table 1, we detect inconsistent tuples with a constant RHS rule $\phi(A, B \rightarrow C, (a1, b1 \| c1))$. The inconsistent tuple set is $S = \{t3, t4, t6\}$.

The inconsistent tuples detected by a variable RHS rule ϕ $(R: X \rightarrow A, tp)$ must be grouped into groups of inconsistent tuple sets Gs. For each group G of Gs, we put the tuples with the same LHS and different RHS into different sets of the same group G.

Example 3: In Table 1, we detect inconsistent tuples with a variable RHS rule $\phi(A, B \rightarrow C, (\_, \_\|\_))$, the inconsistent tuples sets groups G contains two groups, $G[1] = \{\{t1\}, \{t3, t6\}, \{t4\}\}$, and $G[2] = \{\{t2\}, \{t5\}\}$.

4   Detecting the Inconsistencies

In this section, we describe our inconsistency detection algorithm. In Sect. 4.1, we introduce the framework of the algorithm, and then explain the four modules RANGE, GROUP, MATCH, and MERGE in the following subsections, respectively.

4.1   Framework

In order to reduce the times of accessing database, we group each attribute by its value. And for the tuples with same value on same attribute, we group their IDs together in a tuple set and build a hash index for retrieving the tuple sets efficiently. Then we get the tuples matching the LHS by calculating a list of intersections of the grouped tuple sets. At last, we obtain the inconsistencies by checking whether the tuple set matching the LHS is the subset of a tuple set matching the RHS. In our detection method, we only require scanning the database once.

The pseudo code of our algorithm is shown in Algorithm 1. The algorithm has following four steps.

In the first step (line 1), we collect the set of the attributes and its value from the rules to reduce the data requiring accessing during detection. This step is introduced in detail in Sect. 4.2.

In the second step (line 2), we group the tuples for each attribute in A-set by its value. This step is introduced in Sect. 4.3.

Algorithm 1. DETECTION

Input: database D and a set Σ of rules.
Output: inconsistencies tuplesets $sets_\phi$ for each rule ϕ in Σ.
1: $A\text{-set} = \text{RANGE}()$;   //Get attribute set A-set from the rules.
2: $S = \text{GROUP}()$;   //Get the data grouped sets.
3: **foreach** constant CFD   /*detecting with constant CFDs*/
4:      **foreach** attribute $A \in \text{LHS}(\phi)$ and its value $a = tp[A]$
5:          $left\_set \cap = tuple\_set_{A=a}$;
6:      $right\_set \leftarrow tuple\_set_{RHS(\phi)=tp[B]}$;
7:      $sets_{\phi 0} \leftarrow left\_set \setminus right\_set$;
8: **foreach** variable RHS rule   /*initial the unprocessed left_set LS_ϕ of the variable RHS rules*/
9:      $LS_\phi \leftarrow \{ A \in \text{LHS}(\phi)|S[A]\}$;
10: **foreach** FD ϕ /*detecting with FDs*/
11:     $i \leftarrow 0$;
12:     $left\_sets \leftarrow \text{MATCH}(\phi)$;
13:     **foreach** $lset$ in $left\_sets$
14:         **foreach** $rset$ in the tuple sets of $S[RHS(\phi)]$
15:             **If** $|lset \cap rset| < |lset|$ then
16:                 $sets_{\phi i} += lset \cap rset$;
17:     $i \leftarrow i+1$;
18: **foreach** variable CFD ϕ /*detecting with variable CFDs*/
19:     $temp \leftarrow \text{MERGE}(LS_\phi[0], LS_\phi[1])$;
20:     **foreach** $LS[i]$ and $i > 1$
21:         $temp \leftarrow \text{MERGE}(temp, LS_\phi[i])$;
22:     $i \leftarrow i+1$;
23: $LS \leftarrow temp$;

In the third step (line 3–7), we detect the violation of the constant RHS rules. For one constant RHS rule $\phi(R : X \rightarrow A, tp)$, we get the tuple sets from the result of the second step with the attributes in $attr(\phi)$ and the values in tp. The tuples matching the LHS of ϕ are in $left\_set$, and for each X_i in X, $left\_set \cap = tupleset_{X_i = tp[X_i]}$. Likewise, $right\_set = tupleset_{A=tp[A]}$. The inconsistencies can be calculated easily by $left\_set \setminus right\_set$.

In the fourth step (line 8–23), we detect inconsistencies violating the variable RHS rules. In this step, we first consider the FDs. The MATCH procedure returns tuple sets in which the tuples agree on the LHS of ϕ. After getting $left\_sets$ from MATCH module and $right\_sets$ from $S[RHS(\phi)]$, we detect the violations with both $left\_sets$ and $right\_sets$. The tuples violating the variable RHS rules are not independent. We know that the tuples in the same $left\_sets_i$ and $right\_sets_j$ have the same attribute values in the LHS and RHS of ϕ, respectively. For each $left\_sets_i$, if it is the subset of $right\_sets_i$, there is no violation in $left\_sets_i$. Otherwise, there exist violations in $left\_sets_i$. If $|left\_sets_i| > |left\_sets_i \cap right\_sets_i| \neq 0$, then the tuples in $left\_sets_i \cap right\_sets_i$ must be consistencies. After all the FDs are detected, we start checking the variable CFDs. For each variable CFD, if the embedded FD of the CFD is detected before, we only need to check the consistencies detected before with the constant in the CFD. Otherwise, we detect the consistencies just in the way that we detect violations of FDs.

Example 4: Consider the data in Table 1 and the rules in Example 1 again.

First, we obtain the following *A-set* with the algorithm RANGE which will be explained in detail in Sect. 4.2. $A - set = \{(A, \_), (C, \_), (D, \_)\}$.

Second, we get the tuple set grouped with the algorithm GROUP which will be explained in detail in Sect. 4.3. The grouped tuple sets are in Table 4.

The tuple set contains numeric tuple IDs, and the attribute value is an index of the grouped records, so the list of records is much smaller than the origin database. With help of the list, we can easily detect inconsistencies with some intersection and difference operations. Then, we detect inconsistencies with the records in Table 4. The detection with CFD: $\phi2(R : A, C \rightarrow D, (a1, c2 \| d2))$ is transformed into the following calculating. $(tupleset_{A=a1} \cap tupleset_{C=c2}) \backslash tupleset_{D=d2} = \{t3\}$.

Table 4. Grouped Records

tupleset $_{A=a1} = \{t1, t2, t3, t4, t5, t6\}$
tupleset $_{C=c1} = \{t1, t2\}$
tupleset $_{C=c2} = \{t3, t5, t6\}$
tupleset $_{C=c3} = \{t4\}$
tupleset $_{D=d1} = \{t1, t2, t3\}$
tupleset $_{D=d2} = \{t4, t5, t6\}$

The detection with FD: $\phi1(R : A, C \rightarrow D, (\_, \_\|\_))$ is a little more complicated than CFD, since the RHS D is not a constant value. We have to find the tuples matching the FD on the LHS, and no matching on the RHS. First, the tuples matching the LHS can be calculated with two intersections:

(1) intersection $tupleset_1$: $(tupleset_{A=a1} \cap tupleset_{C=c1}) = \{t1, t2\}$
(2) intersection $tupleset_2$: $(tupleset_{A=a1} \cap tupleset_{C=c2}) = \{t3, t5, t6\}$.

If the intersection tuple sets contain more than one tuple, we use the results to check that whether the intersection is a subset of a RHS tuple set or not.

(1) $tupleset_1 \cap tupleset_{D=d1} = \{t1, t2\} = tupleset_1$. That is, $tupleset_1$ is the subset of $tupleset_{D=d1}$. Thus, there is no violation in $tupleset_1$.
(2) $tupleset_2 \cap tupleset_{D=d1} = \{t3\} \neq tupleset_2$, and $\{t3\} \neq \emptyset$. $tupleset_2$ is not the subset of any tuple set of attribute D. Hence, $\{t3\}$ must cause violation.
(3) $tupleset_2 \cap tupleset_{D=d2} = \{t5, t6\} \neq tupleset_2$, and $\{t5, t6\} \neq \emptyset$.

As the result, we obtain the inconsistencies of the rules:
The violation of $\phi2$: $\{t3\}$. And the violation of $\phi1$: $\{t3\}$ conflicts with $\{t5, t6\}$.

4.2 Range

This module aims to get the attributes and values involved in the rules. We get the set of the attributes and its value from the rules to reduce the data to be checked.

In the attributes set $S = \sum(attributeA, valuesetV)$, each attribute A is involved in the set $attr(\sum\phi)$. If there is no constant value in all $\sum\phi$ for attribute A, the value set V only contains a "_" (Line 3–6). Otherwise, we store the distinct value of A in the value set (Line 7–8). If the attribute A is not in $\sum$, we then store A and its value in $\sum$ (Line 10–11). The space complexity of this module is $O(|attr(\sum\phi)|)$ and the time complexity of this module is $O(S)$, where S is the number of the rules.

Algorithm 2. RANGE

Input: a set of rules.
Output: attribute values involved in the rules.

1: **foreach** rule ϕ
2:      **foreach** attribute $A \in attr(\phi)$
3:          **if** A is not exist in A-set
4:              A-set $+= A$;
5:              $valueset_A += tp[A]$;
6:          **else if** $valueset_A != $ "_"
7:              **if** $tp[A]$ is "_"
8:                  $valueset_A \leftarrow \{$"_"$\}$;
9:              **else if** $tp[A] \notin valueset_A$
10:                 $valueset_A += tp[A]$;

Example 5: In Table 1, we detect inconsistencies with following three rules:

$\phi1(R : A, C \rightarrow D, (\_, \_||\_))$,
$\phi2(R : A, C \rightarrow D, (a1, c2||d2))$,
$\phi3(R : B \rightarrow E, (b1||e2))$.

The attributes set, A-set of $\phi1$ and $\phi2$, contains four pairs of attribute and value set.
$A - set = \{(A, \_), (B, \{b1\}), (C, \_), (D, \_), (E, \{e2\})\}$.

4.3   Group

In the range module, we get the attribute set (*A-set*). For each attribute A, there is a value set (*V-set*) which consists of both the value of A and the tuple set (*T-set*), and all the tuples in the *T-set* agree on the same value in A. It is easy to get the tuples with certain value of certain attribute from the result.

The pseudo code of this step is shown in Algorithm 3. We store the attributes with constant values in *A-set* and insert its value in the *V-sets*. If an attribute has no constant value, we only store the attribute in *A-set* (line 1–5). We create a hash for each attribute involved in the rules. Then we scan each tuple in database. If the value does not exist in the hash index of this attribute, we store the value in *V-set* and insert the tuple id into its *T-set* (line 6–13). Meanwhile, we insert the value into the bucket in the hash table of the attribute.

Algorithm 3. GROUP

Input: database D and the attribute set(A-set)
Output: the data set S grouped by value
//Initial the data set S
1: **foreach** attribute A and its value set V in A-set do
2:      store A in S.
3:      **if** the value set of A is not "_"
4:          **foreach** value a in the value set of A do
5:              V-set_A += a;
6: **foreach** tuple $t \in D$ do
7:      **foreach** $A \in A$-set
8:          **if** the value set V of A is "_" or $t[A] \in V$ then
9:              **if** $t[A]$ is not in V-set_A then
10:                 V-set_A += $t[A]$;
11:                 T-$set_{A=t[A]}$ += t;
12:             **else if** t is not in T-$set_{A=t[A]}$ then
13:                 T-$set_{A=t[A]}$ += t;

The time complexity of this module is $O(|T| \cdot |A|)$, where $|T|$ is the number of tuples in the database, and $|A|$ is the number of attributes involved in rules. $O(1)$ is the cost of accessing one value in an attribute. We only scan the values of the attributes involved in the A-set for just once. That is, we only require scanning a subset of the database. So the time complexity of this module is bounded by $O(|T|)$, where $|T|$ is the number of tuples in database.

This module requires the only one pass of the database scan in our detection algorithm. After this module, we get the intermediate result S. S consists of a list of A-$sets$, each A-set consists of an attribute and a V-set, and each V-set consists of a list of pairs ($value$, T-set). $S = \sum A-set(attribute, V-set)$, and a V-set $V = \sum(value, T-set)$. For efficiently retrieving values from S, we build a hash index for attributes and values in S respectively. The result of this module can return the tuple set containing the tuples with a certain attribute value without accessing the original database.

4.4   Match

As discussed in Sect. 4.1, the MATCH module returns tuple sets in which the tuples agree on the LHS of ϕ. In the MATCH module, we consider a strategy of processing the FDs. We store the intermediate results which will be used for the detection with other rules in order to avoid redundant calculation.

As shown in Algorithm 4, if LHS(ϕ) and LHS(ϕ') share the same subset, where ϕ and ϕ' are different rules, we store the merged results in a list named *processed*, since that will be used to detect violations of other rules (Line 3–8). Otherwise, we just merge the LHS of the rule (Line 10–14). The function MERGE used in this module will be explain in detail in Sect. 4.5

Algorithm 4. MATCH

Input: a FD ϕ
Output: the tupleset in which tuples agree on the LHS of ϕ
1: i = processed.size();
2: **while** $|LS_\phi| >= 2$ **do**
3:     **if** a subset $\{A,B\}$ in LS_ϕ exists in $LS_{\phi'}$ and ϕ' is unprocessed
4:         $processed[i]$ $\leftarrow$ MERGE(A,B);
5:         **foreach** ϕ' unprocessed
6:             **if** $\{A,B\}$ is the sebset of $LS_{\phi'}$
7:                 $LS_{\phi'}$ $\leftarrow$ $LS_{\phi'}- \{A, B\} + \{processed[i]\}$;
8:         $i\leftarrow i+1$;
9:     **else**
10:        $temp$ $\leftarrow$ MERGE(LS_ϕ [0], LS_ϕ [1]);
11:        **for** $1< i < |LS_\phi|$ **do**
12:            $temp$ $\leftarrow$ MERGE($temp, LS_\phi$ [i])
13:            $i\leftarrow i+1$;
14:        LS_ϕ $\leftarrow$ $temp$;
15: **return** LS_ϕ

Example 6: Consider the data in Table 1, and the FD $\phi1$ in example 4. We add another FD $\phi4$:

$$\phi1(R : A, C \rightarrow D, (\_, \_||\_));$$

$$\phi4(R : A, C, E \rightarrow B, (\_, \_, \_||\_));$$

First, we got the grouped tuple sets in the GROUP module. The data set is as follows.

$S = \{(A, V\text{-}set_A),(B, V\text{-}set_B),(C, V\text{-}set_C), (D, V\text{-}set_D), (E, V\text{-}set_E)\}$
$V - set_A = \{(A = a1, \{t1, t2, t3, t4, t5, t6\})\}$
$V - set_B = \{(B = b1, \{t1, t3, t4, t6\}), (B = b2, \{t2, t5\})\}$
$V - set_C = \{(C = c1, \{t1, t2\}), (C = c2, \{t3, t5, t6\}), (C = c3, \{t4\})\}$
$V - set_D = \{(D = d1, \{t1, t2, t3\}), (D = d2, \{t4, t5, t6\})\}$
$V - set_E = \{(E = e1, \{t1, t2\}), (E = e2, \{t3, t4, t5\}), (E = e3, \{t6\})\}.$

Then we detect tuple sets which contain tuples matching the FD on the LHS. We denote the tuple set (*T-set*) in which tuples agree on the same condition set (*C-set*) as $T\text{-}set_{C\text{-}set}$. The tuple sets matching $\phi1$ on the left side are as follows.

(1) $T\text{-}set_{A=a1,C=c1} : (T\text{-}set_{A=a1} \cap T\text{-}set_{C=c1}) = \{t1, t2\}$
(2) $T\text{-}set_{A=a1,C=c2} : (T\text{-}set_{A=a1} \cap T\text{-}set_{C=c2}) = \{t3, t5, t6\}.$

As we know, the attributes A and C in the LHS of $\phi1$ also exist in LHS of $\phi4$, so we store the tuple sets matching the LHS of $\phi1$ in the *processed* set in order to avoid redundant calculation. The results can be used in detecting the tuple sets matching $\phi4$ in LHS. We get the intersections which contain more than one tuple as follows.

(1) $T-set_{A=a1,C=c1,E=e1} : (T-set_{A=a1,C=c1} \cap T-set_{e=e1}) = \{t1,\ t2\}$

(2) $T-set_{A=a1,C=c2,E=e2} : (T-set_{A=a1,C=c2} \cap T-set_{e=e2}) = \{t3,\ t5\}.$

In this example, we do not calculate tuple set containing only one tuple likes $T-set_{A=a1,C=c2} \cap T-set_{e=e3}$, since one tuple cannot violate a constant RHS rule alone. This strategy will be explained in detail in Sect. 4.5.

4.5 Merge

This module is the most important part in the algorithm for detecting inconsistencies violating variable RHS rules. And this module dominates cost when detecting inconsistencies violating variable RHS rules.

This block is a sub-module of the match block. Merging two attributes A and B means to find the tuples agree on both A and B. The results contain a list of tuple sets. Each tuple set has a condition set. All the tuples in the same tuple set agree on same condition. For example, given a tuple set $\{t1, t2, t3\}$ and its condition set $\{A = 1, B = 2\}$, we mean $t1[A] = t2[A] = t3[A] = 1$, $t1[B] = t2[B] = t3[B] = 2$. To get the result merged by attribute A and B, we need to check each value a for A and each value b for B from the result of the GROUP module. $T-set_{A=a,B=b}$ is merged by $T-set_{A=a}$ and $T-set_{B=b}$. $T-set_{A=a,B=b} = T-set_{A=a} \cap T-set_{B=b}$.

The pseudo code of the merge algorithm is shown in Algorithm 5. We get two tuple sets from the two lists of tuple sets of the two attributes separately. Then we check the size of the two tuple sets. If both of them and their intersection contain more than one tuple, we store the intersection tuple set in the list of tuple sets which is prepared as the output of this algorithm. Meanwhile, we store the conditions of the two tuple sets together as the condition set of the intersection (Line 5–6). The loop continues until all the tuple sets in one list make intersection with all the tuple sets in the other list.

Algorithm 5. MERGE

Input: two lists of tuplesets according to two different attributes sets M, N
Output: a list of tuplesets matching on all attributes in the input two lists
1: $i = 0$;
2: **foreach** $m \in V\_set_M$
3:      **foreach** $n \in V\_set_N$
4:          **if** $|tuple\_set_{M=m}| > 1$ and $|tuple\_set_{N=n}| > 1$ and $|tuple\_set_{M=m} \cap tuple\_set_{N=n}| > 1$
5:              $C\_set\_i \leftarrow \{M=m, N=n\}$;
6:              $T\_set_{C\_set\_i} \leftarrow tuple\_set_{M=m} \cap tuple\_set_{N=n}$;
7:              $i \leftarrow i+1$;
8: **return** a list consists of $T\_set_{C\_set\_k}$ ($k=0,1,...i$)

Example 7: We use the grouped result in Table 4. And we merge attribute A and C:

(1) $T-set_{A=a1,C=c1} = \{t1, t2\}$,

(2) $T-set_{A=a1,C=c2} = \{t3, t5, t6\}.$

We do not need to calculate the intersection $(T-set_{A=a1} \cap T-set_{C=c3})$, since $T-set_{C=c3}$ contains only one tuple and it is impossible to cause violation.

Then we analyze the cost of this block MERGE. As the tuples in $T-set_{A=a}$ and $T-set_{B=b}$ are ordered by the tuple id, the cost of $(T-set_{A=a} \cap T-set_{B=b})$ is $O(M+N)$, where $M = |T-set_{A=a}|$, and $N = |T-set_{B=b}|$. We assume that there are p values $\{a_1, a_2, ..., a_p\}$ in attribute $valueset_A$ with $|T-set_{A=a_i}| > 1$ and q values $\{b_1, b_2, ..., b_p\}$ in attribute $valueset_B$ with $|T-set_{B=b_j}| > 1$. We only care about the tuple set with size > 1. If there is only one tuple t in the tuple set, it means that there exists no tuples agree with tuple t on the same value. That is, t will not be involved in the inconsistency.

$C(Merge(A, B))$ denotes the cost of merge($attribute$ A, $attribute$ B).

$$C(Merge(A, B)) = \sum_{\substack{1 \leq i \leq p \\ 1 \leq j \leq q}} cost(T-set_{A=a_i} \cap T-set_{B=b_j})$$

$$= \sum_{1 \leq i \leq p} q * |T-set_{A=a_i}| + \sum_{1 \leq i \leq q} p * |T-set_{B=b_j}|$$

$$= q * \sum_{1 \leq i \leq p} |T-set_{A=a_i}| + p * \sum_{1 \leq i \leq q} |T-set_{B=b_j}|$$

$$\leq q * tuplesnum + p * tuplesnum = (p+q) * tuplenum$$

It shows that $C(Merge(A,B)) \leq (p + q) * tuplesnum$, where p is the number of distinct values appear in A in more than one tuple, and q is that in B, $tuplesnum$ is the number of tuples in the database. If both the value sets of A and B that we got from the Range module are "_", $C(Merge(A,B)) = (p + q) * tuplesnum$. The cost is affected by both the $tuplesnum$ and the size of value sets of the attributes that need to be merged. We prove the following proposition and propose a conception of redundancy to help us analyze the complexity of this module.

Proposition: The number of distinct values of one attribute with the $|T-set|>1$ is no more than the $tuplesnum/2$.

Proof: We assume k as the number of distinct values of one attribute A with the $|tupleset|>1$. It means that there is a set S containing k tuplesets and the size of each tupleset is at least 2. As we know, each tuple in the tupleset is different with the tuples in both the same tupleset and other tuplesets. The sums of all the $|tuplesets|$ of attribute A equals $tuplesnum$, then $\sum_{1 \leq i \leq k} |S[i]| \leq tuplesnum$. If $k > tuplesnum/2$, then $\sum_{1 \leq i \leq k} |S[i]| > 2k > tuplesnum$ which conflicts with the result above. The assumption is not supposed and the proposition is true.

Then we can give the upper bound of the p and q mentioned before. $p \leq tuplenum/2$, and $q \leq tuplenum/2$. $C(Merge(A, B)) \leq (p+q) * tuplenum \leq (tuplenum)^2$. Thus the cost of this module $C(Merge(A, B))$ is bounded by both the tuplesnum and the number of distinct values in A, B. It also has an upper bound $O(|T|^2)$, the worst-case complexity, which is impossible in reality.

Definition 6 (Redundancy): We define the redundancy(RDD) of an attribute with the rate $RDD(attribute A) = (1 - \frac{D_A}{T})$, where D_A is the number of distinct values on attribute A, and T is the number of tuples in database. The redundancy of a database R is defined as $(databaseR) = (1 - \frac{1}{T} \times \sum_{A_i \in attr(R)} D_{A_i})$.

According to the definition of redundancy, we can learn that database with high RDD must have a small number of distinct values. The RDD of some attributes in the real world may be very high. For example, in a relation of personal information, the RDD of an attribute "gender" must be very high, since there are only two distinct values "male" and "female". And in the real world, we only use FDs as constraints for the attributes with high RDD. There is no significance to use a FD for attributes such as "ID" whose values are all distinctive. In real world, the RDD of the dataset which can be restricted by FDs may be very high even close to 1.

With the help of redundancy, we show the complexity of this module. The time complexity of this module is $O(|D| \cdot |T|)$ where $|T|$ is the number of tuples in database, and $|D|$ is the number of distinct conditions preparing to be merged. After the definition of redundancy, we can use RDD to express $|D|$, $|D| = (|T| \text{-} RDD \cdot |T|)$. As in real world big data RDD is very likely to be high and even close to 1, meanwhile, $|D|$ will be much smaller than $|T|$, then the cost of this module can be performed in $O(T)$ time.

In this module, we just use the grouped result S instead of the original database. We can get the tuple set with certain attribute value in constant time with the help of the hash index. The cost of one tuple set retrieve is $O(1)$. The cost of retrieving all tuple sets for the detection is bounded by $O(|\sum attr(\phi)|)$, where $|\sum attr(\phi)|$ is the number of attributes in the rules. This cost is much smaller than the cost of scanning the whole database. That is also a reason for the efficiency of our detection method.

5 Experimental Study

In this section, we used two real-world datasets to evaluate the performance of our consistency detection algorithm experimentally.

Dataset 1: This dataset is a relation about call logs from Heilongjiang Liantong Company, which has 14 attributes, and we use 30 M tuples of it. The size of the dataset is 5.67 GB. The limitation of data size is caused by the capability limitation of SQL engineer on a single machine, which is used to run our competitor.

Dataset 2: This dataset is the 1990 US Census data[1] which has 2458285 tuples and 68 attributes. The description of these attributes can be found from the website.

We conducted all the experiments on a Windows 7 machine with a 3.10 GHz Intel CPU and 4 GB of Memory. Each experiment was run 5 times, and the average time is reported. The first experiment is conducted on both dataset 1 and dataset 2, the rest experiments are conducted on the dataset 1. The goal of experiments is to test the impact of five aspects on the detection time. They are (1) the number of tuples (T-SZ), (2) the kind of rules (RK), (3) the number of rules (R-SZ), (4) the redundancy of the attribute values (RDD) and (5) the percentage of dirty tuples.

---

[1] The description of this dataset can be found from following website. http://archive.ics.uci.edu/ml/machine-learning-databases/census1990-mld/USCensus1990-desc.html.

EXP 1: The Impact of T-SZ and RK: In this experiment, we test the impact of T-SZ and RK on detection time. For the first dataset, T-SZ ranges from 5 M to 30 M, in 5 M increments. We use two lists of rules for detection. One consists of constant RHS rules (cRHS) and the other consists of variable RHS rules(vRHS). Both the R-SZs of the two lists are 300. There is a FD in the second rule list, and the embedded FDs of all the rules are the same. The number of attributes of the embedded FD is 4. For the second dataset, we also use two lists of rules, each list contains 15 rules. Figure 1(a) and (c) show the run time of our method and the SQL-based method when detecting with cRHS rules on the two datasets. The result shows that our detection outperforms the previous method, and the detection time increases with T-SZ. Figure 1(b) and (d) show the run time of detection when detecting vRHS rules on the two datasets, it also increases with T-SZ. Both the results in detection on the two datasets indicate that our detection method is more efficiency than SQL-based method.

EXP 2: The Impact of R-SZ and RK: In this experiment, we test the impact of R-SZ and RK on detection time. T-SZ is 10 M, and the R-SZ of rules ranges from 50 to 300, in 50 increments. We still take two lists of rules used in EXP 1 for detection.

In this experiment, we add another rule list consists of 300 variable CFDs. There is no traditional FD in this list. In Fig. 1(e) and (f), we observe that the R-SZ has little effect on the detection time for the variable RHS rules which contains a traditional FD. However, it affects the run time for detecting with cRHS rules and vRHS rules containing no traditional FD. If the embedded FD of this rule has been detected before, we can get the inconsistencies of this rule from the detection result. That is the reason why the R-SZ has little impact on detection time for variable RHS rules including one FD. If the embedded FD has not been detected, we merge the LHS of the rule and start with the attributes with constant values in order to decrease the time complexity of MERGE. The result shown in Fig. 1(f) indicates that our strategy is effective when the size of rules is small. We learn from the figure that the line of vRHS rules with no FD is below the line of vRHS rules with one FD for most time. With the increasing of the rules size, the detection time is more close to that with the FD embedded.

EXP 3: The Impact of RDD: In this experiment, we test the impact of RDD on detection time. T-SZ set to be 5 M, and a rule list consists of both the two list used in EXP 1. We observe from Fig. 1(g) that, it takes more time for detection with low data redundancy. As the tuple sizes of data in this experiment are the same, the one with higher RDD has less distinct values for each attribute. And as we estimated in the MERGE module in Sect. 4, the number of distinct values in each attribute affects the performance of our detection method.

EXP 4: The Impact of Noise: In this experiment, we test the impact of noise on detection time. T-SZ is fixed 10 M. The rule list consists of variable RHS rules in EXP 1, and the noise ranges from 5% to 9%, in 1% increments. We observe from Fig. 1(h) that the noise of the database has little effect on the detection time.

Summary of Experiments: In summary, the experiments have following results.

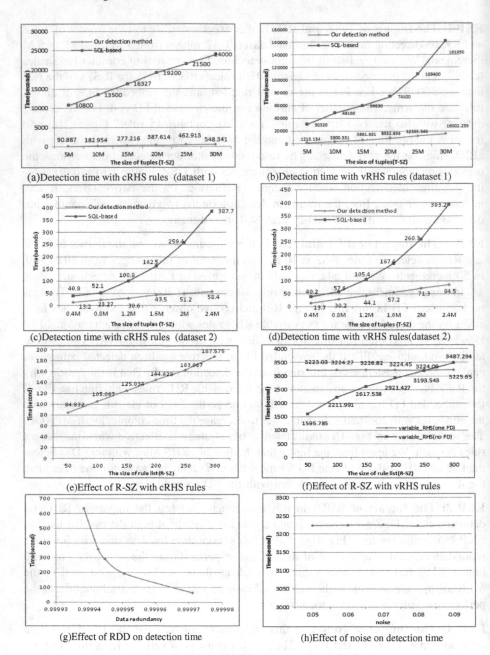

(a)Detection time with cRHS rules (dataset 1)

(b)Detection time with vRHS rules (dataset 1)

(c)Detection time with cRHS rules (dataset 2)

(d)Detection time with vRHS rules(dataset 2)

(e)Effect of R-SZ with cRHS rules

(f)Effect of R-SZ with vRHS rules

(g)Effect of RDD on detection time

(h)Effect of noise on detection time

Fig. 1. Experimental results

(1) Our detection method is much more efficient than the SQL-based method and the detection time is nearly linear when detecting the inconsistency of database with high data redundancy.
(2) The detection time of our method is affected by both the number of tuples and the data redundancy. The number of rules has impact on the results when detecting with constant RHS rules and variable RHS rules without FDs, and it has little effect when detecting with variable RHS rules containing FD.
(3) The noise of database has little impact on the detection time.

6 Conclusion and Future Work

We study the inconsistency detection problem in this paper. We proposed a method to detect consistencies with only one pass scan of data. Our method outperforms the SQL-based method when detecting database with high redundancy. Detecting inconsistencies with high efficiency benefits the database management for improving data quality. And we consider cell as the unit of consistencies instead of tuple, so that the detection result can be clear.

Our future study focuses on two aspects. The first one is extending our method for detection for more kind of rules such as denial constraints [6] and conditional inclusion dependencies. The second aspect is repairing the inconsistencies in big data. We aim to develop efficient methods to restore the consistency in big data.

Acknowledgment. This paper was supported by NGFR 973 grant 2012CB316200, NSFC grant U1509216,61472099,61133002 and National Sci-Tech Support Plan 2015 BAH10F01.

References

1. Wayne, W.E.: Data quality and the bottom line: achieving business success through a commitment to high quality data. In: TDWI report (2004)
2. Bohannon, P., Fan, W., Geerts, F., et al.: Conditional functional dependencies for data cleaning. In: ICDE, pp. 746–755 (2007)
3. Chen, W., Fan, W., Ma, S.: Analyses and validation of conditional dependencies with built-in predicates. In: Bhowmick, S.S., Küng, J., Wagner, R. (eds.) DEXA 2009. LNCS, vol. 5690, pp. 576–591. Springer, Heidelberg (2009)
4. Cong, G., Fan, W., Geerts, F., Jia, X., Ma, S.: Improving data quality: consistency and accuracy. In: VLDB, pp. 315–326 (2007)
5. Fan, W., Geerts, F., Tang, N., et al.: Inferring data currency and consistency for conflict resolution. In: ICDE, pp. 470–481 (2013)
6. Bohannon, P., Fan, W., Flaster, M., et al.: A cost-based model and effective heuristic for repairing constraints by value modification. In: SIGMOD, pp. 143–154 (2005)
7. Chu, X., Ilyas, I.F., Papotti, P.: Holistic data cleaning: putting violations into context. In: ICDE, pp. 458–469 (2013)
8. Kolahi, S., Lakshmanan, L.V.S.: On approximating optimum repairs for functional dependency violations. In: ICDT, pp. 53–62 (2009)

9. Yakout, M., Elmagarmid, A.K., et al.: Guided data repair. In: PVLDB, pp. 279–289 (2011)
10. Korn, F., Muthukrishnan, S., Zhu, Y.: Checks and balances: monitoring data quality problems in network traffic databases. In: VLDB, pp. 536–547 (2003)
11. Xiong, H., Pandey, G., Steinbach, M., et al.: Enhancing data analysis with noise removal. In: TKDE, pp. 304–319 (2006)
12. Fan, W., Geerts, F.: Foundations of Data Quality Management, Synthesis Lectures on Data Management, pp. 71–82 (2012)
13. Chiang, F., Miller, R.J.: Discovering data quality rules. In: VLDB, pp. 1166–1177 (2008)
14. Golab, L., Karloff, H., Korn, F., Srivastava, D., Yu, B.: On generating near-optimal tableaux for conditional functional dependencies. In: VLDB, pp. 1161–1172 (2008)
15. Fan, W., Geerts, F., Li, J., Xiong, M.: Discovering conditional functional dependencies. In: TKDE, pp. 683–698 (2011)
16. Geerts, F., Mecca, G., Papotti, P., Santoro, D.: The LLUNATIC data-cleaning framework. In: PVLDB, pp. 625–636 (2013)
17. Bertossi, L., Bravo, L., et al.: The complexity and approximation of fixing numerical attributes in databases under integrity constraints. In: Information Systems, pp. 407–434 (2008)
18. Fan, W., Li, J., Ma, S., et al.: Towards certain fixes with editing rules and master data. VLDB 3, 173–184 (2010)
19. Talukder, N., Ouzzani, M., Elmagarmid, A.K., et al.: Detecting inconsistencies in private data with secure function evaluation. Technical report, Purdue University (2011)
20. Demsky, B., Rinard, M.: Automatic detection and repair of errors in data structures. In: SIGPLAN Notices, pp. 78–95 (2003)

Entity Identification

Domain-Specific Entity Linking via Fake Named Entity Detection

Jiangtao Zhang[1]($\boxtimes$), Juanzi Li[1], Xiao-Li Li[2], Yao Shi[1], Junpeng Li[3], and Zhigang Wang[1]

[1] Department of Computer Science and Technology,
Tsinghua University, Beijing 100084, China
zhang-jt13@mails.tsinghua.edu.cn, lijuanzi@tsinghua.edu.cn,
wangzigo@gmail.com
[2] Institute for Infocomm Research, A*STAR, Singapore 138632, Singapore
xlli@i2r.a-star.edu.sg
[3] Software School, Xidian University, Xian 710126, China
jpl_xd@163.com

Abstract. The traditional named entity detection (NED) and entity linking (EL) techniques cannot be applied to domain-specific knowledge base effectively. Most of existing techniques just take extracted named entities as the input to the following EL task without considering the interdependency between the NED and EL and how to detect the Fake Named Entities (FNEs). In this paper, we propose a novel approach to jointly model NED and EL for domain-specific knowledge base, facilitating mentions extracted from unstructured data to be accurately matched to uniquely identifiable entities in the given domain-specific knowledge base. We conduct extensive experiments for movie knowledge base by a data set of real-world movie comments, and our experimental results demonstrate that our proposed approach is able to achieve 84.7 % detection precision for NED and 87.5 % linking accuracy for EL respectively, indicating its practical use for domain-specific knowledge base.

Keywords: Entity linking · Named entity detection · Fake named entity · Domain-specific knowledge base · Joint model

1 Introduction

Entity linking (EL), determining the identity of entities mentioned in text, is the key issue in bridging *unstructured* textual data with *structured* knowledge bases (KBs) [7]. It has been widely used in diverse applications such as question answering, information integration and KB construction [19]. Significant portion of recent research in this area focus on linking named entities in text to *general* knowledge bases, such as Wikipedia based KBs or WordNet based KBs.

Recently, establishing *domain-specific* KBs has been found more effective and useful to manage and query knowledge within a specific domain. For example,

S.B. Navathe et al. (Eds.): DASFAA 2016, Part I, LNCS 9642, pp. 101–116, 2016.
DOI: 10.1007/978-3-319-32025-0_7

IMDB[1] contains more concrete and comprehensive movie knowledge than *general* knowledge base Wikipedia or Baidu Baike. Therefore, *domain-specific* EL techniques become more and more important, with the increasing demand for constructing and populating *domain-specific* KBs. An Entity Discovery and Linking (EDL) task is introduced by KBP 2014[2]. In particular, given an unstructured document, the EDL task aims to automatically extract mentions (i.e. Named Entity Detection, or NED), link them to a general KB, e.g. Wikipedia (i.e. Entity Linking or EL), and identify NIL mentions that do not have corresponding KB entries [9]. However, traditional EL methods are ineffective for *domain-specific* EL tasks, due to the different characteristics between *domain-specific* area and *general* area. Specifically, we observe there are two unsolved key **challenges** in *domain-specific* EL problem:

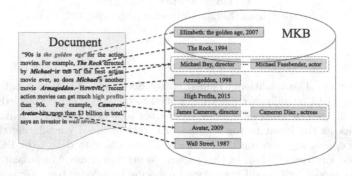

Fig. 1. An illustration for the task of domain-specific entity linking

1. Fake Named Entity: Given a document and a domain-specific KB, there exist many common phrases in the document which could likely be linked to entities in the given KB. However, not all these common phrases should be linked. As an example shown in Fig. 1, the mention *"the golden age"*, *"The Rock"*, *"high profits"* are all common phrases in general domain. For a *domain-specific* Movie-Knowledge-Base (MKB)[3], however, these mentions are the titles of entities/movies in MKB. As such, these mentions in the document are quite likely to be linked to MKB. Nevertheless, according to their context, except *"The Rock"* is *true* named entity, both *"the golden age"* and *"high profits"* are just common phrases that should not be recognized as named entities. We denote these mentions which should not be linked to entities in KB as **Fake Named Entities** (FNEs). Traditional methods do not consider the FNE issue and thus will not work well for the *domain-specific* EL task due to the fact that there exist many FNEs in a *domain-specific* area. In this paper, we propose a novel technique for FNE detection from the given unstructured text.

---

[1] http://www.imdb.com/.

[2] http://nlp.rpi.edu/kbp2014/.

[3] MKB is constructed by knowledge engineering laboratory of department of computer science and technology, Tsinghua University, Beijing.

2. Interdependency: Existing techniques typically treat NED and EL as two separated tasks and use a pipeline/sequencial architecture [10–12,17,21] that simply takes extracted named entities as the input to the following EL task, without considering the interdependency between NED and EL tasks. As such, the errors, i.e. FNEs, occurred in the NED task will inevitably affect the performance of the subsequent EL task. For example in Fig. 1, we could mistakenly treat FNEs *"the golden age"*, *"wall street"* and *"high profits"* as true mentions and link them to entities in a domain-specific knowledge base (e.g. MKB). However, if we consider NED and EL tasks jointly, such FNE errors could be fixed because we can update the confidence of mentions iteratively. For example, the linked entities/movies of above FNEs are not action movies while the main thread of the text is talking about action movies and all other TNEs in the text is related to action movies. Such context information is the result of the EL and thus can in turn be used to lower the confidence of these FNEs. Furthermore, such information is also useful for the ranking of *"Michael"* and *"Cameron"*, which are both famous directors of action movies. Therefore, the errors of FNEs can be fixed because the two tasks EL and NED are inherently coupled. Different from traditional methods, our proposed technique will leverage their interdependency iteratively making both NED and EL tasks more robust.

In summary, detecting the FNEs and linking the true/correct mentions are two significant challenges, because textual mentions in a specific domain could be potentially far more ambiguous than those in general domain. Therefore, we propose a new technique to detect FNEs in a specific domain via jointly modeling named entity detection and entity linking.

Contributions. The main contributions of this paper are summarized as follows.

- We are among the first to explore the problem of joint NED and EL with the domain-specific knowledge base. To the best of our knowledge, our research is the first to define the important concept Fake Named Entities (FNEs), which is critical for *domain-specific* EL task.
- We proposed an effective technique that jointly models NED and EL by iteratively enhancing the confidence of entity extraction and certainty of entity linking. Particularly, we leverage the entity linking result to increase the confidence of true named entities (TNEs) and thus lower the confidence of FNEs. Conversely, this enhancement of extraction/detection confidence improves the performance of the entity linking/disambiguation. This process can be repeated iteratively until convergence, as long as there is an improvement in the extraction and disambiguation.
- To evaluate the effectiveness of our proposed approach, we conducted extensive experiments on a manually annotated data set of real world movie comments and a real *domain-specific* knowledge base. The experimental results show that our proposed approach outperforms baseline methods significantly.

2   Preliminaries

In this section, we first introduce some fundamental concepts for our problem and subsequently define the task of linking named entities in a specific domain.

Domain-Specific Knowledge Base. A domain-specific knowledge base defines a set of representational primitives to model domain knowledge from different perspectives, which can be defined as DSKB= $\{C, E, P, R\}$, where C represents a set of *concepts* in the domain such as actors, movies and producers; $E = \{e_1, e_2, ...e_{|E|}\}$ is the *entities* of concepts such as Steven Spielberg – a movie director; P denotes a set of *properties* to describe attributes of concepts or entities such as actor names, movies' production time; R means the set of triples, each of them describes the *relation* between entities *or* between entity and concept, which can be defined as $\{s, p, o\}$, where $s \in E \cup C, p \in P$, $o \in E \cup C \cup L$, and L is the set of literals. In this paper, we choose domain-specific Movie-Knowledge-Base (MKB) as the target DSKB for our task. The MKB is a high quality knowledge base about movies, TV series and celebrities which integrates several English and Chinese movie data sources from Baidu Baike and Douban, and it contains 23 concepts, 91 properties, more than 700,000 entities and 10 million triples.

Mentions and Linked Entities. We define a *mention* as a textual phrase (e.g., the *"the golden age"* in Fig. 1) which can potentially be linked to some entities in DSKB. We consider every possible n-gram (e.g. $n \leq 5$) as a candidate mention. Given a document d, we define $M = \{m_1, m_2, ..., m_{|M|}\}$ as the set of candidate mentions. In addition, let $E(m) = \{e_1, e_2, ..., e_{|E(m)|}\} \subseteq E$ denote the set of candidate entities which a candidate mention $m \in M$ might be linked to. For a mention m, we define the correct entity $e_m \in E(m)$ which m should actually be linked to as *linked entity* (i.e. ground-truth mapping entity). For example, in Fig. 1, the set of entities that mention *"Cameron"* could be linked to is $E("Cameron") = \{$ *"James Cameron"*, *"Cameron Diaz"*$\}$ and the *linked entity* is *"James Cameron"*.

Fake Named Entity. We define $M_F = \{m_{f1}, m_{f2}, ..., m_{|M_F|}\} \subseteq M$ that should not be linked to any entity in E, which should only be treated as common textual phrases as *Fake Named Entities*(FNEs). We also define the *True Named Entities* (TNEs) as $M_T = \{m_{t1}, m_{t2}, ..., m_{|M_T|}\} \subseteq M$, denoting the mentions that should be linked to entities in E. Obviously, $M_F \cup M_T = M$. As the example shown in Fig. 1, the set TNEs is $M_T = \{$ *"The Rock"*, *"Michael"*, *"Armageddon"*, *"Cameron"*, *"Avatar"*$\}$, while the set FNEs is $M_F = \{$ *"the golden age"*, *"high profits"*, *"wall street"*$\}$.

Context Mention and Entity. For a given mention m in a document d, We define all the other candidate mentions $C_M(m) = \{m_{c1}, m_{c2}, ..., m_{|C_M(m)|}\} \subseteq M$

in the same document or in a certain size window as *Context Mentions*. In our experiments, we employ the window size, which is set as 50, following the experimental setting in literature [15]. Notice that each context mention $m_c \in C_M(m)$ could be ambiguous as we do not know its *linked entity* in E. As such, we define *Context Entities* $C_E(m) = \{e_{c1}, e_{c2}, ..., e_{|C_E(m)|}\} \subseteq E$ as the set of most possible *linked entities* for each context mention for the time being.

Task Definition. Given an unstructured document d in a specific domain and a DSKB pertaining to the same domain, our task is to extract TNEs and filter out FNEs in d, and to develop a function $\sigma : M \to E$ which maps each extracted TNE $m \in M_T$ to its *linked entity* $e \in E(m)$ in DSKB (e.g., MKB). Specifically, our task consists of two parts, namely *Named Entity Detection* (NED) (i.e. *Mention Extraction*) and *Entity Linking* (EL) (i,e. *Disambiguation*). NED is the task of detecting FNEs and extracting TNEs. EL is the task of linking an extracted TNE to a specific definition or instance of an entity in DSKB. The output of our task is the set of *mention* and *entity* mapping pairs: $\{\langle m, \sigma(m) \rangle |\ \forall m \in M_T\}$.

3 Our Proposed Approach

In this section, we propose a novel approach that jointly models NED and EL iteratively to link all the TNEs in an unstructured document to uniquely identifiable entities in DSKB. The main idea of our approach is as follows: in each iterative step, we gradually improve the confidence of TNEs while reduce the confidence of FNEs. Specifically, by leveraging the interdependency of NED and EL, we use the results of EL (linking certainty) to provide the feedback for NED and thus could potentially improve the performance of NED via updating the weights of some features in NED. On the other hand, the results of NED (detection confidence) could also enhance EL process via updating the weights of some features in EL.

3.1 Framework Overview

The framework of our proposed model is shown in Fig. 2. From the figure, we can see that first we train EL and NED models independently based on a manually annotated data set (refer to experiment section) to learn two weighted vectors $\overrightarrow{W}_{el} = \{w_1^{el}, w_2^{el}, w_3^{el}, w_4^{el}\}$ and $\overrightarrow{W}_{ned} = \{w_1^{ned}, w_2^{ned}, w_3^{ned}, w_4^{ned}\}$ for two constructed feature vectors $\overrightarrow{F}_m(e) = \{f_1^{el}, f_2^{el}, f_3^{el}, f_4^{el}\}$ and $\overrightarrow{F}(m) = \{f_1^{ned}, f_2^{ned}, f_3^{ned}, f_4^{ned}\}$ respectively. Next, we apply the two learned models on an input document iteratively to predict TNEs and their *linked entities* in DSKB. The results of NED model $Ned(m)$ show the confidence of a candidate mention m being a TNE while the results of EL model $El_m(e)$ indicate the confidence of a mention m being linked to a candidate entity e. In addition, we take the results of each model to update the weights of some features of the other. That is to say, we can apply the two models mutually and iteratively. With the increase of the number of iterations, the detection confidence and linking certainty of TNEs will increase.

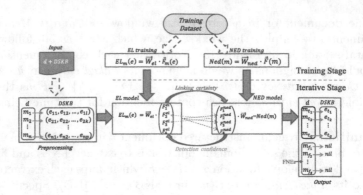

Fig. 2. Framework of our proposed iteratively joint model

Preprocessing. In this subsection, we briefly present how to generate the candidate mentions and entities for a given document d.

First, we build a dictionary D that contains various surface forms of the named entities. The detailed construction method is introduced in [19]. The dictionary D is in the form of $\langle key, value \rangle$ mapping, where the column of the *key* is a list of surface forms and the column of the mapping *value* is the set of entities which can be referred to by the *key*. Next, we consider every possible n-gram (e.g. $n \leq 5$) in d existing in the *key* column of dictionary D to generate a high-recall candidate mentions M, to avoid missing possible real mentions. Then, For each mention $m \in M$, we search for m in the column of *key* in D and add the set of entities *value* to the candidate entities $E(m)$.

3.2   Model Training

As the strategy used by existing studies [13, 20, 24], we can model NED and EL into two binary classifiers. First we need to construct the training set for these two classifiers based on a manually annotated data set. In EL model, the sample in the training set is a pair $(m, e), e \in E(m)$. Let $label(m, e) \in \{0, 1\}$ indicate positive sample or negative sample, which is determined as follows:

1. For each TNE m, if its true *linked entity* in the candidate entity set $E(m)$ is e_m, then (m, e_m) is a *positive* sample. For all other entities $\forall e_i \in E(m), e_i \neq e_m$, the (m, e_i) are regarded as *negative* samples.
2. For each FNE m, $\forall e \in E(m)$, (m, e) is treated as *negative* sample as well.

In NED model, on the other hand, the sample is m. Also let $label(m) \in \{0, 1\}$ be the indicator of positive sample and negative sample. Obviously, the set of TNEs M_T which are manually annotated is the set of *positive* samples while M_F is the set of *negative* samples.

Next, we learn two weight vectors $\overrightarrow{W}_{el}$ and $\overrightarrow{W}_{ned}$ for two constructed feature vectors $\overrightarrow{F}_m(e)$ and $\overrightarrow{F}(m)$ respectively which will be elaborated in next subsection by supervised machine learning technique on training data set – in our experiments, we employ state-of-art classification model SVM due to its good performance. Then these two models can be formulated as: $El_m(e) = \overrightarrow{W}_{el} \cdot \overrightarrow{F}_m(e)$, $Ned(m) = \overrightarrow{W}_{ned} \cdot \overrightarrow{F}(m)$. Specifically, for EL model, we will rank all entities in the candidate entity set $E(m)$ and select the entity with highest score $El_m(e)$ as the most possible *linked entity* in DSKB, and for NED model we classify a mention m into a TNE or FNE based on the result score $Ned(m)$.

Obviously the results of EL $El_m(e)$ show the certainty (strength) of m linking to e, while the results of NED $Ned(m)$ indicate the confidence level of m being a TNE. Therefore, we can use the results of each model to calculate features of the other iteratively, which will benefit the performance of both two models. In next section, we will introduce our constructed features of our two models and explain how these features are used to interact between two models in an iterative manner to improve performance.

3.3   Features in EL Model

Popularity. In the domain-specific area, taking movie comments as an example, people tend to review more popular and classic movies. Therefore, we choose *popularity* as an important context-free feature. We define the popularity via leveraging the count information from Baidu Baike as follows:

$$Pri_m(e) = \frac{count_m(e)}{\sum_{e \in E(m)} count_m(e)} \tag{1}$$

where $count_m(e)$ is defined as the number of times that mention m links to entity e in Baidu Baike. Notice that we can calculate this feature in advance.

Context Relatedness. Intuitively, one would expect that *mentions* which co-occur in the same document are related to one or a few topics, or have certain semantic relatedness [19]. In Fig. 1 both *"The Rock"* and *"Armageddon"* are similar action movies and thus people tend to talk/compare them together. Therefore, we propose a second feature: the *context relatedness*. Specifically, we calculate the average value of the semantic relatedness between all context entities $e_c \in C_E$ and the candidate entity $e \in E(m)$ to get the context relatedness.

$$ConRel_m(e) = \frac{\sum_{e_c \in C_E} SmtRel(e_c, e)}{|C_E|} \tag{2}$$

where $SmtRel(e_c, e)$ is semantic relatedness of a context entity e_c and the candidate entity e. However, there are two problems of above calculation:

1. Context mention $m_c \in C_M$ could also be ambiguous when performing the linking of current mention m. Thus, its corresponding context entity e_c is also unknown in current stage. As such, in the follow-up iteration, we use the results of EL model in last iteration to choose the best context entity with highest score, denoted as $e_{top}(m_c)$, for the relatedness calculation.

2. FNEs in the Context Mentions C_M could damage the performance of the following entity linking task. As mentioned above, the results of NED model $Ned(m)$ indicate the true confidence level of a mention m being a TNE. Therefore, we use $Ned(m_c)$ to denote the confidence level of m_c being a FNE. In other words, if a context mention m_c is a FNE with high probability, then the value of $Ned(m_c)$ will thus be small and has less impact to EL tasks.

Therefore, the context relatedness will be the weighted average value of the semantic relatedness of all Context Entities $e_c \in C_E$ and the candidate entity e, i.e.,

$$ConRel_m(e) = \frac{\sum_{m_c \in C_M} Ned(m_c) * SmtRel(e_{top}(m_c), e)}{|C_M|} \tag{3}$$

$$e_{top}(m_c) = \arg\max_{e_c \in E(m_c)} (El_m(e_c))$$

Next, we adopt two techniques: Wikipedia Link-based Measure (WLM) and Jaccard distance to calculate the semantic relatedness to get two kinds of context relatedness, denoted as $ConRel1_m(e)$ and $ConRel2_m(e)$, respectively.

WLM: The WLM is based on the Wikipedia hyperlink structure [13]. Given two entity e_i and e_j, we define the semantic similarity between them as $WLM(e_i, e_j) = \frac{\log(\max(|E_i|,|E_j|)) - \log(|E_i \cap E_j|)}{\log(|W|) - \log(\min(|E_i|,|E_j|))}$, Where E_i and E_j are the sets of entities that link to e_i and e_j respectively in MKB, and W is the set of all entities in MKB.

Jaccard Distance: We first extract the content of two entity e_i and e_j to compose two bag-of-words representations S_i and S_j respectively, and subsequently calculate Jaccard distance between S_i and S_j.

Content Similarity. It has been an effective way to use the context information to perform entity disambiguation. Therefore, we introduce out last feature for the EL model: *content similarity*. We define the content similarity as the similarity between the context around a candidate mention m and its candidate entity e, i.e., $Consim_m(e)$, which also calculated by Jaccard distance.

3.4   Features in NED Model

Link Probability. Link probability introduced in [12] is a proven feature which indicates how often a mention links to an entity in a knowledge base. For example shown in Fig. 1, for the mention *"wall street"*, the probability that it links to a certain entity is much less than that just treat it as a common phrase, because in most cases when people mention *"wall street"*, they refer it to a location rather than the movie in 1987. Therefore, the link probability is an important feature and is helpful to filter out FNEs. We define the link probability $LP(m)$ as follows.

$$LP(m) = \frac{\sum_{e \in E(m)} count_m(e)}{count(m)} \tag{4}$$

where $count_m(e)$ is defined as the number of times that mention m links to entity e and $count(m)$ is the total occurrence number of m in Baidu Baike. We can also calculate this feature in advance.

Linking Certainty. The results of EL model $El_m(e)$ provide a certainty score whether a mention m correctly links to an entity e. Therefore, we use it as a feature of NED model, to indicate the linking certainty level of a mention. Obviously, the higher the linking certainty that a mention m links to an entity e, the higher the probability that the mention m is not a FNE. We use the value of the highest score entity $e_{top}(m)$, denoted as $El_m(e_{top}(m))$, as the value of linking certainty $LC(m)$ of the mention m. That is to say, we use the results of the EL model as a feature of NED model.

$$LC(m) = El_m(e_{top}(m)) = \max\{El_m(e)|e \in E(m)\}. \tag{5}$$

Coherence. As mentioned above, entities occurring in a given document d are likely to be topically coherent, i.e. they are semantic related. So, we can exploit this topic *coherence* between entities in the document d to define the coherence feature for each candidate mention m, which is defined as the average semantic relatedness between m and all context mentions $m_c \in C_M$. Nevertheless, we still need to know the *linked entity* e for m to calculate the semantic relatedness. Therefore, for each m (and m_c), we also choose the highest score entity $e_{top}(m)$ (and $e_{top}(m_c)$) returned by the EL model as the *linked entity*.

$$Coh(m) = \frac{\sum_{m_c \in C_M} Ned(m_c) * SmtRel(e_{top}(m_c), e_{top}(m))}{|C_M|} \tag{6}$$

We can also calculate two kinds of semantic relatedness for two coherence features $Coh1(m)$ and $Coh2(m)$ as depicted in the Context Relatedness subsection.

3.5 Iterative Process

After training stage, we have learned two weight vectors $\overrightarrow{W}_{el}$ and $\overrightarrow{W}_{ned}$ for our two models. Here, we illustrate our iterative process as follows.

$$
\begin{aligned}
\mathbf{El_m(e)} &= \overrightarrow{W}_{el} \cdot \overrightarrow{F}_m(e) \\
&= w_1^{el} * Prim_m(e) + w_2^{el} * ConRel1_m(e) \\
&\quad + w_3^{el} * ConRel2_m(e) + w_4^{el} * ConSim_m(e) \\
&= w_1^{el} * Prim_m(e) + w_2^{el} * \frac{\sum_{m_c \in C_M} \mathbf{Ned(m_c)} * WLM(\mathbf{e_{top}(m_c)}, e)}{|C_M|} \\
&\quad + w_3^{el} * \frac{\sum_{m_c \in C_M} \mathbf{Ned(m_c)} * Jac(\mathbf{e_{top}(m_c)}, e)}{|C_M|} + w_4^{el} * ConSim_m(e)
\end{aligned}
\tag{7}
$$

$$\mathbf{Ned}(\mathbf{m}) = \vec{W}_{ned} \cdot \vec{F}(m)$$
$$= w_1^{ned} * LP(m) + w_2^{ned} * LC(m) + w_3^{ned} * Coh1(m) + w_4^{ned} * Coh2(m)$$
$$= w_1^{ned} * LP(m) + w_2^{ned} * El_m(\mathbf{e_{top}(m)}) +$$
$$w_3^{ned} * \frac{\sum_{m_c \in C_M} \mathbf{Ned}(\mathbf{m_c}) * WLM(\mathbf{e_{top}(m_c)}, \mathbf{e_{top}(m)})}{|C_M|}$$
$$+ w_4^{ned} * \frac{\sum_{m_c \in C_M} \mathbf{Ned}(\mathbf{m_c}) * Jac(\mathbf{e_{top}(m_c)}, \mathbf{e_{top}(m)})}{|C_M|}$$

$$(8)$$

For any iterative process, one of the most important issue is the convergence. Here, we define *iteration deviation* as the maximal value of difference of $Ned(m)$ of two successive iterations for all $m \in M$. Then we set the condition to complete the iteration is that *iteration deviation* is less than a predefined threshold ε, namely *iteration deviation threshold*, i.e.,

$$\max_{m_i \in M}(Ned(m_i^{(j)}) - Ned(m_i^{(j-1)}) \leq \varepsilon \qquad (9)$$

From extensive experiments, we found that with the increase of the number of iterations, *iteration deviation* gradually decreases and iterative process usually stops after some iterations. We will discuss the convergence issue in experiment section.

The detailed iterative algorithm is given in Algorithm 1.

Input: $M; \forall m \in M, E(m); \vec{W}_{el}; \vec{W}_{ned}$
Output: $M_T; \forall m \in M_T, \langle m, Ned(m) \rangle, \langle m, e_{top}(m), El_m(e_{top}(m)) \rangle$
repeat
  for each $m \in M$ **do**
    for each $e \in E(m)$ **do**
      $El_m(e) = \vec{W}_{el} \cdot \vec{F}_m(e);$
    end
    $e_{top}(m) = \arg\max_{e \in E(m)}(El_m(e));$
  end
  for each $m \in M$ **do**
    $Ned(m) = \vec{W}_{ned} \cdot \vec{F}(m)$
  end
until convergence;

Algorithm 1. Algorithm of the iterative process

---

4   Experiments and Evaluation

To fairly evaluate the effectiveness of our proposed approach, we have conducted extensive experimental studies to compare it with existing methods. All the programs were implemented in Python and all the experiments were conducted on a server (with four 2.7 GHz CPU cores, 1024 GB memory, Ubuntu 13.10).

Table 1. Statistical data of the user data set

Documents	$|FNEs|$	$|TNEs|$	CEs	$\overline{|M|}$	$\overline{|E(m)|}$
843	2529	11848	42105	17.05	2.92

4.1 Data Preparation

To the best of our knowledge, there is no publicly available benchmark data set for the domain-specific EL task. Thus, we manually create a first-of-its-kind gold-standard data set for our task. Since there could be subjective in the manual annotation process, in order to avoid introducing bias in the annotation task, we organized annotators into three groups and each group has several members. The first and second groups annotate the same data set *independently*, while the third group checks the annotation results and annotates those inconsistent named entities. The final results are determined by majority voting. Obviously, the annotation task is very time consuming and labour intensive.

In this paper, we focus to perform the entity linking from user movie comments/documents to the knowledge base MKB (i.e. Movie-Knowledge-Base). Specifically, we crawl user comments on movies from established Websites in China, including Sina, Sohu, 163, and Tianya, with 1 year time span, from 1/1/2014 to 12/31/2014. Finally, we obtained 843 documents forming the gold-standard data set. Table 1 lists the size of the data set and some statistical information about the data set, where $|FNEs|$ and $|TNEs|$ denote the numbers of Fake/True named entities respectively, CEs means the total number of candidate entities, $\overline{|M|}$ represents the average number of mentions per document, and $\overline{|E(m)|}$ shows average number of candidate entities per mention.

4.2 Experimental Results

In this subsection, we study the effectiveness of our proposed approach under different configurations, and compare them with some baseline methods.

Baseline Methods. Since most of joint NED and EL frameworks deal with short text linking to general KB based on high complexity algorithms, which could not apply directly on MKB and our data set, and furthermore these works are lack of publicly available APIs, we created two baselines in this paper, both of which employed the traditional pipeline architecture that takes extracted named entities as the input to the following EL task.

1. Prior Probability-based method (*POP*). First, in NED process, we only used the link probability for detection. Particularly, we set a threshold as 0.2 (which produces the best performance for *POP* method) and retain the mentions whose link probabilities are higher than the pre-set threshold. In the EL process, on the other hand, we used entity popularity for ranking. In other words, the entities with the highest popularity among all the candidate entities is considered as the *linked entity* for this mention.

2. Vector Similarity-based method (*VSim*). We constructed a context vector for each extracted mention and a profile vector for each candidate entity based on standard TF-IDF representation. Then we measure the similarity between these two vectors by computing their *Cosine* distance. Finally, the entity with the highest similarity is considered as the *linked entity* for the mention. For the NED process, we set a threshold as 0.087 (which gives the best results) and only retain the mention whose highest score of vector similarity is larger than it.

Parameter Setting. In our approach, there exists an important parameter, i.e. *iteration deviation threshold*, which needs to be determined.

Iteration Deviation Threshold. Figure 3 shows the curve of the *iteration deviation* versus the number of iterations for eight documents randomly chosen from our data set. From the results we observe that for each document the *iteration deviation* gradually decreases with the increase of the number of iterations. When the number of iterations exceeds 10, the *iteration deviation* flattens out gradually (≤ 0.001). As such, we can set the *iteration deviation threshold* $\varepsilon = 0.001$. This empirically proves that our iterative process converges, and also verifies our proposed approach that NED and EL contribute to each other within limited iterations.

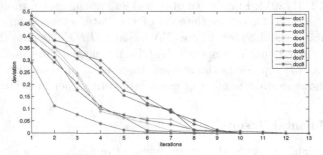

Fig. 3. Convergence of iteration

Our Model and Evaluation Metrics. Now we study the effectiveness of our proposed approach, which is configured into 4 different settings:

- *No Training + No Iterating (NoT+NoI):* we do not use the machine learning method to train the weight of the NED and EL. We assume that all features have the same weight, that is, for NED and EL models, the weight of all features is 0.25. Furthermore, we do not perform the iteration.
- *Training + No Iterating (T + NoI):* we use the machine learning method to learn the weight of the NED and EL, but we do not perform the iteration. Notice that $T + NoI$ is actually the traditional machine learning based approach.
- *No Training + Iterating (NoT + I):* we set the weight of NED and EL manually, but we perform the iteration.
- *Training + Iterating (T + I):* we use the machine learning method to train the NED and EL models and we perform the iteration.

In our proposed approach, EL is performed over noisy NED output and participates to the final decisions about extractions. Therefore, we evaluate NED, EL and their combination by employing the following evaluation metrics:

- **NED:** precision, recall and F1-measure;
- **EL:** accuracy over correctly recognized named entities. Notice here we do not use precision, recall and F1-measure due to the fact that if the correct extracted mentions are given, then precision=recall=F1-measure=accuracy [19] and most EL systems simply use accuracy to assess their performance.
- **Overall NED+EL:** precision, recall and F1-measure, where precision/recall is computed as the product of the NED precision/recall by the EL accuracy.

Table 2. Comparison of experiment results

Approach	NED			EL	Overall NED + EL		
	Precision	Recall	F1	Accuracy	Precision	Recall	F1
POP	0.776	0.643	0.703	0.792	0.615	0.509	0.557
VSim	0.724	0.715	0.719	0.825	0.597	0.590	0.594
NoT+NoI	0.761	0.738	0.749	0.849	0.646	0.627	0.636
T+NoI	0.808	0.754	0.780	0.864	0.698	0.651	0.674
NoT+I	0.826	0.748	0.785	0.852	0.704	0.637	0.669
T+I	0.847	0.788	0.816	0.875	0.741	0.690	0.714

Result and Analysis. Table 2 shows the comparison of our proposed approach and the other two baseline methods. From the results, we can see that 4 different configurations of our proposed approach all significantly outperform the two baseline methods, which demonstrates the effectiveness of our proposed approach.

In general, the performance of EL is higher than NED, for both our approach and baseline methods. That is because the calculation of the EL accuracy is based on the correct results of the NED, i.e., it doesn't take FNEs into consideration. Clearly, we can see that our proposed approach achieves 5.7–8.3 % higher accuracy across all configurations, indicating our approach is very effective for EL task.

Further, for the assessment of the *POP* baseline, obviously, for a mention with high prior probability, the probability of it being a TNE is high. However, due to the fact that *POP* uses the method of simply setting a threshold to exclude the mention with small prior probability, it gets a high precision but low recall. For the *Vsim* baseline, on the other hand, because it considers context rather than prior probability, it is able to get higher recall but lower precision (as it also introduces the FNEs) than *POP*.

Additionally, for our proposed approach with the configuration of *NoT+NoI*, we observe that both NED and EL outperform the two baseline methods because four features are considered. The performance of $T + NoI$ improves as it introduces the machine learning method that takes the importance of different features into consideration. Meanwhile, key point of the $NoT+I$ is to investigate the influence of the iteration to the FNEs. The results indicate that the precision has been further improved due to the fact that iterations exclude FNEs effectively. However, because there is no training in this configuration, the performance of recall falls as all features are treated equally.

Moreover, we can see that the EL accuracy of $T + NoI$ is higher than that of $NoT + I$, which demonstrates that the contribution of iterations to the EL is small as it does not take FNEs into consideration, while the contribution of training to the EL is bigger as training considers the importance of different features.

Finally, as expected, because both the feature importance and the iterations are included in the $T + I$, it is able to achieve the highest performance both for NED precision and EL accuracy, which is consistent with our intuition, since our proposed approach can obtain more related knowledge about candidate entities.

5   Related Work

The problem of EL and NED has been addressed by many researchers starting from papers [1,5]. However, most of existing approaches [3,4,7,8,23,24] focus on the general-purpose knowledge bases and cannot be applied to the domain-specific knowledge base, as we have discussed before.

In addition, many previous systems have employed a pipeline frameworks [10–12,17,21]. Our work is different as we provide high-quality candidate mentions and entity links by jointly modeling NED and EL tasks iteratively. Recently, work [6,16,22] were proposed to perform named entity detection and entity linking jointly to make these two tasks reinforce each other. But their techniques are best-suited for short microblog text (e.g., tweets), while our techniques are better suited for longer documents. In addition, a key difference is that they link mentions to general knowledge base, while our technique links mentions to domain-specific knowledge bases that become more crucial for many domain specific real-world applications.

From above discussion, we can see that considerable approaches have been proposed for general-purpose knowledge bases. Although there are several existing works [2,14,18] addressing domain-specific NED and EL, this area deserves much deeper exploration by research communities as none of above work consider the issue of FNEs, which is essential in the domain-specific area. In this paper, we jointly use the results of named entity detection and entity disambiguation to detect FNEs for domain-specific knowledge base.

6   Conclusion

The current state-of-the-art entity linking research primarily focus on *general* knowledge bases, instead of potentially very useful *domain-specific* knowledge bases. As such, they do not consider two critical problems that we have identified for *domain-specific* knowledge bases, namely *fake named entities* and the *interdependency* between the named entity detection (NED) and entity linking (EL). In this paper, we have proposed a novel approach that dedicates to address the two issues by jointly modeling NED and EL iteratively. We observe from our experimental results that our proposed approach is highly effective comparing with existing baseline methods, indicating it is very promising to be used for many *domain-specific* real-world applications.

Acknowledgements. The work is supported by 973 Program (No. 2014CB340504), NSFC-ANR (No. 61261130588), Tsinghua University Initiative Scientific Research Program (No. 20131089256), Science and Technology Support Program (No. 2014BAK04B00), and THU-NUS NExT Co-Lab.

References

1. Bunescu, R., Pasca, M.: Using encyclopedic knowledge for named entity disambiguation. In: Proceedings of the 11th Conference of the European Chapter of the Association for Computational Linguistics (EACL 2006), pp. 9–16 (2006)
2. Dalvi, N., Kumar, R., Pang, B.: Object matching in tweets with spatial models. In: Proceedings of the Fifth ACM International Conference on Web Search and Data Mining, pp. 43–52 (2012)
3. Finkel, J.R., Grenager, T., Manning, C.: Incorporating non-local information into information extraction systems by Gibbs sampling. In: ACL 2005, pp. 363–370 (2005)
4. Gottipati, S., Jiang, J.: Linking entities to a knowledge base with query expansion. In: EMNLP 2011, pp. 804–813 (2011)
5. Grishman, R., Sundheim, B.: Message understanding conference-6: a brief history. In: Proceedings of the 16th Conference on Computational Linguistics, vol. 1, pp. 466–471 (1996)
6. Guo, S., Chang, M.W., Kiciman, E.: To link or not to link? a study on end-to-end tweet entity linking. In: HLT-NAACL, pp. 1020–1030 (2013)
7. Han, X., Sun, L.: A generative entity-mention model for linking entities with knowledge base. In: Proceedings of the 49th Annual Meeting of the Association for Computational Linguistics: Human Language Technologies, vol. 1, pp. 945–954 (2011)
8. Han, X., Sun, L.: An entity-topic model for entity linking. In: EMNLP-CoNLL 2012, pp.105–115 (2012)
9. Heng, J., Joel, N., Ben, H.: Overview of tac-kbp2014 entity discovery and linking tasks. In: Proceedings of Text Analysis Conference (2014)
10. Lin, T., Mausam, E.O.: Entity linking at web scale. In: AKBC-WEKEX 2012, pp. 84–88 (2012)
11. Mendes, P.N., Daiber, J., Jakob, M., Bizer, C.: Evaluating dbpedia spotlight for the tac-kbp entity linking task. In: Proceedings of the TAC-KBP 2011 Workshop (2011)
12. Mihalcea, R., Csomai, A.: Wikify!: linking documents to encyclopedic knowledge. In: Proceedings of the Sixteenth ACM Conference on Conference on Information and Knowledge Management, pp. 233–242 (2007)

13. Milne, D., Witten, I.H.: Learning to link with wikipedia. In: Proceedings of the 17th ACM Conference on Information and Knowledge Management, pp. 509–518 (2008)
14. Pantel, P., Fuxman, A.: Jigs and Lures: associating web queries with structured entities. In: Proceedings of the 49th Annual Meeting of the Association for Computational Linguistics: Human Language Technologies, vol. 1, pp. 83–92 (2011)
15. Pedersen, T., Purandare, A., Kulkarni, A.: Name discrimination by clustering similar contexts. In: Gelbukh, A. (ed.) CICLing 2005. LNCS, vol. 3406, pp. 226–237. Springer, Heidelberg (2005)
16. Pu, K.Q., Hassanzadeh, O., Drake, R., Miller, R.J.: Online annotation of text streams with structured entities. In: CIKM, pp. 29–38 (2010)
17. Ratinov, L., Roth, D., Downey, D., Anderson, M.: Local and global algorithms for disambiguation to wikipedia. In: HLT 2011, pp. 1375–1384 (2011)
18. Shen, W., Han, J., Wang, J.: A probabilistic model for linking named entities in web text with heterogeneous information networks. In: Proceedings of the 2014 ACM SIGMOD International Conference on Management of Data, pp. 1199–1210 (2014)
19. Shen, W., Wang, J., Jiawei, H.: Entity linking with a knowledge base: Issues, techniques, and solutions. In: IEEE Transactions on Knowledge and Data Engineering, pp. 443–460 (2014)
20. Shen, W., Wang, J., Luo, P., Wang, M.: Linden: linking named entities with knowledge base via semantic knowledge. In: Proceedings of the 21st International Conference on World Wide Web, pp. 449–458 (2012)
21. Sil, A., Cronin, E., Nie, P., Yang, Y., Popescu, A.M., Yates, A.: Linking named entities to any database. In: EMNLP-CoNLL 2012, pp. 116–127 (2012)
22. Sil, A., Yates, A.: Re-ranking for joint named-entity recognition and linking. In: CIKM 2013, pp. 2369–2374 (2013)
23. Zhang, W., Sim, Y.C., Su, J., Tan, C.L.: Entity linking with effective acronym expansion, instance selection and topic modeling. In: IJCAI 2011, pp. 1909–1914 (2011)
24. Zhang, W., Su, J., Tan, C.L., Wang, W.T.: Entity linking leveraging: automatically generated annotation. In: COLING 2010, pp. 1290–1298 (2010)

CTextEM: Using Consolidated Textual Data for Entity Matching

Qiang Yang, Zhixu Li[✉], Binbin Gu, An Liu, Guanfeng Liu, Pengpeng Zhao, and Lei Zhao

School of Computer Science and Technology, Soochow University, Suzhou, China
qiangyanghm@hotmail.com, gu.binbin@hotmail.com,
{zhixuli,anliu,gfliu,ppzhao,zhaol}@suda.edu.cn

Abstract. Entity Matching (EM) identifies records referring to the same entity within or across databases. Existing methods using structured attribute values (such as digital, date or short string values) only may fail when the structured information is not enough to reflect the matching relationships between records. Nowadays more and more databases may have some unstructured textual attribute containing extra Consolidated Textual information (CText for short) of the record, but seldom work has been done on using the CText information for EM. Conventional string similarity metrics such as edit distance or bag-of-words are unsuitable for measuring the similarities between CTexts since there are hundreds or thousands of words with each CText, while existing topic models either can not work well since there is no obvious gaps between the various sub-topics in CText. In this paper, we work on employing CText in EM. A baseline algorithm identifying important phrases with high IDF scores from CTexts and then measuring the similarity between CTexts based on these phrases does not work well since it estimates the similarity in one dimension and neglects that these phrases belong to different topics. To this end, we propose a novel cooccurrence-based topic model to identify various sub-topics from each CText, and then measure the similarity between CTexts on the multiple sub-topic dimensions. Our empirical study on two real-world data set shows that our method outperforms the state-of-the-art EM methods and Text Understanding models by reaching a higher EM precision and recall.

Keywords: Entity Matching · Consolidated textual data · CTextEM · IDF score · Interaction · Sub-topic

1 Introduction

As the data explode for decades, the redundancy and inconsistency between records become more and more serious within and across databases. Entity Matching (EM), also known as record linkage or duplicate detection, aims at finding out records referring to the same entity within or across relation tables.

So far, plenty of work has been done on EM according to the similarities [15] or correlations [18] between various kinds of structured attribute values such as

© Springer International Publishing Switzerland 2016
S.B. Navathe et al. (Eds.): DASFAA 2016, Part I, LNCS 9642, pp. 117–132, 2016.
DOI: 10.1007/978-3-319-32025-0_8

Table 1. Example "House Renting Information" table with CTexts, where r_1, r_2, r_3, and r_5 refer to the same apartment, while r_4, r_7 and r_8 refer to another.

	Residence community	Location (District)	Type	Size	Floor	General supplimental description
r_1	Eastern District Court	Canglang-Xujiang	Residence	75 m^2	3/15	1. Community Planning, unique warmth, flowers and trees patchwork, like a garden, world without dispute, furniture and appliances equipped well. 2. refined decoration, gentle color, facing south.
r_2	Eastern District Court	Canglang-Xujiang	Residence	75 m^2	3/15	1. Community Planning, unique warmth, flowers and trees patchwork, furniture and appliances equipped well. 2. fine decoration, mild color, facing south
r_3	Eastern District Court	Canglang-Xujiang	Residence	-	3/15	1. Community Planning, flowers and trees patchwork, without dispute, furniture and appliances equipped well. 2. refined decoration, color matching gentle, facing south
r_4	Oak Bay Garden	Xiangcheng-Yuanhe	Apartment	100 m^2	25/29	1. general decoration, south facing, nice view, good lighting, air conditioning and water heaters and closed kitchen equiped, 2. free of parking, free of property charges
r_5	Eastern District Court	Canglang-Xujiang	Residence	75 m^2	3/15	1. Unique warmth, community planning well, flowers and trees patchwork, furniture and appliances equipped well. 2. fine decoration, relaxing at ease, world without dispute, gentle color, facing south.
r_6	Eastern District Court	Canglang-Xujiang	Residence	75 m^2	3/15	1. Community Planning, flowers and trees patchwork. 2. good decoration, furniture and appliances equipped well, color matching gentle, facing east
r_7	Oak Bay Garden	Xiangcheng-Yuanhe	Apartment	100 m^2	25/29	1. naive decoration, south facing, good lighting, air conditioning and water heaters and washing machines proved, free of property charges
r_8	Oak Bay Garden	Xiangcheng-Yuanhe	Apartment	100 m^2	-	1. ordinary decoration, south, nice view, air conditioning, water heaters, washing machines, refrigerators, closed kitchen and other necessities 2. free of parking, bag check

digital, date or short string values (see [8] for a survey). However, EM based on structured information only may easily fail for a lack of enough information for EM.

Nowadays there are usually some long free-text descriptions about entities, such as those second-hand goods (like cars, houses, or furniture) for selling online (see Table 1 for example), which have limited structured information but with a "General Supplemental Description" attribute containing some extra information like "orientation", "virescence", "type of decoration" etc. Given that long free-text description which contains various information on several sub-topics, we call it *Consolidated Textual Information* (or **CText** for short) in this paper. So, why do not we use the information of CTexts for better EM? However, conventional string similarity metrics such as edit distance or bag-of-words are unsuitable for measuring the similarities between CTexts since there are usually hundreds or even thousands of words with each CText where much noisy information is mixed with useful information.

There have been some efforts on using CText for EM. For instance, Ektefa et al. [7] calculate both a string similarity score and a semantic similarity score between CTexts. However, the string similarity is simply calculated by Jaccard and the semantic similarity is simply defined by several general "fields" (such as Address, City, Phone, Type) in the WordNet, which only works well on some specific data sets. Gao et al. [9] put forward a semantic features based method, which defines a semantic feature vector like {*time, location, agentive, objective, activity*} for every CText, and then train a classifier to identify duplicate records based on their feature vectors. However, this method is also limited in the dimensions of the features they employed, and thus can not be easily applied to the other data sets.

Essentially, our problem is also very similar to Text Understanding [5], which focuses on understanding the information contained in unstructured text. Some classical topic models such as LDA [2], LSA [16] and PLSA [12] could identify topics from free texts such as the topics of news like "education", "financial", "sports" or "music" etc. However, as a general description/metadata about an entity, the topics in an CText can be seen as sub-topics of a general topic, thus they share many topic words and there is no clear gap between these topics. On the other hand, a sub-topic in CText can be very short (like several words), thus we can hardly learn any sub-topic words as we could do with previous topic models.

Given the above, we propose a novel algorithm that works on mining sub-topics from CText, and then calculating the similarity between CTexts on all sub-topic dimensions. Intuitively, if two phrases are always mentioned in the same sentences, it is quite possible that there exists the association relationship between the two phrases. Based on this intuition, we will build up a Phrase Coocurrence Graph to denote the cooccurrence relationships among all phrases in the CTexts. By doing proper partitioning on the graph, we expect to divide the graph into partitions, each of which corresponds to a sub-topic of the CText. We finally measure the similarity between two entities on all sub-topic dimensions.

A challenge here lies on how we perform the graph partitioning to let each partition closely corresponds to a sub-topic. We first model the problem into an optimization problem and then analyze in theory that this optimization problem is a NP-hard one. To solve the problem, we employ a so-called Phrase Association Degree to measure the similarity between two records on corresponding sub-topic dimensions, and then propose a greedy algorithm that always selects the edge with the minimum Phrase Association Degree as the point of partition. We demonstrate with experiments that the partitions generated by this greedy algorithm can be closely correspond to sub-topics, and the EM results based on this algorithm reach a high precision and recall.

We summarize our contributions as follows: We work on a novel EM problem that uses CText information for EM, and we put forward a cooccurrence-based

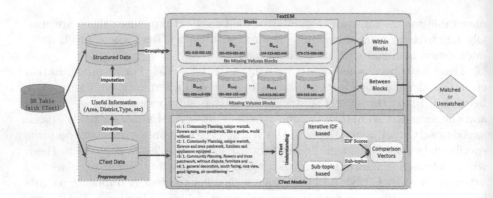

Fig. 1. Workflow overview of CTextEM

sub-topic analytics model that be able to acquire information on multiple sub-topics from the CText for more accurate EM. In addition, we also design a divide-and-conquer workflow that uses structured information to divide all entities into different blocks for EM such that we can greatly minimize the comparison times between entities. Our empirical study on two real-world data set shows that our method outperforms the state-of-the-art EM methods and Text Understanding models by reaching a higher EM precision and recall.

Roadmap. The rest of the paper is organized as follows: We define the CTextEM problem and give our workflow overview in Sect. 2, and then present our algorithms on using CTexts for EM in Sect. 3. After reporting our experimental study in Sect. 4, we cover the related work in Sect. 5. We finally conclude in Sect. 6.

2   Problem Definition

Given a relational table, Entity Matching (EM) identifies all records referring to the same entity within the table. In this paper, we consider tables with both a set of structured attributes (some might be missing) and an unstructured attribute with CText. Particularly, we call the EM task employing CText as **CTextEM**. More formally, we define the CTextEM problem as follows.

Definition 1. *Given a relational table $T = \{r_1, r_2, ..., r_n\}$ under the schema $S = \{[A_1, A_2, ..., A_m], A_U\}$, where m, n are positive integers, r_i $(1 \leq i \leq n)$ denotes a record, A_j $(1 \leq j \leq m)$ denotes an attribute with structured data, and A_U denotes the attribute with CText. $\forall r_i, \forall r_j (1 \leq i, j \leq n, i \neq j)$ in relation table T, CTextEM problem aims at finds a function $\mathcal{F}(r_i, r_j, S)$ and a threshold θ, if and only if: $\mathcal{F}(r_i, r_j, S) \geq \theta$, they are a pair of linked instances referring to the same entity. Otherwise, they are not matched instances.*

In this paper, we employ both structured attributes and CTexts for EM. The basic workflow can be depicted in Fig. 1: We first rely on the structured attribute values to group all records into different blocks, and then use the information in CText to do further EM within or between blocks.

(1) Grouping Records into Blocks: We find a set of structured attributes A_s which satisfy that: two records can not be matched if they do not have the same attribute values under A_s. We put those records sharing the same A_s values into one block. A special case here is the records with missing values under A_s. We put those records having missing values under the same attributes in A_s while sharing the same values under the other attributes in A_s into one block.

(2) EM within or between Blocks: For records within one block, we perform EM between every pair of records by employing the CText. Also, we say the records between two blocks B_1 and B_2 should also be compared pair-wisely, if the two blocks share the same codes under all not-"null" attributes.

For performing EM either within block or between blocks, the key challenge lies on how we acquire useful information from CText for the EM task. In the next section, we will mainly focus on introducing how we mine the information in CTexts for EM between records.

3 Using CTexts for EM

We first present a baseline algorithm based on the IDF scores of phrases, and then put forward a cooccurrence-based sub-topic analytics model for detecting sub-topics from CTexts.

3.1 Baseline: Iterative IDF-Based CTextEM

A baseline algorithm supposes that a set of phrases with the highest IDF scores in a CText can approximately represent the CText. Thus, our similarity function for calculating the similarity between two CTexts will be calculating the similarity between the phrase sets of the two CTexts.

1. Basic Workflow. Particularly, given a CText of a record, we consider all 2-6 word-length phrases from CText as candidate phrases after removing the stop-words. Next, we calculate IDF scores of these phrases and then select phrases to build up the comparison vectors. After that, we calculate the similarity between CTexts, and compare the result with a reasonable threshold. More details are given below:

(a) *Building the Comparison Vectors.* We first calculate the IDF score of every phrase. Note that the IDF score of a phrase is calculated within each block. Then we sort these phrases based on their IDF score in an ascend way and only use top-ranked phrases to represent a database record. Finally, we collect all different phrases from all records into a global phrase set $P_g = \{w_1, w_2, ..., w_g\}$, according to which we can build a boolean vector $v_i = \{bool(r_i, w_1), bool(r_i, w_2), ..., bool(r_i, w_n)\}$ for each record $r_i (1 \leq i \leq n)$, where

$$bool(r_i, w_j) = \begin{cases} 1, & \text{if } w_j \text{ exists in the phrase set of } r_i \\ 0, & \text{otherwise.} \end{cases} \qquad (1)$$

(b) *Computing the Similarity.* Given the comparison vectors v_i and v_j for r_i and r_j ($1 \le i, j \le n$) respectively, we compute the cosine similarity between the two CTexts as follows:

$$sim(r_i, r_j) = \frac{v_i \times v_j}{||v_i|| \cdot ||v_j||} = \frac{\sum_{p=1}^{g} bool(r_i, w_p) bool(r_j, w_p)}{\sqrt{\sum_{p=1}^{g} bool(r_i, w_p)^2} \cdot \sqrt{\sum_{q=1}^{g} bool(r_j, w_q)^2}}. \qquad (2)$$

(c) *Adjusting Blocks.* Let θ to denote the predefined similarity threshold. If $sim(r_i, r_j) > \theta$ and the two instances (r_i, r_j) used to be in the same block, they will be merged into one record in the block. Otherwise, if $sim(r_i, r_j) > \theta$ but the two instances (r_i, r_j) used to be in different blocks, we will move r_j from the original block to r_i's block, and merge it with r_i, assuming that r_j's block is the block with missing values.

2. Iterative Updating IDFs. The intuition of the iterative updating is derived from the fact that: (1) as more matching entities are found, more relevant documents can be utilized for calculating or updating the IDF scores, (2) as more correlative CText are in the same blocks, we can find more matched entities. Thus we will iteratively update the IDF scores of all phrases and then repeat the above three steps, until the IDF scores become stable.

3.2 A Cooccurrence-Based Sub-Topic Analytics Model

The baseline algorithm measures the similarity between two CTexts in one dimension only. However, as a consolidated data, there are actually information of different sub-topics in each CText. Different from those topics such as "sports", "music" and "education" etc., the sub-topics can be taken as various aspects of the same topic. For instance, in the house renting information there are several aspects of the information on "direction", "greening", "property", "traffic" and so forth for describing the situation of an apartment.

In this subsection, we introduce a novel algorithm that works on mining sub-topics from CText, and then calculating the similarity between CTexts on all sub-topic dimensions. Intuitively, if two phrases are always mentioned in same sentences, it is quite possible that there exists some association relationship between the two phrases. Based on this intuition, we will build up a Phrase Cooccurrence Graph (PC-Graph for short), and then employ the so-called Phrase Association Degree (PAD for short) to measure the similarity between two records on corresponding sub-topic dimensions.

1. Constructing the PC-Graph. Give a CText ct, we divide it into a set of segments $t_1, t_2, ..., t_n$ according to the separators such as ",", ".", "?", stopping words etc. We then employ the Longest-Cover method [13] to segment each segment for getting the longest terms in the given vocabulary after filtering the stop words. Next, we add edges with weights between every pair of phrases if the

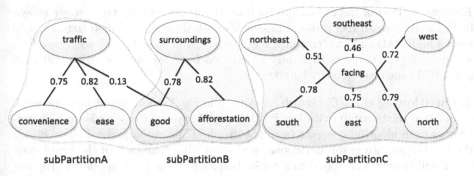

Fig. 2. An example PC-Graph with expected three partitions

two phrases have co-occurred in the same segment, where the weight of an edge between two phrases p_i and p_j can be calculated with the following formula:

$$freq(ct, p_i, p_j) = e^{-gap_{ct}(p_i, p_j)} \cdot bool(p_i, p_j) \qquad (3)$$

where $gap_{ct}(p_i, p_j)$ presents the distance between p_i and p_j in the CText and $e^{-gap_{ct}(p_i, p_j)}$ is to penalize the long distance between two phrases, and $bool(p_i, p_j)$ is used to reduce the influence of similar phrases in the same CText:

$$bool(p_i, p_j) = \begin{cases} 1, & \text{if } sim(p_i, p_j) \leq \theta \\ 0, & \text{otherwise} \end{cases} \qquad (4)$$

where the function $sim(\cdot, \cdot)$ computes the string similarity (e.g., edit similarity) between two phrases, and θ is the string similarity threshold.

Next, we count up the total frequencies of the cooccurence between the phrase pair (p_i, p_j) on all the CTexts in the training set denoted by T as follows:

$$Freq(p_i, p_j) = \sum_{ct \in T} freq(ct, p_i, p_j) \qquad (5)$$

according to which we can calculate the PAD value of an edge linking p_i to p_j with the following formulation:

$$PAD(p_i, p_j) = \frac{Freq(p_i, p_j)}{\sum_{p \in P_g} Freq(p_i, p)} \cdot \log \frac{|P_g|}{|Adj(p_j)| - 1} \qquad (6)$$

where $\frac{Freq(p_i, p_j)}{\sum_{p \in P_g} Freq(p_i, p)}$ calculates the percentage of the degree between p_i and p_j among the total degree of p_i, and $\log \frac{|P_g|}{|Adj(p_j)| - 1}$ is used to penalize a general phrase that always co-occur with other phrases, $Adj(p_j)$ is a phrase set whose elements are always occur with phrase p_j, $|\cdot|$ gets the size of a set.

Example 1. Part of the PC-Graph built on the house renting data set are shown in Fig. 2. As we can see, those phrase pairs that always mentioned together will have a high PAD such as "convenience", "ease", and "southwest" with "traffic", while some phrases pairs that only mentioned together once or twice will have a low PAD such as "good" with "traffic".

2. Partitioning the PC-Graph. As shown in Fig. 2, there might be some weak association relationship (with low PAD scores) between phrase nodes, which prevent us from identifying topics from the graph. Thus, we now consider to divide the PC-Graph into graph partitions, with expecting that each of the graph partition will closely correspond to a topic. Inspired by the work/model in [10], our problem is translated into the following optimization problem: (1) maximizing the sum of PAD scores within each graph partition; while (2) reducing the PAD scores across graph partitions. More formally, our problem is to maximize the following formular:

$$Maximize \sum_{p_1 \in P_g, p_2 \in P_g, p_1 \neq p_2} \frac{PAD(p_1, p_2)}{dis(p_1) + dis(p_2) + \alpha} \qquad (7)$$

$$where \begin{cases} dis(p_1) = Max_{p \in Adj(p_1)} PAD(p_1, p) - Min_{p \in Adj(p_1)} PAD(p_1, p) \\ dis(p_2) = Max_{p \in Adj(p_2)} PAD(p_2, p) - Min_{p \in Adj(p_2)} PAD(p_2, p) \end{cases} \qquad (8)$$

where α is equilibrium factor to prevent the denominator being zero.

Theorem 1. *Finding the optimal solution for Objective 7 is a NP-hard.*

Proof. We prove that the optimal solution is NP-hard even if the number of micro-topics is given. We then prove it by reduction from the balanced maxskip partitioning problem [20]. Given a set V of binary vectors, where $|V|$ is a multiple of p, find a partitioning $\mathcal{P}$ over V such that the following total cost $\mathcal{C}(\mathcal{P})$ is maximized:

$$\mathcal{C}(\mathcal{P}) = \sum_{p_i \in \mathcal{P}} C(P_i) \qquad (9)$$

where $C(P_i) = |P_i|$ is the cost of a graph partition P_i. In our case, we denote the cost of a graph partition P_i as

$$C(P_i) = \sum_{p_1 \in P_g, p_2 \in P_g, p_1 \neq p_2} \frac{PAD(p_1, p_2)}{dis(p_1) + dis(p_2) + \alpha} = \sum_{p_1 \in P_g, p_2 \in P_g, p_1 \neq p_2} 1 - \Delta(P_i) \qquad (10)$$

where $\Delta(P_i)$ is similar to $\bar{v}(P_i)j$ in the blanced maxskip partitioning problem. Thus, the Objective 7 is equivalent to maximizing the total cost of $\mathcal{P}$, i.e. finding the optimal solution for Objective 7 is NP-hard. Hence, Theorem 1 is proved. □

As described in Theorem 1, it is hard to solve this non-linear optimization problem. In the following, we employ a greedy algorithm to solve the problem.

Intuitively, we always greedily select the edge with the minimum PAD as the place to perform the partition.

We define a so-called cohesion score(CScore for short) of every graph partition G_{par}, which can be calculated with the following equation:

$$CScore(G_{par}) = \frac{\sum\limits_{(p_1,p_2) \in P_{G_{par}}} PAD(p_1,p_2)}{\max\limits_{(p_1,p_2) \in P_{G_{par}}} PAD(p_1,p_2) - \min\limits_{(p_1,p_2) \in P_{G_{par}}} PAD(p_1,p_2) + \alpha} \quad (11)$$

where α is equilibrium factor to prevent the denominator being zero, and $\boldsymbol{P_{G_{par}}}$ denotes the set of phrases in the partition G_{par}. Assume that the graph partition G_{par} will be divided into two sub-graph partition G_{par1} and G_{par2} at the edge with the minimum PAD. If this partition operation satisfies the following conditions Eq. 12, we will carry out the partition operation.

$$\begin{cases} CScore(G_{par}) \leq CScore(G_{par1}) + CScore(G_{par2}) \\ |CScore(G_{par1}) - CScore(G_{par2})| \leq \min\limits_{(p_1,p_2) \in P_{G_{par}}} PAD(p_1,p_2) \\ |G_{par1}| > 1, |G_{par2}| > 1 \end{cases} \quad (12)$$

For each graph partition, we iteratively select an edge with the minimum PAD to divide the partition until no more edges satisfy the condition listed in Eq. 12.

3. Acquiring Sub-Topics and Weights. We now acquire sub-topics from the graph partitions. For every graph partition, we calculate an average PAD score for every node in the partition, and then select the one with the highest average PAD score as the sub-topic phrase. Then we take all the other phrases that cooccur with the sub-topic phrase as the sub-topic values.

Assume we get K sub-topics denoted in a vector $< subT_1, subT_2, ..., subT_K >$ from the PC-Graph, where each $subT_i$ $(1 \leq i \leq K)$ denotes a sub-topic. For every dimension, we employ domain knowledge to set the weight of different sub-topics for matching, whose identification degree is in the form of a weight vector $< w_1, w_2, ..., w_K >$. Initially, we set $w_k = 1(1 \leq k \leq K)$, but the weights will be updated iteratively as the entity matching results changing. According to the entity matching result after an iteration, we update the weight w_i as described in Eq. 13. We iteratively update the weight vector until it becomes stable.

$$w_k = \frac{Pos_{subT}(k)}{Pos_{subT}(k) + Neg_{subT}(k)} \quad (13)$$

where $Pos_{subT}(k)$ is the number of all entity pairs (r_i, r_j) satisfying that: if $r_i[k] = r_j[k]$, the entity pair (r_i, r_j) is linked in the current iteration, while $Neg_{subT}(k)$ is the number of all entity pairs (r_i, r_j) satisfying that: if $r_i[k] = r_j[k]$, the entity pair (r_i, r_j) is not linked in the current iteration.

4. Matching Entities on Sub-Topics. Initially, we identify the sub-topic for every CText segment of every record. We then calculate the similarity between every record pair in one block (r_i, r_j) with the adjusted cosine similarity function [19] as follows:

$$Sim(r_i, r_j) = \frac{\sum_{k=1}^{K} w_k^2 \cdot sim(r_i[k], r_j[k])}{\sum_{k=1}^{K} [w_k \cdot sim(r_i[k], r_j[k])]^2} \tag{14}$$

Table 2. The house renting information and the used-car information

	City					Website	
	Beijing	Chengdu	Suzhou	Shenzhen	Tianjin	Ganji	Home of used-car
Attribute number	22	22	22	22	22	12	12
Record number	5.6 k	8.6 k	10.8 k	17.1 k	13.5 k	5.6 k	5.2 k

However, it may happens that there is no obvious sub-topic phrase in a CText segment, which lead us fail to directly identify the sub-topic of the segment. In this case, we employ a probabilistic model to deduce the probability of which topic it belongs to. Let $P(t)$ to denote the set of phrases identified in the segment t, we use the following Eq. 15 to calculate the probability that t belongs to a sub-topic $subT$ according to the *law of total probability*.

$$Pr(subT|P(t)) = \sum_{p \in P(t)} \frac{Pr(p|subT) \cdot Pr(subT)}{\sum_{subT} Pr(p|subT) \cdot Pr(subT)} \tag{15}$$

where $subT$ is a sub-topic, and $Pr(p|subT)$ is the probability occurring the phrase p on the condition of topic $subT$, which can be calculated by our prior knowledge.

After calculating the probability that t belongs to every different sub-topics, we treat the sub-topic with the maximum probability as the sub-topic of the segment, and then still use Eq. 14 to calculate the similarity between two records.

4 Experiments

In this section, we report our experimental study results on two real-world data sets collected from the Web.

- **House.** This database contains the house renting information collected from three house renting information websites, *Ganji, Anjuke, 58tongcheng* of five large-medium cities of China: *Beijing, Shenzheng, Tianjing, Chendu, Suzhou*. The property of the database is given in Table 2.
- **Car.** This database contains second-hand car for selling crawled from Ganji website and "The home of used-car" website, which contains the information of second-cars including structured data and CText information. The property of this data set is also given in Table 2.

We basically use three metrics to evaluate the effectiveness of the methods: **Precision:** the percentage of correctly linked instance pairs among all linked instance pairs, **Recall:** the percentage of correctly linked instance pairs among all instance pairs that should be linked, and $F1$, **Score:** a combination of precision and recall, which is calculated by $F1 = \frac{2*precision*recall}{precision+recall}$. We use the time cost of an algorithm for evaluating the efficiency of a method.

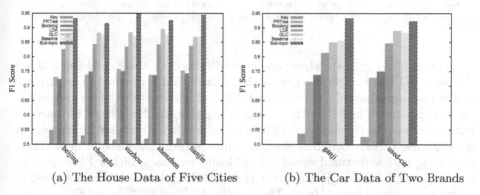

(a) The House Data of Five Cities   (b) The Car Data of Two Brands

Fig. 3. Comparing with previous methods on F1

4.1 Comparison with Previous Methods

In this section, we compare the effectiveness of our two CTextEM algorithm, i.e., *Baseline* and *Sub-topic* methods, with several state-of-the-art EM methods and also CText-based EM methods by using other classical topic-models.

- The *Key-Based EM* method integrates many state-of-the-art techniques based on key values for reducing the comparison cost, such as Q-gram [1] and Inverted indices [4].
- The *Blocking-Based EM* method [3] selects some attributes with high identification to create hash buckets for matching entities. The entities in the same buckets are likely to be the same, while the entities with different hash codes can not be the same.
- The *PRTree-based EM* method [23] builds up a probabilistic rule-based decision tree based on all attributes such that they can perform efficient and effective EM with both key and non-key attributes.
- The *LDA-based EM* method relies on the LDA topic model [2] to mine the hidden variables named topics from CTexts to build up topic vectors for calculating the similarities.
- The *GLC-based EM* method relies on the GLC topic model [21] to understand the information in CTexts and then builds up topic vectors for calculating the similarities.

Table 3. Comparing with previous methods on precision and recall on house

Methods	Beijing		Chengdu		Suzhou		Shenzhen		Tianjin	
	Precision	Recall	Precision	Recall	Precision	Recall	Precision	Recall	Precision	Recall
Key	0.6994	0.4512	0.7116	0.4212	0.7254	0.3998	0.7059	0.4105	0.7142	0.4093
PRTree	0.7504	0.7125	0.7542	0.7239	0.7556	0.7582	0.7694	0.7081	0.7562	0.7472
Blocking	0.7452	0.7028	0.7645	0.7332	0.7583	0.7425	0.7467	0.7259	0.7556	0.7293
LDA	0.8472	0.8066	0.8616	0.8253	0.8438	0.8241	0.8527	0.8320	0.8455	0.8302
GLC	0.8801	0.8625	0.8964	0.8693	0.9045	0.8632	0.9366	0.8590	0.8847	0.8526
Baseline	0.8966	0.8437	0.9059	0.8498	0.8891	0.8524	0.9105	0.8447	0.8725	0.8639
Sub-topic	0.9688	0.8974	0.9472	0.8836	0.9802	0.9163	0.9650	0.8892	0.9823	0.9089

As shown in Fig. 3(a), relying on key attributes only, the Key-Based EM has the lowest F1 scores. The effect of Blocking-Based EM is discounted greatly due to the missing values of the structured data, which leads to the occurrences of *the false-positive*. The PRTree EM works better than the Key-Based EM method but worse than the Baseline, since PRTree uses non-key structured attributes but does not use CText. Our Baseline algorithm extracts information from CText combining with structured data to do EM thus reaches a higher F1 score. The accuracy of LDA-based EM is lower than Baseline, because it is not good at learning sub-topics from CTexts. The Baseline EM and GLC-based EM are neck and neck, but they are both worse than our Sub-Topic method since our sub-topic method uses the CText information in an advanced way.

For more comprehensive comparison, we compare the Precision and Recall of these methods on the house data set. As listed in Table 3, the sub-topic EM also reaches the highest precision and recall among all methods, while GLC-based EM reaches the second highest precision and recall. The effect of the Baseline method is similar to the GLC-based EM, but the LDA-based EM is the worst of the four methods using CTexts.

4.2   Evaluating the Results Extracted from CTexts

We compare the key information extracted from CText with different topic models and our methods. As shown in Table 4, our sub-topic model can acquire more accuracy information than others with the aid of the sub-topic vectors we generated. However, the LDA model only gets some information roughly as shown in the table which is not accurate enough for EM. As can be observed in the table, some important phrases such as "Community Planning" is divided into two phrases. The GLC model either can not get sub-topics and sub-topic phrases well. For example, the phrases "floor" and "twenty" are mixed together. The results of Baseline EM are similar to GLC-based EM. To summarize, the sub-topic model is more suitable than the other models for understanding the information of CTexts.

We also list the weights of different sub-topics on the house data set in Table 5. As can be observed, the sub-topic "floor" has a higher weight than the others since it can better decide the matching results on the data set. It is consistent

Table 4. Comparing with previous methods on the extracted information

Methods	Example	
	1. Community Planning well, unique warmth, flowers and trees patchwork, like a garden, furniture and appliances equipped well, refined decoration, facing south right, twenty floor	1. south facing, good lighting, two air conditioning, water heaters and washing machines equipped, free of property charges
LDA	**Community, Planning,** warmth, flowers, trees, garden, furniture, appliances, decoration, south, floor	**south, facing,** lighting, air, conditioning, water, heaters, washing, machines, property, charges
GLC	**Community Planning,** warmth, flowers and trees, garden, furniture and appliances, refined, decoration, south, **twenty,** floor	**south,** lighting, **two,** air conditioning, water heaters, washing machines, free, property charges
Baseline	**Community Planning,** well, warmth, flowers and trees, garden, furniture and appliances, refined, decoration, south, facing, floor	**facing, south,** lighting, air conditioning, water heaters and washing machines, property charges
Sub-topic	**Community Planning,** warmth, flowers, trees, furniture, appliances, decoration, well-groomed, facing, floor	**facing,** lighting, air conditioning, water heaters, washing machines, property charges

with our exception that the micro-topic with a higher identification degree owns a larger weight than the others.

4.3 Scalability Evaluation

We compare the F1 score and the time cost of the Baseline and sub-topic based EM methods with previous topic models like LDA and GLC. As illustrated in Fig. 4(a), as the records number increasing from 100 to 10000, the F1 Score of the sub-topic based EM method is very stable and is always higher than the other methods. In Fig. 4(b), we can see that the time cost of sub-topic EM is also always less than the Baseline and the other topic models.

5   Related Work

So far, plenty of work has been done on EM based on the string similarities [15], correlations [18], or semantic similarity [6] between various kinds of structured

Table 5. An example of weights for different sub-topics on house data

CText	1. Community Planning well, unique warmth, flowers and trees patchwork, like a garden, furniture and appliances equipped well. 2. Hardcover house, well-groomed room very much, matching color, facing south right, twenty floor.								
Phrases	Community planning	warmth	flowers and trees patchwork	furniture and appliances	decoration	color	facing	floor	...
Weight	0.56	0.31	0.43	0.75	0.69	0.44	0.85	0.89	...

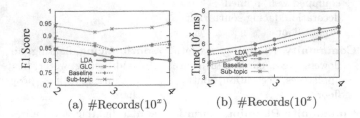

(a) #Records(10^x)   (b) #Records(10^x)

Fig. 4. Scalability: F1 scores and time cost comparison with several topic models.

attribute values of the records such as digital values, date values or short string values in EM (see [8] for a survey). However, EM based on structured information only may easily fail when the structured information is not enough to identify the matching relationships between records.

As a complementary to structured information, we often have some unstructured textual information with each record, which we call as CText for short. Since there can be dozens of sentences (or thousands of words) with each CText, the conventional string similarity metrics can not be applied directly. To utilize the information in CText for EM, the key is to identify useful information from noises, a big challenge is how to identify the key information [22]. Recently, some work has been done for unstructured information. A model based on unstructured text are present in [11], which arrives at a good precision and recall demonstrated with DBWorld posts. However, it needs the support of a special ontology largely. What is more, Ektefa et al. [7] considers a combination of string similarity and semantic similarity between two records, but the measure is not robust since the semantic similarity is simply defined by several general "fields" (such as Address, City, Phone, Type) in the WordNet, which only works well on some specific data sets.

There are also some researches on Text Understanding. Zhang et al. [24] apply deep learning to text understanding from character level inputs all the way up to abstract text concepts, using temporal convolutional networks. They are devoted to learning about the main idea of CText rather than considering the relationship among phrases from CText. Besides, there are some topic models algorithms to discover the main themes for text information in the filed of *NLP*, such as *LDA* [2], *LSA* [16] and *PLSA* [12]. They can get the hidden variables named topic words from text. However, these methods will fail without the obvious topic of text to get the useful information from CText. And some literatures about sub-topic mining have been proposed. Kim et al. [14] propose a method using the co-occurrence of words based on the dependency structure, and anchor texts from web documents to mine sub-topics. But the result of this method is limited by the quality of query and must be supported by external resources. Maowen et al. [17] combine LDA and co-occurrence theory to determine text topics. However, it need to be interpreted by experts to learn about the topics distribution of texts.

6 Conclusions and Future Work

We work on employing CText in EM, i.e., the CTextEM problem, in this paper. To solve the problem, we propose a novel cooccurrence-based topic model to identify various sub-topics from each CText, and then measure the similarity between CTexts on the multiple sub-topic dimensions. Extensive experimental results based on several data collections demonstrate that our proposed Cooccurrence-based Sub-Topic Analytics model can effectively identify sub-topics from CTexts and thus help improve the accuracy of EM on average 10 % of the Iterative IDF-based CTextEM algorithm. As a future work, we would like to extend our work by involving Crowdsouring to help improve the precision and recall of EM. Besides, we would also consider online CTextEM.

Acknowledgements. This research is partially supported by Natural Science Foundation of China (Grant No. 61303019, 61402313, 61472263, 61572336), Postdoctoral scientific research funding of Jiangsu Province (No. 1501090B) National 58 batch of postdoctoral funding (No. 2015M581859) and Collaborative Innovation Center of Novel Software Technology and Industrialization, Jiangsu, China.

References

1. Aizawa, A., Oyama, K.: A fast linkage detection scheme for multi-source information integration. In: Proceedings of International Workshop on Challenges in Web Information Retrieval and Integration, WIRI 2005, pp. 30–39 (2005)
2. Blei, D.M., Ng, A.Y., Jordan, M.I.: Latent dirichlet allocation. J. Mach. Learn. Res. **3**, 993–1022 (2003)
3. Borthwick, A., Goldberg, A., Cheung, P., Winkel, A.: Batch automated blocking and record matching (2011)
4. Christen, P.: A survey of indexing techniques for scalable record linkage and deduplication. Knowl. Data Eng. IEEE Trans. **24**(9), 1537–1555 (2012)
5. Das, Martins, D., A.F.T.: A survey on automatic text summarization. Int. J. Eng (2007)
6. Dhamankar, R., Lee, Y., Doan, A., Halevy, A., Domingos, P.: iMAP: discovering complex semantic matches between database schemas. In: Proceedings of the ACM SIGMOD International Conference on Management of Data, pp. 383–394. ACM (2004)
7. Ektefa, M., Sidi, F., Ibrahim, H., Jabar, M.A., Memar, S., Ramli, A.: A threshold-based similarity measure for duplicate detection. In: IEEE Conference on Open Systems (ICOS), pp. 37–41. IEEE (2011)

8. Elmagarmid, A.K., Ipeirotis, P.G., Verykios, V.S.: Duplicate record detection: a survey. IEEE Trans. Knowl. Data Eng. **19**(1), 1–16 (2007)
9. Gao, C., Hong, X., Peng, Z., Chen, H.: Web trace duplication detection based on context. In: Gong, Z., Luo, X., Chen, J., Lei, J., Wang, F.L. (eds.) WISM 2011, Part II. LNCS, vol. 6988, pp. 292–301. Springer, Heidelberg (2011)
10. Guo, S., Dong, X.L., Srivastava, D., Zajac, R.: Record linkage with uniqueness constraints and erroneous values. Proc. VLDB Endowment **3**(1–2), 417–428 (2010)
11. Hassell, J., Aleman-Meza, B., Arpinar, I.B.: Ontology-driven automatic entity disambiguation in unstructured text. In: Cruz, I., Decker, S., Allemang, D., Preist, C., Schwabe, D., Mika, P., Uschold, M., Aroyo, L.M. (eds.) ISWC 2006. LNCS, vol. 4273, pp. 44–57. Springer, Heidelberg (2006)
12. Hofmann, T.: Probabilistic latent semantic analysis. Proc. Uncertainty Artif. Intell. Uai **25**(4), 289–296 (1999)
13. Kim, D., Wang, H., Oh, A.: Context-dependent conceptualization. In: Proceedings of the Twenty-Third International Joint Conference on Artificial Intelligence, pp. 2654–2661. AAAI Press (2013)
14. Kim, S.-J., Lee, J.-H.: Method of mining subtopics using dependency structure and anchor texts. In: Calderón-Benavides, L., González-Caro, C., Chávez, E., Ziviani, N. (eds.) SPIRE 2012. LNCS, vol. 7608, pp. 277–283. Springer, Heidelberg (2012)
15. Koudas, N., Sarawagi, S., Srivastava, D.: Record linkage: similarity measures and algorithms. In: Sigmod Conference, pp. 802–803 (2006)
16. Landauer, T.K., Foltz, P.W., Laham, D.: An introduction to latent semantic analysis. Discourse Process. **25**(2), 259–284 (1998). Special issue
17. Maowen, W., Dong, Z.C., Weiyao, L., Qiang, W.Q.: Text topic mining based on LDA and co-occurrence theory. In: 7th International Conference on Computer Science & Education (ICCSE), pp. 525–528. IEEE (2012)
18. Parkhomenko, E., Tritchler, D., Beyene, J.: Sparse canonical correlation analysis with application to genomic data integration. Stat. Appl. Genet. Mol. Biol. **8**(1), 1–34 (2009)
19. Sarwar, B., Karypis, G., Konstan, J., Riedl, J.: Item-based collaborative filtering recommendation algorithms. In: Proceedings of the 10th International Conference on World Wide Web, pp. 285–295. ACM (2001)
20. Sun, L., Franklin, M.J., Krishnan, S., Xin, R.S.: Fine-grained partitioning for aggressive data skipping. In: Proceedings of the ACM SIGMOD International Conference on Management of data, pp. 1115–1126. ACM (2014)
21. Garza Villarreal, S.E., Brena, R.F.: Topic mining based on graph local clustering. In: Batyrshin, I., Sidorov, G. (eds.) MICAI 2011, Part II. LNCS, vol. 7095, pp. 201–212. Springer, Heidelberg (2011)
22. Weiss, S.M., Indurkhya, N., Zhang, T., Damerau, F.: Text Mining: Predictive Methods for Analyzing Unstructured Information. Springer Science & Business Media, New York (2010)
23. Yang, Q., Li, Z., Jiang, J., Zhao, P., Liu, G., Liu, A., Zhu, J.: NokeaRM: employing non-key attributes in record matching. In: Dong, X.L., Yu, X., Li, J., Sun, Y. (eds.) Web-Age Information Management. LNCS, vol. 9098, pp. 438–442. Springer, Heidelberg (2015)
24. Zhang, X., LeCun, Y.: Text understanding from scratch. arXiv preprint arXiv: 1502.01710 (2015)

Entity Matching Across Multiple Heterogeneous Data Sources

Chao Kong[1], Ming Gao[1(✉)], Chen Xu[2], Weining Qian[1], and Aoying Zhou[1]

[1] Institute for Data Science and Engineering, ECNU-PINGAN Innovative Research
Center for Big Data, East China Normal University, Shanghai, China
kongchao315@163.com, {mgao,wnqian,ayzhou}@sei.ecnu.edu.cn
[2] Technische Universität Berlin, Berlin, Germany
chen.xu@tu-berlin.de

Abstract. Entity matching is the problem of identifying which entities in a data source refer to the same real-world entity in the others.
Identifying entities across heterogeneous data sources is paramount to
entity profiling, product recommendation, etc. The matching process is
not only overwhelmingly expensive for large data sources since it involves
all tuples from two or more data sources, but also need to handle heterogeneous entity attributes. In this paper, we design an unsupervised approach, called EMAN, to match entities across two or more heterogeneous
data sources. The algorithm utilizes the locality sensitive hashing schema
to reduce the candidate tuples and speed up the matching process. To
handle the heterogeneous entity attributes, we employ the exponential
family to model the similarities between the different attributes. EMAN
is highly accurate and efficient even without any ground-truth tuples.
We illustrate the performance of EMAN on re-identifying entities from
the same data source, as well as matching entities across three real data
sources. Our experimental results manifest that our proposed approach
outperforms the comparable baseline.

Keywords: Entity matching · Exponential family · Locality sensitive
hashing

1 Introduction

Entity matching is the problem of identifying which entities in a data source link
to the same entities in the other data sources. It is a well known and paramount
problem that arises in many research fields, including data cleaning and integration, information retrieval and machine learning. There are many applications
which can benefit from the entity matching task. In the first place, users in a
social network may be the same individuals in the other platforms, but each
user profile and user behavior may be slightly different, e.g., containing different abbreviations, and missing some information. We can improve the product
recommendations after determining the more complete user behaviors. There is
one more point, we should touch on that two websites of second-hand housing

© Springer International Publishing Switzerland 2016
S.B. Navathe et al. (Eds.): DASFAA 2016, Part I, LNCS 9642, pp. 133–146, 2016.
DOI: 10.1007/978-3-319-32025-0_9

may share many house information. After we derive the matched entities across the different websites, we can insight into the more complete house information. Therefore, we have studied the problem of matching entities across two or more heterogeneous data sources in this paper.

However, the entity matching is often a challenging task due to following reasons: (1) it is extremely expensive for the large data sets since the process considers all the tuples as candidates; (2) the process is arduous to compare many heterogeneous attributes for tuple of entities from different data sources; (3) there may be some missing attributes for an entity in some Web applications.

In this paper, we provide an approach, called EMAN, to match entities across two or more heterogeneous data sources. We formulate entity matching task as an unsupervised learning problem. Our proposed approach can result in promising accuracy without ground-truth which are usually arduous and costly to collect in Web applications. For the provided approach, we would like to address the three challenges highlighted earlier. It utilizes the exponential family to combine heterogeneous entity attributes, handle missing data in the unsupervised framework, employ locality sensitive hashing (LSH) to speed up the computation. In summary, our major contributions are as follow.

- We propose an unsupervised method to match entities across two or more heterogeneous data sources. The approach is a unified one which integrates the heterogeneous entity attributes, and employs the distributions from the exponential family to model the similarities of different attributes.
- We employ the locality sensitive hashing (LSH) to block entities. With LSH, EMAN can perform very efficiently but still maintain the promising matching results.
- We illustrate the performance of our algorithm against a comparable baseline on three real data sources. Empirical study results manifest that EMAN outperforms baseline in matching entities across two or more data sources.

The rest of paper is organized as follows. We shortly discuss the related work in Sect. 2. We formally define the problem and describe the overview of our algorithm in Sect. 3. We present the entity matching method in Sect. 4 and report our empirical study in Sect. 5. Finally, we conclude this paper in Sect. 6.

2   Related Work

Entity matching aims at detecting several entities which describe the same entity from given datasets. The study of entity matching problem has become a hot topic in recent years, and some earlier studies can go back to 1950s [1]. However, entity matching is also a wide research problem studied in multiple research communities. In the traditional database community, the problem is described as data matching [2], data deduplication [3], instance identification [4], Merge/Purge [5] and record linkage [6]; In the information retrieval community, the same problem is described as entity resolution [7–11]. The object linkage, object identification [12], and duplicate detection [13–15] are also commonly referred to the same task.

In general, the existing studies of entity matching can be mainly divided into two categories: classification-based and rule-based [17]. For classification-based approaches, they assign labels to a pair after learning the patterns from the training data. Mikhail and Raymond train a classifier by SVM with high accuracy [18]. Dong [19] implements the entity matching algorithm with three crucial features among records based on machine learning technique. The algorithm improves the accuracy of entity matching by merging the attribute values to enrich record information gradually. However, it is not easy to find the exact matched pairs for learning a classifier.

For rule-based approaches, they are deterministic linkage approach [20,21] and judge whether a record pair is matched or not based on rules. Jiannan Wang trains to obtain the most appropriate set of rules on the premise of given rules which is difficult to obtain [17]. Moreover, if the given rules are not sufficient, the deviation of classification results is large and the classification result may not be acceptable. Whang [9] proposed an entity resolution method with evolving rules based on relationship between dynamic semantics and resolution rules to solve the problem of interaction among results. Whang [7] indicates the entity resolution is not an one-time process but incremental process changing constantly. Wang et al. first proposes *how similar is similar problem* in entity matching, and they address the rule-based method to identify the most appropriate similarity functions and thresholds to find entities effectively [17]. Rastogi et al. focus on the scale generic entity matching problem which is implemented with a parallel framework on Hadoop [22]. Lee et al. also attempt to address the scalability problem of entity matching [23]. In that work, they exploit a materialization structure inspired by top-k query processing and develop a scalable entity matching algorithm for evolving rules. However, the rules for linking entities are arduous to learn from data sets.

Fellegi-Sunter's approach is also a rule-based method, and solves the record linkage problem via using an unsupervised and probabilistic linkage approach [16,24]. It works well only when the linkage problem is simple and exact one-to-one matching of username and other user attributes. Sadinle et al. extend Fellegi-Sunter's model to present a probabilistic method for linking multiple data files [25]. Currently, many recent applications generate many data associated with poor quality, including heterogeneous in attributes, error, incomplete and missing values, etc. However, existing entity matching approaches cannot integrate heterogeneous entity attributes and handle missing values. Gao et al. also extend Fellegi-Sunter's approach to link users across different social networks. Their approach can handle heterogeneous user attributes and missing values in user profiles [26]. In addition, Fellegi-Sunter's approach can only link records from two data sources. For linking more than two data sources, the false positive tuples may be large if we link them pair-to-pair by using Fellegi-Sunter's approach. These are the focuses of this work.

3   Entity Matching Approach

In this section, before we overview our proposed approach, we describe a formal definition of the entity matching problem.

3.1   The Problem Definition

We assume that there are N data sources ($N > 1$). Let E_i, $1 \leq i \leq N$, be the set of entities from the i–th source and $\alpha_i(e_{ij})$ represent the observed features of $e_{ij} \in E_i$, i.e., $\alpha_i(e_{ij})$ represents the observed feature vector of e_{ij} from source E_i. Let $\alpha_i(E_i)$ be the set of attribute feature vectors of entities from source E_i. The set of all candidates tuples T can be represented as $\prod_{i=1}^{k} \alpha_i(E_i)$. The entity matching problem is to determine the matched tuples M and unmatched tuples U in T, i.e.,

$$M = \{(\alpha_1(e_{1j_1}), \ldots, \alpha_N(e_{Nj_N})) | e_{1j_1} = \ldots = e_{Nj_N}, e_{ij_i} \in E_i\}, \tag{1}$$

$$U = \{(\alpha_1(e_{1j_1}), \ldots, \alpha_N(e_{Nj_N})) | \exists N_1, N_2, e_{N_1 j_{N_1}} \neq e_{N_2 j_{N_2}}, e_{ij_i} \in E_i\}. \tag{2}$$

When $(\alpha_1(e_{1j_1}), \alpha_2(e_{2j_2}), \ldots, \alpha_N(e_{Nj_N})) \in M$ means that entities e_{ij_i}, $1 \leq i \leq N$, are the same, while $(\alpha_1(e_{1j_1}), \alpha_2(e_{2j_2}), \ldots, \alpha_N(e_{Nj_N})) \in U$ means that at least an entity e_{ij_i} is different from the others. Suppose that each entity can at most match an entity from E_i, M can therefore have at most $min(|E_1|, |E_2|, \cdots, |E_N|)$ matched entity tuples. We may ideally want $T = M \cup U$ but T is usually an extremely large set in real. We therefore consider a smaller T that includes M utilizing some blocking techniques.

3.2   Overview of EMAN

Our proposed entity matching approach consists of four components as following.

Step 1: Candidate tuple generation. The major computational cost is significantly impacted by generating candidate tuples. Blocking methods, such as n-gram indexing and sorted neighborhood, may be the feasible techniques to reduce the number of candidate tuples [27]. However, the number of candidate tuples with N sources of n entities containing in b blocks is $O(\frac{n^N}{b^{N-1}})$. The efficiency and accuracy are significantly impacted by the number of blocks. We therefore utilize the LSH (Locality Sensitive Hashing) schema to speed up the candidate tuple generation since the number of blocks can be arbitrary large (Based on n-gram model, entities can be blocked by utilizing LSH when some string attributes, such as name, address etc., can be represented as a binary vector after shingling).

Step 2: Entity vectorization and similarity computations. We determine a similarity function s^j to evaluate the similarity between the j–th feature of two entities from a candidate tuple. For tuple

$$t_i = (\alpha_1(e_{1i_1}), \alpha_2(e_{2i_2}), \cdots, \alpha_N(e_{Ni_N})),$$

we can compute a similarity vector for m attributes of any two entities of tuple t_i using similarity functions $\{s^j\}_{j=1}^m$. There are $\binom{N}{2}$ similarity vectors between different entity pairs. We take the minimum value of each entry over $\binom{N}{2}$ similarity vectors to model the similarity of tuple t_i, denoted as a m-dimensional vector γ_i. An entity may have multiple attributes. Some of them are individual demographic attributes, e.g., name, location, date, URL and etc. These can be of different data types (e.g., numeric, text, string, categorical, etc.). These attributes can be represented in set and distribution types. Our proposed approach models the similarities between entities to accommodate heterogeneous features using different probability distributions in exponential family.

Step 3: Parameter learning. Given each similarity vector γ_i, EMAN models the similarity values of tuples using two different probability distribution functions, one for matched tuples and another for unmatched ones. The parameter learning step is to infer parameters of the two distributions. More details will be covered in Sect. 4.

Step 4: Tuple scoring and label assignment. For a tuple $t_i \in T$, its score can be computed by $\log \frac{P(t_i \in M | \gamma_i, \hat{\Theta})}{P(t_i \in U | \gamma_i, \hat{\Theta})}$ (for ease of computation). Tuple t_i is more likely to be matched tuple if its score is greater than 0, and otherwise unmatched tuple. Given a threshold, the matched scores of entity tuples are used to judge if they belong to the matched or unmatched tuple sets, i.e., M or U. A tuple is judged as matched tuple if its matched score is larger than the threshold, and otherwise unmatched tuple.

Unlike the earlier Fellegi-Sunter's method, EMAN considers both discrete and continuous similarities as a wider range of probability distributions from the exponential family to model the similarity values of matched and unmatched entity tuples (in Step 1). This is an important extension to handle the heterogenous attribute types, including string, numeric, set, distribution, etc., these exist in the entity matching task.

3.3   EMAN Algorithm

We now present the full EMAN algorithm in Algorithm 1. In this algorithm, PM maintains the set of matched entity tuples. At Lines 2–4, the algorithm employs LSH to block entities from N data sources. At Lines 5–9, it generates all the candidate tuples. A candidate tuple consists of N entities from N data sources, where all entities in a tuple are from the same bucket. In this step, it also computes the similarity vector. At Lines 10–13, it infers the parameters by utilizing the EM-algorithm. Finally, from Lines 14 to 17, it judges whether a tuple belongs to the matched or unmatched one based on the computed matching scores at Line 15.

Algorithm 1. EMAN: Entity matching algorithm

Input: N data sources E_i;
Output: Matched Entities(PM);
1: $T \leftarrow \emptyset; j \leftarrow 0;$
    //*step 1: Candidate tuple generation*
2: **for** $i = 1, \cdots, N$ **do**
3:      employ LSH to block entities from the i−th source;
4: **end for**
    //*step 2: Entity vectorization and similarity computation*
5: **for** each bucket in the LSH **do**
6:      generate $t_i = (\alpha_1(e_{1i_1}), \alpha_2(e_{2i_2}), \cdots, \alpha_N(e_{Ni_N}));$
7:      compute the minimum similarity vector γ_i between entity pairs of t_i;
8:      $T \leftarrow T \cup$ candidate tuples from the bucket;
9: **end for**
    //*step 3: Parameter Learning*
10: **while** parameter set Θ has not converged **do**
11:     E-Step; //handle missing data;
12:     M-Step; //estimate parameters by maximizing the log-likelihood;
13: **end while**
    //*step 4: Tuple scoring and label assignment*
14: **for** $t_i \in T$ **do**
15:     $w_i \leftarrow \log \frac{P(t_i \in M | \gamma_i, \hat{\Theta})}{P(t_i \in U | \gamma_i, \hat{\Theta})};$
16:     according to the score of t_i, keep PM to the top-K candidate entities with
        the largest scores;
17: **end for**
18: **return** PM

---

4   Parameters Inference and Prediction

We employ a generative model to solve the problem defined in previous section. Given the similarity vectors of all candidate tuples, EMAN learns the parameters of similarity distributions for *matched* and *unmatched* tuples based on *exponential family distributions*. In terms of these learned parameters of similarity distributions, EMAN infers whether candidate tuple $t_i \in T$ is *matched* or *unmatched* by estimating probabilities $P(t_i \in M | \gamma_i, \Theta)$ and $P(t_i \in U | \gamma_i, \Theta)$.

4.1   Likelihood

Assume that $P(t_i \in M | \Theta) = p$, i.e., $P(t_i \in U | \Theta) = 1 - p$. Employing Bayes' rule to $P(t_i \in M | \gamma_i, \Theta)$, we can obtain:

$$P(t_i \in M | \gamma_i, \Theta) = \frac{p \times P(\gamma_i | t_i \in M, \Theta)}{P(\gamma_i | \Theta)} \qquad (3)$$

$$P(\gamma_i | \Theta) = p \times P(\gamma_i | t_i \in M, \Theta) + (1 - p) \times P(\gamma_i | t_i \in U, \Theta) \qquad (4)$$

Please note that we do not know the label of a candidate tuple t_i. To represent the joint probability of the observed data, we define a latent variable l_i for candidate tuple t_i. Its value is 1 if tuple t_i is a matched tuple, and otherwise 0. And, we defined $c_i = (l_i, \gamma_i)$ as the *complete data* vector for T. The probability of observation c_j under parameter Θ can be defined as:

$$P(c_i|\Theta) = [P(\gamma_i, t_i \in M|\Theta)]^{l_i}[P(\gamma_i, t_i \in U|\Theta)]^{(1-l_i)}$$
$$= [p \times P(\gamma_i|t_i \in M, \Theta)]^{l_i}[(1-p) \times P(\gamma_i|t_i \in U, \Theta)]^{(1-l_i)} \tag{5}$$

Let $L_i = (l_i, 1 - l_i)$ and thus we obtain the *log-likelihood* for sample $X = \{c_i : i = 1, 2, \cdots, |T|\}$ as:

$$L(\Theta|X) = \sum_{i=1}^{|T|} L_i[logP(\gamma_i|t_i \in M, \Theta), logP(\gamma_i|t_i \in U, \Theta)]'$$
$$+ \sum_{i=1}^{|T|} L_i[logp, log(1-p)]'. \tag{6}$$

4.2 Exponential Family

As mentioned above, most of the probability distributions can be represented by exponential family which is a convenient and widely used family of distributions. Distributions in the exponential family appeal to the machine learning community as some good properties of MLE which is a function of the sufficient statistic and the best unbiased estimator, etc. [22].

In probability and statistics, an *exponential family* is a set of probability distributions and represented by an exponential form which is chosen for mathematical convenience [23]. In other word, an *exponential family* is a set of probability distributions whose PDF and PMF can be expressed in the form as follows:

$$f(x; \theta) = h(x) \exp\left(\theta' S(x) - z(\theta)\right) \tag{7}$$

where θ (may be a vector) is the natural parameter of a distribution. $S(x)$ is a sufficient statistic. Generally, $S(x) = x$. So when the parameter z, h, S are fixed, we will define an exponential family with parameter θ. The exponential family contains as special cases most of the standard discrete and continuous distributions that we use for practical modelling, such as Bernoulli, Multinomial, Poisson, Gamma, Dirichlet, etc.

One of our task is to calculate probabilities $P(t_i \in M|\gamma_i, \Theta)$ and $P(t_i \in U|\gamma_i, \Theta)$. In Eq. 6, we know that the critical step is to calculate $P(\gamma_i|t_i \in M, \Theta)$ and $P(\gamma_i|t_i \subset U, \Theta)$. So we estimate $P(\gamma_i|t_i \in M, \Theta)$ and $P(\gamma_i|t_i \in U, \Theta)$ and assume that γ_i is drawn from a distribution of exponential family, and use the simplifying assumption that the entries of vector γ_i are conditional independent with respect to the state of indicator L_i such as:

$$P(\gamma_i^j|t_i \in M, \Theta) \sim f_{1,j}(\gamma_i^j; \theta_{1,j}), \, for \, j = 1, \cdots, m$$
$$P(\gamma_i^j|t_i \in U, \Theta) \sim f_{0,j}(\gamma_i^j; \theta_{0,j}), \, for \, j = 1, \cdots, m \tag{8}$$

where $f_{\cdot,j}(\cdot;\cdot)$ is the PDF or PMF from the exponential family in Eq. 7.
Next, the log-likelihood in Eq. 6 can be replaced with:

$$
L(\Theta|X) \propto \sum_{i=1}^{|T|} L^i [\sum_{j=1}^{m} \theta'_{1,j} S_{1,j}(\gamma_i^j), \sum_{j=1}^{m} \theta'_{0,j} S_{0,j}(\gamma_i^j)]'
$$

$$
- \sum_{i=1}^{|T|} L^i [\sum_{j=1}^{m} z_{1,j}(\theta_{1,j}), \sum_{j=1}^{m} z_{0,j}(\theta_{0,j})]' + \sum_{i=1}^{|T|} L^i [\log p, \log(1-p)]' .
$$

(9)

4.3   Maximum Likelihood Estimator

Since L^i is a latent vector in Eq. 9, so we estimate the parameter $\Theta = \{p, \theta_{1,j}, \theta_{0,j}, for\, j = 1, \cdots, m\}$ with maximum likelihood estimation using EM algorithm. The EM algorithm begins with initial estimator of unknown parameter Θ and repeat iterative calculation of the expectation(E) and maximization(M) steps until the convergence.

E-step. The objective in E-step is to calculate the conditional expectation of latent variables and estimate the missing data with observed data. Given γ_i and $\Theta^{(k-1)}$ in the k-th iteration, the conditional distribution of l_i is $l_i|\gamma_i, \Theta^{(k-1)} \sim B(1, p_i^{(k)})$ with

$$
p_i^{(k)} = P(l_i = 1|\gamma_i, \Theta^{(k-1)})
$$

(10)

Then $p_i^{(k)}$ will be represented with Eq. 11 as:

$$
p_i^{(k)} = P(l_i = 1|\gamma_i, \Theta^{(k-1)}) = \frac{P(t_i \in M, \gamma_i|\Theta^{(k-1)})}{P(\gamma_i|\Theta^{(k-1)})}
$$

$$
= \frac{p^{(k-1)} \cdot \prod_{j=1}^{m} f_{1,j}(\cdot;\cdot)}{p^{(k-1)} \cdot \prod_{j=1}^{m} f_{1,j}(\cdot;\cdot) + (1-p^{(k-1)}) \cdot \prod_{j=1}^{m} f_{0,j}(\cdot;\cdot)}
$$

(11)

By substituting $p_i^{(k)}$ for l_i, we obtain the expectation function.

M-step. In M-step, we maximize the likelihood after E-step. When we estimate the values of $l_i^{(k)} = p_i^{(k)}$ in E-step, we take derivatives of the log-likelihood to parameters $p, \theta_{1,j}, and\, \theta_{0,j}$ as follows:

$$
\frac{\partial L(\Theta|X)}{\partial p} = \sum_{i=1}^{|T|} (\frac{l_i^{(k)}}{p} - \frac{1 - l_i^{(k)}}{1-p})
$$

(12)

$$
\frac{\partial L(\Theta|X)}{\partial \theta_{1,j}} = \sum_{i=1}^{|T|} l_i^{(k)} (S_{1,j}(\gamma_i^j) - \frac{\partial z_{1,j}(\theta_{1,j})}{\partial \theta_{1,j}})
$$

(13)

$$
\frac{\partial L(\Theta|X)}{\partial \theta_{0,j}} = \sum_{i=1}^{|T|} (1 - l_i^{(k)})(S_{0,i}(\gamma_i^j) - \frac{\partial z_{0,j}(\theta_{0,j})}{\partial \theta_{0,j}})
$$

(14)

Table 1. The MLEs of parameters for both matched and unmatched groups

Distribution	MLE	
	Matched group	Unmatched group
Bernoulli	$p_{1,i}^{(k)} = \frac{\sum_{j=1}^{|T|} l_j^{(k)} \gamma_i^j}{\sum_{j=1}^{|T|} l_j^{(k)}}$	$p_{0,i}^{(k)} = \frac{\sum_{j=1}^{|T|} (1-l_j^{(k)}) \gamma_i^j}{\sum_{j=1}^{|T|} (1-l_j^{(k)})}$
Multinomial	$p_{1,i}^{(k)} = \frac{\sum_{j=1}^{|T|} l_j^{(k)} I_{\gamma_i^j=h}}{\sum_{j=1}^{|T|} l_j^{(k)}}$	$p_{0,i}^{(k)} = \frac{\sum_{j=1}^{|T|} (1-l_j^{(k)}) I_{\gamma_i^j=h}}{\sum_{j=1}^{|T|} (1-l_j^{(k)})}$
Gaussian	$\mu_{1,i}^{(k)} = \frac{\sum_{j=1}^{|T|} l_j^{(k)} \gamma^j}{\sum_{j=1}^{|T|} l_j^{(k)}}$	$\mu_{0,i}^{(k)} = \frac{\sum_{j=1}^{|T|} (1-l_j^{(k)}) \gamma^j}{\sum_{j=1}^{|T|} (1-l_j^{(k)})}$
	$(\sigma_{1,i}^{(k)})^2 = \frac{\sum_{j=1}^{|T|} l_j^{(k)} (\gamma^j-\mu_{1,i}^{(k)})^2}{\sum_{j=1}^{|T|} l_j^{(k)}}$	$(\sigma_{0,i}^{(k)})^2 = \frac{\sum_{j=1}^{|T|} (1-l_j^{(k)})(\gamma^j-\mu_{1,i}^{(k)})^2}{\sum_{j=1}^{|T|} (1-l_j^{(k)})}$
Exponential	$\lambda_{1,i}^{(k)} = \frac{\sum_{j=1}^{|T|} l_j^{(k)}}{\sum_{j=1}^{|T|} l_j^{(k)} \gamma_i^j}$	$\lambda_{0,i}^{(k)} = \frac{\sum_{j=1}^{|T|} (1-l_j^{(k)})}{\sum_{j=1}^{|T|} (1-l_j^{(k)}) \gamma_i^j}$

By calculating, we can infer $\frac{\partial z_{\cdot,j}(\theta_{\cdot,j})}{\partial \theta_{\cdot,j}} = E_{\theta_{\cdot,j}}(S_{\cdot,j}(\gamma^j))^2$ where $\cdot$ can be 1 or 0 and obtain the MLEs of parameters as shown in Table 1. Due to the page limitation, we omit the proofs and computations.

While the distributions of all similarity values were assigned, we can estimate the parameter in the M-step of the k-th iteration. The probability p can be estimated as $p^{(k)} = \frac{\sum_{i=1}^{|T|} l_i^{(k)}}{|T|}$.

4.4 Missing Data

As a result of presence of missing data, we denote the sample X as (X_o, X_m), where X_o represents the observed data and X_m represents the missing data. Let $\Theta^{(0)}$ be the initial value for parameter. The E-step of EM algorithm computes $Q(\Theta; \Theta^{(k-1)}) = E(L(\Theta|X)|X_o, \Theta^{(k-1)})$ during k-$iteration$. Due to the missing data, $S_{1,i}(\gamma_i^j)$ and $S_{0,i}(\gamma_i^j)$ in Eq. 9 are missing. So $Q(\Theta; \Theta^{(k-1)})$ can be calculated by $E(S_{1,j}(\gamma_i^j)|\Theta^{(k-1)})$ and $E(S_{0,j}(\gamma_i^j)|\Theta^{(k-1)})$ for $S_{1,j}(\gamma_i^j)$ and $S_{0,j}(\gamma_i^j)$ in Eqs. 13 and 14 respectively.

4.5 Matching Score Computation

Once parameters Θ are estimated, EMAN determines whether candidate tuple t_i belongs to matched or unmatched one by computing its matching score. To speed up the computation of matching scores for exponential family, we define the match score function as:

$$W_i = log(\frac{P(t_i \in M|\gamma_i, \hat{\Theta})}{P(t_i \in U|\gamma_i, \hat{\Theta})}) \propto \sum_{j=1}^{m} w_i^j \qquad (15)$$

where

$$w_i^j = (\Theta_{1,j}' S_{1,j}(\gamma_i^j) - z_{1,j}(\Theta_{1,j})) - (\Theta_{0,j}' S_{0,j}(\gamma_i^j) - z_{0,j}(\Theta_{0,j})) \qquad (16)$$

142   C. Kong et al.

Table 2. Descriptive statistics of datasets

Data source	# entities	# features
NetEaseHouse[a]	2776	36
AnJuKe[b]	581	34
PingAnFang[c]	630	15

[a]http://house.163.com/.
[b]http://shanghai.anjuke.com/.
[c]http://www.pinganfang.com/.

where $P(t_i \in M|\gamma_i,\hat{\Theta}) > P(t_i \in U|\gamma_i,\hat{\Theta})$ when $W_i > 0$. Alternatively, we can assign t_i to the matched tuple set if $W_i > W_0$ where $W_0 > 0$ is a threshold.

5   Empirical Evaluation

We conduct two experiments to compare the proposed EMAN with the baseline method using three real data sources. First, we manifest the performance of the self-matching problem in which there is a clearly ground truth one-one matching between entities from the identical data source. Secondly, we study the performance of matching three real heterogeneous data sources.

5.1   Experimental Setup

Datasets. We crawled three datasets from the famous estate websites in China for our experiments. The descriptive statistics about the datasets are shown in Table 2. However, the schemes of three data sources are different from each other. We only find 10 useful attributes as shown in the first column of Table 3. In our experiment, we model the similarities of property fees and price with Gaussian distribution, purpose(shop or dwelling) with Bernoulli distribution, and remaining similarities with Exponential distribution as shown in the last column of Table 3.

Comparative Method and Evaluation Measures. We find an unsupervised approach, called Felliegi-Sunter (shorted in **FS**), to be the comparative baseline. Fellegi-Sunter's approach therefore evaluates all attributes by using binary similarity, i.e., the similarity is 1 if two attributes are the same, and otherwise 0. Currently, many data sources are low in quality. The FS approach is too simple to obtain reasonable performance. In our implementation for FS, we therefore set the similarity value to be 1 if the value of similarity of attributes is larger than a tuned threshold, and otherwise 0.

As the mentioned above, we evaluate our method using *Precision@K*, and *Recall@K*. *Precision@K* is the fraction of the matched tuples in the *top-K* result that are correctly matched. *Recall@K* is the fraction of ground truth matched entities that appear among the *top-K* results. To evaluate the scalability of our proposed approach, we also measure the elapsed time in second and the number of candidate tuples.

Table 3. The setup of experiment and parameter estimation for matched and unmatched groups

Feature	Matched	Unmatched	Similarity	Distribution
name	$\lambda_m = 8.36$	$\lambda_u = 33423.12$	LCS	Exponential
address	$\lambda_m = 5.27$	$\lambda_u = 31127.66$	Jaccard	Exponential
developer	$\lambda_m = 23.82$	$\lambda_u = 27.58$	Jaccard	Exponential
construction time	$\lambda_m = 3.92$	$\lambda_u = 4.24$	Jaccard	Exponential
PMC	$\lambda_m = 2.31$	$\lambda_u = 2.38$	Jaccard	Exponential
property fees	$\mu_m = 0.43$	$\mu_u = 0.42$	Euclidean distance	Gaussian
	$\sigma_m = 0.05$	$\sigma_u = 0.04$		
building type	$\lambda_m = 2.76$	$\lambda_u = 2.87$	Jaccard	Exponential
launch date	$\lambda_m = 2.11$	$\lambda_u = 2.20$	Jaccard	Exponential
price	$\mu_m = 0.45$	$\mu_u = 0.43$	Euclidean distance	Gaussian
	$\sigma_m = 0.02$	$\sigma_u = 0.02$		
purpose	$Prob_m = 0.23$	$Prob_u = 0.12$	1,0 for matched or not	Bernoulli

5.2 Self-matching Evaluation

Firstly, we perform our method in self-matching task which is designed such that we know the complete ground truth matched entities, i.e., we match the entities from three replicas of the identical data source. We create two new data sources which are injected some noise into the given data source. For each replica, we randomly inject some noises into string attributes. For each character, it has the same probability to be inserted, deleted or replaced. Take $PingAnFang$ as an example, $PingAnFang_1(\psi)$ is the first replica of $PingAnFang$, where ψ denotes the probability of each character being changed. In our experiment, the value of ψ varies from 10 % to 50 %.

Figure 1(a) and (b) manifest the accuracy of EMAN on $PingAnFang$ by varying ψ from 10 to 50 %. We observe that the accuracy of EMAN is promised. If 10 % noise is injected into the data, almost 90 % matched entities can be found by EMAN. Even 50 % noise is injected into the data, the precision in the top-200 is almost 100 %. Figure 1(c) and (d) illustrate that EMAN outperforms the baseline significantly for self-matching on $PingAnFang$. The result indicates that exponential family is helpful to integrate heterogeneous entity attributes.

5.3 Matching Heterogeneous Data Sources

Scalability of EMAN. In this experiment, we address whether LSH is helpful to speed up EMAN. For entity matching on heterogeneous data sources, we only change the size of $NetEaseHouse$ from 500 to 2,500. In Fig. 2(a) ($EMAN\_L$ is an approach that $EMAN$ employs LSH to block entities), we can find that only less than 1 % candidate tuples are remained after using LSH to block entities. In Fig. 2(b), EMAN associated with LSH detects entities within 2,000 s when the size of data source is almost 2,000. However, the elapsed time of EMAN without

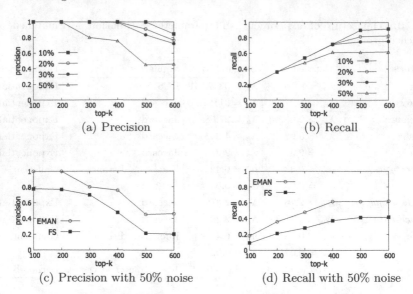

(a) Precision

(b) Recall

(c) Precision with 50% noise

(d) Recall with 50% noise

Fig. 1. Accuracy of EMAN and FS

LSH is more than 12 h. In summary, LSH is helpful to reduce the number of candidate tuples and speed up the computation of EMAN.

Manually Judgement. We now turn to match three heterogeneous data sources, namely complete *NetEaseHouse*, *AnJuKe* and *PingAnFang*. Since the maximum size of matched tuples is less than the minimal number of |*NetEaseHouse*|, |*AnJuKe*| and |*PingAnFang*|, we manually annotated the top-250 matched entities labelled by EMAN and FS. Three entities are judged to be matched tuple when (1) the similar entity name; (2) the similar values in some attributes, such as *address*, *region* and *developer*. The remaining entity triples are assigned the undetermined label. As shown in Fig. 3(a) and (b), we find that the accuracy for the top-250 result of EMAN is more than 70 %, but it is about 60 % for FS. This illustrates that both EMAN and FS are quite good in returning the correctly matched entities for different top-K ranked tuples. EMAN also returns fewer undetermined tuples than FS.

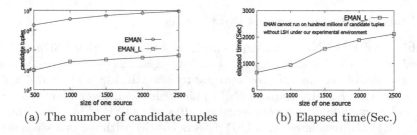

(a) The number of candidate tuples

(b) Elapsed time(Sec.)

Fig. 2. Efficiency of EMAN

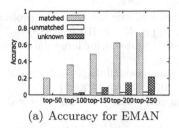
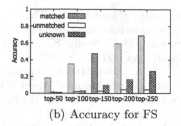

(a) Accuracy for EMAN   (b) Accuracy for FS

Fig. 3. Accuracy for entity matching

For this experiment, we list all parameters learned from our approach. An attribute is more important to match entities if the difference of parameters between matched and unmatched groups is larger. As shown in Table 3, we can also observe that *name* and *address* are two most important attributes to match entities for this task.

6 Conclusion

In this paper, we have studied the problem of entity matching across two or more heterogeneous data sources. It is a challenging task due to the overwhelming expensive, heterogeneous attributes for each entity, and incomplete and missing data. We propose an unsupervised method to deal with the mentioned challenges. We have illustrated our proposed method on three real data sources. Experimental results indicate that EMAN not only outperforms the comparable baseline but also obtains the promising performance.

In our future work, we plan to extend our work to handle some ground-truth tuples with semi-supervised approach, and deploy a distributed algorithm to support more efficient computation.

Acknowledgements. This work is supported by the National Basic Research Program (973) of China (No. 2012CB316203) and NSFC under Grant No. U1401256, 61402177, 61402180 and 61232002. This work is also supported by CCF-Tecent Research Program of China (No. AGR20150114), NSF of Shanghai (No. 14ZR1412600), and a fund of ECNU for oversea scholars, international conference and domestic scholarly visits.

References

1. Newcombe, H.B., Kennedy, J.M., Axford, S.J., James, A.P.: Automatic linkage of vital records. Science **130**(3381), 954–959 (1959)
2. Scannapieco, M., Figotin, I., Bertino, E., Elmagarmid, A.K.: Privacy preserving schema and data matching. In: SIGMOD, pp. 653–664 (2007)
3. Sarawagi, S., Bhamidipaty, A.: Interactive deduplication using active learning. In: ACM SIGKDD, pp. 269–278 (2002)

4. Wang, Y.R., Madnick, S.E.: The inter-database instance identification problem in integrating autonomous systems. In: Data Eng, pp. 46–55. IEEE (1989)
5. Hernandez, M.A., Stolfo, S.J.: The merge/purge problem for large databases. In: SIGMOD, pp. 127–138 (1995)
6. Jin, L., Li, C., Mehrotra, S.: Supporting efficient record linkage for large data sets using mapping techniques. World Wide Web 9(4), 557–584 (2006)
7. Whang, S.E., Garcia-Molina, H.: Incremental entity resolution on rules and data. VLDB J. 23(1), 77–102 (2014)
8. Kolb, L., Thor, A., Rahm, E.: Block-based load balancing for entity resolution with MapReduce. In: CIKM, pp. 2397–2400 (2011)
9. Whang, S., Garcia-Molina, H.: Entity resolution with evolving rules. PVLDB 3(1), 1326–1337 (2010)
10. Getoor, L., Machanavajjhala, A.: Entity resolution: theory practice & open challenges. PVLDB 5(12), 2018–2019 (2012)
11. Singla, P., Domingos, P.: Entity resolution with markov logic. In: ICDM, pp. 572–582 (2006)
12. Tejada, S., Knoblock, C.A., Minton, S.: Learning object identification rules for information integration. Inf. Syst. 26(8), 607–633 (2001)
13. Christen, P.: Data Matching: Concepts and Techniques for Record Linkage, Entity Resolution, and Duplicate Detection. Springer, Heidelberg (2012)
14. Elmagarmid, A.K., Ipeirotis, P.G., Verykios, V.S.: Duplicate record detection: a survey. IEEE Trans. Knowl. Data Eng. 19(1), 1–16 (2007)
15. Winkler, W.E.: Overview of Record Linkage and Current Research Directions. U.S. Census Brueau, Washington (2006)
16. Fellegi, I.P.: A theory for record linkage. J. Am. Stat. Assoc. 64(328), 1183–1210 (1969)
17. Wang, J., Li, G., Yu, J.X., Feng, J.: Entity matching: how similar is similar. PVLDB 4(10), 622–633 (2011)
18. Bilenko, M., Mooney, R.J.: Adaptive duplicate detection using learning string similarity measures. In: ACM SIGKDD, pp. 39–48 (2003)
19. Dong, X., Halevy, A., Madhavan, J.: Reference reconciliation in complex information spaces. In: ACM SIGMOD International Conference on Management of Data, pp. 85–96 (2005)
20. Roos, L.L., Wajda, A.: Record linkage strategies. part I: estimating information and evaluating approaches. Methods Inf. Med. 30(2), 117–123 (1991)
21. Grannis, S.J, Overhage, J,M, McDonald, C.J: Analysis of identifier performance using a deterministic linkage algorithm. In: AMIA (2002)
22. Rastogi, V., Dalvi, N.N., Garofalakis, M.N.: Large-scale collective entity matching. PVLDB 4(4), 208–218 (2011)
23. Lee, S., Lee, J., Hwang, S.-W.: Scalable entity matching computation with materialization. In: CIKM, pp. 2353–2356 (2011)
24. DuVall, S.L., Kerber, R.A., Thomas, A.: Extending the Fellegi-Sunter probabilistic record linkage method for approximate field comparators. J. Biomed. Inform. 43(1), 24–30 (2010)
25. Sadinle, M., Fienberg, S.E.: A generalized Fellegi-Sunter framework for multiple record linkage with application to homicide record systems. J. Am. Stat. Assoc. 108(502), 385–397 (2013)
26. Gao, M., Lim, E.-P., Lo, D., Zhu, F., Prasetyo, P.K., Zhou, A.: C.N.L.: Collective network linkage across heterogeneous social network. In: ICDM (2015)
27. Christen, P.: A survey of indexing techniques for scalable record linkage and deduplication. IEEE TKDE 24(9), 1537–1555 (2011)

Data Mining and Machine Learning

Probabilistic Maximal Frequent Itemset Mining Over Uncertain Databases

Haifeng Li$^{(\boxtimes)}$ and Ning Zhang

School of Information, Central University of Finance and Economics,
Beijing 100081, China
mydlhf@cufe.edu.cn

Abstract. We focus on the problem of mining probabilistic maximal frequent itemsets. In this paper, we define the probabilistic maximal frequent itemset, which provides a better view on how to obtain the pruning strategies. In terms of the concept, a tree-based index *PMFIT* is constructed to record the probabilistic frequent itemsets. Then, a depth-first algorithm *PMFIM* is proposed to bottom-up generate the results, in which the support and expected support are used to estimate the range of probabilistic support, which can infer the frequency of an itemset with much less runtime and memory usage; in addition, the superset pruning is employed to further reduce the mining cost. Theoretical analysis and experimental studies demonstrate that our proposed algorithm spends less computing time and memory, and significantly outperforms the *TODIS-MAX* [20] state-of-the-art algorithm.

Keywords: Uncertain database · Probabilistic frequent itemset · Data mining · Probabilistic Maximal Frequent Itemset

1 Introduction

Frequent itemset mining is one of the traditional and important fields in data mining, which discovers itemsets whose occurrences are larger than a specified threshold. Many efficient algorithms and methods have been developed in the recent years [1]. In such methods, a very important assumption is, however, the mined transactions are exact no matter they are static or increasingly updated.

When new applications are developed and new requirements are met, uncertainty exists often. As an example, Table 1 shows an animal monitor system, which use the cameras to distinguish 3 pandas with names "PanPan", "Tuan-Tuan", and "YuanYuan" among other animals, as well observe their appearances. Nevertheless, the digital image recognition method is not accurate enough

---

H. Li—This research is supported by the National Natural Science Foundation of China(61100112,61309030), Beijing Higher Education Young Elite Teacher Project(YETP0987), Discipline Construction Foundation of Central University of Finance and Economics, Key project of National Social Science Foundation of China(13AXW010), 121 of CUFE Talent project Young doctor Development Fund in 2014 (QBJ1427).

© Springer International Publishing Switzerland 2016
S.B. Navathe et al. (Eds.): DASFAA 2016, Part I, LNCS 9642, pp. 149–163, 2016.
DOI: 10.1007/978-3-319-32025-0_10

Table 1. An uncertain database example

ID	Panpan	Tuantuan	Yuanyuan
1	0.6	0.4	1
2	0.5		0.8

to recognize each panda. Thus, each panda will annotated by a probability to present the existence [2], and we call it the attribute-uncertainty. This new feature brings us new challenges, which cannot be well addressed by the traditional frequent itemset mining methods. The existing uncertain data mining methods can be categorized into two types. One is to achieve the expected frequent itemsets [6–17], another is to obtain the probabilistic frequent itemsets [18–30].

1.1 Motivation

When mining frequent itemsets over exact databases, it has been presented the frequent itemsets are redundant. Many itemset compression methods have been proposed, such as maximal itemset [3], closed itemset [4] and non-derivable itemset [5]. If the users do not care the support of an itemset, but only want to know whether it is frequent, the maximal frequent itemset is the best choice since it is the most efficient method to represent the frequent itemsets. Similarly, if we want to discover frequent itemsets over uncertain databases, to obtain the maximal frequent itemsets can not only make the mining results easier to use, but also reduce the computing cost and memory size. Therefore, in this paper, we investigate how to efficiently discover the maximal frequent itemsets over an attribute-uncertainty model based uncertain database.

1.2 Challenges and Contributions

To address our proposed problem, an intuitive consideration is to enumerate all the probabilistic frequent itemsets and then to filter the maximal frequent itemsets. This method was proposed in [20] named *pApriori*, which introduced the divide-and-conquer method to mine the vertical databases, and used a one-bound estimating method to evaluate the frequency of itemsets. Then, a new method *TODIS-MAX*[20] was proposed for further improvement. Besides all the techniques used in *pApriori*, *TODIS-MAX* also presented its own optimizations. It used a top-down method, which generated the itemsets from supersets to subsets, and thus can efficiently used the pruning strategy when generating the infrequent itemsets. Also, it proposed a method to compute the probability density function of the subsets from the supersets, which can further reduce the computing cost. To our best knowledge, *TODIS-MAX* is the most efficient algorithm to achieve the maximal frequent itemsets from uncertain databases.

However, there are still some problems for addressing: (1) the top-down method may meet a bottleneck when the count of probabilistic frequent 1-items increases, which will result in an exponentially increase of the probabilistic infrequent itemsets, and this is hard to handle even though the time complexity can be reduced to $O(n)$. (2) The value to decide the probabilistic infrequent itemsets is not tight enough, which makes itemsets in multi-levels have to be recomputed. (3) If an itemset is frequent, the computing cost is still high since the time complexity is $O(nlog^2n)$ in the worst case. Accordingly, new problems are posed: How to reduce the most of exponentially increased itemsets? How to further decrease the computing cost of the frequent itemsets? And how to efficiently achieve the maximal frequent itemsets?

In this paper, we address these problems and make the following contributions.

1. We focus on the problem of probabilistic maximal frequent itemset mining over uncertain databases, and define the probabilistic maximal frequent itemset, which is in line to the traditional definition over exact database, and supplies us a better pruning method.
2. We introduce a compact data structure $PMFIT$ to maintain the information of probabilistic frequent itemsets, which in a bottom-up manner, can efficiently organize the mining results for the itemset search. Then the $PMFIM$ algorithm is proposed to depth-first discover the probabilistic maximal frequent itemsets. In this algorithm, we propose a probabilistic support estimation method, which can compute the upper bound and the lower bound of the probabilistic support with a much low cost. The method when together used with $PMFIT$, yields a significantly better performance. Plus, we use the super pruning strategy to further reduce the mining cost.
3. We compare our algorithm with the $TODIS\text{-}MAX$ [20] on 2 synthetic datasets and 3 real-life datasets. Our experimental results show that our algorithm is much more effective and efficient.

The rest of this paper is organized as follows. In Sect. 2 we present the preliminaries and then define the problem. Section 3 introduces the data structures, and illustrates our algorithm in detail. Section 4 evaluates the performance with theoretical analysis and experimental results. Finally, Sect. 5 concludes this paper.

2   Preliminaries and Problem Definition

2.1   Preliminaries

Given a set of distinct items $\Gamma = \{i_1, i_2, \cdots, i_n\}$ where $|\Gamma| = n$ denotes the size of Γ, a subset $X \subseteq \Gamma$ is called an itemset; suppose each item $x_t (0 < t \leq |X|)$ in X is associated with an occurrence probability $p(x_t)$, we call X an uncertain itemset, which is denoted as $X = \{x_1, p(x_1); x_2, p(x_2); \cdots ; x_{|X|}, p(x_{|X|})\}$, and the probability of X is $p(X) = \Pi_{i=1}^{|X|} p(x_i)$. We call the list $\{p(x_1), p(x_2), \cdots, p(x_{|X|})\}$ the probability density function. An uncertain transaction UT is an uncertain itemset with an ID. An uncertain database UD is a collection of uncertain transactions $UT_s (0 < s \leq |UD|)$. Given an uncertain itemset X, the count it occurs in an uncertain database is called the support, denoted $\Lambda(X)$.

Two definitions of the frequent itemset for uncertain data have been proposed. One is based on the the expected support, another is based on the probabilistic support.

Definition 1. *(Expected Frequent Itemset [6]) Given an uncertain database UD, an itemset X is an λ-expected frequent itemset if its expected support $\Lambda^E(X)$ is not smaller than minimum support λ. Here $\Lambda^E(X) = \sum_{UT \in UD}\{p(X)|X \subseteq UT\}$.*

Definition 2. *(Probabilistic Frequent Itemset [26]) Given the minimum support λ, the minimum probabilistic confidence τ and an uncertain database UD, an itemset X is a probabilistic frequent itemset if the probabilistic support $\Lambda_\tau^P(X) \geq \lambda$. $\Lambda_\tau^P(X)$ is the maximal support of itemset X with probabilistic confidence τ, i.e., $\Lambda_\tau^P(X) = Max\{i|P_{\Lambda(X) \geq i} > \tau\}$.*

2.2   Problem Definition

Definition 3. *(Probabilistic Maximal Frequent Itemset) Given the minimum support λ, the minimum probabilistic confidence τ and an uncertain database UD, an itemset X is a probabilistic maximal frequent itemset if it is a probabilistic frequent itemset and is not covered by the other probabilistic frequent itemsets.*

From Definition 3, one can easily see that the support information of an probabilistic frequent itemset $Y \subset X$ can be estimated from the probabilistic maximal frequent itemset X without having to read from the database anymore.In other words, for a probabilistic maximal frequent itemset X, any itemset Y that $Y \subset X$ satisfy the following statement: The probability of Y's support no smaller than λ is larger than τ.

Problem Statement: Based on the previous definition, we present our addressed problem as follows. Given an uncertain database UD, the minimum support λ, the minimum probabilistic confidence τ, we are required to explore all probabilistic maximal frequent itemsets from UD.

3   Probabilistic Maximal Frequent Itemset Mining Method

3.1   Data Structures

Probabilistic Mining Frequent Itemset Tree. To accelerate the searching and pruning speed, we design a simple but effective index named *PMFIT*(**P**robabilistic **M**aximal **F**requent **I**temset **T**ree), in which each node n_X denotes an itemset X; n_X is a 6-tuple $< item, sup, esup, psup, lb, ub >$, in which *item* denotes the last item of the current itemset X, *sup* is the support, *esup* is the expected support, and *psup* is the probabilistic support. *lb* and *ub* separately represent the lower bound and upper bound of probabilistic support. Except the root node, each node has a pointer to its parent node. *PMFIT* can be

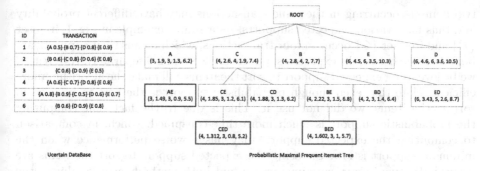

Fig. 1. Probabilistic maximal frequent itemset tree(PMFIT) for $\lambda = 3$, $\tau = 0.1$

constructed by our proposed algorithm in Sect. 3.5. Figure 1 is the *PMFIT* from the 6 transactions. For an example, n_A denotes itemset $\{A\}$ with the support 3, the expected support 1.9, and the probabilistic support 3. As can be seen, 13 nodes are probabilistic frequent itemsets, only 3 nodes are probabilistic maximal frequent itemsets.

Probabilistic Maximal Frequent Itemset Collection. Since the final results do not have to maintain the probabilities of each item, we employed a traditional bitmap based collection to store the probabilistic maximal frequent itemsets, which can help us perform the superset pruning, so that a better performance can be achieved.

3.2 Probabilistic Support Computing

Since the probabilistic density function of itemset X in two transactions T_1 and T_2 can be computed with the convolution between the probabilistic density function in T_1 and the probabilistic density function in T_2, the divide-and-conquer method proposed in [20] is also employed in our paper. That is, the uncertain database will be split into two parts to separately compute the probabilistic density function, and this operation will be recursively conducted until the sub-database has only one transaction. The convolution can be computed with a Fast Fourier Transformer, which, given the size of the uncertain database n, will efficiently reduce the time complexity from $O(n^2)$ to $O(nlog^2(n))$.

3.3 Items Reordering

Bayardo stated that ordering items with increasing support can reduce the search space [3]; together with other pruning strategies, it can further reduce the cost of the support computing. In this paper, we employ the similar heuristic rule with a little differences. That is, the items will be ordered by their expected supports rather their supports. This is due to the instinctive observation that two itemsets occurring in the same transactions may have different probability and thus have various expected supports. As a simple example in Fig. 1, itemset $\{B\}$ and itemset $\{C\}$ occur in four transactions, which means their supports are both 4, nevertheless, the expected supports are 2.8 and 2.6 separately. As a result, we believe using expected support to sort the items will make the algorithm more efficient. Note that even though probabilistic support is the best to be used in sorting the items, we did not use it. This is due to the fact that computing the probabilistic support is much more time-consumed, which, in comparison to computing the expected support, may has a worse performance when the minimum support is low. Choosing the expected support to sort the items are verified effective in our experiments; we find both methods achieve almost the same search space, but computing the expected support is much more efficient.

3.4 Pruning Strategies

We propose pruning strategies to improve the performance. A tight bounds are supplied to infer the range of probabilistic support, or even can ignore the computing of probabilistic support; in addition, a superset pruning method inspired by the traditional mining algorithm is employed.

The Bounds of Probabilistic Support. When mining the probabilistic maximal frequent itemsets over an n-transactions uncertain database, the probabilistic support is not important for the users, so we try to find a method to estimate the frequency of an itemset rather directly compute the probabilistic support.

Theorem 1. *For an itemset X in uncertain database UD, given the minimum probabilistic confidence τ, we can get the lower bound and upper bound of the probabilistic support $\Lambda_\tau^P(X)$, denoted $lb(\Lambda_\tau^P(X))$ and $ub(\Lambda_\tau^P(X))$ as follows.*

$$\begin{cases} lb(\Lambda_\tau^P(X)) = \Lambda^E(X) - \sqrt{-2\Lambda^E(X)ln(1-\tau)} \\ ub(\Lambda_\tau^P(X)) = \frac{2\Lambda^E(X)-ln\tau+\sqrt{ln^2\tau-8\Lambda^E(X)ln\tau}}{2} \end{cases} \tag{1}$$

Proof. For the itemset X, we use ε to denote the expected support $\Lambda^E(X)$, also, we use t to denote the probabilistic support $\Lambda_\tau^P(X)$, which, according to Definition 2, satisfies the following equations.

$$\begin{cases} P_{\Lambda(X)\geq t} > \tau \Leftrightarrow P_{\Lambda(X)>t-1} > \tau \\ P_{\Lambda(X)\geq t+1} \leq \tau \Leftrightarrow P_{\Lambda(X)>t} \leq \tau \end{cases} \tag{2}$$

(1) If we set $t = (1+\xi)\varepsilon$, i.e., $\xi = \frac{t}{\varepsilon} - 1$, where $\xi \geq 0$, that is, $t \geq \varepsilon$, then based on the Chernoff Bound, $P_{\Lambda(X)\geq t} = P_{\Lambda(X)\geq(1+\xi)\varepsilon} \leq e^{-\frac{\xi^2\varepsilon}{2+\xi}}$; based on the first inequality of Eq. 2, we can get $\tau < e^{-\frac{\xi^2\varepsilon}{2+\xi}} = e^{-\frac{(t-\varepsilon)^2}{t+\varepsilon}}$; that is to say, when $t \geq \varepsilon$, $\frac{2\varepsilon-ln\tau-\sqrt{ln^2\tau-8\varepsilon ln\tau}}{2} < t < \frac{2\varepsilon-ln\tau+\sqrt{ln^2\tau-8\varepsilon ln\tau}}{2}$. Since $\frac{2\varepsilon-ln\tau-\sqrt{ln^2\tau-8\varepsilon ln\tau}}{2} \leq \varepsilon$, we can obtain the following inequality.

$$t < \frac{2\varepsilon - ln\tau + \sqrt{ln^2\tau - 8\varepsilon ln\tau}}{2} \qquad\qquad if \quad t \geq \varepsilon \tag{3}$$

(2) If we set $t = (1 - \xi')\varepsilon$, i.e., $\xi' = 1 - \frac{t}{\varepsilon}$, where $\xi' \geq 0$, that is, $t \leq \varepsilon$, then based on the Chernoff Bound, $P_{\Lambda(X)>t} = P_{\Lambda(X)>(1-\xi')\varepsilon} > 1 - e^{-\frac{\xi'^2 \varepsilon}{2}}$; based on the second inequality of Eq. 2, we can get $\tau > 1 - e^{-\frac{\xi'^2 \varepsilon}{2}}$, i.e., $-\sqrt{-\frac{2ln(1-\tau)}{\varepsilon}} < \xi' < \sqrt{-\frac{2ln(1-\tau)}{\varepsilon}}$, then when $t \leq \varepsilon$, $\varepsilon - \sqrt{-2\varepsilon ln(1 - \tau)} < t < \varepsilon + \sqrt{-2\varepsilon ln(1 - \tau)}$. Since $\varepsilon \leq \varepsilon + \sqrt{-2\varepsilon ln(1 - \tau)}$, we can get the following inequality.

$$t > \varepsilon - \sqrt{-2\varepsilon ln(1 - \tau)} \qquad if \quad t \leq \varepsilon \qquad (4)$$

From Eqs. 3 and 4, we can conclude that no matter t is larger or smaller than the ε, it is definitely within the range of $(\varepsilon - \sqrt{-2\varepsilon ln(1 - \tau)}, \frac{2\varepsilon - ln\tau + \sqrt{ln^2\tau - 8\varepsilon ln\tau}}{2})$. Consequently, we can determine the lower bound is $\Lambda^E(X) - \sqrt{-2\Lambda^E(X) ln(1 - \tau)}$, and the upper bound is $\frac{2\Lambda^E(X) - ln\tau + \sqrt{ln^2\tau - 8\Lambda^E(X) ln\tau}}{2}$. ∎

Theorem 1 provides us two pruning strategies. For an itemset X, if the upper bound $\frac{2\varepsilon - ln\tau + \sqrt{ln^2\tau - 8\varepsilon ln\tau}}{2}$ is not larger than the minimum support λ, then X is a probabilistic infrequent itemset. Also, if the lower bound $\varepsilon - \sqrt{-2\varepsilon ln(1 - \tau)} \geq \lambda$, then X is definitely a probabilistic frequent itemset. We can see that for an uncertain database with size n, if the minimum support λ is not within this range, we can successfully hit the target.

Example 1. Using the uncertain dataset in Fig. 1 as the example. If we set the minimum support $\lambda = 1$ and the minimum probabilistic confidence $\tau = 0.1$, then for itemset $\{A\}$, the lower bound is 1.3, which is larger than 1, then itemset $\{A\}$ is a frequent itemset. Further, if we set the minimum support $\lambda = 5$ and the minimum probabilistic confidence $\tau = 0.1$, then for itemset $\{AB\}$, the upper bound is 4.7, which is smaller than 5, and thus itemset $\{AB\}$ is an infrequent itemset.

Superset Pruning. According to our definition, an itemset being probabilistic maximal frequent must satisfy two conditions. (1) the probabilistic support is not smaller than the minimum support; (2) it is not covered by any other probabilistic frequent itemsets. Both computing are needed, then the computing with lower computing cost should be conducted firstly. As the above mentioned, computing the probabilistic support requires $O(nlog^2(n))$ time complexity, which can be improved to $O(n)$ with our method. However, to scan the existing probabilistic maximal frequent itemsets, assuming whose size is m, requires at most $O(m)$ time complexity. Based on the definition of probabilistic maximal frequent itemset, m is much smaller than n. Consequently, for a new generated itemset, we will first decide whether it is cover by a super itemset, then, if not, compute the bounds or the probabilistic support. This strategy can be extended for further pruning. That is, for a probabilistic maximal frequent itemset $X = \{x_1 x_2 \cdots x_n\}$, if they are the last n items in the sorted items list, then items $x_2, \cdots, x_n$ can be pruned directly for further computing. We can see from Fig. 1, since $\{CDE\}$ is an itemset in which the items are the last three ones in the sorted items list, then $\{D\}$, $\{E\}$ will be removed, and the computing for all their descendants can be pruned.

3.5   Algorithm Description

In this section, we propose a depth-first algorithm named PMFIM(**P**robabilistic **M**aximal **F**requent **I**temset **M**ining) to build the bottom-up organized tree, that is, the subsets will be computed first, and then the supersets will be generated if their subsets are all frequent. We discover the itemsets with this manner since the probabilistic frequent itemset also has the apriori property. The algorithm can be conducted in five steps. Algorithm 1 shows the pseudo code of Step 3–5.

Step 1: We get all the distinct items and sort them in an incremental order according to their expected supports before we build the *PMFIT*; during the process, the items with the support or the upper bound lower than the minimum support will be initially pruned.

Step 2: The PMFIT is initialed with only one root node, which represent the null itemset.

Step 3: For a parent node, we begin to generate the child node, and compute the related information to decide whether it is frequent. First, if the child node is covered by one of the maximal frequent itemsets, it is not a maximal but a frequent itemset; specially, if all the items in it are the last ones in the sorted items list, we can determine that all the right nodes are not the final results, and the loop will be ended immediately. Second, after computing the support, the expected support and the support bounds, if the upper bound is not larger than the minimum support, then it is an infrequent itemset; if the lower bound is not smaller than the minimum support, then it is a frequent itemset. Finally, if we can not determine the frequency by the previous value, we need to compute the probabilistic support and compare it to the minimum support.

Step 4: If a child node is frequent, we will recall Step 3 for it; otherwise, it will be pruned.

Step 5: If a node has no children and is not in the final results, it is a probabilistic maximal frequent itemset. We can add it into the probabilistic maximal frequent itemset collection. Because of our depth-first mining manner, there is no need to remove the subset from the collection.

Complexity. The overall time complexity of the *PMFIM* algorithm depends on the database size n, the minimum support λ, the minimum probabilistic confidence τ, and the count of *PMFIT* nodes t. For each new generated node, there are three possible computing cost. The first is $O(m)$ where m is the count of current probabilistic maximal frequent itemsets; the second is $O(n)$; the third is $O(nlog^2n)$. Generally, $m < n < nlog^2n$; thus, the worst time complexity is $O(nlog^2n)$. Nevertheless, as will be demonstrated in our experiments, the count of probabilistic support computing is greatly small, which guarantee that the performance can be improved significantly.

Algorithm 1. PMFIM Algorithm

Require: n_I: node of $PMFIT$ denote itemset I; UD: Uncertain Database; $PMFIC$: Probabilistic Maximal Frequent Itemset Collection; λ: minimum support; τ: minimum probabilistic confidence;

1: **for** each itemset $J(|J| = |I|)$ order larger than I **do**
2:    **if** J is probabilistic maximal frequent itemset and J is the last items in the sorted items list **then**
3:        break;
4:    generate the child node $n_{I \cup J}$ of n_I;
5:    **if** $I \cup J \in PMFIC$ **then**
6:        CALL PMFIM($n_{I \cup J}$, UD, $PMFIC$, λ, τ);
7:    compute $\Lambda^E(I \cup J)$, $\Lambda(I \cup J)$, $lb(\Lambda_\tau^P(I \cup J))$ and $ub(\Lambda_\tau^P(I \cup J))$;
8:    **if** $ub(\Lambda_\tau^P(I \cup J)) \leq \lambda$ **then**
9:        delete $n_{I \cup J}$;
10:       continue;
11:   **if** $lb(\Lambda_\tau^P(I \cup J)) \geq \lambda$ **then**
12:       CALL PMFIM($n_{I \cup J}$, UD, $PMFIC$, λ, τ);
13:   **else**
14:       compute $\Lambda_\tau^P(I \cup J)$;
15:       **if** $\Lambda_\tau^P(I \cup J) \geq \lambda$ **then**
16:           CALL PMFIM($n_{I \cup J}$, UD, $PMFIC$, λ, τ);
17:       **else**
18:           delete $n_{I \cup J}$;
19: **if** n_I has no children and I is not in $PMFIC$ **then**
20:     add I in $PMFIC$;

---

4   Experiments

We conducted the experiments to evaluate the performance of *PMFIM*. The state-of-the-art algorithm *TODIS-MAX*[20], which has been presented much more efficient than *pApriori*, was used as the evaluation method. While *TODIS-MAX* focused on the tuple-uncertainty based databases, we re-implemented it for the attribute-uncertainty based databases. The dataset size $|UD|$, the relative minimum support $\lambda_r(= \frac{\lambda}{|UD|})$, and the minimum probabilistic confidence τ are the main elements that may affect the uncertain data mining, which, as a result, were used to compare the algorithms in runtime and memory cost.

4.1   Running Environment and Datasets

Both algorithms were implemented with C++, compiled with Visual Studio 2010 running on Microsoft Windows 7 and performed on a PC with a 2.90GHZ Intel Core i7-3520M processor and 8GB main memory. We evaluated the algorithms on 2 synthetic datasets generated by the IBM synthetic data generator and 3 real-life datasets [25]. The detailed data characteristics are shown in Table 2. We used the item correlation to show the density of an uncertain database, that is, a smaller correlation value denotes a denser data.

Table 2. Uncertain dataset characteristics

Uncertain DataSet	size of dataset	average. trans. length	minimal trans. length	maximal trans. length	number of items	mean	variance	item Correlation
T25I15D320K	320 002	26	1	67	994	0.87	0.27	38
T40I10D100K	100 000	39	4	77	1000	0.79	0.61	25
KOSARAK	990 002	8	1	2498	41 270	0.5	0.28	5159
ACCIDENTS	340 183	33	18	51	468	0.5	0.58	14
CONNECT4	67 557	43	43	43	129	0.78	0.65	3

4.2 Effect of Relative Minimum Support

We set a fixed minimum probabilistic confidence and compared the performance of the two algorithms when the relative minimum support was changed.

Running Time Cost Evaluation. As can be seen in Fig. 2, the runtime cost of the two algorithms reduced linearly with a decrease of the relative minimum support. Plus, to compare our algorithm to *TODIS-MAX*, we can observe the following results. On the one hand, when the relative minimum support was high, the runtime cost of the *TODIS-MAX* method was a little lower than the *PMFIM* over most of the datasets. This is due to the advantage of the top-down mining manner in *TODIS-MAX*, that is, a faster pruning will be employed if the supersets are not frequent, which is more useful when performing over dense dataset. On the other hand, when the relative minimum support became lower, the performance of *TODIS-MAX* turned worse significantly, which can also be clearly noticed over denser datasets. As shown in the figure, over the densest dataset *CONNECT4*, *PMFIM* achieved almost a thousand time faster than *TODIS-MAX* when the relative minimum support was 0.7, and then, the *TODIS-MAX* can be almost not measured since the runtime cost increased too much; over the *KOSARAK*, the sparsest dataset, our algorithm can also achieved hundreds-fold speedup when the relative minimum support was 0.05. This outperforming further increased when the relative minimum support turned smaller. This is also because of the different mining fashions: *TODIS-MAX* employed the top-down method, which, when the relative minimum support was small, may use more frequent items to generate infrequent itemsets, whose size will exponentially increased along with the count of frequent items. Even though *TODIS-MAX* employed the expected support to prune certain computing, the rest computing was still too large to conduct. In comparison to that, our algorithm was a bottom-up method, i.e., only frequent itemsets were generated; with the help of superset pruning, the computing will not increase sharply when the relative minimum support became lower.

Memory Cost Evaluation. We also compared the maximal memory usages of the two algorithms. As shown in Fig. 3, similar to the runtime cost, the memory usages decreased when the relative minimum support increased. In addition,

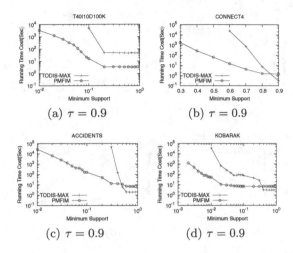

Fig. 2. Running time cost vs relative minimum support

we can see that in a majority of cases *TODIS-MAX* used more memory than our algorithm. Since our algorithm employed the divided-and-conquer method proposed in *TODIS-MAX*, the running memory cost was similar when the relative minimum support was high. Nevertheless, when it became low, the memory usage of *TODIS-MAX* will increased exponentially. Again, it was because of the massively generated infrequent itemsets. Besides, the lattice used in *TODIS-MAX* was another reason that use more memory. Note that we did not show the memory cost of *TODIS-MAX* when the relative minimum support was low, this is because that the runtime cost was too high to perform the *TODIS-MAX* algorithm in limited time.

4.3 Effect of Data Size

We evaluated the scalability of the two algorithms, that is, we performed the algorithms w.r.t. different data sizes, which are shown in Fig. 4. The *T25I15D320K* dataset was used as the evaluation dataset. We separately got the first n(from 20 K to 320 K) transactions to conduct the algorithms. Plus, the relative minimum support and the minimum probabilistic confidence were set to the fixed values. Note even though the relative minimum support was fixed, the minimum support changed since the data size was different. As shown in Fig. 4(a), when the dataset turned larger, the runtime cost of the two algorithms also increased, but the *PMFIM* algorithm was much more stable when the dataset became larger. This presents that our algorithm can accurate estimate most of the probabilistic supports no matter how the data size increased. However, in Fig. 4(b), with the increasing size of transactions, the memory cost of both algorithms increase linearly. It is reasonable since we showed the maximal memory cost during running; thus, once we directly computed the probabilistic support, the memory will be more used for more transactions.

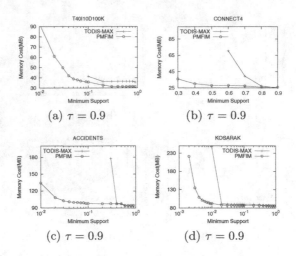

Fig. 3. Memory cost vs relative minimum support

4.4 Effect of Minimum Probabilistic Confidence

Definition 2 shows that a larger minimum probabilistic confidence may result in a less possibility that an itemset becomes frequent, which will reduce the computing cost and the memory usage. We evaluated two algorithms for different minimum probabilistic confidences(from 0.00001 to 0.1) when the relative minimum support was fixed.

Figures 5 and 6 separately presented the runtime cost and the memory usage. To our surprise, we find that when the minimum probabilistic confidence increased, both the runtime and the memory cost of the two algorithms kept almost unchanged. This shows that the minimum probabilistic confidence had little effect on the performance of the algorithms. It is due to the reason that the probabilistic density function was highly sparse when the dataset size was large, which results that most of the minimum probabilistic confidence can slightly change the probabilistic support, and thus can almost not change the types of the itemsets.

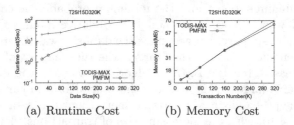

Fig. 4. Effect of data size for $\lambda_r = 0.1$, $\tau = 0.9$

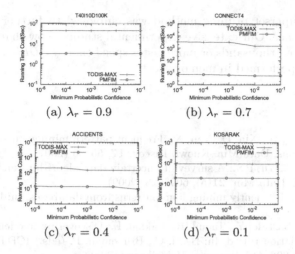

Fig. 5. Runtime cost vs minimum probabilistic confidence

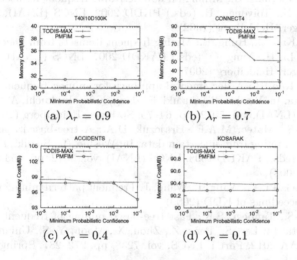

Fig. 6. Memory cost vs minimum probabilistic confidence

5   Conclusions

In this paper we studied the behavior of probabilistic maximal frequent itemset mining over uncertain databases. We defined the probabilistic maximal frequent itemset, which is much reasonable and can supply more pruning considerations. Based on this, an extended enumeration tree named *PMFIT* was introduced to efficiently index and maintain the probabilistic frequent itemsets. A bottom-up algorithm named *PMFIM*, mining in a depth-first manner, was proposed, in which we used the support and the expected support to estimate whether an itemset is frequent, which greatly reduced the computing cost and memory usage;

in addition, a superset pruning method was employed to further improve the mining performance. Our extensive experimental studies show that our *PMFIM* algorithm achieved thousands or more faster speed than *TODIS-MAX*, and also significantly outperformed in memory cost.

References

1. Han, J., Cheng, H., Xin, D., Yan, X.: Frequent pattern mining: current status and future directions. Data Min. Knowl. Discov. **17**, 55–86 (2007)
2. Aggarwal, C.C., Yu, P.S.: A survey of uncertain data algorithms and applications. Trans. Knowl. Data Min. **21**(5), 609–623 (2009)
3. Bayardo, R.J.: Efficiently mining long patterns from databases. In: Proceedings of SIGMOD (1998)
4. Pasquier, N., Bastide, Y., Taouil, R., Lakhal, L.: Discovering frequent closed itemsets for association rulesd. In: Beeri, C., Bruneman, P. (eds.) ICDT 1999. LNCS, vol. 1540, pp. 398–416. Springer, Heidelberg (1998)
5. Calders, T., Goethals, B.: Mining all non-derivable frequent itemsets. In: Elomaa, T., Mannila, H., Toivonen, H. (eds.) PKDD 2002. LNCS (LNAI), vol. 2431, pp. 74–86. Springer, Heidelberg (2002)
6. Chui, C.-K., Kao, B., Hung, E.: Mining frequent itemsets from uncertain data. In: Zhou, Z.-H., Li, H., Yang, Q. (eds.) PAKDD 2007. LNCS (LNAI), vol. 4426, pp. 47–58. Springer, Heidelberg (2007)
7. Chui, C.-K., Kao, B.: A decremental approach for mining frequent itemsets from uncertain data. In: Washio, T., Suzuki, E., Ting, K.M., Inokuchi, A. (eds.) PAKDD 2008. LNCS (LNAI), vol. 5012, pp. 64–75. Springer, Heidelberg (2008)
8. Leung, C.K.-S., Mateo, M.A.F., Brajczuk, D.A.: A tree-based approach for frequent pattern mining from uncertain data. In: Washio, T., Suzuki, E., Ting, K.M., Inokuchi, A. (eds.) PAKDD 2008. LNCS (LNAI), vol. 5012, pp. 653–661. Springer, Heidelberg (2008)
9. Aggarwal, C.C., Li, Y., Wang, J., Wang, J.: Frequent pattern mining with uncertain data. In: Proceedings of KDD (2009)
10. Leung, C.K.-S., Tanbeer, S.K.: Fast tree-based mining of frequent itemsets from uncertain data. In: Lee, S., Peng, Z., Zhou, X., Moon, Y.-S., Unland, R., Yoo, J. (eds.) DASFAA 2012, Part I. LNCS, vol. 7238, pp. 272–287. Springer, Heidelberg (2012)
11. Leung, C.K.-S., MacKinnon, R.K.: BLIMP: a compact tree structure for uncertain frequent pattern mining. In: Bellatreche, L., Mohania, M.K. (eds.) DaWaK 2014. LNCS, vol. 8646, pp. 115–123. Springer, Heidelberg (2014)
12. Leung, C.K.S., Brajczuk, D.A.: Efficient algorithms for the mining of constrained frequent patterns from uncertain data. In: SIGKDD Explorer, vol. 11, No. 2, pp. 123-130 (2009)
13. Calders, T., Garboni, C., Goethals, B.: Efficient pattern mining of uncertain data with sampling. In: Zaki, M.J., Yu, J.X., Ravindran, B., Pudi, V. (eds.) PAKDD 2010, Part I. LNCS, vol. 6118, pp. 480–487. Springer, Heidelberg (2010)
14. Leung, C.K.S., Hao, B.: Mining of frequent itemsets from streams of uncertain data. In: Proceedings of ICDE (2009)
15. Leung, C.K.-S., Jiang, F.: Frequent pattern mining from time-fading streams of uncertain data. In: Cuzzocrea, A., Dayal, U. (eds.) DaWaK 2011. LNCS, vol. 6862, pp. 252–264. Springer, Heidelberg (2011)

16. Nguyen, H.-L., Ng, W.-K., Woon, Y.-K.: Concurrent semi-supervised learning with active learning of data streams. In: Hameurlain, A., Küng, J., Wagner, R., Cuzzocrea, A., Dayal, U. (eds.) TLDKS VIII. LNCS, vol. 7790, pp. 113–136. Springer, Heidelberg (2013)
17. Leung, C.K.-S., Hayduk, Y.: Mining frequent patterns from uncertain data with mapreduce for big data analytics. In: Feng, L., Bressan, S., Winiwarter, W., Song, W., Meng, W. (eds.) DASFAA 2013, Part I. LNCS, vol. 7825, pp. 440–455. Springer, Heidelberg (2013)
18. Zhang, Q., Li, F., Yi, K.: Finding frequent items in probabilistic data. In: Proceedings of SIGMOD (2008)
19. Bernecker, T., Kriegel, H.P., Renz, M., Verhein, F., Zuefle, A.: Probabilistic frequent itemset mining in uncertain databases. In: Proceedings of SIGKDD (2009)
20. Sun, L., Cheng, R., Cheung, D.W., Cheng, J.: Mining uncertain data with probabilistic guarantees. In: Proceedings of KDD (2010)
21. Bernecker, T., Kriegel, H.-P., Renz, M., Verhein, F., Zuefle, A.: Probabilistic frequent pattern growth for itemset mining in uncertain databases. In: Ailamaki, A., Bowers, S. (eds.) SSDBM 2012. LNCS, vol. 7338, pp. 38–55. Springer, Heidelberg (2012)
22. Wang, L., Cheng, R., Lee, S.D., Cheung, D.: Accelerating probabilistic frequent itemset mining: a model-based approach. In: Proceedings of CIKM (2010)
23. Wang, L., Cheung, D., Cheng, R., Lee, S.D., Yang, X.S.: Efficient mining of frequent item sets on large uncertain databases. Trans. Knowl. Data Min. 24(12), 2170–2183 (2012)
24. Calders, T., Garboni, C., Goethals, B.: Approximation of frequentness probability of itemsets in uncertain data. In: Proceedings of ICDM (2010)
25. Tong, Y., Chen, L., Cheng, Y., Yu, P.S.: Mining frequent itemsets over uncertain databases. In: Proceedings of VLDB (2012)
26. Tang, P., Peterson, E.A.: Mining probabilistic frequent closed itemsets in uncertain databases. In: Proceedings of ACMSE (2011)
27. Peterson, E.A., Tang, P.: Fast approximation of probabilistic frequent closed itemsets. In: Proceedings of ACMSE (2012)
28. Tong, Y., Chen, L., Ding, B.: Discovering threshold-based frequent closed itemsets over probabilistic data. In: Proceedings of ICDE (2012)
29. Liu, C., Chen, L., Zhang, C.: Mining probabilistic representative frequent patterns from uncertain data. In: Proceedings of SDM (2013)
30. Liu, C., Chen, L., Zhang, C.: Summarizing probabilistic frequent patterns : a fast approach. In: Proceedings of KDD (2013)

Anytime OPTICS: An Efficient Approach for Hierarchical Density-Based Clustering

Son T. Mai[1(✉)], Ira Assent[1], and Anh Le[2]

[1] Aarhus University, Aarhus, Denmark
{mtson,ira}@cs.au.dk
[2] University of Transport, Hochiminh City, Vietnam
anh@hcmutrans.edu.vn

Abstract. OPTICS is a fundamental data clustering technique that has been widely applied in many fields. However, it suffers from performance degradation when faced with large datasets and expensive distance measures because of its quadratic complexity in terms of both time and distance function calls. In this paper, we introduce a novel *anytime* approach to tackle the above problems. The general idea is to use a sequence of lower-bounding (LB) distances of the true distance measure to produce multiple approximations of the true reachability plot of OPTICS. The algorithm quickly produces an approximation result using the first LB distance. It then continuously refines the results with subsequent LB distances and the results from the previous computations. At any time, users can suspend and resume the algorithm to examine the results, enabling them to stop the algorithm whenever they are satisfied with the obtained results, thereby saving computational cost. Our proposed algorithms, called Any-OPTICS and Any-OPTICS-XS, are built upon this anytime scheme and can be applied for many complex datasets. Our experiments show that Any-OPTICS obtains very good clustering results at early stages of execution, leading to orders of magnitudes speed up. Even when run to the final distance measure, the cumulative runtime of Any-OPTICS is faster than OPTICS and its extensions.

1 Introduction

Clustering algorithms assign unlabeled objects into homogeneous groups (or clusters), usually in terms of a distance measure and have applications in many fields. During the past decades, many clustering algorithms have been introduced, e.g., partition-based, density-based, and grid-based techniques. Among them, density-based clustering is one of the most successful paradigms due to its attractive benefits, e.g., it can detect arbitrarily shaped clusters [6].

In the density-based clustering algorithm DBSCAN [6], an object lies in a dense area, which indicates a cluster, if there are more than μ objects inside its ϵ-neighborhood. However, choosing suitable values for μ and especially ϵ is a non-trivial problem and is a main target of OPTICS [2]. Instead of providing clusters directly, OPTICS starts with arbitrary large value of ϵ, denoted as ϵ^*,

© Springer International Publishing Switzerland 2016
S.B. Navathe et al. (Eds.): DASFAA 2016, Part I, LNCS 9642, pp. 164–179, 2016.
DOI: 10.1007/978-3-319-32025-0_11

and produces a structure called the reachability plot consisting of an ordering of objects. Clusters are then extracted by cutting through the reachability-plot with any value $\epsilon \leq \epsilon^*$. Thus, the reachability plot represents a hierarchical structure of clusters with arbitrary shapes and different densities. Together with DBSCAN, OPTICS has become a state-of-the-art data clustering technique nowadays.

Due to its distance-based scheme, OPTICS suffers from a performance bottle neck when coping with large complex datasets, e.g. trajectories or medical images, and efficient but expensive distance measures, e.g. Dynamic Time Warping (DTW) [16]. Moreover, OPTICS is a batch algorithm. That means it only produces a single result and does not allow interaction with users during its runtime, while interactively exploring the results during execution time has been emerged as a useful approach for analyzing large complex data recently [12,18,20].

In this paper, we introduce an *anytime* [21] approach to tackle these drawbacks of OPTICS. Our algorithms, called anytime hierarchical density-based clustering (Anytime OPTICS (*Any-OPTICS*) and Anytime OPTICS with Xseedlist (*Any-OPTICS-XS*)), employ a sequence of lower-bounding (LB) distances of the true distance measure to produce multiple approximations of the true reachability plot of OPTICS. In the beginning, they quickly produce an approximation result using the first LB distance. Then, they continuously refine the results using the next LB distances and the previous results. As anytime algorithms [21], user can suspend and resume them at any time for examining the results. Thus, they can stop the algorithm whenever they are satisfied with the acquired results to save computational cost. Our algorithms are a general framework that can be applied for many complex datasets. To the best of our knowledge, our algorithms are the first anytime approach for OPTICS proposed in the literature so far.

Contributions. Our contributions are summarized as follows.

- We propose for the first time an anytime approach for hierarchical density-based clustering called *Anytime OPTICS* (Any-OPTICS).
- We introduce an extension of Any-OPTICS called Any-OPTICS-XS which is more efficient than Any-OPTICS when dealing with very expensive distance measures.
- Experiments on real datasets are conducted to evaluate the performance of our algorithms. Any-OPTICS and Any-OPTICS-XS acquire very good clustering results at early stages of execution, thus leading to orders of magnitudes speed up. Even if they run to the end, the total cumulative runtimes of Any-OPTICS and Any-OPTICS-XS are still faster than OPTICS and its extensions.

The rest of this paper is organized as follows. In Sect. 2, we briefly introduce some characteristics of anytime algorithms and some notions of the algorithm OPTICS. The algorithm Any-OPTICS and its extension Any-OPTICS-XS are presented in Sect. 3. We describe the used distance measure and its lower-bounding functions in Sect. 4. Experiments are conducted in Sect. 5. Section 6 is dedicated for discussing related works. Finally, Sect. 7 concludes the paper.

2   Background

2.1   Anytime Algorithms

Anytime algorithms [21] copes with time-consuming problems by trading execution time for quality of results. In contrast to the *batch* ones, during their executions, they can be interrupted to provide a *best-so-far* result and resumed to produce better results at any time. Anytime algorithms have been widely used in many fields, e.g., object recognition [7] and surveillance [17].

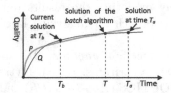

Fig. 1. The progress of two anytime algorithms P and Q.

Characteristics of Anytime Algorithms. According to [20,21], an anytime clustering algorithm should satisfy some important properties such as: (1) The final result should be similar to or better than that of the batch algorithm, (2) The total cumulative runtime of an anytime algorithm should not be much larger than the runtime of the batch algorithm. Figure 1 shows a progress of two anytime algorithms P and Q. If Q is interrupted at T_b, an approximation is returned. At the time T, it acquires the same result as the batch algorithm. If it runs to T_a, its result may be better than that of the batch algorithm. For performance comparison, many researchers consider P better than Q since it acquires better performance in the early stages [20]. Note that the quality of an anytime algorithm is not necessary strictly increased [20].

Anytime Clustering. Though there exist many anytime algorithms in the literature, anytime clustering has not been paid much attention with few proposed algorithm, e.g., [10,12,13,20]. Our work is related to anytime versions of the algorithm DBSCAN [12,13]. We will thoroughly discuss the differences between them and our work throughout the paper and in Sect. 6.

2.2   Hierarchical Density-Based Clustering

In contrast to DBSCAN [6], OPTICS [2] does not produce explicit clustering results. Instead, it produces an order of objects in a dataset, called the *reachability-plot*, that encapsulates the information of all possible clusters wrt. arbitrary values of ϵ that are smaller than a predefined threshold ϵ^*.

Given a set of objects O which contains N objects, a distance function $d : O \times O \rightarrow \mathbb{R}$ and two parameters $\epsilon^* \in \mathbb{R}^+$ and $\mu \in \mathbb{N}^+$.

Definition 1 (ϵ^*-neighborhood). *The neighborhood of an object p, denoted as* $N_{\epsilon^*}(p)$, *is defined as a set of object q so that* $d(p,q) \leq \epsilon^*$.

$$N_{\epsilon^*}(p) = \{q | d(p,q) \leq \epsilon^*\}$$

Definition 2 *(Core-distance). The core-distance of an object p, denoted as core-* $dist_{\epsilon^*,\mu}(p)$, *is defined as follows:*

$$core\text{-}dist_{\epsilon^*,\mu}(p) = \begin{cases} Undef\ (\infty)\ if\ |N_{\epsilon^*}(p)| < \mu \\ \mu\text{-}dist(p)\quad otherwise \end{cases}$$

where μ-dist(p) is the distance between p and its μ-th nearest neighbor.

Definition 3 *(Reachability-distance). The reachability-distance of an object p w.r.t. an object o, denoted as reach-* $dist_{\epsilon^*,\mu}(p,o)$, *is defined as follows:*

$$reach\text{-}dist_{\epsilon^*,\mu}(p,o) = \begin{cases} Undef\ (\infty) & if\ |N_{\epsilon^*}(o)| < \mu \\ max(core\text{-}dist_{\epsilon^*,\mu}(o), d(o,p))\ otherwise \end{cases}$$

OPTICS maintains a sorted seedlist S to expand the order of objects. For constructing the order of objects, OPTICS randomly picks an unprocessed object p and calculates its core-distance. If $core\text{-}dist_{\epsilon^*,\mu}(p) \neq Undef$, all object q inside $N_{\epsilon^*}(p)$ are examined. If q is not inside S, it is inserted into S in an ascending order w.r.t. its reachability-distance from p. If q is already inside S and $reach\text{-}dist_{\epsilon^*,\mu}(q,p)$ is smaller than current reachability-distance of q, then q is assigned a new reachability-distance and S is reordered w.r.t. the new change. The whole process is repeated until S is empty and there is no unprocessed object. Clusters can then be extracted by using a threshold ϵ to cut through the reachability plot and determining the objects' labels. Every object that has $reach\text{-}dist$ larger than ϵ is classified either as an outlier or a border object depend on its $core\text{-}dist$. Otherwise, it is a core object of a cluster. Different thresholds ϵ provide different clustering results. For readability, we drop the terms ϵ^* and μ from the Definitions 2 and 3 in the rest of the paper unless otherwise stated.

3   Anytime Approach for OPTICS

Assume that there exists a sequence Γ of n distance functions $l_i : O \times O \to \mathbb{R}$ so that l_i lower-bounds d for all $1 \leq i < n$ and l_n equals d.

$$\Gamma = \{l_i | \forall p,q \in O : l_i(p,q) \leq d(p,q) \wedge l_n(p,q) = d(p,q)\}$$

The general idea of our *anytime* approach for OPTICS, called Any-OPTICS, is that it runs in multiple levels from L_1 to L_n. At each level L_i, the distance function l_i is used to produce the reachability plot. However, such a naive approach is extremely inefficient since the results from previous levels are totally not exploited to reduce the runtime. Thus, in this Section, we propose a unique approach to connect the results at L_i and L_{i+1}, thus allowing an efficient clustering scheme at each level.

Definition 4 *(Distance at L_i). The distance between two objects p and q at level L_i, denoted as $d_i(p,q)$, is defined as the maximal distance of $l_j(p,q)$ where $1 \leq j \leq i$.*

$$d_i(p,q) = max_{1 \leq j \leq i}\{l_j(p,q)\}$$

At each level L_i, the distance function d_i is now used to produce a reachability plot with OPTICS instead of l_i. By forcing the used distance function at each level to be monotone increased as described in Definition 4, we can have some important properties as described below.

Lemma 1 (Monotonicity of core-distance). *The core-distance of object p at level L_i, denoted as $core\text{-}dist^{L_i}(p)$, is monotone increasing, i.e., $core\text{-}dist^{L_i}(p) \leq core\text{-}dist^{L_{i+1}}(p)$.*

Proof (Sketch). Since $d_i(p,q) \leq d_{i+1}(p,q)$ (due to Definition 4), we have $\mu\text{-dist}^{L_i}(p) \leq \mu\text{-dist}^{L_{i+1}}(p)$. Thus, Lemma 1 holds.

A reachability plot P of OPTICS consists of an order $\omega = \{\omega_1, \cdots, \omega_n\}$ of objects where $\omega_i = o_k \in O \ \wedge \ \omega_i \neq \omega_j \ \forall i,j \in [1,n]$ together with core-dist and reach-dist for each object. A subplot P_{uv} of P is then defined as follows.

Definition 5 (Subplot). *A subplot P_{uv} of P consists of a subsequence $\omega_{uv} = \{\omega_u, \omega_{u+1}, \cdots, \omega_v\}$ of ω so that $(reach\text{-}dist(\omega_u) = Undef \wedge core\text{-}dist(\omega_u) \neq Undef \wedge reach\text{-}dist(\omega_v) \neq Undef)$ and $(reach\text{-}dist(\omega_{v+1}) = Undef \vee v = n)$.*

Obviously, a subplot P_{uv} of P is a subsequence of P which starts and ends at u and v respectively so that if we cut P with a $\epsilon = \epsilon^*$ then all objects ω_i ($u \leq i \leq v$) belong to a same cluster. And for every object ω_j ($j < u \vee j > v$), ω_i and ω_j belong to different clusters. Let P_{L_i} and $P_{L_{i+1}}$ be the reachability plots acquired at levels L_i and L_{i+1}, respectively.

Lemma 2 (Monotonicity of subplot). *For every subplot P_{uv} at level L_{i+1}, denoted as $P_{uv}^{L_{i+1}}$, there exist a subplot P_{kl} at level L_i so that $\forall r \ (u \leq r \leq v)$: $core\text{-}dist^{L_{i+1}}(\omega_r^{L_{i+1}}) \neq Undef \Rightarrow \exists t \ (k \leq t \leq l) : \omega_t^{L_i} = \omega_r^{L_{i+1}}$, i.e., if an object ω_r belongs to $P_{uv}^{L_{i+1}}$ and $core\text{-}dist^{L_{i+1}}(\omega_r) \neq Undef$ then ω_r belongs to $P_{kl}^{L_i}$.*

Proof (Sketch). Assume that $\exists g_1, g_2 \in [u,v]$ so that $\omega_{g_1}^{L_{i+1}}$ and $\omega_{g_2}^{L_{i+1}}$ belong to different subplots $P_{k_1 l_1}^{L_i}$ and $P_{k_2 l_2}^{L_i}$, respectively. Let $h_1 \in [k_1, l_1]$ and $h_2 \in [k_2, l_2]$ be the position of $\omega_{g_1}^{L_{i+1}}$ and $\omega_{g_2}^{L_{i+1}}$ in $P_{k_1 l_1}^{L_i}$ and $P_{k_2 l_2}^{L_i}$, respectively ($\omega_{g_1}^{L_{i+1}} = \omega_{h_1}^{L_i}$ and $\omega_{g_2}^{L_{i+1}} = \omega_{h_2}^{L_i}$). Since $core\text{-}dist^{L_{i+1}}(\omega_{g_1}^{L_{i+1}}) \neq Undef$ and $core\text{-}dist^{L_{i+1}}(\omega_{g_2}^{L_{i+1}}) \neq Undef$, we have $core\text{-}dist^{L_i}(\omega_{h_1}^{L_i}) \neq Undef$ and $core\text{-}dist^{L_i}(\omega_{h_2}^{L_i}) \neq Undef$ (Lemma 1). Moreover, we have, $d_i(\omega_{h_1}^{L_i}, \omega_{h_2}^{L_i}) \geq \epsilon^*$. And there is no chain of object $c = \{c_1, c_2, \cdots, c_p\}$ so that $core\text{-}dist^{L_i}(c_j) \neq Undef$ and $d_i(c_j, c_{j+1}) < \epsilon^*$ and $d_i(c_1, \omega_{h_1}^{L_i}) < \epsilon^*$ and $d_i(c_p, \omega_{h_2}^{L_i}) < \epsilon^*$ to connect $\omega_{h_1}^{L_i}$ and $\omega_{h_2}^{L_i}$. Otherwise, $\omega_{h_1}^{L_i}$ and $\omega_{h_1}^{L_i}$ must belong to the same subplot at L_i. Since $d_i(p,q) \leq d_{i+1}(p,q)$, we have $d_{i+1}(\omega_{g_1}^{L_{i+1}}, \omega_{g_2}^{L_{i+1}}) \geq \epsilon^*$. Also, there exists no such chain like c to connect $\omega_{g_1}^{L_{i+1}}$ and $\omega_{g_2}^{L_{i+1}}$ (Lemma 1 and Definition 4). Thus, $\omega_{g_1}^{L_{i+1}}$ and $\omega_{g_2}^{L_{i+1}}$ must belong to different subplots at L_{i+1}. This leads to a contradiction.

Generally speaking, a subplot P_{uv} at level L_i might be broken into some smaller parts at level L_{i+1} and there is no merging of two subplots at L_i into one at L_{i+1}. However, some border objects might be added to broken parts of P_{uv} at L_{i+1} due to the nature of OPTICS [2].

3.1   The Proposed Algorithm Any-OPTICS

Lemmas 1 and 2 suggest a more efficient way to perform clustering rather than naively re-running OPTICS with distance function d_i at each level L_i. First, if *core-dist*(p) at L_i is *Undef*, we do not need to recalculate it again at L_{i+1} since it is also *Undef* following Lemma 1. It can help to reduce the total number of distance calculations at L_{i+1}, thus speeding up the algorithm. Second, at L_{i+1}, instead of running OPTICS for re-ordering the whole dataset, we only need to re-run OPTICS on every subplot P_{uv} of L_i and simply merge the results together to produce a final ordered list of objects at L_{i+1}. Note that all objects with *Undef core-dist* and *reach-dist* do not belong to any subplot and are regarded as outliers. We only need to put them into the head or tail of the ordered list without further examination. This local updating scheme helps to significantly reduce the runtime for updating the order of objects at each level. Due to space limitation, we only briefly describe the algorithm above.

Lemma 3 (Correctness of Any-OPTICS). *The final result of Any-OPTICS is equivalent to that of OPTICS with an exception on some border objects.*

Proof (Sketch). Since l_i lower-bounds d, we have $d_n(p,q) = d(p,q)$. The distance update phase at L_i only ignores objects with *core-dist* $= \infty$ at L_i which also have *core-dist* $= \infty$ at L_j where $j > i$ (Lemma 1). Thus, the final *core-dists* of object p of Any-OPTICS and OPTICS are identical. Due to Lemma 2, the reachability plot at L_n is preserved at lower levels for all objects q with determined *core-dists* at L_n. Thus, the final result of Any-OPTICS is equivalent to that of OPTICS with only a minor difference on some border objects.

Similar to the naive approach, the time complexity of Any-OPTICS is $\sum_{i=1}^{n} O(\varphi_i N^2)$ where φ_i is the time complexity of l_i. However, since Any-OPTICS can reduce the total number of distance calculations at each level and has efficient local update scheme, it is much more efficient than the naive approach.

3.2   The Proposed Algorithm Any-OPTICS-XS

Though the monotonicity of core-distance can help to reduce the total number of distance calculations at each level, Any-OPTICS still incurs many redundant distance calculations due the ordering scheme of OPTICS. In [3], the authors propose a novel approach built upon a data structure called Xseedlist[1] and a LB distance d_{lb} of d for reducing redundant distance calculations. The general idea of their algorithm is using the LB distances to guide the clustering process and calculating the true distances only when it is necessary. Theoretically, we can employ the Xseedlist for building the reachability plot at each level L_i. However, it is obviously a non-trivial process. Directly using l_i in Xseedlist is not possible since l_i does not necessary lower-bound l_{i+1}. The distance functions d_i satisfy the lower-bound requirement of Xseedlist. However, simply using d_i as a lower-bound for d_{i+1} results in an inefficient scheme since all the distance from d_1 to d_{n-1} have to be calculated for every pair of objects. In this Section, we propose an efficient scheme, called Anytime OPTICS with XSeedlist (Any-OPTICS-XS), for intergrating Xseedlist into the anytime scheme of Any-OPTICS.

The general idea of Any-OPTICS-XS is that it considers all the distances d_j $(1 \leq j \leq i)$ as a lower-bound function for d_{i+1} at L_{i+1} instead of d_i only. To do so, we additionally store for each pair of objects (p, q) the level j that $d_j(p, q)$ has been computed so far. At L_{i+1}, d_j is used as a lower-bounding distance for d_{i+1}. When it is necessary, $d_{i+1}(p, q)$ is calculated, and the new level $i + 1$ is set for (p, q). Consequently, for every object p in object list (OL) and object q in the predecessor list of p $(PL(p))$ at level L_i, $Flag(p, q)$ will contain the current level L_j $(j \leq i)$ of (p, q) instead of a boolean value as in [3]. And $Predist(p, q)$ will contain $max(core\text{-}dist^{L_i}(q), d_j(q, p))$ $(j \leq i)$.

In order to reduce the total number of distance function calls, we use another ordering function ϕ of Xseedlist for sorting two tuples (p, q) and (r, s). The function ϕ must emphasize the distance first instead of the Flag as in A-DBSCAN-XS [13] due to the nature of expansion process of OPTICS. In case the distances are equal, the one with higher $Flag$ will be considered in order to reduce the distance calculation from L_j to L_{i+1}.

$$\phi((p, q), (r, s)) = \begin{cases} > & if\ Predist(p, q) > Predist(r, s) \\ < & if\ Predist(p, q) < Predist(r, s) \\ \begin{cases} > & if\ Flag(p, q) < Flag(r, s) \\ = & if\ Flag(p, q) = Flag(r, s) \\ < & if\ Flag(p, q) > Flag(r, s) \end{cases} & otherwise \end{cases}$$

Generally, the algorithm Any-OPTICS-XS consists of five sub-routines. Due to space limitation, we only briefly described them below.

The main routine called Any-OPTICS-XS is generally similar to Any-OPTICS. At each level L_i, we only need to update all subplot P_{uv} locally following Lemma 2 by calling the function Update-Cluster-Order.

---

[1] Xseedlist consists of a list of objects called the object list (OL). Each object is associated with a so-called predecessor list (PL). Each item of the PL contains a tuple $(Id, Flag, Predist)$ where Id is an object id, $Flag$ indicates whether the $Predist$ is a lower-bound or true distance, and $Predist(p, q)$ contains the reachability distance from q to p. PL is sorted in an ascending order of $Predist$. OL is sorted in an ascending order of $Predist$ of the first object in the PL of each object.

For updating cluster order (function Update-Cluster-Order), an object o_1 at the top of object list (OL) is examined. If the distance between o_1 and its first object in $PL(o_1)$, denoted as o_{11}, is not updated to the current level L_i, we update the distance of the pair (o_{11}, o_1) (function Update-Distance) and reorder the XSeedlist. Otherwise, o_{11} is the closest object that can be reached from all processed objects due to the function ϕ. We write o_1 to the order list, remove o_1 from OL, and resort OL. If $core\text{-}dist(o_1)$ is not $Undef$ (function Update-Core-Dist), we insert all objects inside $N_\epsilon(o_1)$ into the Xseedlist (function Update-Xseedlist). The whole process is repeated until OL is empty.

For updating the core-distance of an object q (function Update-Core-Dist), a sorted list is used for reducing distance calculations. We repeatedly extract the first object p that (p, q) is not updated to L_i, and update (p, q) (function Update-Distance) until the first μ objects in the sorted list have their distance to q updated to L_i. Here, we have the $core\text{-}dist(q)$. Lemma 1 ensures that we do not need to re-calculate $core\text{-}dist(q)$ if it is $Undef$ at previous level.

For inserting object into the Xseedlist (function Update-XSeedlist), for an object $o \in N_{\epsilon^*}(q)$, we first calculate the $Predist(o, q) = max(core\text{-}dist(q), d_j(q, o)$. Note that $core\text{-}dist(q)$ must already be calculated at L_i, while the distance (q, o) may not be updated to the current level L_i $(j < i)$. If o is not in OL, we insert an entry $(o : (q, L_j, Predist(o, q)))$ to OL. Otherwise, we insert the tuple $(q, L_j, Predist(o, q))$ to $PL(o)$. OL is then resorted by means of the function ϕ.

Updating distance for a pair of objects (p, q) (function Update-Distance) is the most important part of Any-OPTICS-XS. In [13], (p, q) is updated from L_j to current level L_i in a single step. However, due to the distance-based scheme of OPTICS, it results in many redundant distance calculations since pairs of objects at lowest level tend to be *blindly* updated first. A better choice is only updating (p, q) to one level (L_j to L_{j+1}) and then let the algorithm *think* whether (p, q) is worth for updating further. It leads to dramatic performance improvement in our experiments.

Lemma 4 *(Correctness of Any-OPTICS-XS). The final result of Any-OPTICS-XS is equivalent to that of OPTICS with an exception on some border objects.*

Proof (Sketch). We have $d_n(p, q) = d(p, q)$ (Definition 4). The distance update step at L_i only ignores objects with $core\text{-}dist = \infty$ which also have $core\text{-}dist = \infty$ at L_j where $j > i$ (Lemma 1). The sorting scheme of the function Update-Core-Dist ensures the correct μ-dist is found. Thus, the final $core\text{-}dists$ of object p of Any-OPTICS-XS and OPTICS are identical. Also, the sorted scheme of OL guarantees that an object with smallest reachability distance at current level is extracted to ordered file. Thus, the reachability distances are the same between Any-OPTICS-XS and OPTICS. Due to Lemma 2, the reachability plot at L_n is preserved at lower levels for all objects q with determined $core\text{-}dists$ at L_n. Thus, the final result of Any-OPTICS-XS is equivalent to that of OPTICS with only a minor difference on some border objects as described in Lemma 3.

The time complexity of Any-OPTICS-XS is $\sum_{i=1}^{n} O(\varphi_i N^2) + N^2 log N$ where φ_i is the time complexity of l_i and $N^2 log N$ is the operation cost of XSeedlist. Therefore, Any-OPTICS-XS may be worse than Any-OPTICS when dealing with less expensive distance measure like the Euclidean (EU) distance.

4   Distance Function

In order to make Any-OPTICS and Any-OTICS-XS work properly, the distance functions l_i should satisfy two conditions (not strictly): (1) l_i should be faster than l_{i+1} in terms of computation time and (2) l_{i+1} should be generally tighter than l_i in terms of LB quality. We use two different distance measures in our experiments including EU [15] and DTW [5].

Lower-Bounding Distances. For EU distance, the Haar wavelet transform [15] is first employed to transfer objects into coefficient domains. At each level L_i, we simply choose k_i % of wavelet coefficients as features of objects and calculate distances among them based on those features.

DTW is a well-known distance measure which has better accuracy than EU in many data mining tasks [5]. However, its quadratic complexity is a major computation bottleneck. In [11,13], the authors propose a way to construct the LB distances for DTW on high-dimensional trajectory data based on LB_Sakurai [16]. Each trajectory is discretized into segments with length l, and the LB distance is calculated by using segment features. For constructing a sequence of LB distances, we only need to select different segment length at each level L_i.

Range Query with LBs. In [16], the authors propose using a sequence of LB distances from coarser to finer for speeding up the query processing compared with indexing techniques. We therefore use the same *multiple-levels filter-and-refinement* for speeding up the ϵ-range query processing of OPTICS.

5   Experiments

5.1   Evaluation Methodology

All experiments are conducted on a workstation with a 2.8 GHz i7 core CPU, 6 GB of RAM, and Windows 7 OS. Runtime results are averaged over ten runs. All algorithms are implemented in Java. We compare our algorithms Any-OPTICS and Any-OPTICS-XS with OPTICS and its variants including:

- OPTICS-M: OPTICS with the *multiple-level filter-and-refine* process described in Sect. 4. As shown in [16], OPTICS-M is more efficient than using indexing techniques [6] for speeding up range queries.
- OPTICS-XS: OPTICS with Xseedlist, proposed in [3]. We slightly modify it to work with multiple LBs by using the first $n - 1$ levels as filter distances to speed up the range query process. We keep L_{n-1} as the LB distance for d. This scheme is more efficient than using only one L_i ($1 \leq i < n$) as the LB distance for d.

We use Normalized Mutual Information (NMI) [19] as an external measure for comparing clustering results. NMI scores fall in the range $[0, 1]$, where 0 means completely independent results and 1 means complete agreement wrt. the ground truth. Unless otherwise stated, for EU distance, we use a sequence of LBs $\Gamma = \{5, 10, 15, 20, 25, 30, 35, 40, 45, 100\}$, where L_i is the percentage of Haar wavelet coefficients used (100 % means true EU distance). For DTW, we use a sequence of LBs $\Gamma = \{35, 30, 25, 20, 15, 10, 5, 1\}$ where L_i is the length l of each segment ($l = 1$ means original DTW distance). To set the parameters of OPTICS, we run DBSCAN with different parameter combinations to find optimal parameter values for each dataset. Then, we use the found parameter μ and choose an arbitrary value $\epsilon^* > \epsilon$ as input parameters for OPTICS. This ensures that the final results found by OPTICS are among the best ones. At each level of our algorithms, clusters are extracted from a reachability plot by using a single cut-off threshold ϵ that varies from ϵ^* to 0 with a step size of 0.01. At each level, the best clustering result and the cumulative runtime are reported.

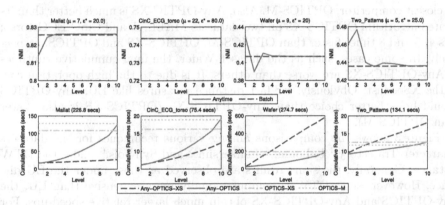

Fig. 2. Performance comparison on some datasets under the EU distance. Parameters μ and ϵ and runtimes of OPTICS are indicated beside the dataset names.

5.2 Performance Analysis

Figure 2 shows the comparison among various techniques on the datasets MAL-LAT, Two_Patterns, CinC_ECG_torso, and Wafer acquired from the UCR Archive.[2] As we can see, our algorithms obtain good clustering results (close to the optimal results) at very early levels. Sometimes the intermediate results are even better than the optimal ones as in the cases of Wafer and Two_Patterns. Therefore, if we stop our algorithms at level 2, for example, we can obtain a dramatic performance acceleration of up to 14× compared to OPTICS and its variants. At the final level, Any-OPTICS almost has the same runtime as that of

---

[2] http://www.cs.ucr.edu/~eamonn/time_series_data/. Note that those datasets are re-interpolated to the length of $2^{\lfloor \log(m) \rfloor + 3}$ (where m is the dimension of each object) to use with the Haar wavelet transform.

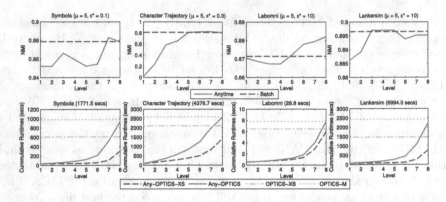

Fig. 3. Performance comparison on some datasets under the DTW distance. Parameters μ and ϵ and runtimes of OPTICS are indicated beside the dataset names.

its closest competitor, OPTICS-M. Also, Any-OPTIC-XS is much better than its related algorithm OPTICS-XS on two datasets MALLAT and CinC_ECG_torso; it is 4, 5 and 8 times faster than OPTICS-M, OPTICS-XS and OPTICS, respectively. In some cases, such as the dataset Wafer, the final cumulative runtimes of Any-OPTICS-XS are worse than others. It is due to the high operation cost of the XSeedlist. Obviously, for fast distance measures like EU, Any-OPTICS should be a better choice than Any-OPTICS-XS (OPTICS-XS is also slower than OPTICS-M).

Figure 3 shows the comparisons among various techniques for the datasets Character Trajectory,[3] Labomni,[4] Lankersim,[5] and Symbols,[6] under the DTW distance. Generally, the same results are observed as those under the EU distance. However, since DTW is computationally more expensive than EU, the Any-OPTICS and Any-OPTICS-XS obtain much larger relative speed-ups. For the dataset Lankersim, we can stop anytime OPTICS at level 1 and obtain a nearly optimal result. At that level, it takes Any-OPTICS-XS only 80 s which is 18, 30, and 88 times faster than OPTICS-XS (1464.6 s), OPTICS-M (2425.6 s) and OPTICS (6994.0 s), respectively. At the end, Any-OPTICS-XS consumes only 772.8 s which is 2, 3, and 9 times faster than the others. Obviously, for expensive distance measures, Any-OPTICS-XS is more efficient than Any-OPTICS due to its pruning power.

5.3   Comparison by Distance Calculation Counts

Figure 4 shows the percentage of distance calculations at each level, relative to level 1 (when no reduction can yet take place). Observe that OPTICS-XS

---

[3] http://archive.ics.uci.edu/ml/.

[4] http://cvrr.ucsd.edu/bmorris/datasets/dataset_trajectory_clustering.html.

[5] http://www.fhwa.dot.gov/publications/research/operations/07029/index.cfm.

[6] http://www.cs.ucr.edu/~eamonn/time_series_data/.

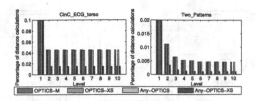

Fig. 4. The total number of distance function calls at each level of different algorithms for the datasets CinC_ECG_torso and Two_Patterns.

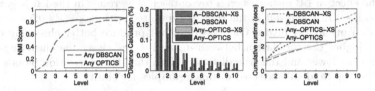

Fig. 5. Comparison between anytime OPTICS and anytime DBSCAN for the dataset COIL20 ($\mu = 5$, $\epsilon^* = \epsilon = 4.75$).

significantly reduces the number of distance calculations at the final level due to the reduction scheme of Xseedlist. Any-OPTICS has slightly fewer distance calculations than OPTICS-M at all levels. Any-OPTICS-XS has the fewest calculations among all methods at every level. This illustrates the efficacy of our distance calculation reduction schemes. When the reduction of more expensive distance calculations in later levels is sufficiently large, the anytime algorithms become faster than the batch ones (as was seen in Figs. 2 and 3, for example).

5.4   Anytime OPTICS Versus Anytime DBSCAN

Figure 5 (right) shows the total numbers of distance calculations of anytime DBSCAN and anytime OPTICS on the dataset COIL20 acquired from the Columbia Object Image Library.[7] Note that, the same results are observed with all other datasets. With the same parameters μ and $\epsilon^* = \epsilon$, A-DBSCAN and Any-OPTICS almost have the same numbers of distance calculations at all levels, while A-DBSCAN-XS consumes less distance calculations than Any-OPTICS-XS. It is due to the fact that OPTICS requires more intensive distance calculations for determining the core-dist and reach-dist of objects than DBSCAN. Consequently, anytime DBSCAN is often faster than anytime OPTICS. However, smaller final runtime does not mean that anytime DBSCAN is more *efficient* than anytime OPTICS in the context of anytime algorithms. Assume that a user wants to have $NMI > 0.7$, she can stop Any-OPTICS at level 1 after 0.95 s, while she has to run A-DBSCAN to level 5 after 1.86 s, which is 2 times slower than Any-OPTICS. Another advantage of anytime OPTICS is that it

---

[7] http://www1.cs.columbia.edu/CAVE/software/softlib/coil-20.php.

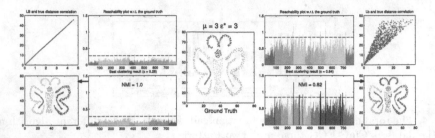

Fig. 6. The effect of LB distances on a toy dataset. The result in the left is generated with $d_{lb1} = d * 0.1$. The result in the right is generated with $d_{lb2} = d * 0.5 + noise$. As we can see, using higher d_{lb} for OPTICS does not produce better clustering quality. If fact, the higher the correlation between d_{lb} and d, the closer the clustering results acquired by the lower and the true distance. Note that the correlation plots are only drawn with 1000 randomly selected distance samples for clarity.

is much less sensitive to the quality of LB distance as shown in Fig. 5 (left). Since the LB distances at levels 1 and 2 are too low, the clustering qualities at levels 1 and 2 of anytime DBSCAN algorithms are consequently very low. However, anytime OPTICS still acquires almost perfect clustering results.

5.5 Effects of the LB Distances

Why does anytime OPTICS produce quite good results at early levels? Fig. 6 shows the results of OPTICS for a toy dataset with different LB distances d_{lb1} and d_{lb2} on the left and right, respectively. Obviously, d_{lb2} is much tighter than d_{lb1} in terms of LB qualities. However, OPTICS acquires much better results on d_{lb1} than on d_{lb2} with $NMI = 1.0$ and $NMI = 0.82$, respectively. The reachability plots on the left also show more consistent orders of objects w.r.t. the ground truth than on the right. Thus, tighter LB does not mean better clustering result. The true reason is the correlation between the LB distance and the true distance instead. As we can see from Fig. 6, d_{lb1} (left) correlates with d better than d_{lb2} (right). As a result, its reachability plot is more consistent with the ground truth since the relationships among objects in terms of pairwise distances are better preserved.

5.6 Impact of Input Parameters

Effects of the Sequence of LBs Γ. Figure 7 shows the performances of anytime OPTICS for different sequences of LBs from Γ_1 to Γ_4 (in an increasing order of LB quality). Here, better LB distances results in better clustering qualities. The runtime of Any-OPTICS generally increases with Γ since the LB distance becomes more expensive at each level. However, for Any-OPTICS-XS, it is more complicated due to its distance reduction scheme. Using coarser LB distances usually leads to fast cumulative runtimes at some initial levels however slower

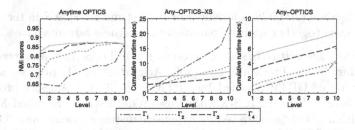

Fig. 7. The effect of the sequence of LBs Γ on the runtimes of Any-OPTICS and Any-OPTICS-XS for the dataset COIL20 ($\mu = 5$ and $\epsilon^* = 6.0$) where $\Gamma_1 = \{1, 2, \cdots, 9, 100\}$, $\Gamma_2 = \{5, 10, \cdots, 45, 100\}$, $\Gamma_3 = \{20, 35, \cdots, 60, 100\}$, and $\Gamma_4 = \{30, 35, \cdots, 70, 100\}$.

results at some last levels as in the case of Γ_1. Conversely, in the case of Γ_4, Any-OPTICS-XS starts much slower but ends much faster than others.

Effects of the Parameter μ ***and*** ϵ^{*8}. Bigger ϵ^* means bigger object's neighborhood and thus more distance calculations and higher runtimes are required at each level. However, the influence of the parameter μ is somewhat unclear since the performance of OPTICS is mostly affected by the neighborhood sizes of objects. The effect of the parameters μ and ϵ^* on the clustering accuracy of OPTICS is still an open research issue and is out of scope of this paper.

6   Related Works and Discussion

Anytime Clustering Algorithms. Recently anytime clustering algorithms have become an emerging research to cope with complex data [8,10,12,14,20].

In [10], the multi-resolution property of the Haar wavelet transform is exploited to cast k-Means into an anytime clustering algorithm. In [20], a lower bounding and a upper-bounding distances of DTW are employed to approximate the distance matrix in the beginning. Then, each entry in the similarity matrix is sequentially selected and updated with its true DTW distances following a predetermined ranking scheme. Kranen et al. [8] propose an anytime algorithm for clustering data stream using an indexing scheme. Act-DBSCAN [14] could also be regarded as an anytime algorithm. However, it is more likely be an active clustering algorithm instead. Among them, anytime DBSCAN algorithms [12,13] are closely related to our approaches in terms of using a sequence of LBs. However, while anytime DBSCAN focuses on producing clusters, anytime OPTICS focuses on providing reachability plots for extracting clusters. Though both approaches exploit the monotonicity properties for reducing computation cost, they differ in the nature: the monotonicity of cluster structures in anytime DBSCAN and the monotonicity of reachability plots in anytime OPTICS. Moreover, as shown

---

[8] Due to space limitation, we only summarize the result here without showing the figure.

in Sect. 5, anytime OPTICS is far superior to anytime DBSCAN in terms of the anytime scheme together with its parameter robustness advantage.

Hierarchical Density-Based Clustering. OPTICS has attracted much research effort in the literature, e.g. [3,4]. Achtert et al. [1] propose an incremental version of OPTICS called IncOPTICS. In [3], the authors propose an algorithm, here called OPTICS-XS, that uses a data structure called XSeedlist together with a LB distance to reduce expensive distance calculations. Our algorithm Any-OPTICS-XS is inspired by this approach. However, with a significant extension including incorporating the monotonicity property of reachability plots and a sequence of LBs, Any-OPTICS-XS dramatically improves the performance compared with OPTICS-XS. Other techniques, e.g., the *data bubble* [4] only approximate the results of OPTICS for enhancing performance. Only some of them focus on complex datasets like our algorithms. Moreover, none of those techniques is an anytime algorithm. To the best of our knowledge, our algorithms Any-OPTICS and Any-OPTICS-XS are the first anytime approach for OPTICS proposed in the literature so far.

Many density-based clustering algorithms rely on the monotonicity property of the cluster structures to enhance the performance under some certain conditions, e.g., subspace projection of data (SUBCLU [9]) or a sequence of LBs (A-DBSCAN [12]). Our anytime OPTICS however is built upon the monotonicity of the order of objects inside the reachability plot.

7   Conclusion

In this paper, we introduce the first any approach for the hierarchical density-based clustering algorithm OPTICS. Our algorithms called Any-OPTICS and Any-OPTICS-XS are built upon a sequence of LBs of the true distance function among objects and work in multiple steps wrt. the LBs. By exploiting the monotonicity property of the core-distance and the order of objects inside the reachability plot, our anytime algorithms can significantly reduce the total number of distance calculations at each level as well as the reachability plot update time. Experiments on a wide range of very large datasets show that anytime OPTICS acquires very good clustering results at early levels. Consequently, they enhance the performance of clustering process up to orders of magnitudes. Even if they are run until the end, they are still much faster than other batch algorithms.

Acknowledgement. We would like to thank Sean Chester for his helps during the preparation of the paper. We special thank anonymous reviewers for their very helpful and constructive comments. Part of this research was funded by a Villum postdoc fellowship.

References

1. Achtert, E., Böhm, C., Kriegel, H.-P., Kröger, P.: Online hierarchical clustering in a data warehouse environment. In: ICDM, pp. 10–17 (2005)
2. Ankerst, M., Breunig, M.M., Kriegel, H.-P., Sander, J.: OPTICS: ordering points to identify the clustering structure. In: SIGMOD, pp. 49–60 (1999)
3. Brecheisen, S., Kriegel, H., Pfeifle, M.: Efficient density-based clustering of complex objects. In: ICDM, pp. 43–50 (2004)
4. Breunig, M.M., Kriegel, H.-P., Kröger, P., Sander, J., Bubbles, D.: Quality preserving performance boosting for hierarchical clustering. In: SIGMOD Conference, pp. 79–90 (2001)
5. Ding, H., Trajcevski, G., Scheuermann, P., Wang, X., Keogh, E.J.: Querying and mining of time series data: experimental comparison of representations and distance measures. PVLDB 1(2), 1542–1552 (2008)
6. Ester, M., Kriegel, H.-P., Sander, J., Xu, X.: Adensity-based algorithm for discovering clusters in large spatial databases with noise. In: KDD, pp. 226–231 (1996)
7. Kobayashi, T., Iwamura, M., Matsuda, T., Kise, K.: An anytime algorithm for camera-based character recognition. In: ICDAR, pp. 1140–1144 (2013)
8. Kranen, P., Assent, I., Baldauf, C., Seidl, T.: Self-adaptive anytime stream clustering. In: ICDM, pp. 249–258 (2009)
9. Kröger, P., Kriegel, H.-P., Kailing, K.: Density-connected subspace clustering for high-dimensional data. In: SDM, pp. 246–256 (2004)
10. Lin, J., Vlachos, M., Keogh, E.J., Gunopulos, D.: Iterative incremental clustering of time series. In: Bertino, E., Christodoulakis, S., Plexousakis, D., Christophides, V., Koubarakis, M., Böhm, K. (eds.) EDBT 2004. LNCS, vol. 2992, pp. 106–122. Springer, Heidelberg (2004)
11. Mai, S.T., Goebl, S., Plant, C.: A similarity model and segmentation algorithm for white matter fiber tracts. In: ICDM, pp. 1014–1019 (2012)
12. Mai, S.T., He, X., Feng, J., Böhm, C.: Efficient anytime density-based clustering. In: SDM, pp. 112–120 (2013)
13. Mai, S.T., He, X., Feng, J., Plant, C., Böhm, C.: Anytime density-based clustering of complex data. Knowl. Inf. Syst. 45(2), 319–355 (2015)
14. Mai, S.T., He, X., Hubig, N., Plant, C., Böhm, C.: Active density-based clustering. In: ICDM, pp. 508–517 (2013)
15. Chan, K.-P., Fu, A.W.-C.: Efficient time series matching by wavelets. In: ICDE, pp. 126–133 (1999)
16. Sakurai, Y., Yoshikawa, M., Faloutsos, C.: FTW: fast similarity search under the time warping distance. In: PODS, pp. 326–337 (2005)
17. Sofman, B., Bagnell, J., Stentz, A.: Anytime online novelty detection for vehicle safeguarding. In: ICRA, pp. 1247–1254, May 2010
18. Ueno, K., Xi, X., Keogh, E.J., Lee, D.-J.: Anytime classification using the nearest neighbor algorithm with applications to stream mining. In: ICDM, pp. 623–632 (2006)
19. Vinh, N.X., Epps, J., Bailey, J.: Information theoretic measures for clusterings comparison: is a correction for chance necessary? In: ICML, pp. 1073–1080 (2009)
20. Zhu, Q., Batista, G.E.A.P.A., Rakthanmanon, T., Keogh, E.J.: A novel approximation to dynamic time warping allows anytime clustering of massive time series datasets. In: SDM, pp. 999–1010 (2012)
21. Zilberstein, S.: Using anytime algorithms in intelligent systems. AI Mag. 17(3), 73–83 (1996)

Efficiently Mining Homomorphic Patterns
from Large Data Trees

Xiaoying Wu[1], Dimitri Theodoratos[2(✉)], and Zhiyong Peng[1]

[1] State Key Laboratory of Software Engineering, Wuhan University, Wuhan, China
{xiaoying.wu,peng}@whu.edu.cn
[2] New Jersey Institute of Technology, Newark, USA
dth@njit.edu

Abstract. Finding interesting tree patterns hidden in large datasets is a central topic in data mining with many practical applications. Unfortunately, previous contributions have focused almost exclusively on mining induced patterns from a set of small trees. The problem of mining homomorphic patterns from a large data tree has been neglected. This is mainly due to the challenging unbounded redundancy that homomorphic tree patterns can display. However, mining homomorphic patterns allows for discovering large patterns which cannot be extracted when mining induced or embedded patterns. Large patterns better characterize big trees which are important for many modern applications in particular with the explosion of big data.

In this paper, we address the problem of mining frequent homomorphic tree patterns from a single large tree. We propose a novel approach that extracts non-redundant maximal homomorphic patterns. Our approach employs an incremental frequency computation method that avoids the costly enumeration of all pattern matchings required by previous approaches. Matching information of already computed patterns is materialized as bitmaps a technique that not only minimizes the memory consumption but also the CPU time. We conduct detailed experiments to test the performance and scalability of our approach. The experimental evaluation shows that our approach mines larger patterns and extracts maximal homomorphic patterns from real datasets outperforming state-of-the-art embedded tree mining algorithms applied to a large data tree.

1 Introduction

Extracting frequent tree patterns which are hidden in data trees is central for analyzing data and is a base step for other data mining processes including association rule mining, clustering and classification. Trees have emerged in recent years as the standard format for representing, exporting, exchanging and integrating data on the web (e.g., XML and JSON). Tree data are adopted in various application areas and systems such as business process management,

---

X. Wu—The research was supported by the NSF of China under Grant No. 61202035, 61272110, and 61232002.

S.B. Navathe et al. (Eds.): DASFAA 2016, Part I, LNCS 9642, pp. 180–196, 2016.
DOI: 10.1007/978-3-319-32025-0_12

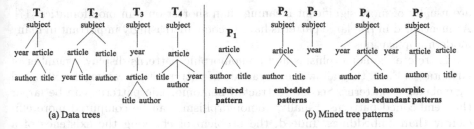

Fig. 1. Different types of mined tree patterns occurring in three of the four data trees.

NoSQL databases, key-value stores, scientific workflows, computational biology and genome analysis.

Because of its practical importance, tree mining has been extensively studied [2,4,6,7,10,11,14,15]. The approaches to tree mining can be basically characterized by two parameters: (a) the type of morphism used to map the tree patterns to the data structure, and (b) the type of mined tree data.

Mining Homomorphic Tree Patterns. The morphism determines how a pattern is mapped to the data tree. The morphism definition depends also on the type of pattern considered. In the literature two types of tree patterns have been studied: patterns whose edges represent parent-child relationships (*child* edges) and patterns whose edges represent ancestor-descendant relationships (*descendant* edges). Over the years, research has evolved from considering isomorphisms for mining patterns with child edges (*induced patterns*) [2,4] into considering embeddings for mining patterns with descendant edges (*embedded patterns*) [10,14,15]. Because of the descendant edges, embeddings are able to extract patterns "hidden" (or embedded) deep within large trees which might be missed by the induced definition [14]. Nevertheless, embeddings are restricted because: (a) they are injective (one-to-one), and (b) they cannot map two sibling nodes in a pattern to two nodes on the same path in the data tree. On the other hand, homomorphisms are powerful morphisms that do not have those two restrictions of embeddings. We term patterns with descendant edges, mined through homomorphisms, *homomorphic* patterns. Formal definitions are provided in Sect. 2. As homomorphisms are more relaxed than embeddings, the mined homomorphic patterns are a superset of the mined embedded patterns.

Figure 1(a) shows four data trees corresponding to different schemas to be integrated through the mining of large tree patterns. The frequency threshold is set to three. Figure 1(b) shows induced mined tree patterns, embedded patterns and non-redundant homomorphic patterns. Figure 1(b) includes the largest patterns that can be mined in each category. As one can see, the shown embedded patterns are not induced patterns, and the shown homomorphic patterns are neither embedded nor induced patterns. Further, the homomorphic patterns are larger than all the other patterns.

Large patterns are more useful in describing data. Mining tasks usually attach much greater importance to patterns that are larger in size, e.g., longer sequences are usually of more significant meaning than shorter ones in bioinfomatics [17]. As mentioned in [16] large patterns have become increasingly important in many modern applications.

Therefore, homomorphisms and homomorphic patterns display a number of advantages. First, they allow the extraction of patterns that cannot be extracted by embedded patterns. Second, extracted homomorphic patterns can be larger than embedded patterns. Finally, homomorphisms can be computed more efficiently than embeddings. Indeed, the problem of checking the existence of a homomorphism of an unordered tree pattern to a data tree is polynomial [9] while the corresponding problem for an embedding is NP-complete [8].

Mining Patterns from a Large Data Tree. The type of mined data can be a collection of small trees [2,4,10,14,15] or a single large tree. Surprisingly, the problem of mining tree patterns from a single large tree has only very recently been touched even though a plethora of interesting datasets from different areas are in the form of a single large tree. Examples include encyclopedia databases like Wikipedia, bibliographic databases like PubMed, scientific and experimental result databases like UniprotKB, and biological datasets like phylogenetic trees. These datasets grow constantly with the addition of new data. Big data applications seek to extract information from large datasets. However, mining a single large data tree is more complex than mining a set of small data trees. In fact, the former setting is more general than the latter, since a collection of small trees can be modelled as a single large tree rooted at a virtual unlabeled node. Existing algorithms for mining embedded patterns from a collection of small trees [14] cannot scale well when the size of the data tree increases. Our experiments show that these algorithms cannot scale beyond some hundreds of nodes in a data tree with low frequency thresholds.

The Problem. Unfortunately, previous work has focused almost exclusively on mining induced and embedded patterns from a set of small trees. The issue of mining *homomorphic patterns* from a *single large data tree* has been neglected.

The Challenges. Mining homomorphic tree patterns is a challenging task. Homomorphic tree patterns are difficult to handle as they may contain redundant nodes. If their structure is not appropriately constrained, the number of frequent patterns (and therefore the number of candidate patterns that need to be generated) can be infinite.

Even if homomorphic patterns are successfully constrained to be non-redundant, their number can be much larger than that of frequent embedded patterns from the same data tree. In order for the mining algorithm to be efficient, new, much faster techniques for computing the support of the candidate homomorphic tree patterns need to be devised.

The support of patterns in the single large data tree setting cannot be any-more the number of trees that contain the pattern as is the case in the multiple small trees setting. A new way to define pattern support in the new setting is needed which enjoys useful monotonic characteristics.

Typically, one can deal with a large number of frequent patterns, by computing only maximal frequent patterns. In the context of induced tree patterns, a pattern is maximal if there is no frequent superpattern [4]. A non-maximal pattern is not returned to the user as there is a larger, more specific pattern, which is frequent. However, in the context of homomorphic patterns, which involve descendant edges, the concept of superpattern is not sufficient for capturing the specificity of a pattern. A tree pattern can be more specific (and informative) without being a superpattern. For instance, the homomorphic pattern P_4 of Fig. 1(b) is more specific than the homomorphic pattern P_5 without being a superpattern of P_5. Therefore, a new sophisticated definition for maximal patterns is required which takes into account both the particularities of the homomorphic patterns and the single large tree setting.

Contribution. In this paper, we address the problem of mining maximal homomorphic unordered tree patterns from a single large data tree. Our main contributions are:

- We define the problem of extracting homomorphic and maximal homomorphic unordered tree patterns with descendant relationships from a single large data tree. This problem departs from previous ones which focus on mining induced or embedded tree patterns from a set of small data trees (Sect. 2).
- We constrain the extracted homomorphic patterns to be non-redundant in order to avoid dealing with an infinite number of frequent patterns of unbounded size. In order to define maximal patterns, we introduce a strict partial order on patterns characterizing specificity. A pattern which is more specific provides more information on the data tree (Sect. 2).
- We design an efficient algorithm to discover all frequent maximal homomorphic tree patterns. Our algorithm wisely prunes the search space by generating and considering only patterns that are maximal and frequent or can contribute to the generation of maximal frequent patterns (Sect. 3).
- Our algorithm employs an incremental frequency computation method that avoids the costly enumeration of all pattern matchings required by previous approaches. An originality of our method is that matching information of already computed patterns is materialized as bitmaps. Exploiting bitmaps not only minimizes the memory consumption but also reduces CPU costs (Sect. 3).
- We run extensive experiments to evaluate the performance and scalability of our approach on real datasets. The experimental results show that: (a) the mined maximal homomorphic tree patterns are *larger* on the average than maximal embedded tree patterns on the same datasets, (b) our approach mines homomorphic maximal patterns up to *several orders of magnitude faster* than state-of-the-art algorithms mining embedded tree patterns when applied to a large data tree, and (c) our algorithm consumes only a *small fraction of the memory space* and *scales smoothly* when the size of the dataset increases (Sect. 4).

2   Preliminaries and Problem Definition

Trees and Inverted Lists. We consider rooted labeled trees, where each tree has a distinguished root node and a labeling function lb mapping nodes to labels. A tree is called *ordered* if it has a predefined left-to-right ordering among the children of each node. Otherwise, it is *unordered*. The *size* of a tree is defined as the number of its nodes. In this paper, unless otherwise specified, a tree pattern is a rooted, labeled, unordered tree.

For every label a in an input data tree T, we construct an inverted list L_a of the data nodes with label a ordered by their pre-order appearance in T. Figure 2(a) and (b) shows a data tree and inverted lists of its labels.

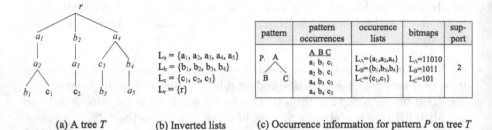

(a) A tree T (b) Inverted lists   (c) Occurrence information for pattern P on tree T

Fig. 2. A tree T, its inverted lists, and occurrence info. of pattern P on T.

Tree Morphisms. There are two types of tree patterns: patterns whose edges represent child relationships (child edges) and patterns whose edges represent descendant relationships (descendant edges). In the literature of tree pattern mining, different types of morphisms are employed to determine if a tree pattern is included in a tree.

Given a pattern P and a tree T, a *homomorphism* from P to T is a function m mapping nodes of P to nodes of T, such that: (1) for any node $x \in P$, $lb(x) = lb(m(x))$; and (2) for any edge $(x, y) \in P$, if (x, y) is a child edge, $(m(x), m(y))$ is an edge of T, while if (x, y) is a descendant edge, $m(x)$ is an ancestor of $m(y)$ in T.

Previous contributions have constrained the homomorphisms considered for tree mining in different ways. Let P be a pattern with descendant edges. An *embedding* from P to T is an injective function m mapping nodes of P to nodes of T, such that: (1) for any node $x \in P$, $lb(x) = lb(m(x))$; and (2) (x, y) is an edge in P *iff* $m(x)$ is an ancestor of $m(y)$ in T. Clearly, an embedding is also a homomorphism. Notice that, in contrast to a homomorphism, an embedding cannot map two siblings of P to nodes on the same path in T. Patterns with descendant edges mined using embeddings are called *embedded* patterns. We call patterns with descendant edges mined using homomorphisms *homomorphic* patterns. In this paper, we consider mining homomorphic patterns. The set of frequent embedded patterns on a data tree T is a subset of the set of frequent homomorphic patterns on T since embeddings are restricted homomorphisms.

Pattern Nodes Occurrence Lists. We identify an occurrence of P on T by a tuple indexed by the nodes of P whose values are the images of the corresponding nodes in P under a homomorphism of P to T. The set of occurrences of P under all possible homomorphisms of P to T is a relation OC whose schema is the set of nodes of P. If X is a node in P labeled by label a, the *occurrence list of* X on T is a sublist L_X of the inverted list L_a containing only those nodes that occur in the column for X in OC.

As an example, in Fig. 2(c), the second and third columns give the occurrence relation and the node occurrence lists, respectively, of the pattern P on the tree T of Fig. 2(a).

Support. We adopt for the support of tree patterns root frequency: the support of a pattern P on a data tree T is the number of distinct images (nodes in T) of the root of P under all homomorphisms of P to T. In other words, the *support* of P on T is the size of the occurrence list of the root of P on T.

A pattern S is *frequent* if its support is no less than a user defined threshold *minsup*. We denote by F_k the set of all frequent patterns of size k, also known as a *k-pattern*.

Constraining Patterns. When homomorphisms are considered, it is possible that an infinite number of frequent patterns of unrestricted size can be extracted from a dataset. In order to exclude this possibility, we consider and define next non-redundant patterns. We say that two patterns P_1 and P_2 are *equivalent*, if there exists a homomorphism from P_1 to P_2 and vice-versa. A node X in a pattern P is *redundant* if the subpattern obtained from P by deleting X and all its descendants is equivalent to P. For example, the rightmost node C of P_3 and the rightmost node B of P_5 in Fig. 3 are redundant. Adding redundant nodes to a pattern can generate an infinite number of frequent equivalent patterns which have the same support. These patterns are not useful as they do not provide additional information on the data tree. A pattern is *non-redundant* if it does not have redundant nodes. In Fig. 3, patterns P_3 and P_5 are redundant while the rest of the patterns are non-redundant. Non-redundant patterns correspond to minimal tree-pattern queries [1] in tree databases. Their number is finite. We discuss later how to efficiently check patterns for redundancy by identifying redundant nodes. We set forth to extract only frequent patterns which are non-redundant but in the process of finding frequent non-redundant patterns we might generate also some redundant patterns.

Maximal Patterns. In order to define maximal homomorphic frequent patterns, we introduce a specificity relation on patterns: A pattern P_1 is *more specific* than a pattern P_2 (and P_2 is *less specific* than P_1) iff there is a homomorphism from P_2 to P_1 but not from P_1 to P_2. If a pattern P_1 is more specific than a pattern P_2, we write $P_1 \prec P_2$. For instance, in Fig. 3, $P_1 \prec P_i$, $i = 2, \ldots, 7$, and $P_2 \prec P_6$. Similarly, in Fig. 1, $P_2 \prec P_1$, $P_5 \prec P_3$, $P_4 \prec P_2$ and $P_4 \prec P_5$. Note that P_4 is more specific than P_5 even though it is smaller in size than P_5.

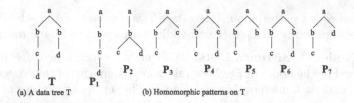

Fig. 3. A data tree and homomorphic patterns.

Clearly, $\prec$ is a strict partial order. If $P_1 \prec P_2$, P_1 conveys more information on the dataset than P_2.

A frequent pattern P is *maximal* if there is no other frequent pattern P_1, such that $P_1 \prec P$. For instance, in Fig. 1, all the patterns shown are frequent homomorphic patterns and P_4 is the only maximal pattern.

Problem Statement. Given a large tree T and a minimum support threshold *minsup*, our goal is to mine all maximal homomorphic frequent patterns from T.

3   Proposed Approach

Our approach for mining embedded tree patterns from a large tree iterates between the candidate generation phase and the support counting phase. In the first phase, we use a systematic way to generate candidate patterns that are potentially frequent. In the second phase, we develop an efficient method to compute the support of candidate patterns.

3.1   Candidate Generation

To generate candidate patterns, we adapt in this section the equivalence class-based pattern generation method proposed in [14,15] so that it can address pattern redundancy and maximality. A candidate pattern may have multiple alternative isomorphic representations. To minimize the redundant generation of the isomorphic representations of the same pattern, we employ a canonical form for tree patterns [5].

Equivalence Class-Based Pattern Generation. Let P be a pattern of size k-1. Each node of P is identified by its *depth-first position* in the tree, determined through a depth-first traversal of P, by sequentially assigning numbers to the first visit of the node. The *rightmost leaf* of P, denoted *rml*, is the node with the highest depth-first position. The *immediate prefix* of P is the sub-pattern of P obtained by deleting the *rml* from P. The *equivalence class* of P is the set of all the patterns of size k that have P as their immediate prefix. We denote the equivalence class of P as $[P]$. Any two members of $[P]$ differ only in their *rmls*. We use the notation P_x^i to denote the k-pattern formed by adding a child node labeled by x to the node with position i in P as the *rml*.

Given an equivalence class $[P]$, we obtain its successor classes by expanding patterns in $[P]$. Specifically, candidates are generated by *joining* each pattern $P_x^i \in [P]$ with any other pattern P_y^j in $[P]$, including itself, to produce the patterns of the equivalence class $[P_x^i]$. We denote the above join operation by $P_x^i \otimes P_y^j$. There are two possible outcomes for each $P_x^i \otimes P_y^j$: one is obtained by making y a sibling node of x in P_x^i, the other is obtained by making y a child node of x in P_x^i. We call patterns P_x^i and P_y^j the *left-parent* and *right-parent* of a join outcome, respectively.

As an example, in Fig. 3, patterns P_1, P_2, P_3, P_5, and P_7 are members of class $[a/b/c]$; P_4 is a join outcome of $P_3 \otimes P_7$, obtained by making the rml d of P_7 a child of the rml c of P_3.

Checking Pattern Redundancy. The pattern generation process may produce candidates which are redundant (defined in Sect. 2). We discuss below how to efficiently check pattern redundancy by identifying redundant nodes. We exploit a result of [1] which states that: a node X of a pattern P is redundant iff there exists a homomorphism h from P to itself such that $h(X) \neq X$. A brute-force method for checking if a pattern is redundant computes all the possible homomorphisms from P to itself. Unfortunately, the number of the homomorphisms can be exponential on the size of P. Therefore, we have designed an algorithm which, given two patterns P and Q, compactly represents all the homomorphisms from P to Q in polynomial time and space[1]. Our algorithm enhances the one presented in [9] which checks if there exists a homomorphism from one tree pattern to another while achieving the same time and space complexity. Its detailed description is omitted here in the interest of space.

During the candidate generation, we cannot however simply discard candidates that are redundant, since they may be needed for generating non-redundant patterns. For instance, the pattern P_5 shown in Fig. 3(b), is redundant, but it is needed (as the left operand in a join operation with P_7) to generate the non-redundant pattern P_6 shown in the same figure. Clearly, we want to avoid as much as possible generating patterns that are redundant. In order to do so, we introduce the notion of *expandable* pattern.

Definition 1 (Expandable Pattern). *A pattern P is* expandable, *if it does not have a redundant node X such that: (1) X is not on the rightmost path of P, or (2) X is on the rightmost path of P and L_X is equal to $L_{X_1} \cup \ldots \cup L_{X_k}$, where $X_1, \ldots, X_k$ are the images of node X under a homomorphism from P to itself.*

Based on Definition 1, if a pattern is non-expandable, every expansion of it is redundant. Therefore, only expandable patterns in a class are considered for expansion.

Finding Maximal Patterns. One way to compute the maximal patterns is to use a postprocessing pruning method. That is, first compute the set S of all frequent homomorphic patterns, and then do the maximality check and eliminate non-maximal patterns by checking the specificity relation on every pair

---

[1] https://web.njit.edu/~dth/HomomorphicTreePattternMining.pdf

of patterns in S. However, the time complexity of this method is $O(|S|^2)$. It is, therefore, inefficient since the size of S can be exponentially larger than the number of maximal patterns.

We have developed a better method which can reduce the number of frequent patterns that need to go through the maximality check. During the course of mining frequent patterns, the method locates a subset of frequent patterns called locally maximal patterns. A pattern P is *locally maximal* if it is frequent and there exists no frequent pattern in the class $[P]$. Clearly, a non-locally maximal pattern is not maximal. Then, in order to identify maximal patterns, we check only locally maximal patterns for maximality. Our experiments show that this improvement can dramatically reduce the number of frequent patterns checked for maximality.

3.2   Support Computation

Recall that the support of a pattern P in the input data tree T is defined as the size of the occurrence list L_R of the root R of P on T (Sect. 2). To compute L_R, a straightforward method is to first compute the relation OC which stores the set of occurrences of P under all possible homomorphisms of P to T and then "project" OC on column R to get L_R. Fortunately, we can do much better using a twig-join approach to compute L_R without enumerating all homomorphisms of P to T. Our approach for support computation is a complete departure from existing approaches.

A Holistic Twig-Join Approach. In order to compute L_X, we exploit a holistic twig-join approach (e.g., $TwigStack$ [3]), the state of the art technique for evaluating tree-pattern queries on tree data. Algorithm $TwigStack$ works in two phases. In the first phase, it computes the matches of the individual root-to-leaf paths of the pattern. In the second phase, it merge joins the path matches to compute the results for the pattern. $TwigStack$ ensures that each solution to each individual query root-to-leaf path is guaranteed to be merge-joinable with at least one solution of each of the other root-to-leaf paths in the pattern. Therefore, the algorithm can guarantee worst-case performance *linear* to the size of the data tree inverted lists (the input) and the size of the pattern matches in the data three (the output), i.e., the algorithm is optimal.

By exploiting the above property of $TwigStack$, we can compute the support of P at the first phase of $TwigStack$ when it finds data nodes participating in matches of root-to-leaf paths of P. There is no need to enumerate the occurrences of pattern P on T (i.e., to compute the occurrence relation OC).

The time complexity of the above support computation method is $O(|P| \times |T|)$, where $|P|$ and $|T|$ denote the size of pattern P and of the input data tree T, respectively. Its space complexity is the $min(|T|, |P| \times heigh(T))$. We note that, on the other hand, the problem of computing an unordered embedding from P to T is NP-complete [8]. As a consequence, a state-of-the-art unordered embedded pattern mining algorithm $Sleuth$ [14] computes pattern support in $O(|P| \times |T|^{2|P|})$ time and $O(|P| \times |T|^{|P|})$ space.

Nevertheless, the *TwigStack*-based method can still be expensive for computing the support of a large number of candidates, since it needs to scan fully the inverted lists corresponding to every candidate pattern. We present below an incremental method, which computes the support of a pattern P by leveraging the computation done at its parent patterns in the search space.

Computing Occurrence Lists Incrementally. Let P be a pattern and X be a node in P labeled by a. Using *TwigStack*, P is computed by iterating over the inverted lists corresponding to every pattern node. If there is a sublist, say L_X, of L_a such that P can be computed on T using L_X instead of L_a, we say that node X can be *computed using L_X on T*. Since L_X is non-strictly smaller than L_a, the computation cost can be reduced. Based on this idea, we propose an incremental method that uses the occurrence lists of the two parent patterns of a given pattern P to compute P.

Let pattern Q be a join outcome of $P_x^i \otimes P_y^j$. By the definition of the join operation, we can easily identify a homomorphism from each parent P_x^i and P_y^j to Q.

Proposition 1. *Let X' be a node in a parent Q' of Q and X be the image of X' under a homomorphism from Q' to Q. The occurrence list L_X of X on T, is a sublist of the occurrence list $L_{X'}$ of X' on T.*

Sublist L_X is the inverted list of data tree nodes that participate in the occurrences of Q to T. By Proposition 1, X can be computed using $L_{X'}$ instead of using the corresponding label inverted list. Further, if X is the image of nodes X_1 and X_2 defined by the homomorphisms from the left and right parent of Q, respectively, we can compute X using the *intersection*, $L_{X_1} \cap L_{X_2}$, of L_{X_1} and L_{X_2} which is the sublist of L_{X_1} and L_{X_2} comprising the nodes that appear in both L_{X_1} and L_{X_2}.

Using Proposition 1, we can compute Q using only the occurrence list sets of its parents. Thus, we only need to store with each frequent pattern its occurrence list set. Our method is space efficient since the occurrence lists can encode in linear space an exponential number of occurrences for the pattern [3]. In contrast, the state-of-the-art methods for mining embedded patterns [14,15] have to store information about all the occurrences of each given pattern in T.

Occurrence Lists as Bitmaps. The occurrence list L_X of a pattern node X labeled by a on T can be represented by a bitmap on L_a. this is a bot array of size $|L_a|$ which has a '1' bit at position i iff L_X comprises the tree node at position i of L_a. Then, the occurrence list set of a pattern is the set of bitmaps of the occurrence lists of its nodes. Figure 2(c) shows an example of bitmaps for pattern occurrence lists.

As verified by our experimental evaluation, storing the occurrence lists of multiple patterns as bitmaps results in important space savings. Bitmaps offer CPU cost saving as well by allowing the translation of pattern evaluation to bitwise operations. This bitmap technique is initially introduced and exploited in [12,13] for materializing tree-pattern views and for efficiently answering queries using materialized views.

3.3   The Tree Pattern Mining Algorithm

We present now our homomorphic tree pattern mining algorithm called *Hom-TreeMiner* (Fig. 4). The first part of the algorithm computes the sets containing all frequent 1-patterns F_1 (i.e., nodes) and 2-patterns F_2 (lines 1–2). F_1 can be easily obtained by finding inverted lists of T whose size (in terms of number of nodes) is no less than *minsup*. The total time for this step is $O(|T|)$. F_2 is computed by the following procedure: let X/Y denote a 2-pattern formed by two elements X and Y of F_1. The support of X/Y is computed via algorithm *TwigStack* on the inverted lists $L_{lb(X)}$ and $L_{lb(Y)}$ that are associated with labels $lb(X)$ and $lb(Y)$, respectively. The total time for each 2-pattern candidate is $O(|T|)$.

The main part of the computation is performed by procedure *MineHom-Patterns* which is invoked for every frequent 2-pattern (Lines 3–4). This is a recursive procedure. It tries to join every $P_x^i \in [P]$ with any other element $P_y^j \in [P]$ including P_x^i itself. Then, it computes the support of each possible join outcome, and adds them to $[P_x^i]$ if they are frequent (Lines 1–6). Once all P_y^j have been processed, it checks if P_x^i is a locally maximal pattern. If so, P_x^i is added to the maximal pattern set $\mathcal{M}$ (Line 7). Then, the new class $[P_x^i]$ is recursively explored in a depth-first manner (Line 8). The recursive process is repeated until no more frequent patterns can be generated.

Once all the locally maximal patterns have been found, the maximality check procedure described in Sect. 3.1 is run to identify maximal patterns among the locally maximal ones and the results are returned to the user (Lines 5–6).

Input: inverted lists $\mathcal{L}$ of tree T and *minsup*.
Output: all the frequent maximal patterns $\mathcal{M}$ in T.

1. $F_1 := \{\text{frequent 1-patterns}\}$;
2. $F_2 := \{\text{classes } [P]_1 \text{ of frequent 2-patterns}\}$;
3. **for** (every $[P] \in F_2$) **do**
4. $MineHomPatterns([P], \mathcal{M} = \emptyset)$;
5. run the maximality checking procedure on $\mathcal{M}$;
6. **return** $\mathcal{M}$;

Procedure $MineHomPatterns([P], \mathcal{M})$
1. **for** (each $P_x^i \in [P]$) **do**
2.   **if** (P_x^i is in canonical form and is expandable) **then**
3. $[P_x^i] := \emptyset$
4.   **for** (each $P_y^j \in [P]$) **do**
5.   **for** (each join outcome Q of $P_x^i \otimes P_y^j$) **do**
6.   add Q to $[P_x^i]$ if Q is frequent;
7.   add P_x^i to $\mathcal{M}$ if none of the members of $[P_x^i]$ is in canonical form;
8. $MineHomPatterns([P_x^i], \mathcal{M})$;

Fig. 4. Homomorphic tree pattern mining algorithm.

4   Experimental Evaluation

We implemented our algorithm *HomTreeMiner* and we conducted experiments to: (a) compare the features of the extracted (maximal) homomorphic patterns with those of (maximal) embedded patterns, and (b) study the performance of *HomTreeMiner* in terms of execution time, memory consumption and scalability.

To the best of our knowledge, there is no previous algorithm computing homomorphic patterns from data trees. Therefore, we compared the performance of our algorithm with state-of-the-art algorithms that compute embedded patterns on the same dataset.

Our implementation was coded in Java. All the experiments reported here were performed on a workstation equipped with an Intel Xeon CPU 3565 @3.20 GHz processor with 8 GB memory running JVM 1.7.0 on Windows 7 Professional. The Java virtual machine memory size was set to 4 GB.

Datasets. We have ran experiments on four real and benchmark datasets (See footnote 1). Due to space limitation, we only present results of our experimental study on one real tree dataset called $Treebank^2$ derived from computational linguistics. The dataset is deep and comprises highly recursive and irregular structures. Its statistics are shown below.

Dataset	Tot. #nodes	#labels	Max/Avg depth	#paths
Treebank	2437666	250	36/8.4	1392231

4.1   Algorithm Performance

We compare the performance of *HomTreeMiner* with two unordered embedded tree mining algorithms *Sleuth* [14] and *EmbTreeMiner* [11]. *Sleuth* was designed to mine embedded patterns from a set of small trees. In order to allow the comparison in the single large tree setting, we adapted *Sleuth* by having it return as support of a pattern the number of its root occurrences in the data tree. *EmbTreeMiner* is a newer embedded tree mining algorithm which, as *HomTreeMiner*, exploits the twig-join approach and bitmaps to compute pattern support.

To the best of our knowledge, direct mining of maximal embedded patterns has not been studied in the literature. We therefore use post-processing pruning which eliminates non-maximal patterns after computing all frequent embedded patterns. For this task, we implemented the unordered tree inclusion algorithm described in [8]. As our experiments show, the cost of this post-processing step is not significant compared to the frequent pattern mining cost.

Execution Time. We measure the total elapsed time for producing maximal frequent patterns at different support thresholds. The total time involves the time to generate candidate patterns, compute pattern support, and check maximality of frequent patterns. To allow *Sleuth*—which is slower—extract some patterns within a reasonable amount of time, we used a fraction of the Treebank dataset which consists of 35 % of the nodes of the original tree. We measured execution times over the entire Treebank dataset in the scalability experiment.

---

2 http://www.cis.upenn.edu/~treebank.

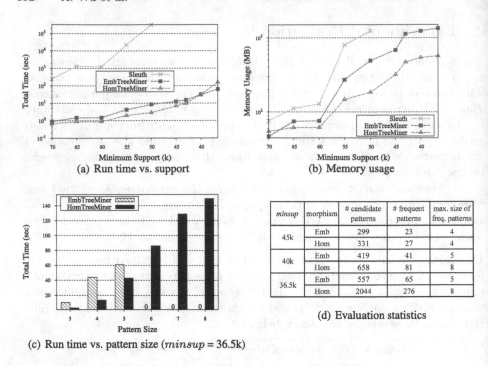

(a) Run time vs. support

(b) Memory usage

(c) Run time vs. pattern size ($minsup = 36.5$k)

$minsup$	morphism	# candidate patterns	# frequent patterns	max. size of freq. patterns
45k	Emb	299	23	4
	Hom	331	27	4
40k	Emb	419	41	5
	Hom	658	81	8
36.5k	Emb	557	65	5
	Hom	2044	276	8

(d) Evaluation statistics

Fig. 5. Performance comparison on a fraction of Treebank.

Figure 5(d) presents evaluation statistics. As one can see, the search space of a homomorphic pattern mining can be larger than that of embedded pattern mining for low support levels. *HomTreeMiner* computes 3.7 times more candidates and produces 4.25 times more frequent patterns than *EmbTreeMiner* at $minsup = 36.5$ k. Since *Treebank* contains many deep, highly recursive paths, the search space of homomorphic patterns becomes large at low support levels.

Figure 5(a) presents the total elapsed time of the three algorithms under different support thresholds. Due to prohibitively long times, we stopped testing *Sleuth* on support levels below 50 k. We can see that *HomTreeMiner* runs orders of magnitude faster than *Sleuth* especially for low support levels. The rate of increase of the running time for *HomTreeMiner* is slower than that for *Sleuth* as the support level decreases. This is expected, since *HomTreeMiner* computes the support of a homomorphic pattern in time linear to the input data size, whereas this computation is exponential for embedded pattern miners (Sect. 3.2). Furthermore, *Sleuth* has to keep track of all possible embedded occurrences of a candidate to a data tree, and to perform expensive join operations over these occurrences. *HomTreeMiner* shows similar or better performance than *EmbTreeMiner* for support levels above 40 K. The large number of candidate homomorphic patterns can negatively affect the time performance of *HomTreeMiner* at low support levels. For example, *HomTreeMiner* is 2.4 times slower than *EmbTreeMiner* in mining frequent patterns at $minsup = 36.5$ k. However, even though the number of (candidate and frequent) homomorphic patterns is always larger than

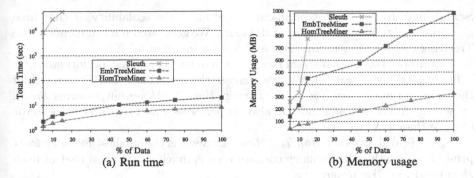

Fig. 6. Scalability comparison on Treebank with increasing size ($minsup = 5.5\%$).

Table 1. Statistics for maximal frequent patterns mined from Treebank.

Dataset	Morphism	# freq. patterns	# loc.max patterns	# max. patterns	%max. over freq. patterns	average #nodes	average height	maximum #nodes	#common max.patterns
Treebank (minsup = 45 k)	Emb	23	n/a	6	26.1	2.8	1.3	4	3
	Hom	27	10	5	16.1	2.8	1.4	4	
Treebank (minsup = 40 k)	Emb	41	n/a	9	22	3.2	1.4	5	4
	Hom	81	43	8	9.8	3.4	1.6	5	
Treebank (minsup = 36.5 k)	Emb	65	n/a	13	20	4.0	1.7	5	1
	Hom	276	90	11	3.9	5.3	2.0	7	

embedded patterns, this difference is not so pronounced in shallower datasets (See footnote 1). As a consequence, *HomTreeMiner* largely outperforms *EmbTreeMiner* in those cases both at higher and low support levels. This is due to its efficient computation of pattern support which does not require the enumeration of pattern occurrences as is the case with *EmbTreeMiner* [11].

Figure 5(c) presents the runtime *HomTreeMiner* and *EmbTreeMiner* need to compute the frequent patterns of a given size varying the pattern size. As we can see, *HomTreeMiner* is more efficient than *EmbTreeMiner* in computing frequent patterns of the same size even though the homomorphic patterns are more numerous.

Memory Usage. We measured the memory footprint of the three algorithms with varying support thresholds. The results are shown in Fig. 5(b). We can see that *HomTreeMiner* has the best memory performance. It consumes substantially less memory than both *Sleuth* and *EmbTreeMiner* in all the test cases, whereas *Sleuth* consumes the largest amount of memory. This is mainly because *Sleuth* needs to enumerate and store in memory all the pattern occurrences for candidates under consideration. In contrast, *HomTreeMiner* avoids storing pattern occurrences by storing only bitmaps of occurrence lists which are usually of insignificant size. Although *EmbTreeMiner* does not store pattern occurrences, it still has to generate pattern occurrences as intermediate results the size of which can be substantial at low support levels.

Scalability. In our final experiment, we studied the scalability of the three algorithms as we increase the input data size. We generated ten fragments of the Treebank dataset of increasing size and fixed *minsup* at 5.5 %.

Figure 6(a) shows that *HomTreeMiner* has the best time performance. It exhibits good linear scalability as we increase the input data size. The growth of the running time of *sleuth* is much sharper. *HomTreeMiner* outperforms *Sleuth* by several orders of magnitude. It also outperforms *EmbTreeMiner* by a factor of more than 2 on average.

Figure 6(b) shows that *HomTreeMiner* always has the smallest memory footprint. The growth of its memory consumption is much slower than that of both *sleuth* and *EmbTreeMiner*.

4.2   Comparison of Mined Maximal Homomorphic and Embedded Patterns

We computed different statistics on frequent and maximal frequent patterns mined by *HomTreeMiner* and *EmbTreeMiner* from Treebank varying the support; the results are summarized in Fig. 5(d) and Table 1. We can make the following observations.

First, *HomTreeMiner* is able to discover larger patterns than *EmbTreeMiner* for the same support level. As one can see in Fig. 5(d) and Table 1, the maximum size of frequent homomorphic patterns and the maximum size and average number of nodes and height of maximum frequent homomorphic patterns is never smaller than that of the embedded patterns for the same support level.

Second, the number of maximal homomorphic patterns is never larger than the number of maximal embedded patterns for the same support (Column 5 of Table 1). Further, the number of homomorphic and embedded frequent patterns is substantially reduced if only maximal patterns are selected (Column 6 of Table 1). However the effect is larger on homomorphic patterns as the number of frequent homomorphic patterns is usually larger than that of embedded patterns for the same support level (Column 3 of Table 1).

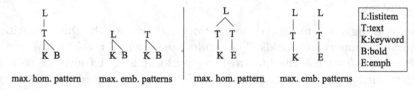

max. hom. pattern   max. emb. patterns | max. hom. pattern   max. emb. patterns

Fig. 7. Examples of maximal patterns mined from XMark at the same support level.

Third, by further looking at the mined maximal patterns we find that the embedded maximal patterns at a certain support level can be partitioned into sets which correspond one-to-one to the maximal homomorphic patterns at the same support level so that all the embedded patterns in a set are less specific than the corresponding homomorphic pattern. Figure 7 shows two pairs of embedded maximal patterns each from the same set in the partition and the corresponding maximal homomorphic pattern. The patterns are extracted from the XMark dataset[3]. Therefore, for a number of applications, maximal homomorphic patterns can offer more information in a more compact way.

5   Related Work

We now discuss, how our work relates to existing literature. The problem of mining tree patterns from a set of small trees has been studied since the last decade. Among the many proposed algorithms, only few mine unordered embedded patterns [10,14].

TreeFinder [10] is the first unordered embedded tree pattern mining algorithm. It is a two-step algorithm. In the first step, it clusters the input trees by the co-occurrence of labels pairs. In the second step, it computes maximal trees that are common to all the trees of each cluster. A known limitation of *TreeFinder* is that it tends to miss many frequent patterns and is computationally expensive.

Sleuth [14] extends the ordered embedded pattern mining algorithm *TreeMiner* [15]. Unlike *TreeFinder*, *Sleuth* uses the equivalence class pattern expansion method to generate candidates. To avoid repeated invocation of tree inclusion checking, *Sleuth* maintains a list of embedded occurrences with each pattern. It defines also a quadratic join operation over pattern occurrence lists to compute support for candidates. The join operation becomes inefficient when the size of pattern occurrence lists is large. Our approach relies on an incremental stack-based approach that exploits bitmaps to efficiently compute the support in time linear to the size of input data.

The work on mining tree patterns in a single large tree/graph setting has so far been very limited. The only known papers are [6,7] which focus on mining tree patterns with only child edges from a single graph, and [11] which leverages homomorphisms to mine embedded tree patterns from a single tree. To the best of our knowledge, our work is the first one for mining homomorphic tree patterns with descendant edges from a single large tree.

6   Conclusion

In this paper we have addressed the problem of mining maximal frequent homomorphic tree patterns from a single large tree. We have provided a novel definition of maximal homomorphic patterns which takes into account homomorphisms, pattern specificity and the single tree setting. We have designed an efficient algorithm that discovers all frequent non-redundant maximal homomorphic tree patterns. Our approach employs an incremental stack-based frequency computation method that avoids the costly enumeration of all pattern occurrences required by previous approaches. An originality of our method is that matching information of already computed patterns is materialized as bitmaps, which greatly reduces both memory consumption and computation costs.

---

[3] http://monetdb.cwi.nl/xml/.

We have conducted extensive experiments to compare our approach with tree mining algorithms that mine embedded patterns when applied to a large data tree. Our results show that maximal homomorphic patterns are fewer and larger than maximal embedded tree patterns. Further, our algorithm is as fast as the state-of-the art algorithm mining embedded trees from a single tree while outperforming it in terms of memory consumption and scalability.

We are currently working on incorporating user-specified constraints to the proposed approach to enable constraint-based homomorphic pattern mining.

References

1. Amer-Yahia, S., Cho, S., Lakshmanan, L.V.S., Srivastava, D.: Minimization of tree pattern queries. In: SIGMOD Conference (2001)
2. Asai, T., Abe, K., Kawasoe, S., Arimura, H., Sakamoto, H., Arikawa, S.: Efficient substructure discovery from large semi-structured data. In: SDM (2002)
3. Bruno, N., Koudas, N., Srivastava, D.: Holistic twig joins: optimal XML pattern matching. In: SIGMOD (2002)
4. Chi, Y., Xia, Y., Yang, Y., Muntz, R.R.: Mining closed and maximal frequent subtrees from databases of labeled rooted trees. IEEE Trans. Knowl. Data Eng. **17**(2), 190–202 (2005)
5. Chi, Y., Yang, Y., Muntz, R.R.: Canonical forms for labelled trees and their applications in frequent subtree mining. Knowl. Inf. Syst. **8**(2), 203–234 (2005)
6. Dries, A., Nijssen, S.: Mining patterns in networks using homomorphism. In: SDM (2012)
7. Goethals, B., Hoekx, E., den Bussche, J.V.: Mining tree queries in a graph. In: KDD (2005)
8. Kilpeläinen, P., Mannila, H.: Ordered and unordered tree inclusion. SIAM J. Comput. **24**(2), 340–356 (1995)
9. Miklau, G., Suciu, D.: Containment and equivalence for a fragment of xpath. J. ACM **51**(1), 2–45 (2004)
10. Termier, A., Rousset, M.-C., Sebag, M.: TreeFinder: a first step towards XML data mining. In: ICDM (2002)
11. Wu, X., Theodoratos, D.: Leveraging homomorphisms and bitmaps to enable the mining of embedded patterns from large data trees. In: Renz, M., Shahabi, C., Zhou, X., Cheema, M.A. (eds.) DASFAA 2015. LNCS, vol. 9049, pp. 3–20. Springer, Heidelberg (2015)
12. Wu, X., Theodoratos, D., Wang, W.H.: Answering XML queries using materialized views revisited. In: CIKM (2009)
13. Wu, X., Theodoratos, D., Wang, W.H., Sellis, T.: Optimizing XML queries: bitmapped materialized views vs. indexes. Inf. Syst. **38**(6), 863–884 (2013)
14. Zaki, M.J.: Efficiently mining frequent embedded unordered trees. Fundam. Inform. **66**(1–2), 33–52 (2005)
15. Zaki, M.J.: Efficiently mining frequent trees in a forest: algorithms and applications. IEEE Trans. Knowl. Data Eng. **17**(8), 1021–1035 (2005)
16. Zhu, F., Qu, Q., Lo, D., Yan, X., Han, J., Yu, P.S.: Mining top-k large structural patterns in a massive network. PVLDB **4**(11), 807–818 (2011)
17. Zhu, F., Yan, X., Han, J., Yu, P.S., Cheng, H.: Mining colossal frequent patterns by core pattern fusion. In: ICDE, pp. 706–715 (2007)

CITPM: A Cluster-Based Iterative Topical Phrase Mining Framework

Bing Li[1], Bin Wang[1], Rui Zhou[2], Xiaochun Yang[1(✉)], and Chengfei Liu[3]

[1] School of Computer Science and Engineering, Northeastern University,
Liaoning 110819, China
libing@stumail.neu.edu.cn, {binwang,yangxc}@mail.neu.edu.cn
[2] Centre for Applied Informatics, College of Engineering and Science,
Victoria University, Melbourne, VIC 3011, Australia
rui.zhou@vu.edu.au
[3] Department of Computer Science and Software Engineering,
Swinburne University of Technology, Melbourne, VIC 3122, Australia
cliu@swin.edu.au

Abstract. A phrase is a natural, meaningful, essential semantic unit. In topic modeling, visualizing phrases for individual topics is an effective way to explore and understand unstructured text corpora. Unfortunately, existing approaches predominately rely on the general distributional features between topics and phrases on an entire corpus, while ignore the impact of domain-level topical distribution. This often leads to losing domain-specific terminologies, and as a consequence, weakens the coherence of topical phrases. In this paper, we present a novel framework CITPM for topical phrase mining. Our framework views a corpus as a mixture of clusters (domains), and each cluster is characterized by documents sharing similar topical distributions. The CITPM framework iteratively performs phrase mining, topical inferring and cluster updating until a satisfactory final result is obtained. The empirical verification demonstrates our framework outperforms state-of-the-art works in both aspects of interpretability and efficiency.

Keywords: Topical phrase · Phrase mining · Document clustering

1 Introduction

Topical phrase mining has been extensively studied for visualizing phrases of individual topics and is of high value to enhance the power and efficiency to facilitate human to explore and understand large amounts of unstructured text data like text corpora. An example is that topical phrases could help a reader quickly find terminologies of different fields from a collection of papers. In another scenario, if researchers could find a research field's phrases appearing with high frequencies in related proceedings in different years, they will be able to have an insight into the academic trend of that research field.

S.B. Navathe et al. (Eds.): DASFAA 2016, Part I, LNCS 9642, pp. 197–213, 2016.
DOI: 10.1007/978-3-319-32025-0_13

Existing approaches can be broadly classified into three categories based on their strategies of handling phrase mining and topical modeling. The first one tries to infer phrases and topics simultaneously, such as Bigram topic model [1], topical N-gram [2], and PDLDA [3]. These methods mostly rely on complex generative models and often suffer from high complexity, and overall demonstrate poor scalability outside small datasets [4]. Another category has been proposed in the hope of overcoming shortcoming of the first category. These approaches perform a post-processing step after inference utilizing Latent Dirichlet Allocation (LDA) model. TurboTopics [5] and KERT [6] belong to this category. However, due to not being regarded as a whole, tokens in one phrase may be assigned to different topics during inferring, which may lead to low recall of phrases. For the third category, a more recent work ToPMine [4] offers a different strategy, first phrase mining then topic modeling. It firstly performs phrase mining then takes *bag-of-phrases* into PhraseLDA to infer topical phrases on the condition that the phrases should share the same latent topic.

Unfortunately, one significant drawback of using any existing approach is that they predominately focus on distributional features between topic and phrases as well as phrase generation method in mining topical phrase, while ignore how domain-level topical distribution impacts on topical phrases. Usually, a corpus is a mixture of different domains, each domain contains its own terminology and the domain can be characterized by documents sharing similar topical distributions. It means that a group of documents which have similar topical distributions can be regarded as belonging to the same domain. There are domain-specific terminologies which show high significance within one domain but low significance in another. For example, a phrase in data mining field `support vector` can be easily mined due to `support` and `vector` showing high significance of co-occurrence in the documents of data mining domain. However, it may be hard for domain-specific phrases to show high significance in the entire corpus where co-occurrence is diluted by the other domains. For example, `support vector` is hard to be regarded as a phrase when surrounded with documents containing `bit vector` in database field and `orthogonal vector` in math field.

In order to effectively mine topical phrases to improve the interpretability of topical phrases even further in both topical phrase mining and topic model inferring by automatically clustering documents according to domain-level topical distributions, we propose a cluster-based iterative topical phrase mining (CITPM) framework. CITPM starts with taking the whole corpus as a cluster, then iteratively update clusters by performing efficient and effective phrase mining and topic inferring until clusters do not change.

The advantages of CITPM framework that differentiate itself from the previous approaches are that: It has high ability to discover domain-specific phrases to improve the accuracy of topical phrase mining. The reason why this is important is that even though some domain-specific phrases are not significant in the entire corpus that fall into oblivion, these phrases can still be mined in our framework, and owing to high quality phrases, the accuracy of topical inferring could be also improved. Moreover, CITPM has higher efficiency compared to the state-of-the-art phrase mining method, TopMine.

The main contributions of this paper are as follows:

- To the best of our knowledge, CITPM is the first effort of applying clustering into topical phrase mining to facilitate accurate topical phrase mining.
- We propose an efficient and accurate frequency counting algorithm and phraseness checking algorithms in phrase mining stage. The frequency counting algorithm has $O(n)$ time complexity in general. The phraseness checking algorithm adopts a pure statistics method which computes a threshold with theoretical guarantee by a given statistical significant level.
- We propose a novel density peaks based document clustering method which is in accord with the underlying hyper distribution of document topics.

The rest paper is organized as follows: Sect. 2 introduces the problem definition. Section 3 discusses every part of the CITPM framework in detail. Section 4 reports the result of empirical verification. Section 5 concludes the paper.

2   Problem Definition

This section introduces some definitions and notations:

Definition 1. *A phrase Pr can be formally represented as a consecutive sequence of tokens from jth position within cluster C_i: $Pr = (w_j^i, ..., w_{j+l}^i), l \geq 0$*

Example 1. [Knowledge Expansion] over [Probabilistic Knowledge Bases]; [Frequent Itemset Mining] from Bursty [Data Streams].

As shown in Example 1, a "good" phrase should be natural and meaningful. In this paper, we use the following three criteria for judging the quality of a phrase:

Frequency: This criterion is based on the observation that a phrase not frequent within a cluster is likely to be not important in the cluster [4]. In Example 1, phrases such as Data Streams and Frequent Itemset Mining tend to be "good" phrases because they have been widely used in data mining domain and have high frequencies. This is also important for statics analysis because lower frequency shows less statical meaning.

Phraseness [7]: If a group of adjacent phrases co-occur more significantly than expected under a given statistical significant level, then these phrases should be merged into a longer phrase. In Example 1, Knowledge and Bases should be regarded as a longer phrase Knowledge Bases because they co-occur frequently.

Completeness: We do not allow a phrase to contain subset phrases. A longer phrase can express a certain semantic more precisely if it also satisfies Frequency and Phraseness criteria. For example, Frequent Itemset Mining is more precise than Frequent Itemset or Itemset Mining.

In this paper, the topic is defined to be a distribution over a fixed vocabulary, and based on this definition, we define topical documents cluster as follows:

Definition 2. *A topical document cluster C_k is a collection of documents $C_k = \{d_0^k, ..., d_{s^k}^k\}$ where all documents have similar document-topic distribution.*

In topic modeling, a document can be represented as a distribution over all topics. If two documents share similar topic distributions, means that they have similar semantic structures. Therefore, we tend to believe that they belong to the same domain. We will introduce how to measure this similarity in Sect. 3.3.

3   CITPM Framework

Figure 1 shows the CITPM framework which is an iterative process and consists of four major stages: preprocessing, phrase mining, topic modeling, and clustering. The first stage preprocessing includes dropping stop-words and stemming. This trivial step can be easily implemented by existing tools [8,9], and therefore we will not expand in this paper. The second phrase mining stage firstly counts all possible phrases' frequency, and finally based on phrases' frequency to decide whether some adjacent phrases should be merged into longer phrases under the guidance by performing statistical independence tests with respect to a given significance level. In stage 3, after performing phrase mining, the presentation of a document can be transformed from a multiset of its words (*bag-of-words*) into a multiset of its phrases (*bag-of-phrases*) and taken as the input of topic modeling. In this paper, we adopts PhraseLDA [4] as our topic model. PhraseLDA is a topic model with the assumption that tokens in the same phrase should share the same latent topic, which is suitable for phrase-centered topic modeling. In the fourth stage, we cluster the documents that share similar document-topic distributions based on a novel density peaks based k-means clustering (DBPK)

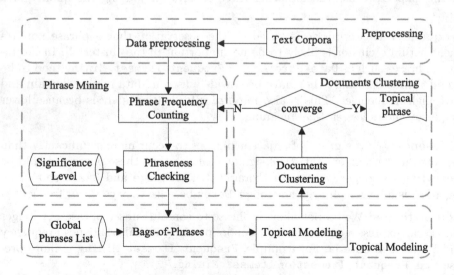

Fig. 1. CITPM framework

method, and take newly clustered documents as the input of phrase mining. CITPM framework will iteratively perform the last three steps until converge.

We next elaborate on the four stages in the following organization: Sect. 3.1 introduces our phrase mining method. Section 3.2 describes the PhraseLDA adopted in our paper. Section 3.3 proposes document clustering method DPBK.

3.1   Phrase Mining

Previous approaches for mining phrases primarily relied on applying a heuristic "importance" ranking or "significance" score to compute the confidence of key phrases [4,10]. However, they are not pure statistics-based and have no theoretical statistical guarantee, and thus often produce low quality phrases. Other approaches adopt external knowledge base or NLP constraints to filter phrases [10,11]. Due to not purely data-driven and requiring specific language knowledge, these methods suffer from low portability that cannot be directly applied to another language. In this paper, we propose an effective phrase mining algorithm by introducing an independence test based method, which is pure statistics-based and has good statistical guarantee.

Our phrase mining algorithm can be divided into two steps: firstly, counting the frequency of each possible phrase in the cluster; secondly, iteratively trying to merge two adjacent phrases to get a longer phrase in a bottom-up manner. We will discuss the two stages in detail in the next two subsections.

Phrase Frequency Counting: A straightforward method that follows from Definition 1 is to enumerate all possible phrases over each document in a cluster C from length 1 to the length of document N_d, and counts the frequency for each possible phrase using a hash counter [4]. The number of all possible phrases in a document is $1 + 2 + ... + N_d = \frac{N_d^2 + N_d}{2}$, and the number of all possible phrases in C is $\sum_{d \in C} \frac{N_d^2 + N_d}{2}$. Although it has an $O(N^2)$ time complexity, since the number of tokens in a document d can be very large, the straightforward method is prohibitively expensive. To address this issue, we propose an one-pass method which draws upon (1) *document segmentation* along with (2) a novel data structure *TrieCounter*.

Document segmentation uses separators of natural language (e.g. comma, period, etc.) to segment document into small partitions. It is an effective property that cannot cross over segments when searching for a potential phrase.

TrieCounter is an effective data structure for reducing the computational complexity for frequency counting as well as retrieving. It has better performance than hash counter due to utilizing a trie structure. An insert operation or retrieval operation on a trie structure has a fixed time cost $O(|Pr|)$ ($|Pr|$ is the length of the phrase) in both worst case and best case, while a hash structure spends $O(N)$ time in the worst case (N is the total length of the cluster and $N \gg |Pr|$), although $O(1)$ time in the best case. For short strings like phrasese (the average length is less than 30), several experimental studies have demonstrated that trie is much faster than hash [12,13]. To be specific, TrieCounter is a trie-structure in which every node stores a token (label) and the frequency

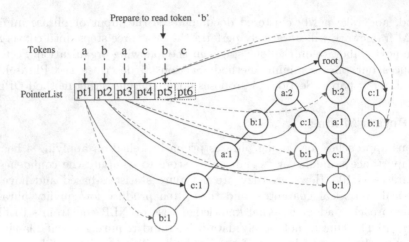

Fig. 2. TrieCounter

of a phrase whose tokens are derived from the path from the root token to the current token (right part of Fig. 2), each node contains a hash index that directs to the child nodes for the sake of the efficiency of retrieval. During its building, TrieCounter dynamically maintains a pointer list where each pointer points to a token of an input sequence and is initialized as pointing to root. Figure 2 shows the process of inserting token sequence a b a c b c to a TrieCounter, the black lines and the blue lines describe the change of pointers in the pointer list before and after reading in token b. With regard to a newly read token b, we firstly add a new pointer (pt5) to the pointer list, then update the whole pointer list as follows: for each pointer, search the child nodes of current pointed node, if the child node is labelled with token b then update pointer to this child node and increase its frequency by 1, otherwise create a new node with label b and frequency 1.

Algorithm 1 shows our phrase frequency counting algorithm. It takes a documents cluster C and a frequency threshold ft as input. Then initializes a Tire-Counter (lines 1 – 2) and updates each node's frequency on each partition of segmented documents (lines 3 – 16) by the process shown in Fig. 2.

Phraseness Checking: One observation is that if two adjacent phrases have significant co-occurrence frequency, they might constitute one longer phrase in a high probability. To measure the co-occurrence of two given phrases Pr_i and Pr_j in a pure statistics manner, our phrase mining method applies a phrase independence test. We firstly consider a null hypothesis H_0 that phrase Pr_i and phrase Pr_j are independent. Under the hypothesis H_0, we have $p(Pr_i \wedge Pr_j) = p(Pr_i) * p(Pr_j)$, therefore, the expected frequency of the co-occurrence can be obtained by $N * p(Pr_i) * p(Pr_j)$, where N is the number of tokens in cluster C. If H_0 is valid, the observation frequency must be similar to the expected frequency. Hence, we constructed a chi-squared test statistic CS in Eq. (1):

Algorithm 1. PHRASE FREQUENCY COUNTING

Input: Cluster C, Frequency threshold ft
Output: Phrase frequency stored in a TrieCounter
1   $TrieCounter \leftarrow \emptyset$;
2   $PointerList \leftarrow \emptyset$;
3   **foreach** $document \; d_i \in C$ **do**
4   Segment d_i into partitions Q;
5   **foreach** $q_j \in Q$ **do**
6   **for** $m \leftarrow 0; m < |q_j|; m + +$ **do**
7   Add a new pointer $pt \leftarrow TrieCounter.root$ to $PointerList$;
8   **foreach** $pt_k \in$ PointerList **do**
9   **if** *Find subnode* $node_t$ *labeled with* $q_j[m]$ *of* pt_k **then**
10            $node_t.counter$ increase by 1;
11            $pt_k \leftarrow node_t$;
12   **else**
13   Create $newNode \leftarrow (q_j[m], 1)$;
14   Add $newNode$ to $TrieCounter$;
15            $pt_k \leftarrow newNode$;
16       $PointerList \leftarrow \emptyset$;

17 Delete the nodes in $TrieCounter$ whose frequency is less than ft;
18 **return** $TrieCounter$;

$$CS = \sum_{k \in categories} \frac{(O_k - E_k)^2}{E_k} \tag{1}$$

where O_k and E_k denote the observed frequency and expected frequency, respectively, of category k within cluster C. There are four categories in our method (Table 1). In Table 1, by $F(Pr_i \wedge Pr_j)$ one refers to the observed co-occurrence frequency of Pr_i and Pr_j, $F(Pr_i)$ and $F(Pr_j)$ denote the observed frequency of Pr_i and Pr_j, respectively. One important advantage is that, retrieving an observed frequency of any phrase in the former stage built TireCounter structure costs only $O(|Pr|)$ time, where $|Pr|$ is the length of the retrieval phrase.

From Pearson's theorem [14], the chi-squared test statistic is drawn from a χ^2 distribution with certain degrees of freedom df (in our method $df = 1$). Hence, we have $CS \xrightarrow{L} \eta \sim \chi^2(1)$. Given a significance level α, a probability threshold below which the null hypothesis will be rejected (commonly selected as $\alpha = 0.05$),

$$\alpha = P\{reject \; H_0 | H_0 \; is \; valid\} = P\{CS > \chi^2_{1-\alpha}(1)\} \tag{2}$$

we can compute the corresponding critical region $\mathcal{W}$ (the region of the possible values of CS under the distribution of χ^2 statistic for which H_0 is rejected) based on Eq. (3)

$$\mathcal{W} = \{CS | \; CS > \chi^2_{1-\alpha}(1)\} \tag{3}$$

Algorithm 2. PHRASNESS CHECKING

Input: $TrieCounter$, Significance-level α
Output: Each document's $bag\text{-}of\text{-}phrases$ form
1 Compute critical region $\mathcal{W}$ by α;
2 **foreach** $document\ d_i \in C$ **do**
3 $\quad$ Segment d_i into partitions Q;
4 $\quad$ **foreach** $q_j \in Q$ **do**
5 $\quad\quad$ **while** $True$ **do**
6 $\quad\quad\quad$ $MaxHeap \leftarrow$ Compute CS for all contiguous phrase pairs by
$\quad\quad\quad\quad$ statical information stored in $TrieCounter$;
7 $\quad\quad\quad$ $best \leftarrow MaxHeap.root$;
8 $\quad\quad\quad$ **if** $CS_{best} \notin \mathcal{W}$ **then**
9 $\quad\quad\quad\quad$ break;
10 $\quad\quad\quad$ **else**
11 $\quad\quad\quad\quad$ Replace original phrases with the $best$;
12 $\quad\quad\quad\quad$ Insert new contiguous phrase pairs which are adjcented to the
$\quad\quad\quad\quad\quad$ $best$ to $MaxHeap$;
13 $\quad\quad\quad\quad$ Do heap adjustment;

14 **return** $document's$ bag-of-phrases $form$

Table 1. Contigency table of phrase Pr_i and phrase Pr_j

	Pr_j	$\neg\ Pr_j$
Pr_i	O_1: $F(Pr_i \wedge Pr_j)$	O_2: $F(Pr_i) - F(Pr_i \wedge Pr_j)$
	E_1: $F(Pr_i) * F(Pr_j)/N$	E_2: $F(Pr_i) * (N - F(Pr_j))/N$
$\neg\ Pr_i$	O_3: $F(Pr_j) - F(Pr_i \wedge Pr_j)$	O_4: $N - F(Pr_i) - F(Pr_j) + F(Pr_i \wedge Pr_j)$
	E_3: $(N - F(Pr_i)) * F(Pr_j)/N$	E_4: $(N - F(Pr_i)) * (N - F(Pr_j))/N$

The algorithm for our phraseness checking can be summarized in Algorithm 2. As seen in Algorithm 2, given a significance level α, our algorithm firstly computes the corresponding critical region (line 1), then iteratively merges the adjacent phrases guided by critical region in a bottom-up manner (lines 5 – 12) until the best CS value is not in the critical region.

3.2 Topic Modeling

In the topic modeling stage, we adopt PhraseLDA [4] as our topic model. PhraseLDA is a variant of LDA model. It builds on the assumption of $bag\text{-}of\text{-}phrases$, and use an undirected clique to model the stronger correlation of tokens in the same phrase. To be specific, the tokens of the g-th phrase of the d-th document form a clique $C_{d,g}$. PhraseLDA assigns the same latent topic to tokens in the same clique. The joint distribution of PhraseLDA can be expressed as:

$$P(Z, W, \Phi, \Theta) = \frac{1}{C} P_{LDA}(Z, W, \Phi, \Theta) \prod_{d,g} f(C_{d,g}) \tag{4}$$

where C is a normalization coefficient, Z denotes latent topics, W is tokens, Φ is multinomial distributions over words, Θ is multinomial distributions over topic. P_{LDA} denotes the joint distribution of classic LDA, $f(C_{d,g})$ is a function to impose variables in a clique takes the same latent topic, it can be computed as Eq. (5)

$$f(C_{d,g}) \begin{cases} 1 & \text{if } z_{d,g,1} = z_{d,g,2} = \cdots = z_{d,g,n} \\ 0 & \text{otherwise} \end{cases} \tag{5}$$

where $z_{d,g,i}$ denotes the assigned topic of i-th token in clique $C_{d,g}$. For the inference, PhraseLDA uses a collapsed Gibbs sampling algorithm to sample latent assignment variables Z from its posterior. The introduction of PhraseLDA is very brief, because it is simply borrowed here. Readers can refer to the work [4] for more details.

3.3   Document Clustering

Topic modeling generates topical phrase as well as distribution matrix between documents and topics (the argument Θ in PhraseLDA model). Each $\theta_i \in \Theta$ is a k-dimensional vector, which represents the distribution of i-th document d_i on k topics.

Traditional approaches for topic document clustering mostly based on the Vector Space Model (VSM) and simply took θ as a space vector and use Euclidean distance, Jaccard similarity, or Cosine similarity [15] to measure the proximity between two documents and tried to minimize the within-class scatter based on k-means clustering [16]. However, there are two drawbacks in the previous approaches: First, distance metrics such as Euclidian and Cosine only reflect the spatial relations between vectors rather than the distributional relations, as a matter of fact, θ itself is a distribution, so it may not be proper to use space distance metrics like Eucilidian or Cosine; Second, the quality of k-means clustering highly depends on how to select the initial centroids. In order to tackle the above issues, we propose a novel document clustering algorithm: density peaks based k-means (DPBK) clustering, an improved k-means which takes Jensen-Shannon divergence as proximity function and selects density peaks as initial centroids.

Formally, given a set of document-topic distribution $\Theta = \{\theta_1, \theta_2, ..., \theta_n\}$, where each distribution is a k-dimensional vector that denotes the distribution of i-th document d_i on k topics. The DPBK clustering aims to partition documents into M ($M < n$) clusters $C = \{C_1, C_2, ..., C_M\}$, and to minimize the total divergence:

$$TotalDivergence = \sum_{i=1}^{M} \sum_{\theta_j \in C_i} JSD(\theta_j, c_i) \tag{6}$$

where c_j is the arithmetic average controid of cluster C_i, and JSD is Jensen-Shannon divergence defined in Eq. (7), where $KL(\theta_i, \theta_j)$ is Kullback-Leibler Divergence [17] between θ_i and θ_j:

$$JSD(\theta_i, \theta_j) = \frac{1}{2}KL(\theta_i, \theta_j) + \frac{1}{2}KL(\theta_j, \theta_i) \tag{7}$$

Choosing the proper initial centroids is the key step of k-means procedure, randomly selected initial centroids often produce poor result quality [18]. Traditional approaches often ignore an important fact that Θ is drawn from a Dirichlet distribution with hyper-parameter $\boldsymbol{\alpha}$, namely, $\Theta \sim Dir(\boldsymbol{\alpha})$. As the example in Fig. 3 suggests, a Dirichlet distribution often contains some density peaks, which arise when some components α_i of argument $\boldsymbol{\alpha}$ have larger $|\alpha_i - 1|$ values. The underlying meaning of these density peaks is that the θs that near the peaks appear significantly. The θs near the peak can be regarded as a cluster and the peak can be taken as the initial centroid. Thus, we propose a novel initial centroid selection approach based on the idea that (1) cluster centroids are characterized by higher densities; (2) centroids are expected to have relatively large distances between each other [19].

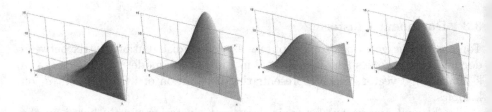

Fig. 3. Four Dirichlet distributions probability density when topic number $k = 3$ for $\alpha = \{(6, 2, 2), (3, 7, 5), (2, 3, 4), (6, 2, 6)\}$. (https://en.wikipedia.org/wiki/Dirichlet_distribution)

Based on the above idea, our approach computes each θ_i's local density ρ_i and distance σ_i with the following two equations Eqs. (8) and (10):

$$\rho_i = \sum_j \varphi(JSD(\theta_i, \theta_j) - r) \tag{8}$$

ρ_i is the local density where $\varphi(x) = 1$ if $x < 0$ and $\varphi(x) = 0$ otherwise, and r is a cut-off distance. In this paper, we define r as:

$$r = \frac{1}{\mu^{|\max_{i \in k}(\alpha_i) - 1|}} \tag{9}$$

where $\mu > 1$ is a constant coefficient. The reason for computing r by this equation is that finding an abrupt density peak needs a small r, while it needs a larger r while find a flat density peak requires a large r.

$$\sigma_i = \min_{j:\rho_j > \rho i} JSD(\theta_i, \theta_j) \tag{10}$$

σ_i is measured by the minimum distance between θ_i and any other θ_j with higher density than θ_i. Based on the two values, the initial centroids are selected as those θs whose σ_i and ρ_i values are anomalously large. It is quantitatively defined as Eq. (11):

$$Score_i = \sigma_i \times \rho_i \qquad (11)$$

Algorithm 3 presents our DPBK clustering algorithm. The algorithm starts with computing the score of every θ by Eq. (11) and selects top-M θs as the initial centroids (lines 1 – 3), then iteratively assigns each document to its closest centroid, and updates the centroids of each cluster until converges (lines 4 – 8).

Algorithm 3. DPBK CLUSTERING

   Input: Document-topic distribution Θ, M
   Output: Clusters C
1 **foreach** $\theta_j \in C_i$ **do**
2   Compute $Score_j$;
3 Select M θs with highest score as initial centroids;
4 **repeat**
5   **foreach** $\theta_j \in C_i$ **do**
6   Assign each θ_i to the nearest centroids;
7   Update the centroids of each cluster by computing arithmetic average centroid;
8 **until** *converge*;
9 **return** C;

There are two conditions to terminate the CITPM algorithm: the number of those documents whose cluster assignments of current iterative round differ from previous round less than a certain threshold, or no new phrase have been mined in current round.

4 Experimental Evaluation

In this section, we firstly introduce the datasets and experimental settings, then evaluate the performance of our CITPM framework by three tasks: (1) demonstrating interpretability by an expert evaluation and a user study *phrase intruder*, (2) showing CITPM's efficiency by comparing it with the existing method, ToPMine, and (3) showing a case study.

Datasets: Table 2 shows the statistics of the datasets we used in our experiments. Four common real datasets are used: (1) **5Conf**[1] has a set of paper titles of conferences related to the areas of artificial intelligence, databases, data mining, information retrieval, machine learning, and natural language processing; (2) **APNews**[2] contains 106K TREC AP news articles (1989); (3) **Titles**[3] is a

---

[1] http://web.engr.illinois.edu/elkishk2/.
[2] http://www.ap.org/.
[3] http://dblp.uni-trier.de/db/.

Table 2. Statics of four datasets

Datasets	5Conf	APNews	Titles	Abstracts
#Documents	44 K	106 K	1555 K	529 K
#Vocabulary	5 K	170 K	96 K	135 K
#Space	2.8 M	229 M	182 M	479 M

full collection of paper titles that extracted from DBLP dataset; and (4) **Abstrtas**[4] contains 529K computer science paper abstracts from DBLP dataset;

Experimental Settings: According to the minimal sample principle, frequency threshold is supposed to ensure $ft \gg v+1$, where v is the number of explanatory variables. Empirically, to guarantee the effectiveness of test statistics, we should have $ft \gg 30$ or $ft \gg 3 * (v + 1)$.

For the setting of M, there are several existing methods to justify the number of clusters M such as the elbow method [20], Akaike information criterion(AIC), Bayesian information criterion(BIC), "jump" method [21]. These methods belong to the field of cluster analysis, and this problem does not fall into the scope of this paper, so we will not expand here. In this paper, we set M as desired clustering of the user.

In our experiments, we set the frequency threshold $ft = \{5, 5, 10, 50\}$ and the number of clusters $M = \{10, 10, 50, 100\}$ for 5Conf, AP news, Titles, and Abstracts, respectively. Let significance level $\alpha = 0.05$ for all datasets, in topic model stage, we set the number of topics as 20 and chose Gibbs sampling iterations as 50 as the arguments for PhraseLDA.

All the algorithms were implemented using Java SE Development Kit 8. The experiments were run on a PC with an Intel Xeon 3.3 GHz 6-Cores CPU X5680 and 24 GB memory with a 1TB disk, running Ubuntu (Linux) operating system.

4.1   Interpretability

To demonstrate CITPM is of high interpretability, we conducted a domain expert evaluation along with a *phrase intruder* test.

An expert evaluation was conducted on topical coherence and phrases quality. For each dataset, we asked five experts (year-2-above Ph.D. candidates of computer science) to rate on a scale of 1 to 10 based on the following two criteria: (1) Topical coherence. The phrases listed in a topic whether in accord with the semantics of this topic. (2) Phrase quality. A phrase is natural, meaningful, complete or not. For convenience of intuitively showing our result, we transformed experts' ratings to standard score (z-score) by $zscore = \frac{\mu}{\sigma}$, where μ is the mean of the ratings and σ is the standard deviation. From Fig. 4 we can see that, in topical coherence evalution, CITPM outperforms ToPMine in all datasets except Titles, and in phrase quality evalution, CITPM is better than ToPMine in all datasets.

---

[4] http://dblp.uni-trier.de/db/.

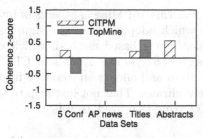

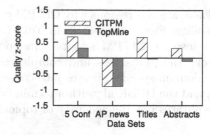

(a) Topical coherence evaluation.   (b) Phrase quality evaluation.

Fig. 4. Expert evaluation of topical coherence and phrase quality.

We also conducted a *Phrase intruder* test. *Phrase intruder* [22] is a commonly used human evaluation criteria for judging the quality of topic separation. A *phrase intruder* test contains a set of questions, and each question contains four phrases which belong to the same topic and one randomly inserted *intruder* phrase which belongs to a different topic. It requires participants to select the right *intruder* or to indicate that they are unable to make a choice.

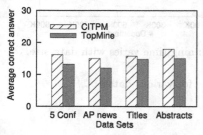

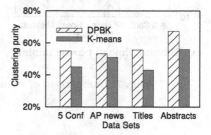

Fig. 5. Phrase intruder tests for various data sets.

Fig. 6. Clustering purity.

We selected 5 knowledgeable participants and set a *phrase intruder* task containing 30 questions. For each question, we randomly selected a topic, sampled 4 phrases from this topic, randomly inserted an *intruder*, and asked the participants to answer it. As seen in Fig. 5, CITPM has a better performance in this task for all datasets.

CITPM not only achieved better performance in the *phrase intruder* test but also got higher scores in the expert evaluation on most datasets compared with ToPMine. The results demonstrate CITPM has a better interpretability.

Figure 6 shows the clustering purity of our document clustering algorithm DPBK vs. k-means. Using DPBK, the initial centroids are selected by a density peaks based method with Jensen Shannon divergence as distance metric, which is better than k-means in accord with the features of specific document-topic distribution.

There are two reasons why CITPM outperforms ToPMine in aspects of interpretability. First, different from ToPMine which adopts a heuristic significant score, instead, CITPM adopts a independence test based method, which can improve the phraseness and completeness. Second, unlike ToPMine, CITPM takes domain-specific phrases into consideration and adopts an iterative framework and the DPBK algorithm to mine these phrases. That not just help to find more phrases but the accuracy of topical inferring improved as well.

4.2   Efficiency

In this section, we examined the efficiency of our method against the state-of-the-art method ToPMine. Figure 7(a) shows the running time of the phrase mining stage in different datasets compared with ToPMine.

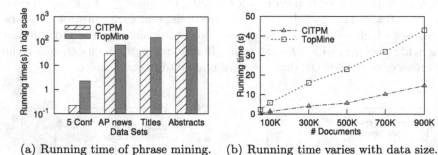

(a) Running time of phrase mining.   (b) Running time varies with data size.

Fig. 7. Comparison of running time for various datasets.

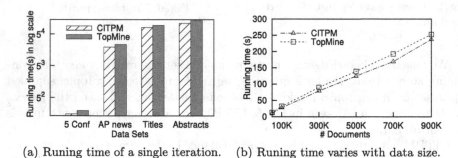

(a) Runing time of a single iteration.   (b) Runing time varies with data size.

Fig. 8. Comparison of running time for various datasets.

As expected, CITPM's TrieCounter structure has better efficiency than ToPMine's HashCounter structure in all four datasets. In the small dataset such as 5Conf, we can see that the time is ten times smaller than ToPMine, while on larger datasets such as Abstracts and APNews, ToPMine takes twice time than CITPM. The reason of the time increase is that a larger dataset may have a larger partition length, which causes a longer retrieval time for TriCounter. Note that the partition length usually has a limited bound in practice, so CITPM still has advantages in a large dataset. Figure 7(b) shows how running time varies with data size, the results demonstrate our phrase mining algorithm has a linear time complexity owing to our efficient one-pass frequency counting algorithm.

Figure 8 shows the total running time of a single iteration comparing to ToPMine. CITPM still has a better time cost than ToPMine. The above two sets of experiments demonstrate that our method is of high efficiency than the best existing method ToPMine. The reason is that, in the phrase mining stage, to repeatedly perform frequency counting and frequency retrieving cause huge time consumption. CITPM's TriCounter structure really beats ToPMine's Hash Counter that help it to achieve a better efficiency.

4.3   Case Study

Table 3 is a sample of three topics' top-10 phrases that mined by CITPM in the 5Conf dataset. From the table we can see that phrases in the same topic is of high cohesion and less related between different topics. For example, phrases in TOPIC 1, TOPIC 2 and TOPIC 3 have close connection within their respective fields, artificial intelligence, database and information retrieval. Each phrase is natural, meaningful and unambiguous, and it is not a simple aggregation of tokens. This demonstrates CITPM works well in practice.

Table 3. A sample of three topics' top-10 phrases.

TOPIC 1	TOPIC 2	TOPIC 3
Machine Learning	Database Systems	Information Retrieval
Artificial Intelligence	Relational Databases	Information Extraction
Support Vector Machines	Knowledge Discovery	Image Retrieval
Logistic Regression	Knowledge Base	Topic models
Manifold Learning	Data Management	Information Systems
Multiple Kernel	Distributed Database Systems	Document Retrieval
Vector Space	Data Base	Probabilistic Models
Metric Spaces	Distributed Database	Mixture Models
Machine Translation	Knowledge Representation	Retrieval System
Version Space	Deductive Databases	Text Retrieval

5   Conclusion and Future Work

We presented a novel framework CITPM for topical phrase mining. This framework characterizes topical domains by document clusters and iteratively performs cluster updating and topical inferring in each iteration. Based on this framework, we designed an efficient phrase frequency counting algorithm and a phraseness checking algorithm. Moreover, we proposed a density-peak based k-means algorithm for topical document clustering. The empirical verification demonstrated our framework is of high interpretability and efficiency. As parts of future work, we will investigate techniques to further improve the performance of CITPM by combining the clustering stage and topic modeling stage.

Acknowledgments. The work was partially supported by the NSF of China for Outstanding Young Scholars under grant 61322208, the NSF of China under grants 61272178, 61572122, the NSF of China for Key Program under grant 61532021, ARC DP140103499, and ARC DP160102412.

References

1. Wallach, H.M.: Topic modeling: beyond bag-of-words. In: Proceedings of the 23rd ICML, pp. 977–984. ACM, Pennsylvania (2006)
2. Wang, X., McCallum, A., Wei, X.: Topical n-grams: phrase and topic discovery, with an application to information retrieval. In: Seventh IEEE International Conference on Data Mining (ICDM), pp. 697–702. IEEE, Nebraska(2007)
3. Lindsey, R.V., Headden III., W.P., Stipicevic, M.J.: A phrase discovering topic model using hierarchical pitman-yor processes. In: Proceedings of the EMNLP, pp. 214–222. ACL, Jeju Island (2012)
4. El-Kishky, A., Song, Y., Wang, C., et al.: Scalable topical phrase mining from text corpora. Proc. VLDB Endowment **8**(3), 305–316 (2014). VLDB Endowment, Hang Zhou
5. Blei, D.M., Lafferty, J.D.: Visualizing topics with multi-word expressions, arXiv preprint arxiv:0907.1013 (2009)
6. Danilevsky, M., Wang, C., Desai, N., et al.: Automatic construction, ranking of topical keyphrases on collections of short documents. In: 2014 SIAM International Conference on Data Mining (SDM). SIAM, Pennsylvania (2014)
7. Wang, C., Danilevsky, M., Desai, N., et al.: A phrase mining framework for recursive construction of a topical hierarchy. In: Proceedings of the 19th ACM SIGKDD International Conference on Knowledge Discovery and Data Mining, pp. 437–445, ACM, Chicago (2013)
8. Porter, M.F.: Snowball: a language for stemming algorithms. Open Source Initiative Osi (2001)
9. Porter, M.F.: An algorithm for suffix stripping. Programming **14**(3), 130–137 (1980)
10. Mihalcea, R., Tarau, P.: Textrank: Bringing order into texts. In: Proceedings of the EMNLP, pp. 275. ACL, Barcelona (2004)
11. Liu, Z., Li, P., Zheng, Y., Sun, M.: Clustering to find exemplar terms for keyphrase extraction. In: Proceedings of the EMNLP, pp. 257–266. ACL, Singapore (2009)
12. Wang, J., Feng, J., Li, G.: Trie-join: Efficient trie-based string similarity joins with edit-distance constraints. Proc. VLDB Endowment **3**(1–2), 1219–1230 (2010). VLDB Endowment, Singapore
13. nedtries Homepage. http://www.nedprod.com/programs/portable/nedtries/

14. Pearson, K.: On the criterion that a given system of deviations from the probable in the case of a correlated system of variables is such that it can be reasonably supposed to have arisen from random sampling. Philos. Mag. **50**(302), 157–175 (1900). Series 5
15. Zengin, M., Carterette, B.: Learning user preferences for topically similar documents. In: Proceedings of the 24th ACM International on Conference on Information and Knowledge Management (CIKM), pp. 1795–1798. ACM, Melbourne (2015)
16. Ding, C., Li, T.: Adaptive dimension reduction using discriminant analysis and k-means clustering. In: Proceedings of the 24th international conference on Machine learning (ICML), pp. 521–528. ACM, Oregon (2007)
17. Kullback, S., Leibler, R.A.: On information and sufficiency. Ann. Math. Stat. **22**(1), 79–86 (1951)
18. Pang-Ning, T., Steinbach, M., Kumar, V.: Introduction to Data Mining. Pearson Addison Wesley, Boston (2006)
19. Rodriguez, A., Laio, A.: Clustering by fast search and find of density peaks. Science **344**(6191), 1492–1496 (2014)
20. Thorndike, R.L.: Who belongs in the family? Psychometrika **18**(4), 267–276 (1953)
21. Sugar, C.A., James, G.M.: Finding the number of clusters in a data set: an information theoretic approach. J. Am. Stat. Assoc. **98**, 750–763 (2003)
22. Chang, J., Gerrish, S., Wang, C., et al.: Reading tea leaves: how humans interpret topic models. In: Advances in neural information processing systems (NIPS), pp. 288–296. NIPS Foundation, Houston (2009)

Deep Convolutional Neural Network Based Regression Approach for Estimation of Remaining Useful Life

Giduthuri Sateesh Babu$^{(\boxtimes)}$, Peilin Zhao, and Xiao-Li Li

Institute for Infocomm Research, A*STAR, Singapore, Singapore
{giduthurisb,zhaop,xlli}@i2r.a-star.edu.sg
http://www.i2r.a-star.edu.sg

Abstract. Prognostics technique aims to accurately estimate the Remaining Useful Life (RUL) of a subsystem or a component using sensor data, which has many real world applications. However, many of the existing algorithms are based on linear models, which cannot capture the complex relationship between the sensor data and RUL. Although Multilayer Perceptron (MLP) has been applied to predict RUL, it cannot learn salient features automatically, because of its network structure. A novel deep Convolutional Neural Network (CNN) based regression approach for estimating the RUL is proposed in this paper. Although CNN has been applied on tasks such as computer vision, natural language processing, speech recognition etc., *this is the first attempt* to adopt CNN for RUL estimation in prognostics. Different from the existing CNN structure for computer vision, the convolution and pooling filters in our approach are applied along the temporal dimension over the multi-channel sensor data to incorporate automated feature learning from raw sensor signals in a systematic way. Through the deep architecture, the learned features are the higher-level abstract representation of low-level raw sensor signals. Furthermore, feature learning and RUL estimation are mutually enhanced by the supervised feedback. We compared with several state-of-the-art algorithms on two publicly available data sets to evaluate the effectiveness of this proposed approach. The encouraging results demonstrate that our proposed deep convolutional neural network based regression approach for RUL estimation is not only more efficient but also more accurate.

Keywords: Multivariate time series analysis · Deep learning · Convolutional neural networks · Supervised learning · Regression methods · Prognostics · Remaining useful life

1 Introduction

Prognostic technologies are very crucial in condition based maintenance for diverse application areas, such as manufacturing, aerospace, automotive, heavy industry, power generation, and transportation. While accessing the degradation

© Springer International Publishing Switzerland 2016
S.B. Navathe et al. (Eds.): DASFAA 2016, Part I, LNCS 9642, pp. 214–228, 2016.
DOI: 10.1007/978-3-319-32025-0_14

from expected operating conditions, prognostic technologies estimate the future performance of a subsystem or a component to make RUL estimation. If we can accurately predict when an engine will fail, then we can make informed maintenance decision in advance to avoid disasters, reduce the maintenance cost, as well as streamline operational activities. This paper proposes a data driven approach to predict RUL of a complex system when the run-to-failure data is available. Existing algorithms in the literature for RUL estimation are either based on multivariate analysis or damage progression analysis [9,15–17,23]. However, it is extremely challenging, if not impossible, to accurately predict RUL without a good feature representation method. It is thus highly desirable to develop a systematical feature representation approach to effectively characterize the nature of signals related to the prognostic tasks.

Recently, a family of learning models has emerged called as deep learning that aim to learn higher level abstractions from the raw data [2,7], deep learning models doesn't require any hand crafted features by people, instead they will automatically learn a hierarchical feature representation from raw data. In deep learning, a deep architecture with multiple layers is built up for automating feature design. Specifically, each layer in deep architecture performs a nonlinear transformation on the outputs of the previous layer, so that through deep learning models the data are represented by a different levels of hierarchy of features. Convolutional neural network, auto-encoders and deep belief network are the mostly known models in deep learning. Depending on the usage of label information, the deep learning models can be learned in either supervised or unsupervised manner. While deep learning models achieve remarkable results in computer vision [11], speech recognition [10], and natural language processing [5]. *To our best knowledge, it has not been exploited in the field of prognostics for RUL estimation.*

Recurrent neural network, a class of deep learning architectures is more intuitive model for time series data [6], however it is suitable for time series future value prediction. In this paper we treat RUL estimation problem as multivariate time series regression and solve it by adapting one particular deep learning model, namely Convolutional Neural Network (CNN) adapted from deep learning model for image classification [1,12,13], which is the first attempt to leverage deep learning to estimate RUL in prognostics. The key attribute of CNN is to conduct different processing units (e.g. convolution, pooling, sigmoid/hyperbolic tangent squashing, rectifier and normalization) *alternatively*. Such a variety of processing units can yield an effective representation of local salience of the signals. Additionally, the deep architecture allows multiple layers of these processing units to be stacked, so that this deep learning model can characterize the salience of signals *in different scales*. Therefore, the features extracted by CNN are task dependent and non-handcrafted. Moreover, these features also own more predictive power, since CNN can be learned under the supervision of target values.

Recently, different CNN architectures are applied on multi-channel time series data for activity recognition problem which is a classification task [24–26]. In [25], a shallow CNN architecture is used consists of only one convolution and one pooling layer, and is restricted to the accelerometer data. In [24,26], deep CNN architectures are used and in these architectures all convolutional and pooling filters are one-dimensional which applied along the temporally over individual sensor time series separately. Different from classification tasks, in the application on RUL estimation which is a regression task, the convolution and pooling filters in CNN are applied along the temporal dimension over all sensors, and all these feature maps for all sensors need to be unified as a common input for the neural network regressor. Therefore, a novel architecture of CNN is developed in this paper. In the proposed architecture for RUL estimation, convolutional filters in the initial layer are two-dimensional which applied along the temporally over all sensors time series and final neural network regression layer employs squared error loss function which makes the proposed architecture is different from the existing CNN architectures for multi-channel time series data [24–26]. In the experiments, the proposed CNN based approach for RUL estimation is compared with existing regression based approaches, across two public data sets. Results clearly demonstrates that the proposed approach is accurately predicts RUL than existing approaches significantly.

This paper is structured as follows: First, Sect. 2 briefly describes the problem settings, including data sets, evaluation metrics and data preprocessing steps that are used to evaluate the effectiveness of different algorithms. Then, Sect. 3 describes our proposed novel deep architecture CNN based regression approach for RUL estimation. Next, Sect. 4 presents the performance comparison of the proposed approach with the standard regression algorithms for RUL estimation. Section 5 summarizes the conclusions from this work.

2 Problem Settings

In prognostics, it is an important problem to estimate the RUL of a component or a subsystem, such as the engine of an airplane. Usually, some sensors, e.g. vibration sensors, are used to collect its information that serve as features to estimate RUL. Formally, assume that d sensors with component index i are employed, so a multivariate time series data $X^i \in \mathbb{R}^{d \times n_i}$ can be obtained, where the j-th column of X^i, denoted as $X_j^i \in \mathbb{R}^d$, is a vector consisting of the signals from the d sensors at the j-th time cycle, and $X_{n_i}^i$ denotes the vector of signals when the component fails and n_i is the useful life time of a component i from the starting. Suppose we have N same category components, e.g., N engines, then we can collect a training set of examples $\{X_j^i | i = 1, \ldots, N; j = n_1, \ldots, n_N\}$. Then the task is to construct a model based on the given training set and to perform RUL estimation on a test set $\{Z^i \in \mathbb{R}^{d \times m_i} | i = 1, \ldots, M\}$, where Z_j^i, $j = 1, \ldots, m_i$ are signals when the component works well. Here RUL for a component i in test set is the number of remaining time cycles it works well from m_i-th time cycle before failure. Now let's introduce two benchmark data sets.

Data Sets: Two data sets chosen in this work, namely the NASA C-MAPSS (Commercial Modular Aero-Propulsion System Simulation) data set and the PHM 2008 Data Challenge data set [19]. The C-MAPSS data set is further divided into 4 sub-data sets as given in Table 1. Both datasets contain simulated data produced using a model based simulation program C-MAPSS developed by NASA [20].

Table 1. Data sets details (Simulated from C-MAPSS)

Data set	C-MAPSS				PHM 2008
	FD001	FD002	FD003	FD004	
Train trajectories	100	260	100	249	218
Test trajectories	100	259	100	248	218
Operating conditions	1	6	1	6	6
Fault conditions	1	1	2	2	2

Both data sets are arranged in an n-by-26 matrix where n corresponds to the number of data points in each component. Each row is a snapshot of data taken during a single operating time cycle and in 26 columns, where 1st column represents the engine number, 2nd column represents the operational cycle number, 3–5 columns represent the three operating settings, and 6–26 columns represent the 21 sensor values. More information about the 21 sensors can be found in [22]. Engine performance can be effected by three operating settings in the data significantly. Each trajectory within the train and test trajectories is assumed to be life-cycle of an engine. While each engine is simulated with different initial conditions, these conditions are considered to be of normal conditions (no faults). For each engine trajectory within the training sets, the last data entry corresponds to the moment the engine is declared unhealthy or failure status. On the other hand, test sets contains data some time before the failure and aim here is to predict RUL in the test set for each engine. For each of the C-MAPSS data set, the actual RUL value of the test trajectories were made available to the public, while the actual RUL value of the test trajectories in PHM 2008 Data Challenge data set is not available.

To fairly compare the estimation model performance on the test data, we need some objective performance measures. In this work, we mainly employ 2 measures: *scoring* function, and Root Mean Square Error (RMSE), which are introduced in details as follows:

Scoring Function: The *scoring* function used in this paper is identical to that used in PHM 2008 Data Challenge. This scoring function is illustrated in Eq. (1), where N is the number of engines in test set, S is the computed score, and $h = (Estimated\ RUL - True\ RUL)$.

$$S = \begin{cases} \sum_{i=1}^{N} \left(e^{-\frac{h_i}{13}} - 1 \right) & for \ \ h_i < 0 \\ \sum_{i=1}^{N} \left(e^{\frac{h_i}{10}} - 1 \right) & for \ \ h_i \geq 0 \end{cases} \tag{1}$$

This scoring function penalizes late predictions (too late to perform maintenance) more than early predictions (no big harms although it could waste maintenance resources). This is in line with the risk adverse attitude in aerospace
industries. However, there are several drawbacks with this function. The most
significant drawback being a single outlier (with a much late prediction) would
dominate the overall performance *score* (pls. refer to the exponential increase in
the right hand side of Fig. 1), thus masking the true overall accuracy of the algorithm. Another drawback is the lack of consideration of the prognostic horizon
of the algorithm. The prognostic horizon assesses the time before failure which
the algorithm is able to accurately estimate the RUL value within a certain confidence level. Finally, this scoring function favors algorithms which artificially
lowers the *score* by underestimating RUL. Despite all these shortcomings, the
scoring function is still used in this paper to provide comparison results with
other methods in literature.

RMSE: In addition to the scoring function, the Root Mean Square Error
(RMSE) of estimated RUL's is also employed as a performance measure. RMSE
is chosen as it gives equal weight to both early and late predictions. Using RMSE
in conjunction with the scoring function would avoid to favor an algorithm which
artificially lowers the *score* by underestimating it but resulting in higher RMSE.
The RMSE is defined as given below:

$$RMSE = \sqrt{\frac{1}{N} \sum_{i=1}^{N} h_i^2} \tag{2}$$

A comparative plot between the two evaluation metrics is shown in Fig. 1. It
can be observed that at lower absolute error values the scoring function results in
lower values than the RMSE. The relative characteristics of the two evaluation
metrics will be useful during the discussion of experimental results in the later
part of this paper.

In addition, to learn a model, we need to perform some data pre-processing
for which the details are given as follows.

Operating Conditions: Several literature [9,16,23], have shown that by plot-
ting the 3 operating setting values, the data points are clustered into six different
distinct clusters. This observation is only applicable for data sets with different
operating conditions, but data points from FD001 and FD003 in C-MAPSS data
set are all clustered at a single point instead – they are single operating condition sub-data sets. These clusters are assumed to correspond to the six different
operating conditions. It is therefore possible to include the operating condition

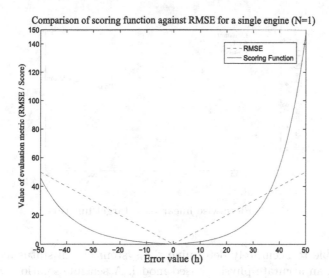

Fig. 1. Comparison of evaluation metric values for different error values

history as a feature. This is done for FD002, FD004 and PHM 2008 Data Challenge data sets by adding 6 columns of data (multiple operating condition data sets), representing the number of cycles spent in their respective operating condition since the beginning of the series [16].

Data Normalization: Due to the 6 operating conditions, each of these operating conditions results in disparate sensor values. Therefore prior to any training and testing, it is imperative to do data normalization so that the data points to be within uniform scale range using Eq. (3). As normalization was carried out within the uniform scale range for each sensor and each operating condition, this will ensure equal contribution from all features across all operating conditions [16]. Alternatively, it is also possible to incorporate operating condition information within the data to take into consideration various operating conditions.

$$Norm(x^{c,f}) = \frac{x^{(c,f)} - \mu^{(c,f)}}{\sigma^{(c,f)}}, \ \forall c, f \qquad (3)$$

where c represents operating conditions; f represents each of the original 21 sensors. $\mu^{(c,f)}$ is the mean and $\sigma^{(c,f)}$ is the standard deviation in c operating condition.

RUL Target Function: In its simplest form prognostic algorithms are similar to regression problems. However, unlike typical regression problems, an inherent challenge for data driven prognostic problems is to determine the desired output values for each input data point. This is because in real world applications,

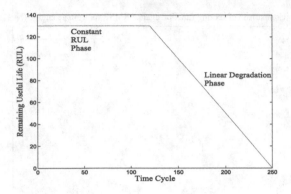

Fig. 2. Piece-wise linear RUL target function

it is impossible to accurately determine the system health status at each time step without an accurate physics based model. A sensible solution would be to simply assign the desired output as the actual time left before functional failure [16]. This approach however inadvertently implies that the health of the system degrades linearly with usage. An alternative approach is to derive the desired output values based on a suitable degradation model. For this data-set a piece-wise linear degradation model has proposed in [9], which limits the maximum value of the RUL function as illustrated in Fig. 2. The maximum value was chosen based on the observations and its numerical value is different for each data-set.

Both these approaches have their own advantages. The piece-wise linear RUL target function is more likely to prevent the algorithm from overestimating the RUL. In addition, it is also a more logical model as the degradation of the system typically only starts after a certain degree of usage. On the other hand, the linear RUL function follows the definition of RUL in the strictest sense which defined as the time to failure. Therefore, the plot of time left of a system against the time passed naturally results in a linear function. However, it should be noted that in cases where knowledge of a suitable degradation model is unavailable, the linear model is the most natural choice to use.

3   Deep Convolution Neural Network for RUL Estimation

This section presents the architecture of deep learning CNN for RUL estimation from multi-variate time series sensor signals. The inputs are normalized sensor signals in addition to the extracted features corresponding to the operating condition history. The target values are the RUL of system at corresponding time cycle. The considered target RUL function is a piece-wise linear function as described in the previous sections.

Convolutional neural networks have great potential to identify the various salient patterns of sensor signals. Specifically, lower layers processing units obtain the local salience of the signals. The higher layers processing units obtain the salient patterns of signals at high-level representation. Note that each layer may

have a number of convolution or pooling operators (specified by different parameters) as described below, so multiple salient patterns learned from different
aspects are jointly considered in the CNN. When these operators with the same
parameters are applied on local signals (or their mapping) at different time segments, a form of translation invariance is obtained [2,7,8]. Consequently, what
matters is only the salient patterns of signals instead of their positions or scales.
However, in RUL estimation we confront with multiple channels of time series
signals, in which the traditional CNN cannot be used directly. The challenges in
our problem include (i) Processing units in CNN need to be applied along temporal dimension and (ii) Sharing or unifying the units in CNN among multiple
sensors. In what follows, we will define the convolution and pooling filters along
the temporal dimension, and then present the entire architecture of the CNN
used in RUL estimation.

3.1 Architecture

We start with the notations used in the CNN. A sliding window strategy is
adopted to segment the time series signal into a collection of short pieces of
signals. Specifically, an instance used by the CNN is a two-dimensional matrix
containing r data samples each sample with D attributes (In case of single operating condition sub-data sets D attributes are taken as d raw sensor signals and
in case of multiple operating condition sub-data sets D attributes includes d raw
sensor signals along with extracted features corresponding to the operating condition history as explained in operating condition subsection in problem settings
section). Here, r is chosen to be as the sampling rate (15 used in the experiments
because one of the test engine trajectories has only 15 time cycle data samples),
and the step size of sliding a window is chosen to be 1. One may choose larger
step size to decrease the amount of the instances for lesser computational cost.
For training data, the true RUL of the matrix instance is determined by the true
RUL of the last record.

In this proposed architecture as shown in Fig. 3, conventional CNN is modified and applied to multi-variate time series regression as follows: On each segmented multi-variate time series we perform feature learning jointly. At the end
of feature learning, we concatenate a normal multi-layer perceptron (MLP) for
RUL estimation. Specifically in this work, we use 2-pairs of convolution layers
and pooling layers, and one normal fully connected multi layered perceptron.
It includes D-channel inputs and length of each input is 15. This segmented
multi-variate time series ($D \times 15$) is fed into a 2-stages of convolution and pooling layers. Then, we concatenate all end layer feature maps into a vector as the
MLP input for RUL estimation. Training stage involves the CNN parameters
estimation by standard back propagation algorithm using stochastic gradient
descent method to optimize objective function, which is cumulative square error
of the CNN model.

Convolution Layer: In the convolution layers, the previous layer's feature
maps are convolved with several convolutional kernels (to be learned in the

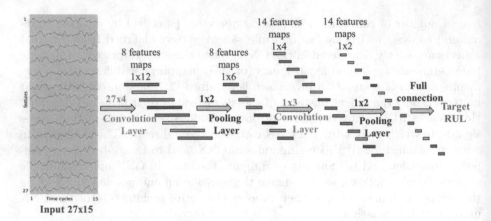

Fig. 3. Proposed CNN architecture for RUL estimation on PHM 2008 Data Challenge data set. This architecture consists of segmented multi-variate time series input, 2 convolutional filtering layers, 2 pooling filtering layers, and one fully connected layer.

training process). The output of the convolution operators added by a bias (to be learned) and the feature map for next layer is computed through the activation function. The output feature map of convolution layer computed as given below:

$$x_j^l = sigm\left(z_j^l\right), \quad z_j^l = \sum_i x_i^{l-1} * k_{ij}^l + b_j^l \tag{4}$$

Where $*$ denotes the convolution operator, x_i^{l-1} and x_j^l are the convolution filter input and output, $sigm()$ denotes the sigmoid function, and z_j^l is the input of non-linear sigmoid function. Sigmoid function is used due to its simplicity. We apply convolution filters of size $D \times 4$ in the first convolution layer. In the second convolution layer we apply convolution filters of size 1×3.

Pooling Layer: In the pooling layers, the input features are sub-sampled by suitable factor such that the feature maps resolution is reduced to increase the invariance of features to distortions on the inputs. We utilize *average* pooling without overlapping for all stage in our work. The input feature-maps are partitioned by the average pooling and results into a set of non-overlapping regions. For each sub-region output is the average value. Pooling layer output feature map is computed as given below:

$$x_j^{l+1} = down\left(x_j^l\right) \tag{5}$$

Where x_j^l is the input and x_j^{l+1} is the output of pooling layer, and $down(.)$ represents the sub-sampling function for *average* pooling. We apply pooling filters of size 1×2 in the first and second pooling layers.

3.2   Training Process

As in traditional MLP training for regression task, we used the squared error loss function in our CNN based architecture defined as: $E = \frac{1}{2}(y(t) - y^*(t))^2$, where $y^*(t)$ is the predicted RUL value and $y(t)$ is the target RUL of the t-th training sample. In the training of our CNN model, we utilize stochastic gradient descent based optimization method for optimal parameters estimation of the network and back propagation algorithm to minimize the loss function [14]. Training procedure includes three cascaded phases of forward propagation, backward propagation and the application of gradients.

Forward Propagation: The objective of the forward propagation is to determine the predicted output of CNN model on segmented multi-variate time series input. Specifically, each layer output feature maps are computed. As mentioned in the before sections, each stage contains convolution layer followed by pooling layer. We compute the output of convolution and pooling layers using Eqs. (4) and (5) respectively. Eventually, a single fully connected layer is connected with feature extractor.

Backward Propagation: Once one iteration of forward propagation is done, we will have the error value, with the squared error loss function. The predicted error propagates back on each layer parameters from last layer to first layer, derivatives chain commonly applied for this procedure.

For the backward propagation of errors in the second stage pooling layer, the $\mathbf{x}_j^{l-1}$'s derivative is calculated by the up-sampling function $up(.)$, it is an inverse operation of the sub-sampling function $down(.)$

$$\frac{\partial E}{\partial \mathbf{x}_j^{l-1}} = up(\frac{\partial E}{\partial \mathbf{x}_j^l}) \tag{6}$$

In the second stage feature extraction layer, $\mathbf{z}_j^l$'s derivative is calculated as same in hidden layer of MLP.

$$\delta_j^l = \frac{\partial E}{\partial \mathbf{z}_j^l} = \frac{\partial E}{\partial \mathbf{x}_j^l}\frac{\partial \mathbf{x}_j^l}{\partial \mathbf{z}_j^l} = sigm'(\mathbf{z}_j^l) \odot up(\frac{\partial E}{\partial \mathbf{x}_j^{l+1}}) \tag{7}$$

In the above equation element wise product is denoted by $\odot$ symbol and bias derivative is calculated by summating all values in δ_j^l as given below:

$$\frac{\partial E}{\partial b_j^l} = \sum_u (\delta_j^l)_u \tag{8}$$

The kernel weight $\mathbf{k}_{ij}^l$'s derivative is calculated by summating all values related the kernel and it is calculated with convolution operation as given below:

$$\frac{\partial E}{\partial \mathbf{k}_{ij}^l} = \frac{\partial E}{\partial \mathbf{z}_j^l}\frac{\partial \mathbf{z}_j^l}{\partial \mathbf{k}_{ij}^l} = \delta_j^l * reverse(\mathbf{x}_i^{l-1}) \tag{9}$$

Where $reverse(.)$ is the function of reversing corresponding feature extractor. At the end, we calculate $\mathbf{x}_i^{l-1}$'s derivative as given below:

$$\frac{\partial E}{\partial \mathbf{x}_i^{l-1}} = \sum_j \frac{\partial E}{\partial \mathbf{z}_j^l} \frac{\partial \mathbf{z}_j^l}{\partial \mathbf{x}_i^{l-1}} = \sum_j pad(\delta_j^l) * reverse(\mathbf{k}_{ij}^l) \tag{10}$$

In the above equation $pad(.)$ denotes the padding function, it pads zeros to δ_j^l at both ends. Specifically, $pad(.)$ function will pad at each end of δ_j^l with $n_2^l - 1$ zeros, where n_2^l is the size of $\mathbf{k}_{ij}^l$.

Apply Gradients: After the calculation of values of parameters derivatives, we can apply them to update parameters. Assume that the cost function that we want to minimize is $E(\mathbf{w})$. Gradient descent tells us to modify weights $\mathbf{w}$ in the direction of steepest descent in E:

$$w_{ij}^l = w_{ij}^l - \eta \frac{\partial E}{\partial w_{ij}^l} \tag{11}$$

Where η is the learning rate, the learning rate is a parameter that determines how much an updating step influences the current value of weights, and if it's too large it will have a correspondingly large modification of the weights w_{ij}. More details about forward propagation, backward propagation and application of gradients can be found in [3,14].

4 Experimental Results

In this section, we have performed extensive experiments for comparison of our proposed CNN based regression model (CNN in short) with three regression algorithms in the state-of-the-art, including Multi-layer Perceptron (MLP) [18], Support Vector Regression (SVR) [4] and Relevance Vector Regression (RVR) [21], on two publicly available data sets. The tunable parameters of all the four techniques, namely CNN, MLP, SVR and RVR, are chosen using standard 5-fold cross-validation procedure based on the *training set* only, where we tune their parameter values for training these models on the randomly selected four folds and choose their final values that give the best results in the last fold.

4.1 Results on C-MAPSS Data Set

The four algorithms were tested on four C-MAPSS sub-data sets (see Table 1). Table 2 illustrates their comparison results across four sub-data sets in terms of RMSE values. It is observed that CNN achieved the lower RMSE values consistently on all the sub-data sets than MLP, SVR and RVR, regardless of the operating conditions, indicating the proposed deep learning method can find more informative features than shallow features and features from naive MLP network. Among the four methods, MLP achieved higher RMSE values on all the four sub-data sets than the remaining methods, signifying that naive deep model can even harm the performance and further verified the necessity to explore modern deep learning techniques. SVR achieved the lower RMSE values than MLP and RVR on single operating condition data sets, i.e. the first and third sub-data sets. Furthermore, RVR achieved the lower RMSE values than MLP and SVR on multiple operating condition data sets, i.e. the second and fourth sub-data sets. This demonstrates that none of the existing traditional methods can beat the others consistently, while our proposed CNN method consistently achieves significantly better results across multiple data sets.

Table 2. *RMSE* for various algorithms on C-MAPSS data set

Algorithms	C-MAPSS data sets			
	FD001	FD002	FD003	FD004
MLP	37.5629	80.0301	37.3853	77.3688
SVR	20.9640	41.9963	21.0480	45.3475
RVR	23.7985	31.2956	22.3678	34.3403
CNN	**18.4480**	**30.2944**	**19.8174**	**29.1568**

Table 3. *Scores* for various algorithms on C-MAPSS data set

Algorithms	C-MAPSS data sets			
	FD001	FD002	FD003	FD004
MLP	$1.7972 * 10^4$	$7.8028 * 10^6$	$1.7409 * 10^4$	$5.6166 * 10^6$
SVR	$1.3815 * 10^3$	$5.8990 * 10^5$	$1.5983 * 10^3$	$3.7114 * 10^5$
RVR	$1.5029 * 10^3$	$1.7423 * 10^4$	$\mathbf{1.4316 * 10^3}$	$2.6509 * 10^4$
CNN	$\mathbf{1.2867 * 10^3}$	$\mathbf{1.3570 * 10^4}$	$1.5962 * 10^3$	$\mathbf{7.8864 * 10^3}$

Similarly, in the same C-MAPSS data sets, Table 3 describes the comparison results for all the four methods in terms of the evaluation *scores*, illustrated as *scoring function* in Fig. 1. It is observed that CNN achieved lower (better) score values than the MLP, SVR and RVR on multi operating condition data sets, i.e. second and fourth sub-data sets, as well as on 1 single operating condition data set, i.e. first sub-data set. Among the four methods MLP achieved higher score values (worst results) on all the four sub-data sets than remaining methods regardless of the operating conditions. CNN achieved slightly higher (worse) scores than the RVR on one single operating condition data set, i.e. third sub-data set, even though the RMSE values are lower. Coupled with the characteristics of each evaluation metric (Fig. 1), it implies that the slightly high score could be caused by certain outliers in predicting the RUL. Based on these observations, we find that performance of the methods for RUL estimation also depends on their operating conditions.

Table 4. Scores for various algorithms on PHM 2008 Data Challenge test data set

Algorithms	Score
MLP	3212
SVR	15886
RVR	8242
CNN	**2056**

4.2   Results on PHM 2008 Data Challenge Data Set

Finally, we also evaluate the performance of the four algorithms on the PHM 2008 Data Challenge test data set. After we execute the 4 algorithms to compute the estimated RULs of 218 engines in the test data set, they were then uploaded to the NASA Data Repository website and a single score was then calculated by the website as the final output.

We can observe from the results in Table 4, our proposed CNN based approach outperforms the existing regression methods based approaches significantly by producing much lower *score* (see Fig. 1), indicating that the predicted failure time from our proposed CNN model is very near to the actual failure time or their ground truth values. Hence, we can conclude that CNN based regression approach is better than the standard shallow architecture based regression methods for RUL estimation.

5   Conclusion

Clearly, accurate estimation of RUL has great benefits and advantages in many real-world applications across different industrial verticals. As the first attempt to adapt deep learning to estimate RUL for prognostic problem, this paper investigated a novel deep architecture CNN based regressor to estimate the RUL of complex system from multivariate time series data. This proposed deep architecture mainly employs the convolution and pooling layers to capture the salient patterns of the sensor signals at different time scales. All identified salient patterns are systematically unified and finally mapped into the RUL in the estimation model. To evaluate the proposed algorithm, we examined its empirical performance on two public data sets and our experimental results shows that it significantly outperforms the existing state-of-the-art shallow regression models that have been utilized extensively for RUL estimation in literature. As in our future study, we would like to further explore novel deep learning techniques to tackle a variety of emerging real-world problems in prognostics field.

References

1. Bengio, Y., Courville, A., Vincent, P.: Representation learning: a review and new perspectives. IEEE Trans. Pattern Anal. Mach. Intell. **35**(8), 1798–1828 (2013)

2. Bengio, Y.: Learning deep architectures for AI. Found. Trends Mach. Learn. **2**(1), 1–127 (2009)
3. Bouvrie, J.: Notes on convolutional neural networks, November 2006. http://cogprints.org/5869/1/cnn_tutorial.pdf
4. Chang, C.C., Lin, C.J.: LIBSVM: a library for support vector machines. ACM Trans. Intell. Syst. Technol. **2**, 27:1–27:27 (2011). http://www.csie.ntu.edu.tw/~cjlin/libsvm
5. Collobert, R., Weston, J.: A unified architecture for natural language processing: deep neural networks with multitask learning. In: Proceedings of the 25th International Conference on Machine Learning, pp. 160–167. ACM (2008)
6. Connor, J.T., Martin, R.D., Atlas, L.E.: Recurrent neural networks and robust time series prediction. IEEE Trans. Neural Netw. **5**(2), 240–254 (1994)
7. Deng, L.: A tutorial survey of architectures, algorithms, and applications for deep learning. APSIPA Trans. Sig. Inf. Process. **3**, 29 (2014)
8. Fukushima, K.: Neocognitron: a self organizing neural network model for a mechanism of pattern recognition unaffected by shift in position. Biol. Cybern. **36**(4), 193–202 (1980)
9. Heimes, F.O.: Recurrent neural networks for remaining useful life estimation. In: International Conference on Prognostics and Health Management, PHM 2008, pp. 1–6, October 2008
10. Hinton, G., Deng, L., Yu, D., Dahl, G.E., Mohamed, A., Jaitly, N., Senior, A., Vanhoucke, V., Nguyen, P., Sainath, T.N., et al.: Deep neural networks for acoustic modeling in speech recognition: the shared views of four research groups. IEEE Sig. Process. Mag. **29**(6), 82–97 (2012)
11. Krizhevsky, A., Sutskever, I., Hinton, G.E.: Imagenet classification with deep convolutional neural networks. In: Advances in Neural Information Processing Systems, pp. 1097–1105 (2012)
12. LeCun, Y., Kavukcuoglu, K., Farabet, C.: Convolutional networks and applications in vision. In: Proceedings of 2010 IEEE International Symposium on Circuits and Systems (ISCAS), pp. 253–256, May 2010
13. LeCun, Y., Bengio, Y.: Convolutional networks for images, speech, and time series. In: Arbib, M.A. (ed.) The Handbook of Brain Theory and Neural Networks, pp. 255–258. MIT Press, Cambridge (1998)
14. LeCun, Y., Bottou, L., Orr, G.B., Müller, K.-R.: Efficient BackProp. In: Orr, G.B., Müller, K.-R. (eds.) NIPS-WS 1996. LNCS, vol. 1524, p. 9. Springer, Heidelberg (1998)
15. Lim, P., Goh, C.K., Tan, K.C., Dutta, P.: Estimation of remaining useful life based on switching kalman filter neural network ensemble. Ann. Conf. Prognostics Health Manag. Soc. **2014**, 1–8 (2014)
16. Peel, L.: Data driven prognostics using a kalman filter ensemble of neural network models. In: International Conference on Prognostics and Health Management, PHM 2008, pp. 1–6, October 2008
17. Ramasso, E., Saxena, A.: Review and analysis of algorithmic approaches developed for prognostics on CMAPSS dataset. Ann. Conf. Prognostics Health Manag. Soc. **2014**, 1–11 (2014)
18. Rumelhart, D.E., Hinton, G.E., Williams, R.J.: Learning representations by back-propagating errors. In: Anderson, J.A., Rosenfeld, E. (eds.) Neurocomputing: Foundations of Research, pp. 696–699. MIT Press, Cambridge (1988). http://dl.acm.org/citation.cfm?id=65669.104451
19. Saxena, A., Goebel, K.: PHM08 challenge data set. NASA AMES prognostics data repository. Technical report, Moffett Field, CA (2008)

20. Saxena, A., Goebel, K., Simon, D., Eklund, N.: Damage propagation modeling for aircraft engine run-to-failure simulation. In: International Conference on Prognostics and Health Management, PHM 2008, pp. 1–9, October 2008
21. Tipping, M.E.: The relevance vector machine. In: Solla, S.A., Leen, T.K., Müller, K.R. (eds.) Advances in Neural Information Processing Systems, vol. 12, pp. 652–658. MIT Press, Cambridge (2000)
22. Wang, P., Youn, B.D., Hu, C.: A generic probabilistic framework for structural health prognostics and uncertainty management. Mech. Syst. Sig. Process. **28**, 622–637 (2012)
23. Wang, T., Yu, J., Siegel, D., Lee, J.: A similarity-based prognostics approach for remaining useful life estimation of engineered systems. In: International Conference on Prognostics and Health Management, PHM 2008, pp. 1–6, October 2008
24. Yang, J.B., Nguyen, M.N., San, P.P., Li, X.L., Krishnaswamy, S.: Deep convolutional neural networks on multichannel time series for human activity recognition. In: Proceedings of the 24th International Conference on Artificial Intelligence, pp. 3995–4001. AAAI Press (2015)
25. Zeng, M., Nguyen, L.T., Yu, B., Mengshoel, O.J., Zhu, J., Wu, P., Zhang, J.: Convolutional neural networks for human activity recognition using mobile sensors. In: 6th International Conference on Mobile Computing, Applications and Services (MobiCASE), pp. 197–205. IEEE (2014)
26. Zheng, Y., Liu, Q., Chen, E., Ge, Y., Zhao, J.L.: Time series classification using multi-channels deep convolutional neural networks. In: Li, F., Li, G., Hwang, S., Yao, B., Zhang, Z. (eds.) WAIM 2014. LNCS, vol. 8485, pp. 298–310. Springer, Heidelberg (2014)

Multiple-Instance Learning with Evolutionary Instance Selection

Yongshan Zhang[1], Jia Wu[2(✉)], Chuan Zhou[3], Peng Zhang[2], and Zhihua Cai[1]

[1] Department of Computer Science,
China University of Geosciences, Wuhan, China
yszhang@2014.cug.edu.cn, zhcai@cug.edu.cn
[2] The Centre for Quantum Computation and Intelligent Systems (QCIS),
University of Technology Sydney, Sydney, Australia
{jia.wu,peng.zhang}@uts.edu.au
[3] Institute of Information Engineering, Chinese Academy of Sciences, Beijing, China
zhouchuan@iie.ac.cn

Abstract. Multiple-Instance Learning (MIL) represents a new class of supervised learning tasks, where training examples are bags of instances with labels only available for the bags. To solve the instance label ambiguity, instance selection based MIL models were proposed to convert bag learning to traditional vector learning. However, existing MIL instance selection approaches are all based on the instances inside the bags. In this case, at the original instance space, those potential informative instances, which do not occur in the bags are discarded. In this paper, we propose a novel learning method, MILEIS (Multiple-Instance Learning with Evolutionary Instance Selection), to adaptively determine the informative instances for feature mapping. The unique evolutionary search mechanism, including instance initialization, mutation, and crossover, ensures that MILEIS can adjust itself to the data without explicit specification of functional or distributional form for the underlying model. By doing so, MILEIS can also take full advantage of those creative informative instances to help feature mapping in an accurate way. Experiments and comparisons on real-world applications demonstrate the effectiveness of the proposed method.

Keywords: Multiple-instance learning · Instance selection · Feature mapping · Evolutionary machine learning · Classification

1 Introduction

Multiple-instance learning (MIL) is a novel type of learning task proposed by Dieterrich et al. [1] during Bio-pharmaceutical activity test. It provides a framework to handle the collections of instances (*i.e.,* bags) instead of individual instances. In the MIL problem, an individual example is called a bag, which contains multiple instances [2,3]. Compared with single instance learning (SIL), the label of a bag is observable, while the label of an instance in the bag is

© Springer International Publishing Switzerland 2016
S.B. Navathe et al. (Eds.): DASFAA 2016, Part I, LNCS 9642, pp. 229–241, 2016.
DOI: 10.1007/978-3-319-32025-0_15

unobservable due to the bag constraint (*i.e.*, a bag is labeled positive if at least one instance inside the bag is positive, and negative otherwise). Therefore, conventional supervised classification methods are not suitable for solving MIL problems because of the instance label ambiguity [4].

Over the past few years, many applications have been formulated as the MIL problems, such as drug activity prediction [5], graph mining [6,7], image classification [8], web recommendation [9], and object detection [10]. For example, in content based image classification, an image can be repressed as a bag and regions inside the images can be represented as instances. A bag is labeled as positive if any region inside the image contains objects interesting to users, *e.g.*, a leopard. To solve these MIL problems, researchers have proposed many approaches, which can be roughly divided into two categories: (a) updating an existing learning algorithm to tackle the label ambiguity problem [11] and (b) developing a learning paradigm specifically for multiple-instance learning [12]. However, in the real-word applications, one potential problem that deteriorates the performance of the above methods is the possible large number of instances in a bag. For example, the content-based image classification data set contains numerous images, with each image consisting of a number of small regions. Accordingly, an image can be presented as a bag of instances (*i.e.*, regions). Therefore, the total number of the instances may be very large. For a region-based image, different regions/instances in a bag make different contributions to the image classification. The more informative instances we have, the more information can be provided to the learning task. In this case, how to select the most informative instances in each bag remains a challenging problem for multiple-instance learning.

Recently, some instance selection based MIL methods have been proposed. According to changes of the feature space occurring in the learning process, they can be generally grouped into two categories: (a) non-feature mapping approaches and (b) feature mapping approaches. For the former approaches (such as arithmetic mean model, geometric mean model and max-min model [13,14]), a bag will be represented by an instance inside the bag. By doing so, all bags are transformed to a set of instances in the same feature space. Thus, traditional learning classifier can be used for classification. This type of approaches provide an intuitive way to directly use one instance for bag representation, and could achieve good performance in some special domains. Nevertheless, the non-feature mapping approaches will lead to the information loss issue, because only a small number of instances are explored for bag representation with the remaining instances being discarded. By contrast, the basic idea of the latter feature mapping approaches is to map each bag into a single instance in the new feature space by using a set of instance prototypes (*i.e.*, IPs). And then, a traditional classifier can be constructed in this new feature space. Notice that all existing feature mapping approaches (such as MILES [15] and MILIS [16]) are based on the instance prototype IPs, which consists of the instances in bags. In this case, in the original instance space, those potential informative instances, which do not occur in the bags are discarded. In other words, the performance of existing MIL feature mapping approaches will be restricted by the available instance prototype IPs selected from training bags. Therefore, how to select a promising set of instance prototypes is vital to feature mapping approaches.

Based on the above observation, we propose a self-adaptive learning framework for instance selection based MIL in this paper. Our method uses evolutionary principles to design an automated search strategy to find the optimal instance prototype IPs for bag mapping. The unique evolutionary computation processes, including initialization, mutation, and crossover, ensure that our method can adjust itself to the data without any explicit specification of functional or distributional form for the underlying model. Compared with the existing feature mapping MIL methods, our proposed method can take full advantage of those creative informative instances to help bag mapping in an accurate way. Experiments and comparisons on two different types of real-world applications (each application contains three benchmark data sets) demonstrate that the proposed MILEIS (Multiple-Instance Learning with Evolutionary Instance Selection) can successfully find optimal instance prototype IPs for further learning.

The remainder of the paper is organized as follows: In Sect. 2, we review the related work on instance selection based multiple-instance learning approaches. Preliminary and problem statement are addressed in Sect. 3. Section 4 presents the implementation of the proposed evolutionary instance selection based MIL. To demonstrate the effectiveness of the proposed method, we report the experimental results in Sect. 5. Finally, we conclude the paper in Sect. 6.

2   Related Work

Multiple-instance learning is a variation of supervised learning [1]. To tackle a variety of real-world applications, numerous MIL approaches are proposed [17]. Recently, a novel type of approaches convert the MIL problem into a standard single instance learning problem and then the MIL issue can be solved by a conventional classifier [14–16], which can be broadly divided into two categories: (a) non-feature mapping methods and (b) feature mapping methods.

2.1   Non-Feature Mapping Based MIL

In non-feature mapping methods, the basic idea is to choose one instance to substitute a bag in a reasonable manner. An intuitive method is to randomly choose an instance for bag representation. Besides, the other three main promising non-feature mapping models are "arithmetic mean model", "geometric mean model" and "max-min model" [13,14]. The first two non-feature mapping models are based on the assumption that each individual instance within a bag contributes independently and equally to the bag label. Therefore, the arithmetic mean model simply calculates the arithmetic mean of the instances for each bag, while the geometric mean model calculates the geometric mean of the instances for each bag. The max-min model records both the minimum and maximum values of each dimension for every bag. After choosing the representative instances from each bag, the MIL problem is converted to a SIL problem, and a conventional classifier can be applied to the new instances for a classification task instead of the bags. However, non-feature mapping methods will lead to the problem of information loss and impair the classification performance, because most information of instances in a bag are discarded.

2.2 Feature Mapping Based MIL

In feature mapping methods, the basic idea is to choose a set of instance prototype IPs to map each bag into a new bag-level feature space. Two representative methods are MILES [15] and MILIS [16]. MILES does not define an explicit mechanism for instance prototypes selection, because the instance prototypes (i.e., IPs) are composed of all instances in the training bags. After that, MILES maps each bag into a feature space defined by IPs via a bag-instance similarity measure. Since negative instances in negative bags can have very general distributions, in MILIS, the most positive instance and the most negative instance are selected as instance prototype from each positive bag and negative bag, respectively. The notion of most negative is reciprocal to most positive and measured by the likelihood of the instance being negative based on the distributions of negative instances. These instance prototypes (IPs) are used to map each bag into a new bag-level feature space, and then any traditional classifier can be directly employed for further learning. Obviously, the IPs generated by MILES and MILIS consist of the existing instances in the training bags. Such instance restriction may result in inferior classification performance, because in the original instance space, those potential informative instances, which do not occur in the bags are discarded.

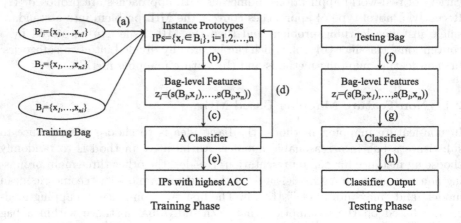

Fig. 1. A conceptual view of multiple-instance learning with evolutionary instance selection (MILEIS). The training phase of MILEIS includes: (a) Initial instance selection; (b) Feature mapping in training data set; (c) SIL classification in training data set; (d) Self-adaptive Instance updating via evolutionary mechanism; and (e) Choosing the optimal IPs with highest learning performance. Meanwhile, the testing phase of MILEIS consists of: (f) Feature mapping with the optimal IPs in testing data set; (g) SIL classification in testing data set; and (h) Obtaining the classification result.

By making full use of the existing instances and the creative informative instances, our proposed self-adaptively evolutionary instance selection based MIL can achieve a more accurate bag mapping.

3   Preliminaries and Problem Statement

3.1   Preliminaries

Suppose that there is a bag set $B = \{B_1, ..., B_n\}$ with n bags, and B_i is the ith bag. Assume that $Y = \{y_1, ..., y_n\}$ is the label set where y_i is the label of B_i. A positive bag's (B_i^+) label can be denoted as $y_i = +1$, while a negative bag's (B_j^-) label can be denoted as $y_j = -1$. For each bag B_i, the number of instances inside the bag can be denoted by n_i, and $x_{i,j}, j = 1, 2, ..., n_i$ indicates the jth instance in B_i. For each instance $x_{i,j}$, the label is unobservable due to the bag constraint.

Algorithm 1. MILEIS: MIL with Evolutionary Instance Selection

---

Input:

   The Size of IPs candidates: L;

   Maximum Evolutionary Generation: $MaxGen$;

   IPs candidate pool: $X = \{x_1, ..., x_L\}$;

   Training Data Set: B_a; Testing Data Set: B_b;

Output:

   The predicted class labels Y_b of the testing bags B_b;

   //**Training Phase:**

1: $X \leftarrow$ Initialize each instance prototype (Sect. 4.1)
2: $t \leftarrow 1$
3: **while** $(t \leq MaxGen)$ **do**
4:     $H_a \leftarrow$ Apply each IPs x_i to map B_a into the bag-level features (Sect. 4.2).
5:     $(X^c)^t \leftarrow x_c^t \leftarrow$ Apply a conventional classifier bulit on H_a to calculate the fitness (classification accuracy) of all IPs $f(x_i^t)$ and find the x_c^t with best fitness (Sect. 4.3);
6:     Apply mutation operator to obtain IPs x_i^c form parent candidate pool.
7:     Apply crossover operator to obtain IPs x_i^m form mutation candidate pool.
8:     $X^{t+1} \leftarrow$ Apply selection operator to obtain the next generation from crossover candidate pool, if $f(v_i) \geq f(x_i)$, keep IPs v_i (generated from crossover operation) to the next generation, otherwise retain IPs x_i.
9:     $t = t + 1$;
10: **end while**
11: $x_c^* \leftarrow x_c^t$.   //The final optimal instance prototype set

   //**Testing Phase:**

12: $H_b \leftarrow$ Apply the optimal IPs x_c^* to map the testing bags B_b into the bag-level features.
13: $Y_b \leftarrow$ Apply a conventional classifier on to H_b predict the labels of bags B_b.

3.2   Overall Framework of MILEIS

Figure 1 outlines the framework of the proposed MILEIS algorithm. MILEIS is separated into the training and testing phase, which *aims* at converting an MIL problem into a SIL problem and then improving its classification performance.

Given a training set B_a containing various positive bags and negative bags, MILEIS first employs Gaussian-kernel-based Kernel Density Estimator (KDE) to perform the initial instance selection (step (a) in Fig. 1), which can find the promising set of instance prototypes (IPs) from the training bags. Bag-level feature mapping is achieved by the selected IPs via a bag-instance similarity measure in the training data set (step (b)). After that, the MIL problem is converted into a SIL problem. A conventional classifier can be applied to address this problem (step (c)). In MILEIS, instance updating is performed in the process of the evolutionary mechanism (step (d)). This alternating optimization framework can guarantee to find a proper set of instance prototypes which can enhance the classification performance. After the training phase, the most proper instance prototypes set is obtained (step (e)) for the further testing phase. In step (f), the testing bags can be mapped into the bag-level features with the most proper IPs as step (b) in the training phase. In step (g), the bag-level testing feature is classified by a conventional classifier as step (c) in the training phase. Finally, the classification results in the testing data set are obtained to validate the effectiveness of the proposed method.

4   Evolutionary Instance Selection Based MIL

Algorithm 1 reports the detailed process of the proposed MILEIS framework. There are four main steps in MILEIS model: (1) Initial instance selection; (2) Bag-level feature representation; (3) Classification; (4) Self-adaptive instance updating. These four main steps constitute the training phase in MILEIS. Subsequently, the details of the four main steps are presented as the following subsections.

4.1   Initial Instance Selection

In MILEIS, instance prototypes selection is the first step to choose a most representative instance from each training bag for the construction of bag-level feature mapping. Here, an efficient approach for instance selection is presented.

Following the MIL assumption, All instances in the negative bag are negative, and at least one positive instance is in the positive bag. In other words, a positive bag may contain positive instances and negative instances, namely *true positive instances* and *false positive instances*. The key problem is how to identify the true positive instances in the positive bags [16]. To achieve this goal, we first apply the Gaussian-kernel-based kernel density estimator (KDE) [18] to model all instances contained in the negative bags:

$$f(x) = \frac{1}{Z \sum_i n_i} \sum_{y_i=-1} \sum_{j=1}^{n_i} \exp(-\gamma \|x - x_{i,j}^-\|); \tag{1}$$

where $x_{i,j}^-$ denotes the jth instance in the ith negative bag, Z is a constant normalization factor which can be ignored in our calculation, and γ is a scale parameter to control the range of influence for training instances. The above equation also defines a normalized probability density function for the negative instances in all negative bags.

To form a promising set of instance prototypes (IPs), we pick an instance with the lowest likelihood value (*i.e.*, the most positive instance) from each positive bag and pick an instance with the highest likelihood value (*i.e.*, the most negative instance) from each negative bag via Eq. (1). The total number of IPs is equal to the number of the training bags. Compared with randomly instance selection mechanism, the instance selection presented in our paper is more robust to noise corruption and outliers in the data set, which has been experimentally verified in [16].

4.2 Bag-Level Feature Representation

Before introducing the bag-level feature representation, we need present the Hausdorff distance, which is a distance metric between bags and instances. To be specific, the distance between bag B_i and instance x can be defined as:

$$d(B_i, x) = \min_{x_{i,j} \in B_i} ||x_{i,j} - x||^2, \tag{2}$$

where $d(B_i, x)$ can be regarded as the distance between instance x and its nearest neighbor in B_i.

Given the above distance metric, the bag-instance similarity measure [15] is derived using an exponential function:

$$s(B_i, x) = \exp(-\lambda d(B_i, x)) = \max_{x_{i,j} \in B_i} \exp(-\lambda ||x_{i,j} - x||^2). \tag{3}$$

By utilizing Eq. (3), we can calculate the similarity between a bag (*i.e.*, a set of instances) and an instance. The bag-level feature vector for bag B_i can be presented by the similarities between bag B_i and each instances prototype in IPs. The mathematical definition of bag-level feature representation can be denoted as follows:

$$z_i = [s(B_i, x_1), ..., s(B_i, x_i), ..., s(B_i, x_{N_a})]; \tag{4}$$

where $\mathbf{x} = \{x_1, ..., x_i, ..., x_{N_a}\}$ is the IPs with N_a instances prototypes. As mentioned in Sect. 4.1, the total number of IPs is equal to the number of the training bags, so N_a is also the size of the training bags.

Hence, the MIL problem is converted into a traditional single instance (SIL) problem. The performance of a classifier is decided by the bag-level features, and the bag-level features are converted by IPs via Eq. (4). There is no doubt that a proper IPs is vital to enhancing the classification performance.

4.3   Classification: Extreme Learning Machine

After mapping each bag into a bag-level feature by the IPs via a bag-instance similarity measure, the MIL problem is converted into a SIL problem. At the moment, a conventional classifier can be employed to perform a classification task. In our paper, self-adaptive instance selection process is the hinge instead of the classifier. Therefore, we can choose a classifier arbitrarily. Here, we apply Extreme Learning Machine (ELM) [19] as a classifier in MILEIS for its better generalization performance and extremely fast learning speed. Different from traditional Neural Networks which adjust the network parameters iteratively, the input weights and hidden biases are chosen arbitrarily while the output weights are calculated analytically by using Moore-Penrose (MP) generalized inverse in ELM.

For a bag-level feature instance set $\{(z_i, t_i)\}_{i=1}^{N_a}$, where each bag-level instance z_i contained N_a inputs corresponds to a label t_i with m outputs. Assume that l is the number of hidden neurons, ϖ is the $l \times N_a$ input weight matrix, b is the $l \times 1$ biases vector and β is the $l \times m$ output weight matrix. More specifically, the ELM network structure can be formulated as:

$$t_i = \sum_{j=1}^{l} \beta_j g(\varpi_j \cdot z_i + b_j), i = 1, 2, ..., N; \tag{5}$$

where ϖ_j and β_j are the input and output weight vectors connecting the jth hidden neurons, $\varpi_j \cdot z_i$ indicates the inner product of ϖ_j and z_i and the activation function $g(\varpi_j \cdot z_i + b_j) = 1/(1 + \exp(-(\varpi_j \cdot z_i + b_j)))$ is the sigmoid function.

Equation (5) can be written as the following compacted form:

$$H\beta = T; \tag{6}$$

where H is called the hidden layer output matrix of the network and T is the output matrix of the whole network.

4.4   Self-Adaptive Instance Updating

After obtaining the ELM classifier for bag-level features, we can validate the selected IPs and update them accordingly. In this section, we propose a self-adaptive evolutionary process to search the optimal IPs for bag-level feature construction. It consists of three major steps: (1) mutation, (2) crossover and (3) selection. The mutation and crossover operations maintain the diversity of IPs and guarantee to find the promising IPs (IPs Diversity), while the selection operation selects good IPs with high fitness (IPs Updating). Because the aim of MILEIS is to maximize the classification performance, a good IPs should correspond to a high classification accuracy (ACC). Thus, the calculation of fitness function (*i.e.*, training ACC) can be presented as: $ACC = \sum_{i=1}^{N_a} \delta(c(z_i), y_i)/N_a$, where $\delta(c(z_i), y_i)$ is one if the prediction label $c(z_i) = y_i$ and zero otherwise, N_a is the size of the training bags. Accordingly, we drive an instance updating process to obtain the optimal IPs based on the highest ACC as follows:

(1) **IPs Initialization:** In the first generation, the selected IPs in Sect. 4.1 are represented as an individual. The rest of the individuals are denoted by other IPs which are randomly generated from each training bag. Assume that L is the size of the candidate IPs, d is the dimension of an instance prototype in IPs. The candidate IPs in tth generation can be represented as:

$$X^t = \begin{bmatrix} x_{1,1}^t & \cdots & x_{1,j}^t & \cdots & x_{1,N_a}^t \\ \vdots & \ddots & \vdots & \ddots & \vdots \\ x_{L,1}^t & \cdots & x_{L,j}^t & \cdots & x_{L,N_a}^t \end{bmatrix} \tag{7}$$

where $\mathbf{x}_i^t = \{x_{i,1}^t, ..., x_{i,j}^t, ..., x_{i,N_a}^t\}$ is the ith individual (IPs set) in the IPs candidate pool, $x_{i,j}^t$ is the jth instance prototype of the ith IPs set, N_a is the size of each IPs (*i.e.*, the size of the training bags).

(2) **IPs Diversity:** In order to maintain the diversity of the IPs, the mutation and crossover operation should be applied [20]. Specifically, for any IPs in the candidate pool, a new variation IPs individual can be generated as follows:

$$\mathbf{v}_i^t = \mathbf{x}_{r_1}^t + F \cdot (\mathbf{x}_{r_2}^t - \mathbf{x}_{r_3}^t) + F \cdot (\mathbf{x}_{r_4}^t - \mathbf{x}_{r_5}^t); \tag{8}$$

where the indicates r_1, r_2, r_3, r_4 and r_5 are mutually exclusive integers randomly selected from the range $[1, L]$, which are different from the index i. F, the mutation rate, is set to 0.5. After the mutation stage, a binominal crossover operation is applied, which forms the final trail vector $\mathbf{u}_i^t = [u_{i,1}^t, ..., u_{i,j}^t, ..., u_{i,N_a}^t]$ can be presented as follows:

$$u_{i,j}^t = \begin{cases} v_{i,j}^t, & if(rndreal(0,1) < CR \ or \ j = j_{rand}) \\ x_{i,j}^t, & otherwise; \end{cases} \tag{9}$$

where j_{rand} is an integer randomly chosen in the range $[1, L]$, and $rndreal\ (0,1)$ is a real number randomly generated in $(0, 1)$. CR, the crossover rate, is set to 0.9.

(3) **IPs Updating:** This process determines whether the variation IPs (generated from step 2) or the target IPs (generated from step 1) can survive to the next generation. In this process, a greedy search strategy is adopted. According to their ACC performance, only if the variation IPs with higher ACC can replace the target IPs and survive to the next generation.

5   Experiments

5.1   Experimental Setting

To evaluate the effectiveness of our MILEIS framework, we use classification accuracy (ACC for short) and area under the ROC curve (AUC for short, which is widely used for other evolutionary machine learning, such as evolutionary Bayesian works [21]) as the evaluation metrics. The definition of ACC is given in

Sect. 4.4. Besides, the AUC of the classifier can be calculated as: $E = (P_0 - t_0(t_0 + 1)/2)/t_0 t_1$, where t_0 and t_1 are the number of negative and positive instances repressively. $P_0 = \sum r_i$, with r_i denoting the rank of the ith negative instance in the ranked list. For real-world data sets used in our experiments[1], we use 10-fold cross validation (CV) to evaluate our proposed MILEIS. All reported results shown in our paper are obtained by 10-fold CV. Besides, the two parameters L and $MaxGen$ in Algorithm 1 are set to 20 and 50, respectively.

5.2   Baseline Approaches

For comparison purposes, we use the following baseline approaches (instance selection methods based MIL) from feature mapping and non-feature mapping perspectives.

Feature Mapping Approaches: (a) **MILES** maps each bag into a feature space by all training instances via a bag-instance similarity measure [15]; (b) **MILIS** applies kernel density estimation (KDE) to select the instance prototypes from each training bag, which are used to map each bag into a bag-level feature [16]; (c) **MILIS_Pos** selects the instance prototypes from the positive training bags instead of the whole training bags for MILIS; and (d) **MILRS** randomly selects an instance from each training bag to form the instance prototype set.

Non-Feature Mapping Approaches: (a) **ARITHMETIC** calculates the arithmetic mean of the instances for each bag to replace each bag without feature mapping [22]; (b) **GEOMETRIC** calculates the geometric mean of the instances for each bag, and then these geometric instances substitute for the bags [14]; and (c) **MAXMIN** records both the minimum and maximum value of each dimension to combine a new instance for each bag [14].

After converting the MIL problem into a SIL problem, a conventional classifier can be applied. For fair comparison, we employ ELM as the classifier for each compared method in our experiments.

Table 1. Classification accuracy (ACC) and area under the ROC curve (AUC) with their standard deviation on drug activity prediction. The best results are indicated in bold typeface.

	ACC			AUC		
	MUSK1	MUSK2	MUTAGENESIS	MUSK1	MUSK2	MUTAGENESIS
MILEIS	**0.8111 ± 0.11**	**0.8100 ± 0.11**	**0.7944 ± 0.07**	**0.9179 ± 0.05**	**0.8975 ± 0.07**	**0.8971 ± 0.06**
MILES	0.7778 ± 0.15	0.7200 ± 0.13	0.7111 ± 0.10	0.8672 ± 0.18	0.8072 ± 0.15	0.7253 ± 0.10
MILIS	0.7889 ± 0.17	0.7800 ± 0.15	0.7333 ± 0.07	0.8617 ± 0.14	0.8678 ± 0.12	0.7018 ± 0.10
MILIS_Pos	0.8000 ± 0.14	0.7700 ± 0.18	0.7222 ± 0.10	0.9039 ± 0.08	0.8591 ± 0.16	0.7086 ± 0.13
MILRS	0.7889 ± 0.12	0.7900 ± 0.12	0.7556 ± 0.10	0.8649 ± 0.13	0.8410 ± 0.17	0.8119 ± 0.16
ARITHMETIC	0.6222 ± 0.15	0.6300 ± 0.16	0.6500 ± 0.13	0.5696 ± 0.27	0.6358 ± 0.18	0.6126 ± 0.09
GEOMETRIC	0.7333 ± 0.12	0.7300 ± 0.09	0.7167 ± 0.16	0.8194 ± 0.17	0.8427 ± 0.14	0.8129 ± 0.09
MAXMIN	0.5556 ± 0.15	0.5800 ± 0.18	0.6167 ± 0.11	0.5367 ± 0.21	0.6223 ± 0.22	0.5652 ± 0.15

---

[1] http://www.miproblems.org/datasets/.

5.3   Drug Activity Prediction

The objective of drug activity prediction is to predict the potency of the drug molecules on certain disease states. The corresponding data sets consist of their descriptions of molecules, where a molecule is represented as a bag and low-energy shapes of the molecule are denoted as instances in the bag. MUSK1 has 92 bags (the total number of the instance is 476), where 47 bags are positive and 45 bags are negative. MUSK2 contains 6359 instances grouped into 102 bags, of which 39 are labeled positive and 63 are negative. Similarly, the instances in MUSK1 and MUSK2 are described by a 166-dimensional feature vector. In MUTAGENESIS data set, there are 188 bags (63 positive and 125 negative). For all 188 bags, the total number of instances is 9444. Each instance is described by a 7-dimensional feature vector.

The results for two types of baselines (feature mapping and non-feature mapping) and proposed MILEIS on three drug activity prediction data sets are reported in Table 1. Feature mapping approaches can achieve better performances than non-feature mapping approaches on both ACC and AUC, which demonstrates that bag-level feature mapping remains the creative information discarded by non-feature mapping approaches for classification. Meanwhile, the proposed MILEIS can achieve much better performance compared other baselines. To be specific, MILEIS achieves 81.11 % ACC, 91.79 % AUC on MUSK1; 81.00 % ACC, 89.75 % AUC on MUSK2; and 79.44 % ACC, 89.71 % AUC on MUTAGENESIS. MILEIS's remarkable performances on drug activity prediction owe to the unique of evolutionary search mechanism, which guarantees that MILEIS can find the most proper IPs for bag-level feature mapping.

Table 2. Classification accuracy (ACC) and area under the ROC curve (AUC) with their standard deviation on region-based image categorization. The best results are indicated in bold typeface.

	ACC			AUC		
	TIGER	ELEPHANT	FOX	TIGER	ELEPHANT	FOX
MILEIS	**0.7400 ± 0.12**	**0.7650 ± 0.05**	**0.6050 ± 0.11**	**0.8827 ± 0.08**	**0.9156 ± 0.04**	**0.7980 ± 0.06**
MILES	0.7200 ± 0.14	0.7250 ± 0.09	0.5850 ± 0.11	0.7832 ± 0.13	0.7801 ± 0.12	0.6330 ± 0.12
MILIS	0.7150 ± 0.15	0.7450 ± 0.12	0.5800 ± 0.11	0.7814 ± 0.15	0.8264 ± 0.11	0.6257 ± 0.12
MILIS_Pos	0.6400 ± 0.10	0.7150 ± 0.08	0.5550 ± 0.13	0.7326 ± 0.10	0.7653 ± 0.10	0.5888 ± 0.14
MILRS	0.7250 ± 0.07	0.7350 ± 0.10	0.5900 ± 0.15	0.8188 ± 0.06	0.8191 ± 0.12	0.6352 ± 0.14
ARITHMETIC	0.5350 ± 0.08	0.5650 ± 0.07	0.5400 ± 0.14	0.5528 ± 0.11	0.5930 ± 0.11	0.5341 ± 0.19
GEOMETRIC	0.6850 ± 0.10	0.7050 ± 0.08	0.5400 ± 0.09	0.7186 ± 0.11	0.7712 ± 0.10	0.5384 ± 0.04
MAXMIN	0.5550 ± 0.11	0.5350 ± 0.11	0.5250 ± 0.11	0.5037 ± 0.17	0.5255 ± 0.12	0.5005 ± 0.14

5.4   Content-Based Image Annotation

In this subsection, we reported MILEIS's performance for image annotation tasks. The content-based image annotation task is to identify the target object within the image or not. The original data are color image from the Corel data set that have been preprocessed and segmented using Blobworld system [23]. In this case, an image (denoted as a bag) contains many regions (presented as instances). In our experiments, we utilized three different data sets (TIGER,

ELEPHANT and FOX), which are the benchmark data sets for testing MIL algorithms. In each data set, there are 100 positive and 100 negative bags. For TIGER, ELEPHANT and FOX data sets, the total numbers of instances are 1096, 1259 and 1474 respectively, and the instances are all described by a 230-dimensional feature vector which represent color, texture and shape of the region.

In Table 2, we report the experimental results of MILEIS with two types of baselines on three different data sets. The results show that non-feature mapping methods are inferior to feature mapping methods except GEOMETRIC. For GEOMETRIC, it can obtain the comparative results on TIGER, ELEPHANT and FOX data sets compared to feature mapping methods. By contrast, MILEIS gains 74.00 % ACC, 88.27 % AUC on TIGER; 76.50 % ACC, 91.56 % AUC on ELEPHANT; and 60.50 % ACC, 79.80 % AUC on FOX respectively. This suggests that MILEIS is also effective on content-based image annotation, mainly because that it can make use those creative/potential instances to help annotation in an accurate way.

6   Conclusions

In this paper, we proposed a novel learning method, MILEIS (Multiple-Instance Learning with Evolutionary Instance Selection), which can adaptively determine the informative instances for bag-level feature mapping. The unique evolutionary search mechanism, including instance initialization, mutation, and crossover, ensures that MILEIS can adjust itself to the data without explicit specification of functional or distribution forms for the underlying model. Experiments on drug activity prediction and content-based image annotation (each application contains three data sets) demonstrated superior performance of the proposed MILEIS in terms of classification accuracy (ACC) and area under the ROC curve (AUC).

Acknowledgments. This work was supported in part by the National Nature Science Foundation of China (No. 61403351), the China Scholarship Council Foundation (No. 201206410056), the key project of the Natural Science Foundation of Hubei province, China under Grant No. 2013CFA004, the Australian Research Council (ARC) Discovery Projects under Grant No. DP140100545, the Self-Determined and Innovative Research Founds of CUG (No. 1610491T05) and the National College Students' Innovation Entrepreneurial Training Plan of CUG (WuHan) (No. 201410491083).

References

1. Dieterich, T.G., Lathrop, R.H., Lozano-Pérez, T.: Solving the multiple instance problem with axis-parallel rectangles. Artif. Intell. **89**(1), 31–71 (1997)
2. Andrews, S., Tsochantaridis, I., Hofmann, T.: Support vector machines for multiple-instance learning. Adv. Neural Inf. Process. Syst. **15**(2), 561–568 (2002)
3. Maron, O., Lozano-Prez, T.: A framework for multiple-instance learning. Adv. Neural Inf. Process. Syst. **200**(2), 570–576 (1998)

4. Ray, S., Craven, M.: Supervised versus multiple instance learning: an empirical comparison. In: ICML, pp. 697–704 (2005)
5. Zhao, Z., Gang, F., Sheng, L., Elokely, K.M., Doerksen, R.J., Chen, Y., Wilkins, D.E.: Drug activity prediction using multiple-instance learning via joint instance and feature selection. BMC Bioinform. **14**(Suppl. 14), 535–536 (2013)
6. Wu, J., Zhu, X., Zhang, C., Yu, P.: Bag constrained structure pattern mining for multi-graph classification. IEEE Trans. Knowl. Data Eng. **26**(10), 2382–2396 (2014)
7. Wu, J., Pan, S., Zhu, X., Cai, Z.: Boosting for multi-graph classification. IEEE Trans. Cybern. **45**(3), 416–429 (2015)
8. Hong, R., Meng, W., Yue, G., Tao, D., Li, X., Wu, X.: Image annotation by multiple-instance learning with discriminative feature mapping and selection. IEEE Trans. Cybern. **44**(5), 669–680 (2014)
9. Zhou, Z.H., Jiang, K., Li, M.: Multi-instance learning based web mining. Appl. Intell. **22**(2), 135–147 (2005)
10. Ali, K., Saenko, K.: Confidence-rated multiple instance boosting for object detection. In: CVPR, pp. 2433–2440 (2014)
11. Zhang, M.L., Zhou, Z.H.: Adapting RBF neural networks to multi-instance learning. Neural Process. Lett. **23**(1), 1–26 (2006)
12. Yuan, H., Fang, M., Zhu, X.: Hierarchical sampling for multi-instance ensemble learning. IEEE Trans. Knowl. Data Eng. **25**(12), 2900–2905 (2013)
13. Xu, X., Frank, E.: Logistic regression and boosting for labeled bags of instances. In: Dai, H., Srikant, R., Zhang, C. (eds.) PAKDD 2004. LNCS (LNAI), vol. 3056, pp. 272–281. Springer, Heidelberg (2004)
14. Dong, L.: A comparison of multi-instance learning algorithms. University of Waikato (2006)
15. Chen, Y., Bi, J., Wang, J.: Miles: multiple-instance learning via embedded instance selection. IEEE Trans. Pattern Anal. Mach. Intell. **28**(12), 1931–1947 (2006)
16. Fu, Z., Robles-Kelly, A., Zhou, J.: Milis: multiple instance learning with instance selection. IEEE Trans. Pattern Anal. Mach. Intell. **33**(5), 958–977 (2011)
17. Amores, J.: Multiple instance classification: Review, taxonomy and comparative study. Artif. Intell. **201**(4), 81–105 (2013)
18. Kim, J.S., Scott, C.D.: Robust kernel density estimation. J. Mach. Learn. Res. **13**(1), 2529–2565 (2012)
19. Huang, G.B., Zhu, Q.Y., Siew, C.K.: Extreme learning machine: Theory and applications. Neurocomputing **70**(1), 489–501 (2006)
20. Wang, H., Rahnamayan, S., Sun, H., Omran, M.: Gaussian bare-bones differential evolution. IEEE Trans. Cybern. **43**(2), 634–647 (2013)
21. Wu, J., Pan, S., Zhu, X., Zhang, P., Zhang, C.: SODE: self-adaptive one-dependence estimators for classification. Pattern Recogn. **51**, 358–377 (2016)
22. Wu, J., Zhu, X., Zhang, C., Cai, Z.: Multi-instance multi-graph dual embedding learning. In: ICDM, pp. 827–836 (2013)
23. Carson, C., Belongie, S., Greenspan, H., Malik, J.: Blobworld: image segmentation using expectation-maximization and its application to image querying. IEEE Trans. Pattern Anal. Mach. Intell. **24**(8), 1026–1038 (2002)

Exploiting Human Mobility Patterns for Gas Station Site Selection

Hongting Niu[✉], Junming Liu, Yanjie Fu, Yanchi Liu, and Bo Lang

State Key Laboratory of Software Development Environment,
Beihang Univeristy, 100191 Beijing, China
{niuhongting,langbo}@buaa.edu.cn,
{jl1433,yanjie.fu,yanchi.liu}@rutgers.edu

Abstract. Advances in sensor, wireless communication, and information infrastructure such as GPS have enabled us to collect massive amounts of human mobility data, which are fine-grained and have global road coverage. These human mobility data, if properly encoded with semantic information (i.e. combined with Point of Interests (POIs)), is appealing for changing the paradigm for gas station site selection. To this end, in this paper, we investigate how to exploit newly-generated human mobility data for enhancing gas station selection. Specifically, we develop a ranking system for evaluating the business performances of gas stations based on waiting time of refueling events by mining human mobility data. Along this line, we first design a method for detecting taxi refueling events by jointly tracking dwell times, GPS trace angles, location sequences, and refueling cycles of the vehicles. Also, we extract the fine-grained discriminative features strategically from POI data, human mobility data and road network data within the neighborhood of gas stations, and perform feature selection by simultaneously maximizing relevance and minimizing redundancy based on mutual information. In addition, we learn a ranking model for predicting gas station crowdedness by exploiting learning to rank techniques. The extensive experimental evaluation on real-world data also show the advantages of the proposed method over existing approaches for gas site selection.

Keywords: Refueling event detection · Gas station distribution · Site selection

1 Introduction

Recent years have witnessed an explosive increasing of both automobile amount and urban population in super cities, and thus there are massive needs of oil consumption for the daily transportation. Besides, the civilization has brought serious problems of environmental pollution to super cities, for example, air pollution and noise pollution. Moreover, it has caused heavy traffic in transforation systems and wasted precious time of people. Therefore, a careful site selection of gas stations becomes critical in urban planning for both governments and

© Springer International Publishing Switzerland 2016
S.B. Navathe et al. (Eds.): DASFAA 2016, Part I, LNCS 9642, pp. 242–257, 2016.
DOI: 10.1007/978-3-319-32025-0_16

residences, because distributing gas station sites can effectively help shorten commute distances optimize road network development and reduce pollutant emission. Essentially, site selection of gas stations aims to rank the candidate sites in terms of their gas refueling crowdedness for developing gas stations.

In the literature, computer scientists and urban planners mainly focus on exploiting urban geography data (e.g., road networks, point of interests (POIs), etc.) and human mobility data (e.g., taxi GPS traces, bus GPS traces, etc.) for tackling the problems, such as recommending taxi trajectory, inferring air pollution, etc., in urban space. Besides,there are several criteria used for site selection, including analytical hierarchy process (AHP) [3] and ordered weighted average [6]. Dimitrios optimized the site selection with linear regression models, a spatial simultaneous autoregressive model and a geographically weighted regression model [5,9]. Similar to the problem of our paper, gas station site selection with multi factor evaluation model [13] was proposed with weights obtained from survey.Moreover, shopping mall location selection with fuzzy MCDM [12] and GIS approach [4] has explored before. For the behavior detection and extraction, Krumm [8] tried to predict previous covered trajectory, and gave a system based on trajectories. Zheng [17] established a model of histories using tree-based hierarchical graph with GPS multiple traces. As to POI ranking, Lavandoski [10] used location based ratings for recommendation. Also, there are some works aiming to select sites for developing gas stations in the petroleum industry [11,15]. However, the performances of classic methods are limited by the availability of mobile data, and many highly related factors, such as traffic situation, point-of-interest and road network, are rarely discussed. For example, works [2,7] only investigated limited aspects that affect the site selection of gas station (e.g., policy guidance of gas station development) and heavily relied on expensive data (e.g., historical prices of oil and gas). In addition, Semih *et al.* [13] take into account the environment protection and balanced the traffic flow, commercial trade, popularity characteristics, road planning and safety etc. These models aim to establish a comprehensive process for selecting gas station sites for government decision making. They therefore integrate both macro and micro factors about government regularization, local population and urban environment. Unfortunately, this also make it unpractical for real world application due to the requirement of expensive data. Indeed, there are some exploratory works on analyzing refueling behavior and gas consumption using mobile data [14,16]. But these methods are not specifically designed for tackling gas station site selection.

With the development of internet, mobile and sensing technologies [1], large amount of mobile data have been collected in a cheaper and faster way to sense the pulse of super cities. For example, we can easily collect the POIs data, road network data, and taxi GPS traces of Beijing using the sensing and GPS techniques. The availability of these mobile data provides the opportunities to explore a novel approach for selecting sites of gas stations in terms of a three stage paradigm: (1) refuelling event detection, (2) influence feature extraction, and (3) gas station ranking. Figure 1 presents the framework for ranking gas stations in terms of crowdedness using human mobility data and POI data.

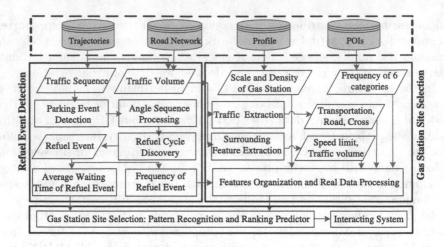

Fig. 1. The architecture of gas station ranking

The three-stage paradigm reflects the three research insights for gas station site selection as following. First, how to effectively identify the refueling events from human mobility data? While work [16] proposes a method for detecting refueling events based on taxi trajectories, the detection accuracy suffers from both the uncertainty of taxi GPS locations and the noises from non-refuel parking. To overcome this challenge, we provide a comprehensive method to discriminate refueling events from parking events by jointly model dwell time, driving direction (angle sequence), location trace (coordinate sequence) and mileage estimation of refueling cycle. Second, how to define and select the optimal set of the features for effective discriminating crowded gas station? Gas station site selection can be collaboratively affected by many factors including the distribution of POIs, the design of road networks, the conditions of traffic flows and the neighborhood profiles in or near gas stations. While it is appealing to extract large amount of features, these features may be intercorrelated and redundant. We therefore select the most informative factors based on the minimal-redundancy-maximal relevance model named mRMR in order to improve the ranking effectiveness. In addition, to prepare the bench mark crowdedness, previous studies typically analyze expensive information (e.g., oil sales, revenue, etc.) for estimating the crowdedness of a given gas station. Unlike classic methods, in this paper, we strategically mine vehicle trajectories in order to extract the refueling average waiting time since this metric is an indicator which contains the behaviors of real world drivers. Third, how to discover the most crowded gas stations for taking further steps and satisfying the needs of refueling requirements. Here we exploit a pairwise learning to rank technique mixed with a neural network to effectively rank the top-K crowded gas stations. Finally, we conduct a comprehensive performance evaluation on real world gas station related data, and the experimental results demonstrate the effectiveness and efficiency of our method.

2   Refueling Event Detection

It is traditionally challenging to collect historical refueling records from publicity. Nowadays, the availability of mobility data is more present for some public services such as taxi and bus etc. We therefore devise a three-stage method to detect the refueling events using vehicle GPS traces by jointly tracking dwelling locations, GPS angles, and refueling cycles. Each refueling event here is represented by a trajectory which indeed is a continuous sequence of GPS records.

2.1   Tracking GPS Locations

We note that the GPS equipment of each taxi typically report a record every minute. This yields a trajectory where each record contains the information about the location of latitude and longitude, the driving direction at the reported time stamp. Base on the above, we denote a trajectory by

$$\mathbf{G} = \{(c_0, \theta_0, t_0), (c_1, \theta_1, t_1), \ldots\}$$

where c_i, θ_i is the coordinate and driving direction of a taxicab at time t_i.

Assuming that this trajectory $\mathbf{G}$ contains a refueling event, our task is to extract a sub trajectory which contains the GPS records reported during the refueling events. In this way, we extract the trajectory candidates for identifying refueling events. Here we propose a simple algorithm in terms of two intuitions of car refueling. First, based on the reported GPS location, we are able to compute the distance between a taxicab and the nearby gas station. If this GPS record is reported in a refueling event, the distance should not exceed the radius of the circle area of a gas station. In Beijing, the largest radius of the circle area of a gas station is around 300 m. In our experiment, we set 300 m are the radius of the circle area of a gas station. Second, the government has impose a regularization that the speed limit in a gas station is 30 m/min. Thus, in the extracted GPS records, the average speed of a refueling taxi should be less than this speed limit.

Formally, the GPS records in a gas refueling event R should be a subset of G whose elements satisfy the following two constraints: 1. $\exists$ a gas station of coordinate G s.t. $\forall c_j \in \mathbf{R}, \|c_j - G\| < 300m$ 2. $\forall c_j, c_{j+1} \in \mathbf{R}, \frac{\|c_{j+1} - c_j\|}{t_{j+1} - t_j} <$ 30 m/min.

2.2   Tracking GPS Angles

In the last subsection, we extract the candidates of refueling events by tracking the distances and the speeds of a targeted taxi when this taxi is approaching a gas station. However, it is possible that the candidates of refueling events may contain the GPS records of low speed driving near gas station under certain transportation scenarios, such as traffic accidents, traffic jams and traffic lights. Thus, to filer out these noise GPS records, we provide a novel algorithm from a perspective of driving directions. Given a taxi is approaching a gas station for refueling, the driving direction in a refueling event is different from most

Algorithm 1. Refilling Behavior filtering based on time series of angles

Input: A time series of taxi coordinates and angles L_i and θ_i, i starts from 1, C is constant;
Output: 1/0 indicating refueling or street parking
Initialization: Right Turn=0; Left Turn=0

1: **define** angle speed: $\omega_i = \frac{\theta_{i+1} - \theta_i}{t_{i+1} - t_i}$
2: **for** i in range(1:length(R)):
3:   **if** $\omega_i > C$:
4:     Right Turn = 1
5:   **else if** $\omega_i > C < -C$:
6:     Left Turn = 1
7: **if** Right Turn*Left Turn=1:
8:   return **1** %a refill event
9: **else**:
10:    return **0** %not a refill event

of the non-refueling events (e.g., left turns, right turns, straight ahead, etc.). To differentiate the non-refueling events, we exploit the power of the angular velocity which describes the speed of rotation and the orientation of the instantaneous axis about a taxi. For these non-refueling events, the angular velocity doesn't change or changes in one direction only. But, in each refueling event, the directions of the angular velocity include a clock-wise direction and a counterclockwise direction during the two phases of entering and leaving gas stations. Figure 2 presents a motivating example of different driving directions for distinct purposes (e.g., refueling, left turn, and right turn). Based on the above intuition, the angle based filtering algorithm is presented in Algorithm 1. In Algorithm 1, we set $C = 5°/s$ to check the vehicle did not change direction in corresponding to angle detection limit, since a slight direction change of a vehicle is permitted when keeping on one lane in the road.

Fig. 2. Driving route comparison between refueling and non-refueling taxis

2.3   Tracking Refueling Cycles

Aside from tracking GPS locations and angles, we propose to track the refueling cycles to filter out hidden noise candidates. As known to all, taxi drivers may visit gas stations for shopping or restrooms. The GPS records of these outlier events are similar to those of true refueling events. Therefore, simply tracking dwelling location, dwelling time, and driving angles are not enough for filtering out these outliers. However, by observing that most of taxies periodically refuel with a period of around 24 h (i.e., 500 km per day), whereas these outlier events (e.g., visit gas station for shopping or rest room) may happen many times aperiodically within one day. Motivated by this observation, we can refine the refueling candidate events by tracking the refueling cycles for each taxi.

From a set of a taxicab refueling events time series $R^k = \{R_1^k, R_2^k, \ldots, R_n^k\}$ we get after tracing GPS locations and angles, where $R_i^k = \{(c_{i,0}^k, \theta_{i,0}^k, t_{i,0}^k), (c_{i,1}^k, \theta_{i,1}^k, t_{i,1}^k) \ldots\}$ is i th refueling event time series of taxicab k, we can get a time series of refueling gas ID of our objective is to select a refined subset RS^k of R^k such that $\forall t_{i,m}^k \in RS^k$, $t_{i,m+1}^k - t_{i,m}^k \approx 24$ h. That is, any time intervals of a taxicab's refueling event should be around 24 h.

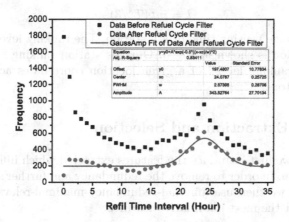

Fig. 3. Refueling events time interval filtering

To solve this problem, we first define an refueling probability function P_i^k of an event R_i^k:

$$P_i^k = \frac{1}{n-1} \sum_{j \neq i}^{n} exp - \frac{|t_{i,0}^k - t_{j,0}^k| - 24}{12}$$

That is, if there are two events which have a time interval of 24 h, they will give a high weight to each other while the weights from aperiodic events are very low. If the taxicab is recorded consecutively as "in serve" for d days, we will choose d events with top d P^k as our final refined refueling events, assuming that a taxicab will refuel one time per day in average. The performance of our strategy is presented in Fig. 3. We can find that before refueling cycle filtering, most events we detected have a time interval less than 2 h because of the aperiodic visiting (see black squares). After filter, taxi refuel time interval forms a normal distribution with a peak at interval of 24.08 h and a full width at half maximum (FWHM) of 2.87 h which agree very well with reality. Thus the refuel events after the three steps are most refined and their statistics is most correspond to reality. So far, we complete entire extraction process for refueling event detection with three steps above. Thus we filter the refueling events and get the refueling volume in gas station, which are used in following tasks.

2.4   Waiting Time Discretization

After extract all refueling traces and their refueling time series, we could calculate the average refuel waiting time of each taxi stations. According to consultations from experts in gas station, the minimum and maximum time period of refuel event is between 2 min and 15 min and a refuel without waiting is less than 3 min. We discretize the waiting time with an interval of 3 min. Thus, rank level of s station l^s is defined as number of l cars are ahead waiting to refuel in gas station S.

$$l = int\,(T^s/3) \qquad (1)$$

where T^s is average waiting time of sth station. The ranking level of current in service gas station is shown in Fig. 5. Our gas station ranking system is built to predict the ranking level l of a given location coordinates according to its surrounding factors.

3   Feature Extraction and Selection

In this section, we first introduce the features extracted which influence gas station site selection. In order to remove the redundancy and further strengthen our training model, we then use minimal-redundancy-maximal-relevance (mRMR) method to select the most informative features.

3.1   Feature Extraction

Rather than simply considering the statistics of urban geography and human mobility, we introduce 14 fine-grained features extracted from geography data and taxi trajectory data surrounding gas stations. These features are extracted from four factors (as shown in Table 1), which are gas station profile, POI, road network, and traffic.

We have to mention that oil price in different gas stations is also a key factor of gas station site selection. However, the oil price all over Beijing is same due to the government regulation. It's no need to consider the price difference in Beijing, however, price difference in other cities can be easily adapted to our system by adding one more feature.

Table 1. Features of gas station site selection and feature selection

Factors	Features	Index	MI Score	Status
Profile	Density	11	0.03963	Remove
	Scale	14	0.21897	Keep
POIs	Entertainment	1	0.13923	Keep
	Restaurants	3	0.24145	Keep
	Restroom	2	0.56484	Keep
	Shopping centers	4	0.09911	Remove
	Sight	5	0.04562	Keep
	Transportation	6	0.16125	Keep
Road network	Crossroad	7	0.21927	Keep
	Road feature number	9	0.09606	Keep
	Road feature volume	10	0.40607	Keep
Traffic	Parking volume	8	0.42740	Keep
	Speed limit around	12	0.68535	Keep
	Traffic volume around	13	0.67557	Remove

3.2 mRMR Feature Selection

In order to resist data noise and strengthen our training model performance, we identify the most characterizing features based on mutual information $I(F; C)$. Given two random variables x, y, their mutual information is defined as:

$$I(x; y) = \int \int p(x, y) log \frac{p(x, y)}{p(x)p(y)} dx dy,$$

where p(x), p(y), p(x,y) are probabilistic density functions. In Fig. 4, we evaluate the mutual information of individual feature f_i and the discretized waiting time l. From the results, we see that although most of the features are informative for our ranking level prediction, some of them may bring redundancy due to the high correlation with each other. To solve this problem and choose our best feature subset F, the minimal-redundancy-maximal-relevance (mRMR) is applied to maximize dependency D and to minimize redundancy R simultaneously:

$$D = \frac{1}{|F|} \sum_{x_i \in S} I(x_i; c)$$

$$R = \frac{2}{|F|^2 - |F|} \sum_{x_i, x_j \in F; i \neq j} I(x_i; c)$$

$$max \; \Phi(D, R) = D - R$$

For simplicity, we start from full set and recursively remove one feature until we reach a 80 % feature subset with maximum Φ. The MI score and their status after feature selection are presented in Table 1. Features after selection is 11.

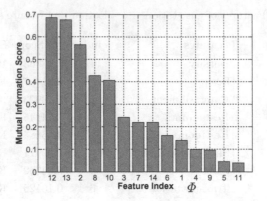

Fig. 4. MI of individual feature and class

4 Gas Station Ranking for Site Selection

After detecting the refueling event and filtering the surrounding factors, we extract effective features and propose a model to give the prediction on which region in Beijing should establish or remove a gas station. Here, we exploit a learning-to-rank model to evaluate gas stations locations in terms of ranking level l.

4.1 Model Descriptions

First, let us introduce the learning-to-rank model. We propose a pair-wise learning-to-ranking model based on a two-layer artificial neural network (ANN). The input of this model is a gas station h represented by a feature vector, $h = (f_1, f_2, \ldots, f_K)$. Here, the f_i list totally 14 as symbol in Table 1. The output is the predicted ranking level y proportional to our discretized waiting time l of refueling process $y = g(h; W) : \mathbb{R}^K \mapsto \mathbb{R}$, which is a typical measurement of throughput and capacity of a gas station. Here, W is the parameter collection and the model g is computed according to Eq. 2,

$$y = \varphi(b'' + \sum_n w_n \varphi(b'_n + \sum_m w'_{mn} \varphi(b_m + \sum_k f_k w_{km}))) \tag{2}$$

where $\varphi(x) = \frac{1}{1+exp(-x)}$ is a Sigmoid function; b_m, b'_n, and b'' are the biases associated with the neuron in the first, second, and third layers respectively; w_{km}, w'_{mn}, and w_n represent the weights associated with the input of different layers. For simplicity, we denote the predicted refueling waiting time by $g(h; W) = g(h)$.

4.2 Model Training

In the training process, we feed a list of gas stations with corresponding average waiting time of the detected refueling events into the model. Instead of employing traditional least square error as an objective function, we use a pair-wise loss function to capture the ranking inconsistencies between predictions and the ground truth. Specifically, the ranked list can be encoded into a directed graph, $G = \{V, E\}$, with the node set V as gas stations and the edge set E as pairwise ranking orders. For instance, $i \rightarrow j$ means a gas station i is ranked higher than j. For a generative model, graph G is generated by model g and parameters W through a probability function $P(G|g, W) = \prod_{i,j \in V, i \rightarrow j \in E} P(i \rightarrow j)$. Here, we adopt the concept of pair-wise likelihood. Each gas station pair has a posterior $P(i \rightarrow j)$ by

$$P(i \rightarrow j) \propto \varphi(g(h_i) - g(h_j)). \tag{3}$$

In other words, the longer taxi drivers have to wait in gas station i than gas station j, the bigger $P(i \rightarrow j)$ should be. On the contrary, the case, in which $i \rightarrow j$ but $g(h_i) < g(h_j)$, will punish the loss function. As a result, we can transfer the ranking problem to find the best estimation of W to maximize the posterior likelihood of the graph, i.e., $P(G|g, W)$. The log of the posterior likelihood is

$$
\begin{aligned}
L(\mathbf{W}|G, g) &\propto \sum_{i,j \in V, i \rightarrow j \in E} ln\varphi(g(h_i) - g(h_j)) \\
&= \sum_{i,j \in V, i \rightarrow j \in E} ln\frac{1}{1 + exp(-(g(h_i) - g(h_j)))}
\end{aligned}
\tag{4}
$$

By using the gradient decent (shown in Eq. 5), we can find an estimation for each parameter in W, where η is a learning rate determining the pace of each step.

$$\mathbf{W}^{t+1} = \mathbf{W}^t - \eta \frac{\partial L(\mathbf{W})}{\partial \mathbf{W}}. \tag{5}$$

5   Experimental Results

To validate the efficiency and effectiveness of our proposed recommender system, extensive experiments are performed on real world data sets collected in Beijing City in 86 days. The bench mark ranking level of 425 existing gas stations are used for training and validation.

5.1   Experimental Data

Table 2 shows our four different data sources, which are collected in Beijing, China. We focus on the urban region inside fifth ring road because gas station in suburban areas are rarely crowded due to the low density of cars.

GPS Trajectories: We use the real-world taxi GPS traces generated by 16,859 taxis in Beijing over a period of 86 days. The mobility traces are the 2.22 billion records of cabs' driving states in consecutive time (1 min interval), with each represented as a tuple (coordinate, driving speed, driving direction, cab status, date and time). Besides, we extract profile and traffic data from GPS Trajectories which is corresponding to gas station features in Table 1.

POIs Densities: A POIs from four categories (entertainment, restaurant, shopping, and transportation) are selected for our study. The densities of these four categories can reflet the population density and consumption level of each area. It is supposed to have a high demand on gas consumption if these POIs are densely distributed.

Road Network: Road network data we used contains information of crossroad, road level, exit and entrance of road etc. We also combine road network data and trajectory data for better road status detection.

Table 2. Statistics of the experimental data

Data sources	Properties	Statistics
GPS Trajectories	Number of taxis	16,859
	Effective days	86
	Trips (million)	2,223
POIs	Entertainment	4,435
	Restaurant	40,543
	Shopping	6,187
	Sight	1,736
	Restroom	5,502
	Transportation	13,973
Road network	Number of crossroad	433,391
	Exit, entrance and service	171,504

5.2 Baseline Algorithms

We evaluate the effectiveness of our model with a set of baselines: (1) RBF Kernel SVM, features and target level are processed in non-linear space; (2) Linear SVM, features and target level are solved in linear space; (3) Decision Tree Learning; and (4) Latent Dirichlet Allocation (LDA).

The algorithms are implemented in Python (feature extraction and learning algorithms) and Matlab (visualization). And all the experiments are performed on a x64 machine with 3.40 GHz Intel i7 CPU (4 cores) and 24 GB RAM. We use 3-fold cross validation for the evaluation.

5.3 Evaluation Metrics

To show the effectiveness of the proposed model, we use the following two metrics for evaluation.

(1) Cross Validation Accuracy. 3-fold cross validation is used here, and the accuracy of each round is defined as:

$$Accuracy = \frac{\sum_{i \in S_1} \delta(l_i - Y(F^i))}{|S_1|} \tag{6}$$

where $Y(F^i)$ is the prediction result of ith testing example based on its feature vector $F^i = (f_1^i, \ldots f_k^i)$ and l_i is the ranking level defined in Eq. (14). In our ANN ranking model, $Y(F^i)$ is a step function of ANN output $g(h, W)$, which is proportional to discretized waiting time l_i.

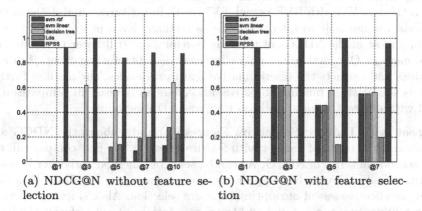

(a) NDCG@N without feature se- (b) NDCG@N with feature selec-
lection   tion

Fig. 5. Performance comparison in terms of NDCG (with and without feature selection)

(2) Normalized Discounted Cumulative Gain. Some urban planners may care more about the accuracy of top-N rankings of the selected sites for gas station rather than overall precision. Here we also use NDCG@N as an evaluation criterion. The discounted cumulative gain (DCG) metric is evaluated over top N gas stations on the ranked gas station list by assuming that low-capacity gas stations should appear on the top of the ranked list. The discounted cumulative gain (DCG@N) is given by

$$DCG[n] = \begin{cases} rel_1, & if\ n = 1 \\ DCG[n-1] + \frac{rel_n}{log_2 n}, & if\ n >= 2 \end{cases} \tag{7}$$

Given the ideal discounted cumulative gain DCG', NDCG at the n-th position can be computed by NDCG$[n]= \frac{DCG[n]}{DCG'[n]}$. The larger NDCG@N is, the higher top-N ranking accuracy is.

5.4   Overall Performances

In this subsection, we report the overall performance of our proposed method (named RPSS) comparing to baseline algorithms on dataset without feature selection and dataset with feature selection in terms of NDCG (5) and CV accuracy (6).

Dataset without Feature Selection. In Fig. 5(a), we present the performance comparison of NDCG in terms of dataset without feature selection. As can be seen in Fig. 5(a), our method achieves 1 NDCG@1, 1 NDCG@3, 0.8401 NDCG@5, 0.8819 NDCG@7, and 0.8747 NDCG@10, which outperform the baseline algorithms with a significant margin. However, as can be seen, three baseline methods, SVM with RBF kernel, SVM with linear kernel, and decision tree methods, are not able to hit the top 1 gas stations. Even in top 7 and top 10 ranking, most of the NDCGs of the four baselines are still lower than our proposed method. Our method not only effectively detects refueling events, but also identifies and extracts the discriminative features. Besides, our method further exploits the pairwise ranking objective, and offers an increase in comparison to SVM with different kernels, decision tree, and LDA methods.

Dataset with Feature Selection. Figures 5(b) and 6 shows the NDCG and CV accuracy in terms of dataset with feature selection. We first compare all the five methods in terms of NDCG. As can be seen, our method and other baseline algorithms all achieve a better performance comparing to Fig. 5(a). This validates the effectiveness of our optimal feature selection. Also, since our method accurately predicts top 1, top 3 and top 5 gas stations, it outperforms the baselines with a significant margin.

In sum, the NDCG and CV accuracy results of our method on both datasets with feature selection and without feature selection validate the effectiveness of our strategy on detecting refueling events, extracting discriminative features, and exploiting ranking objective for gas station site selection.

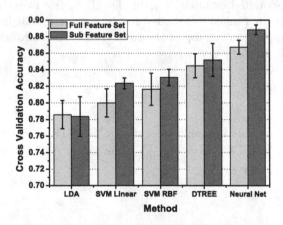

Fig. 6. Performance comparison in terms of 3-fold cross validation accuracy

5.5   Study for Gas Station Suitability

We carry out study to validate the effectiveness of our method even though it is complex. To estimate the accuracy of refueling behavior, we take a survey with 3 gas stations' waiting time sequences as ground truth. Of the 3 gas stations, our waiting time from survey is the same level with our predict results. So positive prediction match facts. We also select 2 newly closed collapsed gas station in Beijing so far from trajectory data collecting time. Because two gas station is closed, so we define that the waiting time from survey is 0. Of the 2 gas stations, ID 406 is effective, which ID 399 is a little bit partial difference. In view of lower level with our predict, it is also some like reasonable for our prediction. We can infer that maybe it is government action with closing this gas station, without considering about the suitability but just environment factors or other unknown and uncertainty factors. Gas station cases list on Table 3.

Table 3. Comparison between survey and prediction

Gas station	Survey	Prediction	Consistency
ID1379	3	3	1
ID666	2	2	1
ID471	4	4	1
ID399	0	1	0
ID406	0	0	1

Urban planning with gas station establishing in 2014 happened, so we can give our method evaluated further. We not only proved our estimate and detection of refuel events is effective, but also proved prediction of gas station's construction and closure by taxicab and POIs pattern is acceptable. According our analysis, we can suggest suitability of gas station site selection by pattern discovering.

6   Conclusion

In this paper, we develop a system for ranking gas stations based on the crowdedness using heterogeneous mobile data (e.g., POIs, vehicle trajectories, road networks, etc.). Our system consists of three stages: (1) refueling event detection, (2) feature extraction of gas stations, and (3) gas station ranking. Specifically, we first detect the refueling events using vehicle GPS traces by jointly tracking dwell times, GPS angles, location sequences, and refueling cycles of the taxicabs. We then define the discriminative features of gas stations from two aspects: (1) profile features and (2) surrounding features. Besides, we select the optimal set of features by simultaneously maximizing relevance and minimizing redundancy based on mutual information, and transform the gas stations into feature vectors. In addition, we learn a gas station ranking predictor by exploiting the neural predictive model and a pairwise ranking objective, and rank gas stations in terms of the crowdedness for gas station site selection. Finally, the experimental results of performance evaluation on real world gas station related data demonstrate the performances of the proposed method.

Acknowledgement. This work is supported by the Foundation of the State Key Laboratory of Software Development Environment (Grant No. SKLSDE-2015ZX-04).

References

1. Bowerman, B., Braverman, J., Taylor, J., Todosow, H., Von Wimmersperg, U.: The vision of a smart city. In: 2nd International Life Extension Technology Workshop, Paris (2000)
2. Chan, T.Y., Padmanabhan, V., Seetharaman, P.: An econometric model of location and pricing in the gasoline market. J. Mark. Res. **44**(4), 622–635 (2007)
3. Chen, C.-F.: Applying the analytical hierarchy process (AHP) approach to convention site selection. J. Travel Res. **45**(2), 167–174 (2006)
4. Cheng, E.W., Li, H., Yu, L.: A GIS approach to shopping mall location selection. Build. Environ. **42**(2), 884–892 (2007)
5. Efthymiou, D., Antoniou, C., Tyrinopoulos, Y.: Spatially aware model for optimal site selection. Transp. Res. Rec. J. Transp. Res. Board **2276**(1), 146–155 (2012)
6. Gorsevski, P.V., Donevska, K.R., Mitrovski, C.D., Frizado, J.P.: Integrating multi-criteria evaluation techniques with geographic information systems for landfill site selection: a case study using ordered weighted average. Waste Manage. **32**(2), 287–296 (2012)
7. Iyer, G., Seetharaman, P.: Too close to be similar: product and price competition in retail gasoline markets. QME **6**(3), 205–234 (2008)
8. Krumm, J., Horvitz, E.: Predestination: inferring destinations from partial trajectories. In: Dourish, P., Friday, A. (eds.) UbiComp 2006. LNCS, vol. 4206, pp. 243–260. Springer, Heidelberg (2006)
9. Kuo, R.J., Chi, S.-C., Kao, S.-S.: A decision support system for selecting convenience store location through integration of fuzzy AHP and artificial neural network. Comput. Ind. **47**(2), 199–214 (2002)
10. Lavandoski, J., Albino Silva, J., Vargas-Sánchez, A.: Institutional theory in tourism studies: evidence and future directions. Technical report, CIEO-Research Centre for Spatial and Organizational Dynamics, University of Algarve (2014)
11. Ma, Y.-F., Gao, R.-X.: A gas station distribution model. Oper. Res. Manage. Sci. **5**, 013 (2005)
12. Önüt, S., Efendigil, T., Kara, S.S.: A combined fuzzy MCDM approach for selecting shopping center site: an example from Istanbul, Turkey. Expert Syst. Appl. **37**(3), 1973–1980 (2010)
13. Semih, T., Seyhan, S.: A multi-criteria factor evaluation model for gas station site selection. J. Glob. Manage. **2**(1), 12–21 (2011)
14. Shang, J., Zheng, Y., Tong, W., Chang, E., Yu, Y.: Inferring gas consumption and pollution emission of vehicles throughout a city. In: Proceedings of the 20th ACM SIGKDD International Conference on Knowledge Discovery and Data Mining, pp. 1027–1036. ACM (2014)

15. Taibi, H., Haibo, H., Xiulan, L., Liangjun, L.: Application analysis and marketing proposals of the lng vehicles and gas filling stations in China. Nat. Gas Ind. **9**, 026 (2010)
16. Zhang, F., Wilkie, D., Zheng, Y., Xie, X.: Sensing the pulse of urban refueling behavior. In: Proceedings of the ACM International Joint Conference on Pervasive and Ubiquitous Computing, pp. 13–22. ACM (2013)
17. Zheng, Y., Li, Q., Chen, Y., Xie, X., Ma, W.-Y.: Understanding mobility based on gps data. In: Proceedings of the 10th International Conference on Ubiquitous Computing, pp. 312–321. ACM (2008)

Exploring the Procrastination of College Students: A Data-Driven Behavioral Perspective

Yan Zhu[1], Hengshu Zhu[2], Qi Liu[1], Enhong Chen[1(✉)], Hong Li[3], and Hongke Zhao[1]

[1] University of Science and Technology of China, Hefei, China
{zhuyan90,zhhk}@mail.ustc.edu.cn, {qiliuql,cheneh}@ustc.edu.cn
[2] Baidu Research-Big Data Lab, Beijing, China
zhuhengshu@baidu.com
[3] Hefei University, Hefei, China
xiaoke_93@126.com

Abstract. Procrastination refers to the practice of putting off impending tasks due to the habitual carelessness or laziness. The understanding of procrastination plays an important role in educational psychology, which can help track and evaluate the comprehensive quality of students. However, traditional methods for procrastination analysis largely rely on the knowledge and experiences from domain experts. Fortunately, with the rapid development of college information systems, a large amount of student behavior records are captured, which enables us to analyze the behaviors of students in a quantitative way. To this end, in this paper, we provide a data-driven study from a behavioral perspective to understand the procrastination of college students. Specifically, we propose an unsupervised approach to quantitatively estimate the procrastination level of students by the analysis of their borrowing records in library. Along this line, we first propose a naive Reading-Procrastination (naive RP) model, which considers the behavioral similarity between students for procrastination discovery. Furthermore, to improve the discovery performance, we develop a dynamic Reading-Procrastination (dynamic RP) model by integrating more comprehensive characteristics of student behaviors, such as semester-awareness and month-regularity. Finally, we conduct extensive experiments on several real-world data sets. The experimental results clearly demonstrate the effectiveness of our approach, and verify several key findings from psychological fields.

1 Introduction

Procrastination refers to the practice of putting off impending tasks to a later time, sometimes to the "last minute" before a deadline, which is usually due to the habitual carelessness or laziness [18]. As a matter of fact, exploring procrastination plays an important role in educational psychology, which can help track and evaluate the comprehensive quality of students.

S.B. Navathe et al. (Eds.): DASFAA 2016, Part I, LNCS 9642, pp. 258–273, 2016.
DOI: 10.1007/978-3-319-32025-0_17

However, psychological fields often focus on analyzing the causes [11, 14] and effects [2, 6] of procrastination, few efforts have been devoted to quantitatively discovering the procrastination of students. Meanwhile, traditional methods for procrastination analysis largely rely on the knowledge and manual labor of domain experts, such as questionnaire and survey. Such self-reported approach has been found that it has only a moderate correlation with observed procrastination [13]. Therefore, it is appealing to develop an approach to automatically uncover the procrastination behavior, which is still under-addressed.

Fortunately, thanks to the rapid development of college information system, a large amount of behavior records of students are captured, which opens a better venue for analyzing students habits. To this end, we introduce a data-driven study from a behavioral perspective to explore the procrastination of college students. Specifically, we propose an unsupervised approach to quantitatively estimate the procrastination level of students through the analysis of their borrowing records in college library. Particularly, based on the research findings in psychological studies, we assume that procrastination is a latent factor which may affect the hold time of borrowed books. Therefore, instead of directly mining procrastination phenomenon, we propose to comprehensively model the borrowing behaviors of students and probe the procrastination through predicting the hold time of borrowed books in library. Along this line, we propose a naive Reading-Procrastination (naive RP) model, which takes consideration of the behavioral similarity between students for procrastination discovery. To improve the discovery performance, we develop a dynamic Reading-Procrastination (dynamic RP) model by integrating more comprehensive characteristics of student behaviors, such as semester-awareness and month-regularity. A unique characteristic of the dynamic RP model is that it can depict the procrastination behavior in a probabilistic and empirical Bayesian perspective. Finally, extensive experiments are carried out on real world data sets collected from a Chinese college. The experimental results demonstrate the effectiveness of our approach, and verify several key findings from psychological fields.

2   Preliminaries

In this section, we first introduce the details of our real world data. Then, we present the basis of procrastination assumption, and finally describe the formulation of the procrastination discovery problem.

2.1   Data Description

The data set used in our study is the library borrowing records provided by a Chinese four-year university. The snapshot of the data set is shown as Table 1. In addition, the data set also includes the profile information of students such as grade, major and sex. In particular, since library usually has limited volumes of each book, students must comply with some restricted borrowing rules made by college library. First, each student can hold at most a limited number of

Table 1. A snapshot of library borrowing records

User id	Book id	Borrowing time	Due time	Return time
U_1	B_5	2010-04-01 09:05:20	2010-07-31 23:59:59	2010-07-12 16:12:16
U_2	B_2	2010-04-01 09:08:22	2010-07-31 23:59:59	2010-05-02 19:25:08
U_2	B_6	2010-04-01 09:08:45	2010-07-31 23:59:59	2010-06-24 10:05:36
...	...	...	...	...
U_3	B_1	2010-04-13 14:16:17	2010-08-12 23:59:59	2010-05-15 12:01:03
U_4	B_2	2010-04-13 14:16:55	2010-08-12 23:59:59	2010-06-44 11:45:56

(e.g. six) books in total. That is to say, students who have already had six books at hand must return at least one book if they want to borrow another one. Second, each book can be held for a limited number of (e.g. 120) days in fairness to all of students. To explore the procrastination behavior of students, we choose the records of students entering school from 2006 to 2010 as study candidates, including 17,531 students and 107,818 books with 1,007,406 borrowing records.

2.2 Procrastination Assumption

A study of academic procrastination reveals that 46 % of subjects reported they nearly always or always procrastinate on writing a paper, while 27.6 % procrastinate on studying for examinations [10]. Furthermore, it is estimated that 80 %–95 % of college students engage in procrastination, approximately 75 % consider themselves procrastinators [11]. Therefore, the procrastination of students is very likely to influence their behaviors in library, and the extent of which is determined by the level of procrastination. Based on the above, here we assume that procrastination is a latent factor which may affect the hold time of borrowed books. Therefore, such factor can be learned through comprehensively modeling the borrowing behaviors of students.

2.3 Problem Formulation

Since the procrastination of students is not an explicit variable, we cannot directly estimate the value of procrastination through supervised approach. Therefore, we need to find some quantitative observations and seek the relationship between the procrastination and them. Based on our procrastination assumption, we can regard procrastination as a latent factor that determines the hold time of borrowed books. Through the modeling of hold time, we can learn the procrastination value for each student. If the hold time of borrowed books can be predicted accurately, it can be verified that procrastination surely is an attribute of college students and undoubtedly makes effect on their hold time of borrowed books. For simplicity, we assume the procrastination value of each student is invariable over the four years in college.

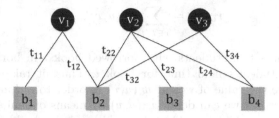

Fig. 1. Example of the B-S bipartite network.

Formally, we define e_u as the procrastination value of student u. The hold time of the book i borrowed by students u at time t is represented by $h_{u,i}(t)$. Our task is to model the causal relationship $e_u \to \{h_{u,i}(t)\}$ from e_u to a set of borrowing records $\{h_{u,i}(t)\}$ of student u. In this way, the procrastination value e_u can be obtained as a result of the hold time modeling.

3   Naive RP Model

Based on procrastination assumption, we can estimate the hold time of borrowed book $h_{u,i}(t)$ through procrastination value e_u. However, intuitively, $h_{u,i}(t)$ is not completely determined by e_u, since different books have different *required reading time*. Therefore, we should first clarify the *required reading time* $r_{u,i}(t)$ for student u and the borrowed book i at time t. The *required reading time* is not equal to the actual hold time of borrowed book, but is just an estimation of reading time that student u may spend in reading book i due to different reading speed and different focused content. Suppose this *required reading time* has been obtained, we can make estimation of hold time $h_{u,i}(t)$ by $r_{u,i}(t) + e_u$. By minimizing the error function, we can learn the procrastination value e_u for all of students. The error function is represented as:

$$\min_{\{e_u|u\in U\}} \sum_{(u,i,t)\in R} (h_{u,i}(t) - r_{u,i}(t) - e_u)^2 + \sum_{u\in U} \lambda e_u^2, \qquad (1)$$

where R is the set of borrowing records and U is the set of students. In order to make a tradeoff between the magnitude of *required reading time* and procrastination factor, we add a regularization term λe_u^2 into the above error function.

3.1   Required Reading Time Estimation

The *required reading time* $r_{u,i}(t)$ is an important component of the hold time $h_{u,i}(t)$ and its estimation also has impact on precise estimation of procrastination level of each student. However, less information about the student can be used to describe this *required reading time*. As for book i, it is impractical to analyze its content for reading time estimation due to the lack of electronic data. Therefore, we propose to leverage the neighborhood approach for estimation and define $r_{u,i}(t)$ through borrowing history of book i:

$$r_{u,i}(t) = \sum_{v \in L(i)} w_i(u,v) h_{v,i}, \tag{2}$$

where $L(i)$ is the set of students who borrowed book i in borrowing history. $h_{v,i}$ is the vth student's hold time for book i. Thus, it raises an issue that how to determine the value of weight $w_i(u,v)$ in order to estimate the *required reading time*. Directly, we can define $w_i(u,v)$ by means of similarity $sim_i(u,v)$ between student u and his neighbors. To this end, we build a book-student (B-S) bipartite network $G = \{V, B, T\}$ as shown in Fig. 1, where $V = \{v_1, ..., v_{|V|}\}$ denotes the set of students in library history, and $B = \{b_1, ..., b_{|B|}\}$ denotes the set of borrowed books. $T = \{t_{vi}\}$ is the edge set, where t_{vi} denotes the time student v borrowed book i previously. For convenience of calculating the similarity between students, in this paper, we let $t_{vi} = (t_{vi}^s, t_{vi}^m, t_{vi}^d)$, where t_{vi}^s, t_{vi}^m and t_{vi}^d represents the semester, month and day when student v borrowed book i. Thus, we can define the similarity $sim_i^1(u,v)$ as:

$$sim_i^1(u,v) = \frac{e^{I(q_u = q_v)}}{1 + |t_{ui}^m - t_{vi}^m|}, \tag{3}$$

where $I(q_u = q_v)$ is the indicator function, q_u and q_v denote the major of student u and v respectively. It is sound that students who majored in same field might focus on the same content of the book. Moreover, the more adjacent month students borrow this book at one year, the more likely they sign up for the same course and the more similar knowledge they need to learn from this book.

However, to depict the similarities of students, these observational similarities are not enough when estimating the *required reading time* of borrowed books. Thus, we need to define the similarities from different point of view. As mentioned in preliminaries section, students must comply with the borrowing volume number constraint. So students who frequently borrow books relatively have less hold time of borrowed books. In this way, we incorporate this borrowing frequency factor into the similarity definition. In this paper, we only consider the borrowing frequency in each semester. Specifically, we first seek out edge set $T_v^s = \{t_{vj} | t_{vj}^s = t_{vi}^s\}$, which is the set of dates student v had borrowing actions in semester t_{vi}^s. Second, we divide semester into several timestamps and count the borrowing number in each timestamp utilizing t_{vi}^m and t_{vi}^d in set T_v^s. Then, we define a vector $\overrightarrow{n_{v,i}}$ to represent the frequency, in which each entry corresponds to the above counting borrowing number in each timestamp. After obtaining vector $\overrightarrow{n_{u,i}}$ and $\overrightarrow{n_{v,i}}$, we can compute similarity conveniently. Motivated by Tanimoto similarity coefficient [15], which is a generalized Jaccard similarity coefficient, and is defined as:

$$T_s(\overrightarrow{x}, \overrightarrow{y}) = \frac{\overrightarrow{x} \cdot \overrightarrow{y}}{|\overrightarrow{x}|^2 + |\overrightarrow{y}|^2 - \overrightarrow{x} \cdot \overrightarrow{y}} = \frac{\overrightarrow{x} \cdot \overrightarrow{y}}{|\overrightarrow{x} - \overrightarrow{y}|^2 + \overrightarrow{x} \cdot \overrightarrow{y}}, \tag{4}$$

we define the similarity $sim_i^2(u,v)$ as:

$$sim_i^2(u,v) = \frac{\cos < \overrightarrow{n_{u,i}}, \overrightarrow{n_{v,i}} >}{|\overrightarrow{n_{u,i}} - \overrightarrow{n_{v,i}}| + \cos < \overrightarrow{n_{u,i}}, \overrightarrow{n_{v,i}} >}. \tag{5}$$

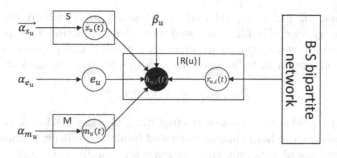

Fig. 2. The graphical representation of dynamic RP model.

In Eq. 5, we consider not only the frequency but also the magnitude in similarity computation. For example, if $\overrightarrow{n_{u,i}} = (2,4)$, and $\overrightarrow{n_{v1,i}} = (1,4)$, $\overrightarrow{n_{v2,i}} = (2,2)$, $\overrightarrow{n_{v3,i}} = (2,4)$, $\overrightarrow{n_{v4,i}} = (4,8)$, then $sim_i^2(u,v1) = 0.4940$, $sim_i^2(u,v2) = 0.3217$, $sim_i^2(u,v3) = 1$ and $sim_i^2(u,v4) = 0.1827$. We can find that $sim_i^2(u,v4)$ is the smallest although $\overrightarrow{n_{v4,i}}$ is proportional to $\overrightarrow{n_{u,i}}$. Taking account of $sim_i^1(u,v)$ and $sim_i^2(u,v)$, we represent $w_i(u,v)$ as:

$$w_i(u,v) = \frac{sim_i^1(u,v) + sim_i^2(u,v)}{\sum_{v=1}^{L_i} sim_i^1(u,v) + sim_i^2(u,v)}.$$

(6)

3.2 Limitation Discussion

Having estimated the *required reading time* $r_{u,i}(t)$ and learned the procrastination value e_u, we can also predict the hold time of future borrowed books for student u. However, the naive RP model assumes that all students share the same regularization coefficient λ, which may be unsuitable for all students. Besides, other factors except procrastination may also affect the hold time of borrowed books, which have impact on the precision of procrastination estimation. To this end, a graphical model named dynamic RP model is developed, where the procrastination value e_u of every student is determined all by its own. In the next section, we will illustrate how dynamic RP model can capture these factors as well as can control estimation accuracy automatically.

4   Dynamic RP Model

In this section, we analyze some potential factors that dynamically influence the hold time of borrowed books. To depict these dynamic characteristics, we develop our dynamic RP model from probabilistic point of view, where the parameters estimations are under the framework of empirical Bayes.

4.1   Hold Time Component Elements

The dynamic RP model is developed on the basis of naive RP model and is shown in Fig. 2, where unshaded variables indicate latent factors that determine the hold time. In Fig. 2, $h_{u,i}(t)$ and $r_{u,i}(t)$ stand for the hold time and the *required reading time* of book i borrowed by students u at time t, respectively. e_u represents the above-mentioned procrastination value and we assume it follows a zero-mean Gaussian distribution with a parameter α_{e_u} for student u:

$$P(e_u|\alpha_{e_u}) = \mathcal{N}(e_u|0, \alpha_{e_u}^{-1}). \tag{7}$$

Naturally, besides the *required reading time*, procrastination is not the only factor that affects the hold time of borrowed books. In college, students making borrowing actions at different time of semester usually have different type of borrowing patterns. Some students tend to borrow books for learning guidance at the beginning of each semester, while others prefer to borrow at the end of each semester just for final examinations. Therefore, the hold time of their borrowed books are influenced by the specific time of each semester. Besides, course teachers assign tasks with various levels in each semester, which also has impact on the hold time of borrowed books. To capture this point, we specify $\overrightarrow{s_u(t)}$ to describe these factors, as shown in Fig. 2. Specifically, we select several time points in each semester as *kernel points*. The more close the borrowing actions take place to a specific kernel, the greater degree the hold time falls into that pattern. This factor $f_s(t)$ is defined formally as:

$$f_s(t) = \frac{\sum_{k=1}^{K} e^{-|t-t_k^0|} \cdot s_{u,k}(t)}{\sum_{k=1}^{K} e^{-|t-t_k^0|}}, \tag{8}$$

where K is the number of *kernel points*, $s_{u,k}(t)$ is the kth entry in $\overrightarrow{s_u(t)}$, t_k^0 stands for the date of kth *kernel point*. Therefore, different dates in the same semester share the same value of $\overrightarrow{s_u(t)}$, and there are totally S semesters. Again, we choose a form of zero-mean Gaussian for $s_{u,k}$ with a parameter $\alpha_{s_{u,k}}$ for student u, in which $\alpha_{s_{u,k}}$ is the kth entry in $\overrightarrow{\alpha_{s_u}}$, and

$$P(s_{u,k}(t)|\alpha_{s_{u,k}}) = \mathcal{N}(s_{u,k}(t)|0, \alpha_{s_{u,k}}^{-1}). \tag{9}$$

Furthermore, it is common that students may be free this month and become busy next month due to various reasons. These unpredictable factors come out of some stochastic events, which can alter students' reading schedule and then determine the hold time of their borrowed books. With this consideration, we choose the variable m_u to denote those month regularity factors. As shown in Fig. 2, the variable $m_u(t)$ is also shared by different dates in the same month and there are M months in total. A zero-mean Gaussian distribution with a parameter α_{m_u} is also available for m_u.

$$P(m_u(t)|\alpha_{m_u}) = \mathcal{N}(m_u(t)|0, \alpha_{m_u}^{-1}). \tag{10}$$

These above discussed factors are the supplementary component elements of the hold time of borrowed books. Adding these elements can not only model-ing the hold time of borrowed books more comprehensively, but also make the estimation of procrastination value more precise. In practice, if students have borrowed very few books previously, it is better to use naive RP model since these dynamic factors can not be captured by dynamic RP model adequately. In contrast, when there are enough borrowing records, we choose dynamic RP model to depict those dynamic characteristics of each student. The model training process is presented as follow.

4.2   Empirical Bayes Framework for Estimation Accuracy Control

Based on the discussion above, we prepare to train dynamic RP model with respect to the procrastination value e_u and the above-mentioned variables. For convenience, we define $\Theta_u = \{e_u, \overrightarrow{s_u(t)}, m_u(t)\}$ and $\Lambda_u = \{\alpha_{e_u}, \overrightarrow{\alpha_{s_u}}, \alpha_{m_u}\}$. The likelihood function is given by:

$$P(\overrightarrow{h_u}|r_{u,i}(t), \Theta_u, \beta_u) = \prod_{(u,i,t)\in R(u)} \mathcal{N}(h_{u,i}(t)|p_{ui}(t), \beta_u^{-1}), \tag{11}$$

$$p_{u,i}(t) = r_{u,i}(t) + e_u + \frac{\sum_{k=1}^{K} e^{-|t-t_k^0|} \cdot s_{u,k}(t)}{\sum_{k=1}^{K} e^{-|t-t_k^0|}} + m_u(t), \tag{12}$$

where $\overrightarrow{h_u}$ is the vector of $h_{u,i}(t)$ in students borrowing set $R(u)$. In Eq. 11, we also assume $h_{u,i}(t)$ is characterized by a Gaussian distribution with a parameter β_u for student u. Due to the conjugate property, the posterior distributions of Θ_u are also Gaussian and they are represented by:

$$P(e_u|\overrightarrow{h_u}, \Theta_u, \alpha_{e_u}) = \mathcal{N}(e_u|\mu_{e_u}, \lambda_{e_u}),$$
$$P(s_{u,k}(t)|\overrightarrow{h_u}, \Theta_u, \alpha_{s_{u,k}}) = \mathcal{N}(s_{u,k}(t)|\mu_{s_{u,k}(t)}, \lambda_{s_{u,k}}), \tag{13}$$
$$P(m_u(t)|\overrightarrow{h_u}, \Theta_u, \alpha_{m_u}) = \mathcal{N}(m_u(t)|\mu_{m_u(t)}, \lambda_{m_u}),$$

where the means and variances of Θ_u are calculated as:

$$\mu_{e_u} = \beta_u \lambda_{e_u}^{-1} \sum_{(u,i,t)\in R(u)} (h_{u,i}(t) - p_{u,i}(t) + e_u),$$

$$\lambda_{e_u} = \alpha_{e_u} + \beta_u |R(u)|,$$

$$\mu_{s_{u,k}(t)} = \beta_u \lambda_{s_{u,k}}^{-1} \sum_{(u,i,t)\in R_s(u,t)} W_k(t)(h_{u,i} - p_{u,i} + W_k(t)s_{u,k}(t)),$$

$$\lambda_{s_{u,k}} = \alpha_{s_{u,k}} + \beta_u \sum_{(u,i,t)\in R_s(u,t)} W_k(t)^2,$$

$$\mu_{m_u(t)} = \beta_u \lambda_{m_u}^{-1} \sum_{(u,i,t)\in R_m(u,t)} (h_{u,i}(t) - p_{u,i}(t) + m_u(t)),$$

$$\lambda_{m_u} = \alpha_{m_u} + \beta_u |R_m(u,t)|,$$

where $R(u)$, $R_s(u,t)$ and $R_m(u,t)$ are the set of borrowing records for student u generated in all four years, in the semester of t and in the month of t, respectively. $W_k(t)$ is given by:

$$W_k(t) = \frac{e^{-|t-t_k^0|}}{\sum_{j=1}^{K} e^{-|t-t_j^0|}}.$$

Suppose the priors Λ_u have been obtained, which are inferred from the data and will be discussed later. By maximizing the posterior distributions of Θ_u, the procrastination value of student u and other dynamic variables can be estimated by the means of Eq. 13. With these obtaining variables, we can also make prediction for new instance of hold time of borrowed book. The predictive distribution of $h_{u,\hat{i}}(t)$ takes the form of:

$$P(h_{u,\hat{i}}(t)|\overrightarrow{h_u}, r_{u,i}(t), \Lambda_u, \beta_u) = \int P(h_{u,\hat{i}}(t)|r_{u,i}(t), \Theta_u, \beta_u)$$
$$\times P(e_u|\overrightarrow{h_u}, \Theta_u, \alpha_{e_u}) P(\overrightarrow{s_u(t)}|\overrightarrow{h_u}, \Theta_u, \overrightarrow{\alpha_{s_u}}) P(m_u(t)|\overrightarrow{h_u}, \Theta_u, \alpha_{m_u}) \mathrm{d}\Theta_u. \tag{14}$$

As stated above, both the conditional distribution of $h_{u,i}(t)$ and the distributions of Θ_u are all Gaussian distribution, which makes it possible to derive the closed form solution. By means of integral operations, the mean of predictive distribution is obtained using the result that substituting the means of posterior distributions of Θ_u into Eq. 12 simply.

Now we discuss how to get appropriate priors Λ_u from the data. The marginal likelihood function is represented by:

$$P(\overrightarrow{h_u}|r_{u,i}(t), \Lambda_u, \beta_u) = \int P(\overrightarrow{h_u}|r_{u,i}(t), \Theta_u, \beta_u)$$
$$\times P(e_u|\alpha_{e_u}) P(\overrightarrow{s_u(t)}|\overrightarrow{\alpha_{s_u}}) P(m_u(t)|\alpha_{m_u}) \mathrm{d}\Theta_u, \tag{15}$$

where $\overrightarrow{\alpha_{s_u}}$ and $\overrightarrow{s_u(t)}$ are K-dimensional vectors that each entry of it corresponds to $\alpha_{s_{u,k}}$ and $s_{u,k}(t)$, respectively. Fortunately, all terms in Eq. 15 are Gaussian, which comes out a closed form when integrating over parameters Θ_u. Thus, we obtain the following results by maximizing Eq. 15.

$$\beta_u = \frac{|R(u)| - \gamma_{e_u} - \sum_{k=1}^{K} \gamma_{s_{u,k}} - \gamma_{m_u}}{\sum_{(u,i,t) \in R(u)} (h_{u,i}(t) - p_{ui}(t))^2},$$

$$\alpha_{e_u}^{-1} = e_u^2 \gamma_{e_u}^{-1}, \qquad \gamma_{e_u} = \frac{\lambda_{e_u} - \alpha_{e_u}}{\lambda_{e_u}},$$

$$\alpha_{s_{u,k}}^{-1} = \frac{\gamma_{s_{u,k}}^{-1}}{|R_s(u,t)|} \sum_{(u,i,t) \in R_s(u,t)} s_{u,k}(t)^2, \qquad \gamma_{s_{u,k}} = \frac{\lambda_{s_{u,k}} - \alpha_{s_{u,k}}}{\lambda_{s_{u,k}}},$$

$$\alpha_{m_u}^{-1} = \frac{\gamma_{m_u}^{-1}}{|R_m(u,t)|} \sum_{(u,i,t) \in R_m(u,t)} m_u(t)^2, \qquad \gamma_{m_u} = \frac{\lambda_{m_u} - \alpha_{m_u}}{\lambda_{m_u}}.$$

Afterwards, we can substitute above results into Eq. 13 and alternate between maximizing posterior distributions of Θ_u and using the above results to update prior parameters Λ_u and β_u until convergence criterion is satisfied.

The framework of empirical Bayes guarantees the precision of these learning variables. In this case, we need to explain why we do not use *fully Bayesian* treatment. As we discussed above, all variables are assumed to be Gaussian, which renders a closed form when making inferences and predictions. However, if we adopt the framework of fully Bayesian treatment, both inferences and predictions are analytically intractable. Therefore, we need to resort to approximate inference like variational methods [5] or MCMC-based methods [8]. Variational methods typically scale well to large applications and may produce inaccurate results for our estimation problem, while MCMC-based methods are too time-consuming for training our model. Considering this, we employ the framework of empirical Bayes, which can save plenty of time for inferences and predictions.

5   Experimental Results

In this section, we comprehensively evaluate our procrastination discovery approach based on several real-world data sets. First, we evaluate the effectiveness of our models through predicting the hold time of borrowed books. Then, we empirically verify our learned procrastination value from psychological fields.

5.1   Prediction Performance

Experimental Setup. In our experiments, we empirically extracted 812,506 records from 10,035 students, who borrowed more than 30 books throughout the four years in college, as evaluation data set. Furthermore, we randomly selected 144,590 records as test set and the remaining records were used for training models. To the best of our knowledge, there is no existing work for the hold time of borrowed books prediction. Therefore, we exploit several intuitive but state-of-the-art baselines for evaluation. First, we propose to use the the average hold time of previous borrowed books of the given student for prediction (i.e., *Average*), which is based on the intuition of habitual momentum. Second, from the preference factorization perspective, we use the Probabilistic Matrix Factorization (i.e., *PMF*) [7] with 30 latent factors for predicting the hold time. Third, we choose some supervised machine learning techniques for prediction. The selected features for these methods are listed in Table 2.

In our experiments, we exploited Weka to conduct the above baselines, which is an open source software under the GNU General Public License[1]. The parameters of these machine learning methods and the value of λ in our naive RP model are set according to 10-fold cross validation. Besides, we empirically set K to be 3 in dynamic RP model. The time of *kernel points* were set to September 15th, November 15th and January 15th for the first semester, while March 15th, May 15th and July 15 for the second semester, since they correspond to the start, middle and end of each semester, respectively.

---

[1] http://www.cs.waikato.ac.nz/ml/weka/.

Table 2. Selected prediction features

For student features:
• Borrowing number previously
• Minimum hold days of borrowed books previously
• Maximum hold days of borrowed books previously
• Average hold days of borrowed books previously
• Hold days of borrowed book last time
• Number of times coming to library previously
For book features:
• Number of times been borrowed
• Minimum hold days been borrowed
• Maximum hold days been borrowed
• Average hold days been borrowed
For interaction features:
• Borrowing semester
• Borrowing month

Evaluation Metrics. For predicting the hold time of borrowed books, we can treat it as a regression problem or classification problem. The regression performance can be evaluated by the Root Mean Squared Error (RMSE), which is defined as $RMSE = \sqrt{\frac{\sum_{(u,i) \in C}(h_{u,i}(t) - \hat{h_{u,i}}(t))^2}{|C|}}$, where $\hat{h_{u,i}}(t)$ is the predicted hold time of the real value $h_{u,i}(t)$, and C denotes the test set. As for classification performance, we can separate the records into two classes for prediction, where one represents those returned back within a month (i.e., 30 days) and the other represents those longer than a month. We regard the books returned back within 30 days as positive class and leverage the classic evaluation metric *F-measure* for evaluation, which is harmonic mean of metric *Precision* and *Recall*.

Performance and Discussion. Fig. 3 shows the RMSE performance and F-measure performance with respect to students who borrowed different number of books totally. From the two figures, we can observe that our dynamic RP model has the best prediction performance. As for the naive RP model, it has comparative performance with the machine learning methods. Particularly, more attention should be paid to the performance of *PMF*, which has the highest value of RMSE, even higher than the *Average* method. This indicates that the relationship between students and books based on the latent preferences are not important for books hold time.

Based on the above experimental results, we can obtain the following conclusions. First, our two proposed models are effective in predicting the hold time of books borrowed by college students, especially the dynamic RP model. This guarantees the validity of our procrastination assumption. Although there still

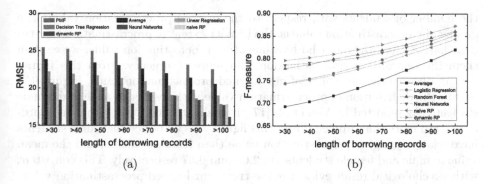

Fig. 3. Performance with respect to students who borrowed different number of books. (a) Regression performance. (b) Classification performance.

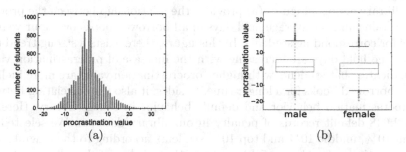

Fig. 4. Distribution of the learned procrastination value. (a): The number of students with respect to the procrastination value. (b): The distribution of procrastination value with to male and female students.

exists some predictive bias, two reasons can explain this result. One is that the events happened in a short period of time for each student are hardly captured, which can also affect the hold time of borrowed books more or less. The other is that there are still some errors in the estimation of *required reading time*, which is also difficult to seek the optimal estimation. Second, some machine learning methods such as *Decision Tree Regression* and *Neural Networks* have nearly same regression performance with the naive RP model. This implies that these machine learning methods are very likely to discover the knowledge about the *required reading time* and procrastination factors from data, while it is difficult for them to capture more dynamic factors. Moreover, the lack of the ability to discover the procrastination of students is the crucial limitation, which is the superiority of our models.

5.2   Procrastination Verification

In this subsection, we validate our learned procrastination value of students.

Distribution Evidences. Figure 4 shows the distribution of the procrastination value of all 10,035 students in our data set. Specifically, Fig. 4(a) illustrates the number of students with respect to the learned procrastination value. The greater the procrastination value is, the heavier extent of procrastination the student has. As for students who have negative procrastination value, we explain them the opposite of procrastination, the sense of urgency. From this figure, we can observe the normality of our learned procrastination value. Figure 4(b) reveals the distribution of procrastination value with respect to male and female students. As reported by Van et al. [17], it is slightly more likely that men procrastinate more than women. From this figure, we can notice that male students have relatively higher procrastination value than female students, and the mean value of male and female students are 2.05 and 0.57 respectively. This consistent with psychological result evidences the truth our learned procrastination value.

Library Actions Evidences. To illustrate how procrastination makes impact on the library actions. Figure 5(a) provides the relationship between the procrastination value and the average hold days of all borrowed books over four years in college for corresponding students. In this figure, there exists a strong trend that the average hold days are increasing with the increase of procrastination value. This indicates that students with higher procrastination value are more inclined to hold borrowed books for a longer time. Besides, it also has correlation between the procrastination behavior and default behavior in college library. Here, we have 32,648 default records of penalty in our library data set. We selected the bottom 10 %, middle 10 % and top 10 % students according to the magnitude of procrastination value. Figure 5(b) presents the number of default records with these students. In this figure, apparently, the top 10 % students with the highest procrastination value have much more default records, whereas the bottom 10 % students have much less. It is nature that those top 10 % students procrastinate to return their borrowed books, then gradually evolve to forget it and exceed the due return time. According to the above statistics, the rationality of our learned procrastination value can be verified. Besides, as an application of library service, students with high procrastination level will be reminded to return the borrowed books in case these books are in urgent need by other students.

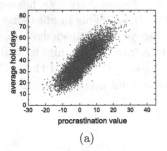
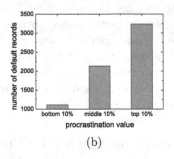

(a)   (b)

Fig. 5. The statistics for the procrastination value with library actions. (a): The average hold days of all borrowed books over four years in college for students with corresponding procrastination value. (b): The number of default records in library with the bottom 10 %, middle 10 % and top 10 % procrastination value.

Association Evidences. To further verify our learned procrastination value, we take the association results between procrastination value and other characteristics of students with psychological fields conclusions for comparison. As Culnan et al. [3] revealed that procrastination may influence college freshmen weight change, we first test our learned procrastination value on students' body weight data. Figure 6 shows the average standard deviation of students weight over four years in college with the bottom 10%, middle 10% and top 10% procrastination

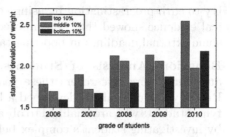

Fig. 6. The average standard deviation of weight over four years in college with the bottom 10%, middle 10% and top 10% procrastination value, grouped by entering college year.

value. Here, we group students according to their entering college year. From this figure, we can find that the weight of students with highest procrastination value fluctuates more greatly over four years in college than that of students with lowest procrastination value. This statistical result is consistent with Culnan's conclusion, which evidences the reliability of our learned procrastination value. As for academic performance, Steel et al. [12] noted that procrastination may not have contributed significantly to poorer grades and students who completed all of the practice exercises tended to perform well on the final exam no matter how much they delayed. Therefore, we test on the scholarship data. There are 1,267 students winning a scholarship, and the number of students having procrastination value in the bottom 10%, middle 10% and top 10% account for 58, 70 and 76 respectively. We find that the gap between top 10% and bottom 10% or middle 10% is not obvious, it seems that the number of students with higher procrastination value is even more. This consistent statistics also gives more evidence to the validity of our learned procrastination value.

6 Related Work

Generally, the related work of this paper can be grouped into two categories, i.e., procrastination in psychological research and behaviors analysis of students.

Procrastination in Psychological Research. Procrastination has been studied for a long time in the field of psychology. One of research directions is the study of impact brought by procrastination behavior. Tice et al. [16] found that some negative associations are linked to procrastination, such as depression, anxiety, irrational behaviour and low self-esteem. Another research direction focuses on the exploration of the possible causes of procrastination. For example, Steel et al. [11] revealed that task aversiveness, task delay, self-efficacy, and impulsiveness, as well as conscientiousness and its facets of self-control, distractibility, organization, and achievement motivation are strong and consistent predictors of procrastination. Study also involves in result of the remedy of procrastination behavior. For example, Ariely et al. [2] specially studied people who strategically try to curb procrastination by using costly self-imposed deadlines. Their empirical evidence showed that self-imposed deadlines are not always as effective as some external deadlines in boosting task performance.

Behaviors Analysis of Students. The behaviors analysis of students is attracting more researchers these years due to the increasing available data. Recently, Guan et al. [4] developed a learning framework Dis-HARD for identifying students who are qualified to obtain the financial funding support in college by investigating student's complex behaviors within campus. Agrawal et al. [1] proposed solutions for grouping students who exhibit different ability level into sections so that the overall gain for students is maximized. To examine students' learning process and improve their study performance, researches also covered in education, emerging educational data mining (EDM) [9].

Above all, however, no existing work has explored the procrastination from college library data, which comes out our procrastination exploration work.

7   Concluding Remarks

In this paper, we introduced a data-driven study from a behavioral perspective to explore the procrastination of college students. To this end, we proposed an unsupervised approach to quantitatively estimate the procrastination level of students through the analysis of their borrowing records in college library. Specifically, we first propose a naive Reading-Procrastination (naive RP) model, which takes consideration of the behavioral similarity between students for procrastination discovery. Furthermore, to improve the discovery performance, we develop a dynamic Reading-Procrastination (dynamic RP) model by integrating more comprehensive characteristics of student behaviors, such as semester-awareness and month-regularity. A unique characteristics of the dynamic RP model is it can depict the procrastination behavior in a probabilistic and empirical Bayesian perspective. Finally, we conducted extensive experiments on several real-world data sets collected from a Chinese college. The experimental results clearly demonstrated the effectiveness of our approach, and verified several key findings from psychological fields.

Acknowledgments. This research was partially supported by grants from the National Science Foundation for Distinguished Young Scholars of China (Grant No. 61325010), the Science and Technology Program for Public Wellbeing (Grant No. 2013GS340302). Qi Liu gratefully acknowledges the support of the Youth Innovation Promotion Association of CAS and acknowledges the support of the CCF-Intel Young Faculty Researcher Program (YFRP).

References

1. Agrawal, R., Golshan, B., Terzi, E.: Grouping students in educational settings. In: SIGKDD 2014, pp. 1017–1026. ACM (2014)
2. Ariely, D., Wertenbroch, K.: Procrastination, deadlines, and performance: self-control by precommitment. Psychol. Sci. **13**(3), 219–224 (2002)
3. Culnan, E., Kloss, J.D., Grandner, M.: A prospective study of weight gain associated with chronotype among college freshmen. Chronobiol. Int. **30**(5), 682–690 (2013)
4. Guan, C., Lu, X., Li, X., Chen, E., Zhou, W., Xiong, H.: Discovery of college students in financial hardship. In: ICDM. IEEE (2015)
5. Jordan, M.I., Ghahramani, Z., Jaakkola, T.S., Saul, L.K.: An introduction to variational methods for graphical models. Mach. Learn. **37**(2), 183–233 (1999)
6. Klassen, R.M., Krawchuk, L.L., Rajani, S.: Academic procrastination of undergraduates: low self-efficacy to self-regulate predicts higher levels of procrastination. Contemp. Educ. Psychol. **33**(4), 915–931 (2008)
7. Mnih, A., Salakhutdinov, R.: Probabilistic matrix factorization. In: Advances in neural information processing systems, pp. 1257–1264 (2007)
8. Neal, R.M.: Probabilistic inference using markov chain monte carlo methods. Technical report CRG-TR-93-1 (1993)
9. Romero, C., Ventura, S.: Data mining in education. Wiley Interdisc. Rev.: Data Min. Knowl. Discov. (DMKD) **3**(1), 12–27 (2013)
10. Solomon, L.J., Rothblum, E.D.: Academic procrastination: frequency and cognitive-behavioral correlates. J. Couns. Psychol. **31**(4), 503 (1984)
11. Steel, P.: The nature of procrastination: a meta-analytic and theoretical review of quintessential self-regulatory failure. Psychol. Bull. **133**(1), 65 (2007)
12. Steel, P.: The Procrastination Equation: How to Stop Putting Things Off and Start Getting Stuff Done. Random House Canada, Toronto (2010)
13. Steel, P., Brothen, T., Wambach, C.: Procrastination and personality, performance, and mood. Personality Individ. Differ. **30**(1), 95–106 (2001)
14. Sub, A., Prabha, C.: Academic performance in relation to perfectionism, test procrastination and test anxiety of high school children. Psychological Studies (2003)
15. Tan, P.N., Steinbach, M., Kumar, V., et al.: Introduction to Data Mining, vol. 1. Pearson Addison Wesley, Boston (2006)
16. Tice, D.M., Baumeister, R.F.: Longitudinal study of procrastination, performance, stress, and health: the costs and benefits of dawdling. Psychol. Sci. **8**, 454–458 (1997)
17. Van Eerde, W.: A meta-analytically derived nomological network of procrastination. Personality Individ. Differ. **35**(6), 1401–1418 (2003)
18. Wikipedia: Procrastination – wikipedia, the free encyclopedia (2015). http://en.wikipedia.org/wiki/Procrastination. Accessed 1 May 2015

Recommendation

Joint User Attributes and Item Category in Factor Models for Rating Prediction

Jiang Wang[1], Yuqing Zhu[2], Deying Li[1(✉)], Wenping Chen[1], and Yongcai Wang[1]

[1] School of Information, Renmin University of China, Beijing 100872, China
{jiangw,deyingli,wenpingchen,ycw}@ruc.edu.cn
[2] Department of Computer Science, California State University at Los Angeles, Los Angeles, CA 90032, USA
yzhu14@calstatela.edu

Abstract. One important problem of recommender system is rating prediction. In this paper, we use the movie rating data from Movie-Lens as an example to show how to use users' attributes to improve the accuracy of rating prediction. Through data analysis, we observe that users having similar attributes tend to share more similar preferences and users with a special attribute have their own preferred items. Based on the two observations, we assume that a user's rating to an item is determined by both the user intrinsic characteristics and the user common characteristics. Using the widely adopted latent factor model for rating prediction, in our proposed solution, we use two kinds of latent factors to model a user: one for the user intrinsic characteristics and the other for the user common characteristics. The latter encodes the influence of users' attributes which include user age, gender and occupation. On the other hand, we jointly use user attributes or item category information and rating data for calculating similarity of users or items. The similarity calculating results are used in our proposed latent factor model as a regularization term to regularize users or items latent factors gap. Experimental results on MovieLens show that by incorporating users' attributes influences, much lower prediction error is achieved than the state-of-the-art models. The prediction error is further reduced by incorporating influences from item category popularity and item popularity.

Keywords: Recommendation · Rating prediction · Matrix factorization

1 Introduction

With the rapid increasing of web information, recommender system is becoming more and more popular since it is good for both consumers and service or product provider. On one hand, it helps consumers to alleviate the information overload problem. On the other hand, it brings better sales to product or service provider. Diversity recommender systems have been designed to meet the needs of different platforms in recent years. Such as commodities recommendation for

© Springer International Publishing Switzerland 2016
S.B. Navathe et al. (Eds.): DASFAA 2016, Part I, LNCS 9642, pp. 277–296, 2016.
DOI: 10.1007/978-3-319-32025-0_18

E-commerce websites, friends recommendation for social network platform and point-of-interest recommendation for location-based social networks (LBSN).

Depending on the application, different recommendation problems have been defined and studied. The top-N item recommendation and rating prediction are two most widely studied categories of recommendation problems. On one hand, top-N item recommendation tasks aim to recommend a user a list of items that she may be interested in. For example, in movie websites, movie recommendation aims to recommend unseen movies to users. Rating prediction, on the other hand, is to predict the preference rating of a user to a product or service that she has not rated before. The products or services with high predicted ratings are recommended to users. In this study, we are interested in the movie rating prediction problem with user attribute data from MovieLens.

In recent years, social information is widely used in the recommendation system and the performance has been improved a lot. One popular method using social information in recommendation systems is SocialMF [1]. SocialMF is proposed by adding the social regularization term to the matrix factorization objective function. The additional social regularization term ensures that the distance of the latent vectors of two users or items will become closer if two users or items are more similar. Commonly, users' or items' similarity are computed based on user-item ratings, which is not applicable for new users or items. In our dataset, both user attributes and item categories information are available. For item rating prediction, an interesting question here is: Are the user attributes and item categories information helpful to similarity calculation and can they improve the accuracy of rating prediction?

To answer the question above, we conducted data analysis on MovieLens's movie rating data. We observe that users having similar attributes tend to share more similar user preferences and users with certain attribute have their own favorite items, while other users without that attribute may not be interested in those items. Based on these observations, we incorporate user attributes influences into our movie rating prediction model which is based on the widely adopted latent factor model realized by Matrix Factorization (MF). Together with influences from other factors including item category, user neighbor, item neighbor, item popularity and item category popularity, we show that the proposed model outperforms state-of-the-art baselines including Biased MF, RSVD and Social MF [1,2], measured by both Mean Absolute Error (MAE) and Root Mean Square Error (RMSE). To the best of our knowledge, this is the first study exploiting user attributes as latent factor in item rating prediction. To summarize, the main contributions arise from this study are as follows:

- We conduct data analysis and observe that users having similar attributes tend to share more similar user preferences and users with certain attribute have their own favorite items, while other users without that attribute may not be interested in those items. This is an important observation that could be useful for not only item rating prediction, but also related studies, e.g., item recommendation and user attribute analysis from rating data.

- We directly model the influence of user attributes into movie rating prediction using matrix factorization. Specifically, for each user, we use two different latent factors to represent his or her intrinsic and common characteristics respectively. Every user's intrinsic latent factors are unique and determined by himself or herself. While common latent factors of a user are determined by his or her attributes' type and every attribute has a latent factors. On the whole, a user's latent factors are modeled as the linear combination of his or her intrinsic latent factors and all attributes' latent factors. In our recommender system, we have also considered other factors including item popularity and item category popularity. To the best of our knowledge, this is the first study that models user attribute influence as latent factors into rating prediction, while previous studies only take a attribute as a bias.
- We conducted extensive experiments to evaluate the effectiveness of incorporating influence of user attributes and other factors in item rating prediction and compared the prediction accuracy of the proposed model with an array of strong baseline models. The experiment results show that using both user or item profile information and rating sequence information for selecting appropriate neighbor can further improve the performance and different user attributes have imparity status in item rating prediction.

The rest of this paper is organized as follows. We review the related work in Sect. 2. The data analysis is reported in Sect. 3. The details of the proposed item rating prediction model is presented in Sect. 4, Experimental evaluation is showed in Sect. 5. Finally, Sect. 6 concludes this paper.

2 Related Work

The collaborative filtering (CF) technique has become more and more popular in personalized recommender task [3–6] since CF methods are domain independent and only require the past activities history of users, i.e. user-item rating matrix, to make recommendations. According to different means of utilizing the user-item rating matrix, collaborative filtering approaches are usually divided into two kinds of categories [7]: memory-based CF and model-based CF.

Memory-based CF, also called as neighbor-based methods, use the history user rating behaviors data for recommendations. The basic idea of memory-based CF is to find similar users or items for target user or item by using similarity measures. Once the neighborhoods are formed, memory-based CF usually takes a weighted sum of ratings given by their neighbors (target user' neighbors or target item' neighbors) as a prediction. user-based CF [7,8] and item-based CF [9,10] are two typical kinds of memory-based method. User-based CF predicts the ratings based on the opinions of target user's neighbors, which have similar item preferences with target user. On the other hand, item-based CF provide prediction based on the ratings of items that is similar to target item.

In contrast with memory-based CF, which utilize entire user-item matrix to provide recommendations for target users, model-based CF make use of statistical and machine learning techniques to learn a predictive model from training data. The predictive model can characterize the rating behaviors of target users. Then model-based CF use the trained model to predict the unobserved ratings, rather than directly utilize the entire user-item matrix to compute predictions. Latent factor model is one of the most successful CF models, in which users and items are jointly mapped into a shared latent space of much lower dimensionality. As the most successful realization of latent factor model, matrix factorization (MF) [1,11,12] has been successfully applied to various recommender systems including location rating prediction for Jiepang [13] and Yelp [14], event recommendation for Meetup [15] and Douban [16], and personalized tweet tag recommendation [17].

Among various MF models proposed, SVD++ [11] is probably one of the most successful models. This model integrates the implicit feedback information from a user to items (e.g., based on user' s purchase history or browsing history). More specifically, the user vector of latent factors in this model is complemented by the latent factors of the items to which the user has provided implicit feedback. Recently, Ma [1] proposed a social regularization MF method, named Social MF, to employ the similar and dissimilar relationships between users and items to improve recommendation accuracy. The similarity between items is measured based on their ratings using Pearson' s correlation or cosine similarity. It is usually difficult to reflect the real similarity of two users or items that only using the rating information to measure similarity. In many literatures, many other user-/item-specific attributes are introduced to calculating user similarity or item similarity. In contrast with rating data, user attribute information, for example, age, gender, occupation for a user, can represent the characteristics more accurately. Item has a similar situation with user. In our experiments, user attribute information and item category information are introduced to calculating similarity, and the proposed models achieve better rating prediction accuracy than Social MF [1] indicating that adding the influence from attributes neighborhood is more effective than the influence only from neighbors chosen by rating-based similarity measures.

Based on MF, influence from other aspects of users or items besides the ratings can be flexibly and easily modeled. For example, Koenigstein et al. [18] incorporated rich item bias into MF model to capture the taxonomy information of music. Each music has multiple types of information such as track, album, artist and genre. In their proposed model, MF was extended by adding shared bias parameters for items linked by common taxonomy. Moreover, some other work has also shown that popularity is helpful in improving the recommendation accuracy [19,20]. To the best of our knowledge, user or item attributes are introduced to most of literatures as biases. However, few work focus on taking user or item attributes information as latent factors to improve the quality of recommendation. In our problem setting, we consider the attributes from a user further elaborates about the rating. We model users' attributes as latent factors and incorporate the attributes into rating prediction. Next, we conduct data analysis of the MovieLens data.

3 Data Analysis

3.1 Dataset

Our study is based on the popular released MovieLens Dataset[1]. The data was collected through the MovieLens web site[2] during the seven-month period from September 19th, 1997 through April 22nd, 1998. It contains 100,000 user-item ratings (scale from 1 to 5) rated by 943 users on 1,682 items.

This data has been cleaned up - users who had less than 20 ratings or did not have complete demographic information were removed from this data set. A user contains a unique id, name, age, gender and occupation. This data have 21 occupations and a user only has one of them. An item has a unique id, name, release date, URL and its categories. This data have 19 categories and an item may have more than one category. However, we don't use released date and URL in our study. A rating tuple contains user id, item id, rating from 1 to 5 stars and date. In this study, we do not use the date feature for its less relevance to our research problem. Table 1 reports the minimum, maximum, and average number of ratings per user and per item respectively.

Table 1. Statistics of dataset movieLens

Statistics	User	Item
Min. Num. of Ratings	20	1
Max. Num. of Ratings	737	583
Avg. Num. of Ratings	106.04	59.45

3.2 Observations

To gain a better understanding of users' rating behaviors, we study the rating data introduced above. More specifically, we study the influence of user attributes for rating behavior at two levels: item-level and rating-level. We make two observations from the data analysis: (a) Users have similar attributes tend to share more similar user preferences; and (b) Users with certain attribute have their own favorite items, while other users without that attribute may not be interested in those items. These patterns are independent of any particular user. We now report the details of the data analysis.

Item Level. To better study users rating behaviors at rating level, we cluster the users into groups by user attributes. Users having same attribute value are clustered in same group. As an example, we cluster users into female group and male group by gender. In order to observe the item level rating behaviors more intuitively, we measure the Rating Count Difference(RCD) of each item i for

---

[1] http://www.grouplens.org/system/files/ml-100k.zip.
[2] movielens.umn.edu.

two user groups us_j and us_k, these two user groups have different attribute value a_j and a_k, respectively. Considering the user number gap of those two user groups, using Absolute Rating Count Difference(ARCD) is unfair. Hence we use Normalized Rating Count Difference(NRCD) to measure RCD, defined as follows.

$$NRCD_i(us_j, us_k) = \frac{\frac{\sum_{t \in I} |us_{kt}|}{\sum_{t \in I} |us_{jt}|} \cdot | us_{ji} | - | us_{ki} |}{\max_{t \in I}(\frac{\sum_{t \in I} |us_{kt}|}{\sum_{t \in I} |us_{jt}|} \cdot | us_{ji} | - | us_{ki} |)} \quad (1)$$

Where $NRCD_i(us_j, us_k)$ is normalized rating count difference of user group us_j and user group us_k for item i, I is the set of all items, us_{ji} and us_{ki} denote the two sets of users who have rated item i, and $| \bullet |$ denotes the cardinality of a set.

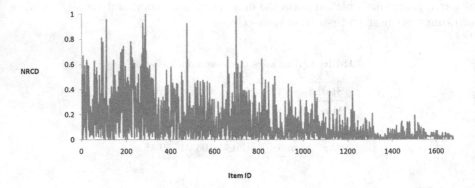

Fig. 1. Normalized Rating Count Difference(NRCD) of female and male

Figure 1 plots the rating count difference(NRCD) of female and male. As shown in Fig. 1, female's rating count for each item shows great difference to male's. Specifically, about 13 % items' NRCD is over 0.3 and average of NRCD is about 0.13. Another interesting phenomenon is about 45 % items' NRCD less than 0.05. These two phenomena indicate that female and male share similar preferences for most of items, while for some particular items, their preferences show huge gap. In other words, users' gender has a certain degree of influence for users' rating behaviors. For age (3 age groups) and occupation (21 occupations), Rating count difference have similar circumstance with gender. As a result of the limitation of space, we omit the NRCD figures of age groups and occupations. Table 2 lists top-5 ARCD items of young adults and mid adults. Item 286 has the max ARCD of 213, while the max NRCD of items is 515. This is because NRCD and ARCD are two measure methods which have a certain relation, but inequitable. Any way, these two measure methods both indicate user attributes have influence of user's rating behaviors in another way.

Table 2. Top-5 ARCD items of Young Adults(YA) and Mid Adults(MA)

Item ID	RC YA	RC MA	NRCD	ARCD
286	87	300	0.67	**213**
269	54	206	0.55	152
515	13	151	**0.83**	138
211	19	152	0.74	133
483	37	169	0.57	132

Table 3. Statistics of Student(S) and Educator(E)

Statistics	Item id	Group	Diff
Max. of Avg Ratings Difference	919	-	1.46
Min. of Avg Ratings Difference	257,etc.	-	0
Avg. of Avg Ratings Difference	-	-	0.32
Max. of Avg Ratings	170	S	4.93
Max. of Avg Ratings	48	E	4.69

Rating Level. At the rating level, we analyze the average of each item's rating for diverse user groups. Without loss of generality, we use student group and educator group as an example. These two user groups clustered according occupation. To avoid the influence of sparse data, we don't consider items which are less than 5 user rated in either student or educator. After the preprocessing, only 500 items remind. But we surprising discover that about 30 % items' average difference are larger than 0.5. Especially, the max gap reaches 1.46 for item 919. More detailed statistics of these two groups are summarized in Table 3.

We also compare others user group pairs, they have similar circumstance with student and educator group. From these analysis in rating level, we can get a similar conclusion with item level — user attributes have certain influence for user rating behaviors.

Motivated by the two observations, we propose incorporating the above attribute characteristics for rating prediction, so as to improve item recommendation accuracy.

4 Rating Prediction

Rating prediction is a basic problem in recommendation system and has been widely studied in literature. In this paper, we use r_{ui} to denote the rating that user u gives to item i (i.e., a movie), and r_{ui} is in the range of 1 to 5 stars with more stars indicating higher preference. Given the existing ratings made by M users to N items, the task is to predict the unknown rating $\hat{r}_{ui}$, if user u has not

Table 4. Notations and semantics

N_i	Set of neighbors of item i
K	Set of (u,i) pairs with known r_{ui} ratings
T	Set of (u,i) pairs using test
s_{uf}	Similarity of user(item) u and user(item)f
s_{uf}^a	Attribute similarity of user(item) u and user(item)f
s_{uf}^r	Rating similarity of user(item) u and user(item)f
r_{ui}	Observed ratings of user u to item i
$\hat{r}_{ui}$	Predicted ratings of user u to item i
μ	Mean of all known r_{ui} ratings
b_u	Bias parameters for user u
b_i	Bias parameters for item i
$\mathbf{p}_u$	Latent factors of user u
$\mathbf{q}_i$	Latent factors of item i
$\mathbf{a}_{ua}$	Latent factors of user u's age group
$\mathbf{g}_{ug}$	Latent factors of user u's gender
$\mathbf{o}_{uo}$	Latent factors of user u's occupation
γ	Similarity weighting parameter
ρ_i	Normalized popularity of item i
γ_i	Popularity weighting parameter for item i
τ_c	Normalized popularity of item category c
γ_c	Popularity weighting parameter for item category c

rated item i before. In the following, we first briefly introduce matrix factorization and then present our proposed model by incorporating various influences into the prediction. Table 4 lists the notations used in this paper.

4.1   Matrix Factorization

Our proposed method is based on the latent factor model realized by matrix factorization. Matrix factorization map users and items into a joint latent space with dimension $f << min(M; N)$. The inner product of a user vector $\mathbf{p}_u \in R^{f \times 1}$ and an item vector $\mathbf{q}_i \in R^{f \times 1}$ is used to approximate the user's preference to the item (see [21] for a detailed introduction of matrix factorization). Accordingly, the predicted rating of user u to item i is computed using

$$\hat{r}_{ui} = \mathbf{p}_u^T \mathbf{q}_i, \tag{2}$$

where $\mathbf{p}_u$ and $\mathbf{q}_i$ can be learned from the user-item rating matrix with known ratings. However, users may have certain degree of biases: some users are more lenient and some are very strict about ratings. Similarly, items may also have some degree of biases because of location or branding for example. To achieve more accurate rating prediction, Biased MF extends the basic matrix factorization by considering the biases,

$$\hat{r}_{ui} = \mu + b_u + b_i + \mathbf{p}_u^T \mathbf{q}_i, \tag{3}$$

where μ is the average rating of all known ratings, b_u and b_i are the user bias and item bias, respectively. Learning the unknown parameters $\mathbf{p}_u$, $\mathbf{q}_i$, b_u and b_i is an optimization problem to minimize the regularized squared error on the set of known ratings K.

$$\min_{p^*, q^*, b^*} \sum_{(u,i) \in K} (r_{ui} - \hat{r}_{ui})^2 + \lambda_1(\| \mathbf{p}_u \|^2 + \| \mathbf{q}_i \|^2) + \lambda_2(b_u^2 + b_i^2)$$

In this equation, λ_1 and λ_2 are regularization parameters used to avoid overfitting. Both stochastic gradient descent (SGD) and alternating least squares (ALS) algorithms can be used to solve the optimization function and learn the parameters [2,21]. In this paper, we adopt SGD to learn the parameters following the algorithm presented in [2].

4.2   Incorporating Neighborhood Influence

Inspired by [1,22], user or item neighbors information can help rating prediction. In the case of missing explicit neighbor information in MovieLens, we use top-N similar users of target user instead of his or her neighbors information and then plug in those similar users to the aforementioned matrix factorization framework. There are several methods we can borrow in the literature to compare the similarity between two users. In this paper, we jointly adopt rating similarity and attribute similarity, which is defined as:

$$s_{uf} = \gamma \cdot s_{uf}^a + (1 - \gamma) \cdot s_{uf}^r$$

$$s_{uf}^r = \frac{\sum_{k \in I(u) \cap I(f)} (r_{uk} - \bar{r}_u) \cdot (r_{fk} - \bar{r}_f)}{\sqrt{\sum_{k \in I(u) \cap I(f)} (r_{uk} - \bar{r}_u)^2} \cdot \sqrt{\sum_{k \in I(u) \cap I(f)} (r_{fk} - \bar{r}_f)^2}}$$

$$s_{uf}^a = \frac{1}{3}(s_{uf}^{age} + s_{uf}^{gender} + s_{uf}^{occupation})$$

where s_{uf} is the similarity between user u and f, s_{uf}^r indicates the rating similarity and we adopt Pearson Correlation Coefficient(PCC) to calculate it. $I(u)$ is a set of items that rated by user u, and $\bar{r}_u$ represents the average rate of user u. s_{uf}^a represents the attribute similarity and it's a linear combination of age, gender and occupation similarity. Age similarity s_{uf}^{age} is a normalized value of the age difference of user u and f. Gender similarity s_{uf}^{gender} is equal to 1 if the gender of user u and f are the same, otherwise it is a small smooth value. The calculation method of occupation similarity $s_{uf}^{occupation}$ is similar to s_{uf}^{age}. And parameter γ

is introduced to balance the importance between rating and attribute similarity. Incorporating neighborhood influence, we proposed the matrix factorization objective function as follow:

$$\min_{p^*,q^*,b^*} \sum_{(u,i)\in K} (r_{ui} - \hat{r}_{ui})^2 + \beta_1 \sum_{f\in N(u)} s_{uf} \parallel \mathbf{p}_u - \mathbf{p}_f \parallel^2 + \lambda_1(\parallel \mathbf{p}_u \parallel^2 + \parallel \mathbf{q}_i \parallel^2) + \lambda_2(b_u^2 + b_i^2)$$

where β_1 is the regularization parameter, and $N(u)$ represents user i's neighbors. In this method, the neighbors information is employed in designing the neighbor regularization term to constrain the matrix factorization objective function. The neighbor regularization term also indirectly models the difference of users' tastes. More specifically, if user u has a high similarity with user f, this regularization term actually indirectly minimizes the distance between latent vectors $\mathbf{p}_u$ and $\mathbf{p}_f$.

From the above, since we define the implicit user social information as the similar users, we can naturally extend this idea to take advantages of the implicit item social information, which can be found through the similar items. The similarity calculation method of items are similar to users, and also includes rating similarity and attribute similarity. Attribute similarity is calculated by cosine similarity of two items' category vector. The Social Regularization method described above is a very general approach, and it can be easily extended to incorporate the item social information. The objective function can be formulated as:

$$\min_{p^*,q^*,b^*} \sum_{(u,i)\in K} (r_{ui} - \hat{r}_{ui})^2 + \beta_1 \sum_{f\in N(u)} s_{uf} \parallel \mathbf{p}_u - \mathbf{p}_f \parallel^2 + \beta_2 \sum_{f'\in N(i)} s_{if'} \parallel \mathbf{q}_i - \mathbf{q}_{f'} \parallel^2$$
$$+\lambda_1(\parallel \mathbf{p}_u \parallel^2 + \parallel \mathbf{q}_i \parallel^2) + \lambda_2(b_u^2 + b_i^2)$$

4.3 Incorporating User Profile Influence

Based on our observations in Sect. 3.2, user's attributes have influence of item rating. These observations suggest that considering user attributes influence may improve the accuracy of item rating prediction.

In this paper, to model users' rating behavior, we first assume that a user's rating preference is determined by the user intrinsic characteristics and the common characteristics. Each user' intrinsic characteristics are unique, while common characteristics are shared by all users. Limited by the data set, common characteristics using in this paper contain age, gender and occupation characteristics. For a user u, we use $\mathbf{p}_u$, $\mathbf{a}_{ua}$, $\mathbf{g}_{ug}$ and $\mathbf{o}_{uo}$ to model its intrinsic, age, gender and occupation characteristics, respectively. Next, we use user profile for rating prediction in our proposed methods.

Age Influence. Analyzed in Sect. 3.2, users' age span in MovieLens is very big, the youngest is only 7 years old, while the oldest is 73. But users with similar ages show close preferences. In our model, we introduce age latent factors to exploit age groups for more accurate item rating prediction. For an age group

Table 5. Objective functions for incorporating neighborhood influence, age influence, gender and other factors

$$
\min_{p\star,q\star,a\star,b\star} \sum_{(u,i)\in K} (r_{ui} - \hat{r}_{ui})^2 + \beta_1 \cdot \sum_{f\in N(u)} s_{uf} \parallel \mathbf{p}_u - \mathbf{p}_f \parallel^2 + \beta_2 \cdot \sum_{f'\in N(i)} s_{if'} \tag{9}
$$
$$
\parallel \mathbf{q}_i - \mathbf{q}_{f'} \parallel^2 + \lambda_1 (\parallel \mathbf{p}_u \parallel^2 + \parallel \mathbf{q}_i \parallel^2) + \lambda_2 (b_u^2 + b_i^2) + \lambda_3 \parallel \mathbf{a}_a \parallel^2
$$

$$
\min_{p\star,q\star,a\star,g\star,b\star} \sum_{(u,i)\in K} (r_{ui} - \hat{r}_{ui})^2 + \beta_1 \cdot \sum_{f\in N(u)} s_{uf} \parallel \mathbf{p}_u - \mathbf{p}_f \parallel^2 + \beta_2 \cdot \sum_{f'\in N(i)} s_{if'} \tag{10}
$$
$$
\parallel \mathbf{q}_i - \mathbf{q}_{f'} \parallel^2 + \lambda_1 (\parallel \mathbf{p}_u \parallel^2 + \parallel \mathbf{q}_i \parallel^2) + \lambda_2 (b_u^2 + b_i^2) + \lambda_3 (\parallel \mathbf{a}_a \parallel^2 + \parallel \mathbf{g}_g \parallel^2)
$$

$$
\min_{p\star,q\star,a\star,g\star,o\star,b\star} \sum_{(u,i)\in K} (r_{ui} - \hat{r}_{ui})^2 + \beta_1 \cdot \sum_{f\in N(u)} s_{uf} \parallel \mathbf{p}_u - \mathbf{p}_f \parallel^2
$$
$$
+ \beta_2 \cdot \sum_{f'\in N(i)} s_{if'} \parallel \mathbf{q}_i - \mathbf{q}_{f'} \parallel^2 + \lambda_1 (\parallel \mathbf{p}_u \parallel^2 + \parallel \mathbf{q}_i \parallel^2) + \lambda_2 (b_u^2 + b_i^2) \tag{11}
$$
$$
+ \lambda_3 (\parallel \mathbf{a}_a \parallel^2 + \parallel \mathbf{g}_g \parallel^2 + \parallel \mathbf{o}_o \parallel^2)
$$

$$
\min_{p\star,q\star,a\star,g\star,o\star,\gamma\star,b\star} \sum_{(u,i)\in K} (r_{ui} - \hat{r}_{ui})^2 + \beta_1 \cdot \sum_{f\in N(u)} s_{uf} \parallel \mathbf{p}_u - \mathbf{p}_f \parallel^2
$$
$$
+ \beta_2 \cdot \sum_{f'\in N(i)} s_{if'} \parallel \mathbf{q}_i - \mathbf{q}_{f'} \parallel^2 + \lambda_1 (\parallel \mathbf{p}_u \parallel^2 + \parallel \mathbf{q}_i \parallel^2) \tag{12}
$$
$$
+ \lambda_2 (b_u^2 + b_i^2 + \gamma_i^2 + \sum_{c\in C_i} \gamma_c^2) + \lambda_3 (\parallel \mathbf{a}_a \parallel^2 + \parallel \mathbf{g}_g \parallel^2 + \parallel \mathbf{o}_o \parallel^2)
$$

a, it is associated with a latent vector $\mathbf{a}_a \in R^{f\times 1}$. Let u_a be the age group of user u and $\mathbf{a}_{ua}$ be the age factor of user u. By incorporating the age influence, the predicted rating $\hat{r}_{ui}$ is now defined in Eq. 4, where $\alpha_1 \in [0, 1]$ is a parameter that controls the importance of age influence in rating prediction. The objective function is updated accordingly, see Eq. 9 in Table 5.

$$
\hat{r}_{ui} = \mu + b_u + b_i + (\mathbf{p}_u + \alpha_1 \cdot \mathbf{a}_{ua})^T \mathbf{q}_i, \tag{4}
$$

Gender Influence. Similar with age, we study the relationship between user gender and their rating behavior in Sect. 3.2. Users in same gender have smaller preference difference than those in opposite genders. To better use gender information in our model, we introduce gender latent factors to explore the influence of gender in item recommendation. Female or male is associated with a latent vector $\mathbf{g}_g \in R^{f\times 1}$. Let u_g be the gender of user u and $\mathbf{g}_{ug}$ be the gender factor of user u. By incorporating the gender influence, the predicted rating $\hat{r}_{ui}$ is now defined in Eq. 5, where $\alpha_2 \in [0, 1]$ is a parameter that controls the importance of gender influence in rating prediction. The objective function is updated accordingly, see Eq. 10 in Table 5.

$$
\hat{r}_{ui} = \mu + b_u + b_i + (\mathbf{p}_u + \alpha_1 \cdot \mathbf{a}_{ua} + \alpha_2 \cdot \mathbf{g}_{ug})^T \mathbf{q}_i, \tag{5}
$$

Occupation Influence. In Movielens, a user belongs to one of 21 occupations. Based on our observations in Sect. 3.2, the occupation of a user may reflects the characteristics of a user, e.g., an artist may care about items' artistic quality more, while a writer may attach more importance to items' literariness. In other words, users often consider items in their own professional perspective. Intuitively, users with same occupation show similar item preference. Inspired by the above observations, we also introduce occupation latent factors in our model for getting better rating prediction. For each occupation o, it maps to a latent vector $\mathbf{o}_o \in R^{f \times 1}$. Let u_t be occupation of user u and o_{ut} be the occupation factor of user u. By incorporating the occupation influence, the predicted rating $\hat{r}_{ui}$ is now defined in Eq. 6, where $\alpha_3 \in [0, 1]$ is a parameter that controls the importance of occupation influence in rating prediction. The objective function is shown in Eq. 11 in Table 5.

$$\hat{r}_{ui} = \mu + b_u + b_i + (\mathbf{p}_u + \alpha_1 \cdot \mathbf{a}_{ua} + \alpha_2 \cdot \mathbf{g}_{ug} + \alpha_3 \cdot \mathbf{o}_{uo})^T \mathbf{q}_i, \tag{6}$$

4.4 Item Popularity and Category Influences

The aforementioned methods are standing in the user's perspective. Next we convert perspective to item, discuss two features that have been widely used in traditional collaborative filtering method, namely *popularity* and item *category popularity*. For simplicity, we model both item popularity and category popularity as a rating bias z.

$$z = \gamma_i \cdot \rho_i + \gamma_c \cdot \tau_c \tag{7}$$

In above equation, $\rho_i \in [0, 1]$ is the normalized popularity of item i, $\tau_c \in [0, 1]$ is the normalized popularity of category c. The two parameters γ_i and γ_c are the popularity weighting parameters for item i and category c respectively, both are learned from the training data. With rating bias z, the predicted rating is shown in Eq. 8. The objective function considering both item popularity and category popularity is shown in Eq. 12 in Table 5.

$$\hat{r}_{ui} = \mu + b_u + b_i + z + (\mathbf{p}_u + \alpha_1 \cdot \mathbf{a}_{ua} + \alpha_2 \cdot \mathbf{g}_{ug} + \alpha_3 \cdot \mathbf{o}_{uo})^T \mathbf{q}_i \tag{8}$$

4.5 Parameter Estimation

All the objective functions (e.g., Eqs. 9, 10, 11 and 12) in the proposed models share the same form. Next, we detail the parameter estimation for Eq. 12 (where $z = \gamma_i \cdot \rho_i + \gamma_c \cdot \tau_c$) as an example using Stochastic Gradient Descent (SGD) algorithm [2]. Let e_{ui} be the error associated with the prediction $e_{ui} = r_{ui} - \hat{r}_{ui}$. The parameters are learned by moving in the opposite direction of the gradient with a learning rate η in an iterative manner. The details of iterative formula is as follows:

$$b_u \leftarrow b_u + \eta \cdot (e_{ui} - \lambda_2 \cdot b_u)$$
$$b_i \leftarrow b_i + \eta \cdot (e_{ui} - \lambda_2 \cdot b_i)$$
$$\gamma_i \leftarrow \gamma_i + \eta \cdot (e_{ui} \cdot \rho_i - \lambda_2 \cdot \gamma_i)$$
$$\forall c \in C_i \ : \ \gamma_c \leftarrow \gamma_c + \eta \cdot (e_{ui} \cdot \tau_c - \lambda_2 \cdot \gamma_c)$$
$$\mathbf{p}_u \leftarrow \mathbf{p}_u + \eta \cdot (e_{ui} \cdot \mathbf{q}_i - \beta_1 \cdot \sum_{f \in N(u)} s_{uf} \cdot (\mathbf{p}_u - \mathbf{p}_f) - \lambda_1 \cdot \mathbf{p}_u)$$
$$\mathbf{q}_i \leftarrow \mathbf{q}_i + \eta \cdot (e_{ui} \cdot (\mathbf{p}_u + \alpha_1 \cdot \mathbf{a}_{ua} + \alpha_2 \cdot \mathbf{g}_{ug} + \alpha_3 \cdot \mathbf{o}_{uo})$$
$$- \beta_2 \cdot \sum_{f' \in N(i)} s_{uf'} \cdot (\mathbf{q}_i - \mathbf{q}_{f'}) - \lambda_1 \cdot \mathbf{q}_i)$$
$$\mathbf{a}_{ua} \leftarrow \mathbf{a}_{ua} + \eta \cdot (e_{ui} \cdot \alpha_1 \cdot \mathbf{q}_i - \lambda_3 \cdot \mathbf{a}_{ua})$$
$$\mathbf{g}_{ug} \leftarrow \mathbf{g}_{ug} + \eta \cdot (e_{ui} \cdot \alpha_2 \cdot \mathbf{q}_i - \lambda_3 \cdot \mathbf{g}_{ug})$$
$$\mathbf{o}_{uo} \leftarrow \mathbf{o}_{uo} + \eta \cdot (e_{ui} \cdot \alpha_3 \cdot \mathbf{q}_i - \lambda_3 \cdot \mathbf{o}_{uo})$$

5 Experiments

We now conduct experiments on the MovieLens dataset to evaluate the proposed models and compare the proposed models with state-of-the-art baselines.

5.1 Experimental Setting

Dataset. We use the MovieLens dataset that has been studied in Sect. 3.1 in our experiments. For each user, we randomly select 80 % of ratings for training, and the remaining 20 % for testing. As the result, we have 80, 000 ratings to build the matrix factorization model for the prediction of the remaining 20,000 ratings. The data sparsity is 93.7 %.

Evaluation Metric. We adopt two popular evaluation metrics, namely, Mean Absolute Error (MAE) and Root Mean Square Error (RMSE). The smaller MAE or RMSE value means better rating prediction accuracy. In the following equations, T is the set of user-item rating pairs (u, i) used in testing.

$$MAE = \frac{1}{|T|} \sum_{(u,i) \in T} | r_{ui} - \hat{r}_{ui} |$$

$$RMSE = \sqrt{\frac{1}{|T|} \sum_{(u,i) \in T} (r_{ui} - \hat{r}_{ui})^2}$$

Baseline Methods. We compare the proposed models with the following 8 baseline methods. All the experiments were conducted on a server with Intel Xeon E5310 1.60 GHz CPU (8 cores) and 20 G memory. We implemented the algorithms in C++ with the support of Eigen libary[3] for fast vector/matrix manipulations.

1. *Global Mean:* this method predicts an unknown rating to be the average of all known ratings, *i.e.*, $\hat{r}_{ui} = u$.
2. *User Mean:* this method utilizes the mean rating of each user to predict the missing values for the corresponding user.
3. *Item Mean:* this method uses the mean rating of each item to predict the missing values for the corresponding item.
4. *RSVD:* this is the Regularized SVD method. It is equivalent with the method proposed by Salakhutdinov and Minh in [23]. The underlining distribution is assumed as Gaussian distribution. The details of this method are also introduced in Sect. 4.1.
5. *Social MF:* this model considers implicit social information between items and/or users. The implicit social information can be derived from most similar and dissimilar users/items using Pearson's correlations or cosine similarity of ratings. As our model considers neighbors influences from not only rating information but also attribute information, for a fair comparison, this method only includes rating information from most similar items in Social MF. More detailed discussion about neighbors selection is presented in Sect. 5.2.
6. *Biased MF:* this is the MF model with user and item biases briefly described in Sect. 4.1. Biased MF is widely used as a baseline in recommender systems.
   Proposed Methods. We extended Biased MF to incorporate influences from multiple factors: user age (A), user gender (G), user occupation (O), item popularity (P), and item category popularity (C). The proposed methods are denoted using the letters in parentheses to indicate the influences considered in each method.
7. *NA-MF:* this method incorporates both implicit neighborhood and user age influence (Sect. 4.3, Eq. 5).
8. *NAG-MF:* this method incorporates both implicit neighborhood, user and gender influence (Sect. 4.3, Eq. 6).
9. *NAGO-MF:* this method incorporates implicit neighborhood, user age, user gender and user occupation influence (Sect. 4.3, Eq. 7).
10. *NAGOP-MF:* this method incorporates implicit neighborhood, user age, user gender, user occupation and item popularity influence, by setting $z = \gamma_i \cdot \rho_i$ in Eq. 8 (Sect. 4.4).
11. *NAGOPC-MF:* this model incorporates all factors: implicit neighborhood, user age, user gender, user occupation, item popularity and item category popularity influence, by setting $z = \gamma_i \cdot \rho_i + \gamma_c \cdot \tau_c$ (Sect. 4.4).

---

[3] http://eigen.tuxfamily.org.

We also evaluate another two methods: *AGOP-MF* and *AGOPC-MF*. These two methods do not incorporate implicit neighborhood influence but incorporate influences from other factors (i.e., *A*, *G*, *O*, *P* and *C*) indicated by the method names.

Parameter Setting. We performed 5-fold cross-validation on the training set to empirically set the hyperparameters. The number of latent factors f = 20, the weight coefficient of similarity $\gamma = 0.8$, the relative importance of age, sex and occupation influences are set to $\alpha_1 = 1$, $\alpha_2 = 1$, $\alpha_3 = 1$. The neighborhood regularization parameters are set to $\beta_1 = 0.002$, $\beta_2 = 0.002$. The regularization parameters: $\lambda_1 = 0.1$, $\lambda_2 = 0.1$, $\lambda_3 = 0.1$. The latent factors are learnt by SGD with initial learning rate $\eta = 0.01$, which decreases by a factor of 0.95 after each 10 iterations. The same parameters are used in all methods for fair comparison for all our proposed methods and the baseline methods whenever applicable. For example, the number of latent factors is also set to 20 in baseline methods *Biased MF*, *RSVD* and *Social MF*. For neighborhood influence, by default, the proposed methods use the 10 nearest neighbors for each user or item. For all the methods based on matrix factorization, the reported results are averaged over 5 runs to avoid the impact of initialization in parameter learning.

5.2   Experimental Results

We first compare the proposed methods with baseline methods and then search the best weight coefficient of similarity. Lastly, we evaluate the importance of user attributes in our proposed methods.

Method Comparison. The prediction errors measured by MAE and RMSE of all methods are reported in Table 6 with best results highlighted in boldface. We make four observations from the results.

First, incorporating attributes influence into item rating prediction greatly reduces prediction errors measured by both MAE and RMSE. All the proposed methods with attributes influence (i.e., methods 7–13) outperform all baseline

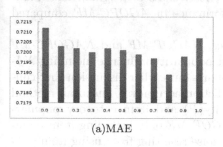
(a)MAE

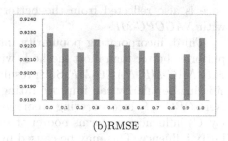
(b)RMSE

Fig. 2. Comparisons of different similarity weight of *NAGOP-MF* on MAE and RSME metrics.

Table 6. MAE and RMSE of all methods, the lower the better.

ID	Method	MAE	RMSE
1	Global Mean	0.9435	1.1256
2	User Mean	0.8362	1.0437
3	Item Mean	0.8163	1.0229
4	RSVD	0.7492	0.9496
5	Social MF	0.7372	0.9304
6	Biased MF	0.7418	0.9468
7	NA-MF	0.7228	0.9234
8	NAG-MF	0.7205	0.9213
9	NAGO-MF	0.7193	0.9204
10	NAGOP-MF	**0.7189**	**0.9199**
11	NAGOPC-MF	0.7190	0.9201
12	AGOP-MF	0.7275	0.9321
13	AGOPC-MF	0.7275	0.9322

methods (methods 1–6). The best prediction accuracy is achieved by *NAGOP-MF* which considers neighborhood(N), user age(A), user gender(G), user occupation(O) and item popularity (P). With attribute age influence alone, *NA-MF* outperforms all baselines including state-of-the-art methods *Bias MF* and *Social MF*. This result suggests that using user attributes is of great help to item rating prediction. Further considering factors like user gender(G) and user occupation(O) leads to relatively small additional reduction in prediction errors.

Second, without incorporating neighborhood influence, *AGOP-MF* performs poorer than most methods with neighborhood influence including *NA-MF*, *NAG-MF*, *NAGO-MF*, *NAGOP-MF* and *NAGOPC-MF*. The poorer performance of *AGOP-MF* against *NA-MF* suggests that the neighborhood influence is more effective than the combination of the three factors (G, O, and P) in item rating prediction. On the other hand, the effectiveness of neighborhood influence is also reflected from the better performance of *AGOPC-MF* compared with *NAGOPC-MF*.

Third, incorporating popularity influence, *NAGOP-MF* and *NAGOPC-MF* performs better than most methods without popularity influence including *NA-MF*, *NAG-MF*, *NAGO-MF*. Such result supports our earlier discussion that considering popularity can improve rating prediction accuracy. Unnatural, method *NAGOPC-MF* incorporating item popularity(P) and item category popularity(C) influences performs poorer than *NAGOP-MF* only including item popularity influence. This may be caused by the noise resulting from using item popularity(P) and item category popularity(C) in one prediction model. Because item category popularity have certain direct relationship with item popularity.

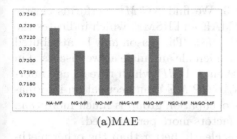
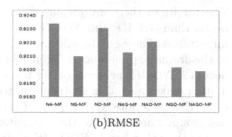

(a)MAE   (b)RMSE

Fig. 3. Comparisons of different types of methods on MAE and RSME metrics.

Last, among the 6 baseline methods, *Social MF*, the state-of-the-art methods, perform the best evaluated by both MAE and RMSE. While *Global Mean* perform the worst.

Impact of Different Similarity Weight. As described in Sect. 4.2, γ is a weight coefficient to balance the importance between rating sequence similarity and attribute similarity. Bigger γ represents attribute similarity is more important and vice versa. To explore the impact of γ on the rating prediction accuracy, we select 11 different γ from 0 to 1 to use in the *NAGOP-MF* method. The *NAGOP-MF* method is selected as the method for evaluation because it has the best performance.

Figure 2(a) and (b) respectively plot MAE and RMSE of the *NAGOP-MF* method by using different γ ranging from 0 to 1. From the figure, $\gamma = 0.8$ gives the best prediction accuracy by considering both MAE and RMSE, and only using rating similarity ($\gamma = 0$) or attribute similarity ($\gamma = 1.0$) gets worse results. These show that both rating similarity and attribute similarity are help for rating prediction, and attribute similarity are more important than rating similarity. This may be caused by that users with similar attributes can better reflect the users' close preference than those users with similar rating sequence.

Analysis of Each Attribute Factor. Age(A), gender(G), and occupation(O) factor are three main factors for rating prediction in this task. Although the results of *NA-MF*, *NG-MF*, and *NO-MF* shown in Fig. 3 can indicate their effectiveness for the task alone, the contribution of each factor to *NAGOP-MF* should also be explored. This is because combining multiple latent factors to form a unified model does not mean the results of the new model is the performance summarization of each factor.

To better understand the importance of each attribute factor, we also adopt the strategy of combining two factors from user attributes, i.e., *NAG-MF*, *NAO-MF* and *NGO-MF*. we keep using neighborhood(N) in this section because of user attribute also using in neighborhood selection, while item popular(P) is moved for the reasons that it has no relationship with user attribute. Specifically, we test the strategy of all three user attributes factors using *NAGO-MF*.

The results of them are displayed in Fig. 3. We find *NG-MF* performs clearly better than *NA-MF* and *NO-MF* both in MAE and RSME, which indicates the importance of user gender information to the task. The reason may be attributed to the dramatic preference difference between female and male as we discussed in Sect. 3.2. *NO-MF* achieves better results than *NA-MF*, which reveals age information makes a smaller contribution to *NAGOP-MF* than occupation information. This is because the number of user's occupation group is much more than age group, which leads to the occupation factors more personalized.

On the other hand, *NGO-MF* performs clearly better than the other methods with two attribute factors, which again indicates the top importance of gender, occupation take the second and age is the last. Lastly, when all the three attributes are combined together (*NAGO-MF*), the performance is further improved, which indicates each attribute is good for rating prediction.

6 Conclusion

In this work we proposed a new, simple, and efficient way to incorporate user attributes and item category on ratings prediction in several methods commonly used for matrix factorization. We firstly analyze the MovieLens dataset and find that a user's rating behavior is certainly correlated with his or her attribute type. Based on this observation, we model a user with two vectors of latent factors one for its intrinsic characteristics and the other for its common characteristics determined by his or her attribute type. The experimental results on real dataset have shown that our model is effective and outperforms several alternative methods. Other factors like user neighbor, item neighbor, item popularity and item category popularity can further improve the rating prediction accuracy. Nevertheless, using both item popularity and item category popularity in one method may bring noise and lead to bad results.

Acknowledgments. This research was supported in part by the National Natural Science Foundation of China under Grant 91124001 and the Fundamental Research Funds for the Central University, and the Research Funds of Renmin University of China under grant 2015030273.

References

1. Ma, H.: An experimental study on implicit social recommendation. In: Proceedings of the 36th International ACM SIGIR Conference on Research and Development in Information Retrieval, pp. 73–82. ACM (2013)
2. Koren, Y.: Factorization meets the neighborhood: A multifaceted collaborative filtering model. In: KDD, pp. 426–434. ACM (2008)
3. Adomavicius, G., Tuzhilin, A.: Toward the next generation of recommender systems: A survey of the state-of-the-art and possible extensions. TKDE **17**(6), 734–749 (2005)

4. Koenigstein, N., Dror, G., Koren, Y.: Yahoo! music recommendations: Modeling music ratings with temporal dynamics and item taxonomy. In: RecSys, pp. 165–172. ACM (2011)
5. Verstrepen, K., Goethals, B.: Unifying nearest neighbors collaborative filtering. In: Proceedings of the 8th ACM Conference on Recommender Systems, RecSys 2014, pp. 177–184. ACM (2014)
6. Natarajan, N., Shin, D., Dhillon, I.S.: Which app will you use next?: Collaborative filtering with interactional context. In: RecSys, pp. 201–208. ACM (2013)
7. Breese, J.S., Heckerman, D., Kadie, C.: Empirical analysis of predictive algorithms for collaborative filtering. In: Proceedings of the Fourteenth Conference on Uncertainty in Artificial Intelligence, pp. 43–52. Morgan Kaufmann Publishers Inc., (1998)
8. Jin, R., Chai, J.Y., Si, L.: An automatic weighting scheme for collaborative filtering. In: SIGIR, pp. 337–344 (2004)
9. Deshpande, M., Karypis, G.: Item-based top-n recommendation algorithms. ACM Trans. Inf. Syst. **22**(1), 143–177 (2004)
10. Sarwar, B., Karypis, G., Konstan, J., Riedl, J.: Item-based collaborative filtering recommendation algorithms. In: WWW, pp. 285–295. ACM (2001)
11. Koren, Y.: Factorization meets the neighborhood: a multifaceted collaborative filtering model. In: Proceedings of the 14th ACM SIGKDD International Conference on Knowledge Discovery and Data Mining, pp. 426–434. ACM Press (2008)
12. Koren, Y., Bell, R., Volinsky, C.: Matrix factorization techniques for recommender systems. Computer **42**(8), 30–37 (2009)
13. Lian, D., Zhao, C., Xie, X., Sun, G., Chen, E., Rui, Y.: GeoMF: joint geographical modeling and matrix factorization for point-of-interest recommendation. In: Proceedings of KDD 2014, pp. 831–840. ACM (2014)
14. Hu, L., Sun, A., Liu, Y.: Your Neighbors affect your ratings: on geographical neighborhood influence to rating prediction. In: Proceedings of the 37th International ACM SIGIR Conference on Research & Development in Information Retrieval, pp. 345–354. ACM (2014)
15. Pham, T.-AN., Li, X., Cong, G., Zhang, Z.: A general graph-based model for recommendation in event-based social networks. In: ICDE, Seoul, Korea, 13–17 April 2015
16. Zhang, W., Wang, J.Y.: A collective bayesian poisson factorization model for cold-start local event recommendation. In: Proceedings of the 21th ACM SIGKDD International Conference on Knowledge Discovery and Data Mining, pp. 1455–1464. ACM Press (2015)
17. Feng, W., Wang, J.: We can learn your #hashtags: Connecting tweets to explicit topics. In: ICDE, Chicago, USA, 31 March-4 April 2014
18. Koenigstein, N., Dror, G., Koren, Y.: Yahoo! music recommendations: Modeling music ratings with temporal dynamics and item taxonomy. In: RecSys, pp. 165–172. ACM (2011)
19. Cremonesi, P., Koren, Y., Turrin, R.: Performance of recommender algorithms on top-n recommendation tasks. In: RecSys, pp. 39–46. ACM (2010)
20. Steck, H.: Item popularity and recommendation accuracy. In: RecSys, pp. 125–132. ACM (2011)
21. Koren, Y.: Collaborative filtering with temporal dynamics. In: KDD, pp. 447–456. ACM (2009)

22. Ma, H., Zhou, D., Liu, C., Lyu, M.R., King, I.: Recommender systems with social regularization. In: Proceedings of the Fourth ACM International Conference on Web Search and Data Mining, WSDM 11, Hong Kong, China, pp. 287–296 (2011)
23. Salakhutdinov, R., Mnih, A.: Probabilistic matrix factorization. In: Proceedings of Advances in Neural Information Processing Systems, NIPS 2007 (2007)

Expert Recommendation in Time-Sensitive Online Shopping

Ming Han[1,2(✉)] and Ling Feng[1,2]

[1] Department of Computer Science and Technology, Tsinghua National Laboratory
for Information Science and Technology (TNList), Beijing, China
hanm13@mails.tsinghua.edu.cn, fengling@mail.tsinghua.edu.cn
[2] Centre for Computational Mental Healthcare Research,
Institute of Data Science Tsinghua University, Beijing 100084, China

Abstract. Like offline shopping mode, when shopping online for the
time-sensitive products, people also would like to consult online shopping experts, which can save their time and avoid risk. In this paper,
we propose a generalized framework to discover and recommend experts
in time-sensitive online shopping. First, we derive the order pattern of
users' purchases in order to establish their relationship. Then we quantify users' purchase ability and influence acceptance from their historical
purchase log. With this knowledge we select accessible users as expert
candidates and compute their reputation. Finally, we filter and recommend the experts to users. The experiments on real-world dataset and
users show that (1) the accuracy of existing recommendation systems
can be improved by embedding the expert discovery process while preserving good coverage; (2) our method achieves a better performance in
matching and ranking experts compared with baselines.

Keywords: Expert discovery · Social influence and time-sensitive online
shopping

1 Introduction

Motivation. When shopping or investing in an unfamiliar field, people usually
want to consult the experts for advice in order to save their time and avoid risk.
For example, some people would like to pay a *fashion buyer* to pick out fashion
clothing for them and some companies are willing to pay a Certified Public
Procurement Officer (CPPO) or Certified Professional Public Buyer (CPPB) to
get advice for the device purchasing.

With the booming of web 2.0, more and more people prefer shopping online.
Like the traditional offline mode, when shopping or investing online, some people
also want to get help from the experts or the expert-driven approaches. Many
online products and investments are *time-sensitive*, that is, their value changes
over time and they usually have a purchase deadline, such as crowdfunding and
online auctions. Since people feel more pressure when the purchase deadline is

© Springer International Publishing Switzerland 2016
S.B. Navathe et al. (Eds.): DASFAA 2016, Part I, LNCS 9642, pp. 297–312, 2016.
DOI: 10.1007/978-3-319-32025-0_19

coming soon, the expert consultant is valuable for their decision making process. As far as we know, little work has been done on this problem since it is time-consuming to discover and certify the online experts manually in order to meet the consultant demand. Thus, we think it is necessary to find a way to discover and recommend the experts to online users automatically. Besides, the recommendation can also enhance the activeness of the shopping community.

Background. There is a lot of research on trust and influence in recommendation systems as well as social network and they are related to expert discovery.

Recommendation systems are designed to help users with the *information overload* problem, but they can not meet the demand discussed above. These systems usually only generate lists of items for users. While in some fields, such as financial products, people usually want more reliable information based on the rich experience of experts rather than item lists. Besides, traditional recommendation systems rely on the *taste similarity* between users, while this strategy fails to meet the needs of users who want to get advice from higher-level users.

By including the trust network, people proposed the Trust-Aware Recommender Systems (TARSs). However, the scope of their application is very limited and can not be adopted to discover experts. First and foremost, it takes plenty of effort to get the accurate *explicit trust* since most of the online shopping websites don't support the mechanism for people to express their trust to others. We argue that TARSs ignore the important issue: *How does the trust form?* Most of them suppose that the trust value is *given* beforehand. Besides, we think that the trust between users are topic-sensitive which is not considered in the TARSs. For example, you may trust a computer expert when buying a new laptop, while you may not trust him when picking out a fashionable T-shirt. And we argue that the purchase order is also important for the trust computation which is also lacking in TARSs. For example, you may get more valuable information from the one who often buys the same item earlier than you instead of the latter one.

In social network, many researchers focus on the problem of *information diffusion*, especially on the *influence maximization* problem. They propose many methods to efficiently find the "seed" users who have the most significant influence onc community. However, we can not apply these models to discover experts directly since there is a lack of explicit "links" of relationship in online shopping community such as *following*. The "links" in online shopping community are much weaker and implicit: instead of through friends, people are more often influenced by strangers who also bought the same items.

Our Work. We propose a generalized framework to discover and recommend experts in time-sensitive online shopping. In this paper, we show (1) how to leverage users' accessibility derived from the order pattern of their purchases in order to generate candidates; and (2) how to quantify their purchase ability and influence acceptance to compute their reputation in online shopping community. Our model has following features:

- **Topic Sensitive.** We take topics into consideration when computing users' purchase ability and reputation, because we argue that a person is only good

at shopping on limited categories and a user may trust another user on a certain category.
- **High Coverage.** Because we use the accessibility propagated through the community network to select candidates, our model can decrease the impact of data sparsity and noisy data.
- **Attack Resistant.** Since we combine users' purchase history with their network status to compute their reputation, we can filter out users who want to raise their reputation by deceitful activities such as *fake purchases*.

The rest of this paper is structured as follows: In Sect. 2, we present the related work. In Sect. 3 we elaborate our model step by step. Next in Sect. 4, we describe the experiments we conducted. Finally, we conclude this study and propose the future work in Sect. 5.

2   Related Work

Behavior Analysis of Online Shopping and Shopping Aid. In this field, researchers classify consumer behavior into two categories: *getting information* behavior and *purchasing* behavior. They think users' behavior is influenced by various factors. For example, Pavlou et al. studied factors that influence consumer's acceptance of e-commerce including trust, perceived usefulness, ease of use and risk [14]. Besides, people also proposed various tools based on behavior analysis to help users with online shopping. For example, An et al. designed a model to recommend investors to crowdfunding projects based on analysis of investors' behavior patterns [1]. Das et al.proposed a novel recommendation system called Shopping Advisor [6] which generates question flowcharts automatically to users and helps them to figure out what they want.

Trust in Recommender System. In order to overcome drawbacks of traditional recommendation systems such as the *cold start* problem, people brought trust into recommendation. Some researchers derive trust from rating set and model trust propagation process. For example, Donovan et al. proposed a computational model to compute profile-level and item-level trust from rating data in [13]. Guha et al. developed a framework of trust and distrust propagation schemes in [8]. On the other hand, some researchers used explicit trust score between users to improve recommendation performance. For example, Massa et al. proposed the Trust-aware Recommender System which improves collaborative filtering with a trust matrix [12].Jamali et al. proposed a random walk model for combining trust-based and item-based recommendation [9].

Influence Diffusion. Due to dramatic development of social network, many researchers focus on the topic of *influence diffusion*. Domingos et al. led to propose the *influence maximization* (IM) problem of social network in [7]. Then Kempe et al. proposed two discrete influence propagation models [10], which are widely accepted. Later on, people realized that by utilizing topics of influence they can improve the performance of their models significantly. Therefore

Barbieri et al. used history activities on social network to compute topic proportion of relationship strength among vertices and proposed the Topic-Aware Influence Cascade (TIC) model [3]. To support topic-aware influence maximization, Barbieri et al. proposed a similarity-based model called INFLEX [2].

Though above research performs well on discovering users' shopping behavior pattern and their influence to the social network, we can not apply their methods directly to solve our problem, because the application context is very different and there lacks a generalized model to integrate various factors to discover experts in time-sensitive online shopping.

3   Model Framework

Problem Definition. The input of our problem $F = (\mathbf{S}, \mathbf{H})$ contains two parts: (1) static feature set of products $\mathbf{S} = \{(i, S^l)\}_{i,l}$, where S^l is the lth static feature of product i; and (2) activity history set $\mathbf{H} = \{(v, i, G_f, t)\}_{v,i,f,t}$, where v stands for a customer and G_f stands for an activity of type f. Given the input F, a query user u, a query topic c and a number of expert k, our goal is to recommend a k-expert list for u on c.

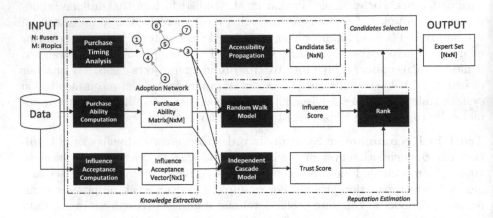

Fig. 1. The framework of expert discovery and recommendation.

Our model framework includes three parts: knowledge extraction, candidate selection and reputation estimation (Fig. 1). In our model, a user has two roles: *producer* and *acceptor* of influence. We first derive the order patterns of users' purchases to establish their relationship. Then we quantify users' purchase ability and acceptance of influence from their activity history. With this knowledge, we propagate accessibility through the network in order to find accessible candidates. And then we quantify the trust they get from others and the influence they bring to others. Finally, we combine these two factors to get the reputation of candidates and use it in order to filter candidates to obtain a final expert set.

3.1   Knowledge Extraction

In this section, we present how to extract the knowledge about users including adoption network, individual purchase ability and influence acceptance.

Adoption Network. As [16] indicates, as long as there are consistent time-based patterns where two users purchase or access the same item sequentially, the influence can be modeled as information flow from early adopters to latter adopters. So to capture the order pattern of users' purchases and establish their relationship, we construct a network called *adoption network* based on the order of users' purchases.

We use a graph to formulate this process where nodes represent users and edges represent the relationship of their purchase orders. Specifically, given the purchase history of N users including timestamps, if two users u and v have purchased the same item, we link them with a directed edge $e(u, v)$. Then we compare the timestamps of their purchases and use the times of u purchasing common items earlier than v as the weight of $e(u, v)$ and set its direction $u \to v$. Finally, we divide each edge weight by the weight sum of edges from its starting node. In the following context, we use an adjacent matrix $B_{N \times N}$ to represent this graph.

Purchase Ability. We evaluate the purchase ability of users by analyzing their activity history. Specifically, for every user, we first evaluate whether his purchase is *successful* and *satisfying*. Then we compute its weight with the amount of effort the user has given. Finally, by synthesizing all his purchases in the query category c, we get his topic-sensitive purchase ability on c.

Definition 1. *We define a purchase of u about item i is a **successful purchase** when it satisfies following constraints:*

- *u has paid some money for i;*
- *i meets its pre-setting sale goal (if any);*
- *there is no disputes after u purchasing i;*
- *the sentiment of comments from u about i (if any) is positive[1].*

We use a sign function $SP_u(i)$ to map every purchase into 1 if it satisfies above constraints, otherwise -1.

Then we evaluate the weight of purchase by following features:

- $SE_u(i)$: average sentiment value of u's comments about i;
- $M_u(i)$: expenditure of the purchase ;
- $L_u(i)$: whether u *like* i (by clicking *like* button);
- $S_u(i)$: whether u *share* i on other websites (by clicking *share* button).

---

[1] In this study, we apply a Chinese text processing python package, SnowNLP to do sentiment analysis.

In order to decrease the impact of different value ranges, we use $x' = \frac{x-min}{max-min}$ to scale first two features and map last two features into 1 for *like* or *share*, otherwise 0. We linearly combine them to get purchase weight $W_u(i) = \delta \cdot SE_u(i) + \eta \cdot M_u(i) + \mu \cdot L_u(i) + \theta \cdot S_u(i)$. Inspired by the observation of our dataset and the work of [5], we argue that ability of users fit the normal distribution when the number of users is large enough. Thus by fitting the normal distribution we can estimate parameters involved above to get $\delta = 0.3, \eta = 0.6, \mu = 0.05$ and $\theta = 0.05$ in this study.

With variables computed above, we can give the formulation of u's purchase ability on category c:

$$P(u,c) = sigmoid(\sum_{i \in D_c} SP_u(i) \cdot W_u(i)) \tag{1}$$

where D_c stands for the set of u's purchases on c. Since the sum $\sum_{i \in D_c} SP_u(i) \cdot W_u(i)$ may be negative while we want the sign to represent "whether a purchase is successful or not", we adopt the sigmoid function which is wildly used in other fields to map weight sum into the range of $(0, 1)$. Intuitively, if a user has a large ratio of *successful purchases* on category c, he is more likely good at purchasing on category c and have a high value of $P(u, c)$.

Influence Acceptance. We use a positive variable A_u to evaluate acceptors' acceptance of influence. It is derived from the activity log of users and the information of products they have bought. The features used to quantify it are as following:

- **Relative Time of Purchasing.** Intuitively, the latter u buys a product, the more chance he wants to wait and see market situation. So we think users' acceptance of influence has positive correlation to the relative time of purchase which is represented as $X_i(t) = \frac{t - T_{start}^i}{T_{end}^i - T_{start}^i}$, where T_{start}^i and T_{end}^i are start time and end time of i.
- **Stimuli Amount.** If there are more stimuli when purchasing, the user is more prone to rely on these stimuli to make his purchase decision. So we think the amount of stimuli $Y_i(t)$ is also relative to users' acceptance of influence. In order to measure it, we weightedly sum various stimuli including the number of *purchase* ($N_i^p(t)$), *share*($N_i^s(t)$), *comment*($N_i^c(t)$) *and like*($N_i^l(t)$) at time t: $Y_i(t) = a * N_i^p(t) + b * N_i^s(t) + c * N_i^c(t) + d * N_i^l(t)$.
- **Item Popularity.** If u often buys unpopular products when there are many other hotter ones in the same category, he may be more independent from others' opinion. Thus we think users' acceptance of influence is negatively correlative to product popularity which can be represented as $Z_i(t) = \frac{1}{Rank_i(t)}$, where $Rank_i(t)$ is the order of i measured by the value of $Y_i(t)$ in the same category.

We apply the *z-score* standardization to scale features above to get $X_i(t), 'Y_i(t)'$ and $Z_i(t)'$. Since the expenditure $M_u'(i)$(also be scaled) represents the degree he is influenced by others' opinions, we also include it into the formula to measure users' acceptance of influence:

$$A_u = sigmoid(\sum_{i=1}^{n} \sum_{t \in T_i} (\alpha * X_i(t)' + \beta * Y_i(t)' + \omega * Z_i(t)') \cdot \frac{M'_u(i)}{n \cdot |T_i|}) \quad (2)$$

where n is the number of purchases and T_i is the set of purchase time for product i. We also adopt the sigmoid function to map the sum result into positive range.

Similarly, we adopt the normal distribution to estimate parameters above and get $a = 0.7$, $b = 0.1$, $c = 0.1$, $d = 0.1$, $\alpha = 0.6$, $\beta = 0.2$, $\omega = 0.2$.

3.2 Expert Candidates Selection

The static knowledge presented above builds the base of our model, with them we can generate expert candidates.

In practice, people usually consult someone nearby that they can get in touch with, so the influence from them is larger than those far away from their social circle. Inspired by this observation, we use the accessibility to find the people whose purchases can be accessed by the query user as expert candidates.

In fact, the adoption network $B_{N \times N}$ already contains some information of accessibility between users: the neighbors directly connected to the query user are his accessible users by *one step propagation*. However, people can get in touch with others in an indirect manner. For example, you may decide to buy an item since you know *a friend of your friend* has bought it. So in order to explore such relationships we propagate accessibility through adoption network. Specifically, the goal of this part is to produce a matrix F from which we can get accessibility of any two users. Then we use accessible users as the expert candidates of query user u.

In this study we adopt the propagation model of [8] to derive the closed set of accessibility. First we formalize *atomic propagations* on adoption network B, which can be seen as the "basis set". Then we use them to extend a conclusion by a constant-length sequence of forward steps. On the other hand, any inference regarding accessibility should be expressed as a combination of elements in this basis set. In this study, we use four types of atomic propagations:

- **Direct Propagation**: If i access j and j access k, then we infer that i can access k. The operator of this propagation is B itself.
- **Co-citation**: If i_1 accesses j_1 and j_2, and i_2 accesses j_1, then i_2 should access j_2 as well. Its operator is $B^T B$.
- **Transpose Access**: If i accesses j then accessing j should imply accessing i. Its operator is B^T.
- **Access Coupling**: If both i and j access k, then accessing i should imply accessing j. Its operator is BB^T.

Given the basic adoption matrix B and parameters $\alpha_1, \alpha_2, \alpha_3, \alpha_4$ as weights for combining atomic propagation schemes, all the atomic propagations can be captured into a combined matrix $C_{B,a}$ as follows:

$$C_{B,a} = \alpha_1 B + \alpha_2 B^T B + \alpha_3 B^T + \alpha_4 BB^T \qquad (3)$$

Then we can derive multi-step propagation with $C_{B,a}$. Let a positive integer k be the propagation step number, we express the multi-step propagation as a matrix powering operation:

$$F = \sum_{k=1}^{K} \gamma^k \cdot C_{B,a}^k \qquad (4)$$

where γ is a discount factor to penalize the lengthy of propagation steps.

After the process above, we get matrix F in continuous domain. We need to convert it into discrete ones by the "majority rounding" method [8]. Specifically, let J be users whom user u can directly access and i is the user we want to predict his accessibility to $u (i \notin J)$. We order J with i according to the entries of $F_{uj'}$ where $j' \in J \cup i$. Then we set $F_{ui} = 1$ to represent u can access i if most of its q nearest neighbors can be accessed by u. We loop this process until all the users' accessibility to u has been set. Finally, we use set $I = \{i | F_{ui} = 1\}$ as expert candidates for user u. The parameters involved in this section is set by the cross validation of the recommendation (see Sect. 4.1 for details) and they are: $\alpha_1 = 0.3, \alpha_2 = 0.5, \alpha_3 = 0.1, \alpha_4 = 0.1, \gamma = 0.5, q = 4$.

3.3   Reputation Computation

In this section, we present the computation of reputation which is used to filter candidates to the final expert set. We think the reputation of producers comes from two parts: influence he brings to community and trust he gets from community.

Influence to Community. The purchase ability of producers can influence acceptors to some extent: higher purchase ability means more influence. However, we argue it is not enough since it is only based on the purchase history of users while no information about his status in the community network is included. What's more, since a user can cheat as an expert by manipulating many "successful and satisfying" purchases deliberately, including network status can also resist this attack.

In order to combine the network status of users, we use the random walk model to derive their influence on community. Let I_c be the influence vector of users on category c, and $1 - \phi$ be the damping factor, the influence computation can be expressed as following:

$$I'_c = (1 - \phi)B \times I_c + \phi p \qquad (5)$$

where $p_u = \frac{P(u,c)}{\sum_{v \in I} P(v,c)}$. This iterative process runs until it converges and the final I_c is the influence values of producers.

Trust from Community. Since people usually consult experts they *trust* for advice, experts should be those who receive more trust than ordinary users. So we argue that trust, which a producer gets from community, is another source of reputation. Though online shopping websites usually lack the mechanism for users to express their trust to others, we can derive it from the knowledge we have extracted. Specifically, first we formulate local trust among direct-connected users, then we propagate it through the network to get global trust and compute the *influence spread* of each user.

Based on the observation of our dataset, we argue that there are three factors related to trust computation: (1) purchase ability of influence producers, (2) acceptance of influence acceptors, and (3) distance between them in adoption network. Specifically, a acceptor i is more likely to trust a producer j with high purchase ability. And the higher influence acceptance i has, the more probability he trust others. Since the distance $Dis(i,j)$ between them reflect the accessibility of trust, it is also relative to the computation. To sum up, we get following formula for the trust computation from user i to j on the category c:

$$Trust(i,j,c) \propto \frac{P(j,c) \times A_i}{Dis(i,j)} \tag{6}$$

Though there may be various optional formulations of the relationship between trust and these three factors, in this study, we just use the simplest formulation $Trust(i,j,c) = \frac{P(i,c) \times A_i}{Dis(i,j)}$. Because we just use the value to rank candidates, the order rather than accuracy value is more important. After that, we use the normalization $x' = \frac{x-min}{max-min}$ to map trust into [0,1].

Adding the trust computed above to adoption network, we can construct trust network. Then we apply the Independent Cascade Model(ICM) [10] which is wildly used in information diffusion to generate the user set who are influenced by a candidate. Specifically, suppose the state of a candidate i is "active" and the rest users are "non-active". On the first iteration of propagation, i has a single chance to active each of its directly connected neighbor j with probability $Trust(i,j,c)$. If i succeeds, then j will become "active" in this step and no matter whether i succeeds or not, it can not active j any more. Then j replaces i to be the activator to continue this iteration until no more activation is possible. The set of active nodes at the end of this process, denoted as $\rho(i)$, is the set that trust i. Therefore we use its relative size $\frac{|\rho(i)|}{\sum_{j \in I} |\rho(j)|}$ to measure the trust he get from community under category c.

Rank and Filtering. With the influence and trust computed above, we can get the final reputation $R(i,c)$ to evaluate expert candidates:

$$R(i,c) = \psi \times \frac{|\rho(i)|}{\sum_{j \in I} |\rho(j)|} + (1-\psi) \times \boldsymbol{I}_c(i) \tag{7}$$

Finally, we order candidates by their reputation and keep top-k users as final experts to recommend to the query user. The parameters included in this

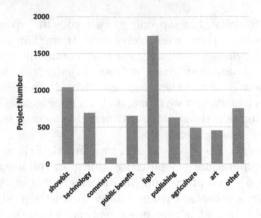

Fig. 2. Category distribution of zhongchou dataset

section are also determined by the cross validation of recommendation and they are: $\phi = 0.85, \psi = 0.4$.

4   Experimental Results

In this paper, we adopt reward-based crowdfunding as application background since it is a representative field of time-sensitive online shopping and investment. To evaluate our model's effectiveness, we conduct two experiments: (1) first we embed the expert discovery into an existing recommender system to explore whether its recommendation result can be improved; (2) then we conduct a user study by surveying real-world users.

4.1   Recommendation

Since trust-aware recommender systems have the selection process of neighbors based on *similarity* measure and *trust* measure, we can replace *trust measure* with expert discovery process to see how much its performance is improved compared with traditional recommender systems. In this way, we can evaluate our work indirectly.

Zhongchou.com Dataset. The dataset we use to evaluate our work is derived from Zhongchou.com which is the largest crowdfunding website in China. Since we can not access its data through API, we built up a web crawler to get its public data. By doing so, we construct a three-year (from February, 2013 to March, 2015) dataset (Table 1) including 6,477 projects which funded by 117,698 investors with a total number of 184,514 pledges. The projects have nine different categories (Fig. 2) which can be used to evaluate the topic-sensitive feature of our model. Among these projects, 1,883 projects were successfully funded, 4,091 (63.1 %) failed to achieve their pledging goals and the rest were still ongoing by the end time of crawling.

Table 1. Statistics for the zhongchou dataset.

	Successful	Failed	Ongoing	Total
Projects	1,883	4,091	503	6,477
Proportion	29.1 %	63.1 %	7.8 %	100 %
Investors	-	-	-	117,698
Initiators	-	-	-	4605
Pledges	146,566	23,620	14,328	184,514
Pledged (CNY)	65,710,159	3,346,806	2,005,515	71,062,481
Likes	369,902	183,202	21,855	574,959
Shares	26,562	14,317	9,282	50,161
Updates	6,288	2,387	628	9,303
Time span	-	-	-	736 days

Experiment Setup. Because there is no "ratings" in the Zhongchou.com dataset, we need to derive them from original features. Since the money people paid can reflect their evaluation of projects, we use it to infer "rating" set. First, we divide pledge options of projects equally into 5 bins. And we use the order of the bin that pledge falls into as the first reference value of rating, denoted as R_i. Then we divide user's investment range equally into 5 bins and use the order of the bin that pledge falls into as the second reference, donated as R_j. Finally, we get the inferred rating R by combining R_i and R_j: $R = \frac{R_i + R_j}{2}$.

We use the five-fold cross validation to evaluate recommendation performance and the measurements for accuracy are Mean Absolute Error (MAE) and Mean Absolute User Error (MAUE) [12]. Since ratings are hard to be predicted for sparse dataset, we also adopt measurement *coverage* including *rating coverage* and *user coverage*.

We embed our model into the existing trust-aware recommender system [12] by replacing the computation of trust matrix with expert discovery. Since the candidate selection in our model has iterative steps, we conducted two experiments with different iteration steps. We call the algorithm with few propagation steps as TRED1 (k = 10) and the one with more propagation steps as TRED2 (k = 20). Apart from the trust-aware recommender system, we also adopt other popular recommendation algorithms as comparison including SVD [11], user-based and item-based collaborative filtering [4, 15].

In order to capture relative merits of these methods in various situation, we also classify projects and users to report results in different views [12]: *cold start users*, who pledged from 1 to 4 projects; *heavy buyers*, who pledged more than 10 projects; *opinionated users*, who pledged more than 4 projects and whose standard deviation is greater than 1.5; *black sheep*, users who pledged more than 3 projects and for which the average distance of their rating on project i with respect to mean rating of project i is greater than 0.5; *niche projects*, which were pledged by less than 5 users; *controversial projects*, which received ratings whose standard deviation is greater than 0.5.

Results of Experiment. The result on ratings is shown in the left part of Table 2. From the first row we can see that TRED2 exceeds over others in terms of both MAE and coverage, it lowers the MAE by 63.03 % while preserving the coverage. It is worth noting that, for cold start users, TREDs exceeds the best of others by 43.31 % on MAE and 47.49 % on coverage, which means TREDs can handle the *cold start* problem well. The fact that TRED2 does better than TRED1 demonstrates that in a reasonable scope, a larger candidate set produces a better result. On cold start users and niche projects, TREDs still remain better effectiveness while others suffer a lot from data sparsity. In this experiment we also find that TREDs do not exceed over collaborative filtering for controversial items and opinionated users. We think this is because the expert discovery process depends on the opinion of minority, thus it suffers a little bit from controversial input. Similarly, for black sheep, it also gives up some coverage to get a better result on MAE.

Table 2. Accuracy and coverage measures on ratings and users, for different RS algorithms on different views.

Views	MAE/Ratings coverage (%)					MAUE/Ratings coverage (%)				
	CF_{user}	CF_{item}	SVD	TRE1	TRE2	CF_{user}	CF_{item}	SVD	TRE1	TRE2
All	0.600	0.638	0.633	0.406	**0.234**	0.750	0.656	1.283	0.495	**0.222**
	1.73	8.23	22.86	20.9	**22.9**	3.03	15.15	16.30	36.30	**37.23**
Cold	0.501	1.602	1.222	0.416	**0.284**	0.333	1.603	1.880	0.359	**0.293**
users	3.15	3.14	10.99	14.15	**16.21**	2.68	4.46	7.86	31.73	**33.62**
Heavy	0.376	0.349	0.615	0.357	**0.346**	0.453	0.413	0.515	0.397	**0.366**
buyers	12.57	22.02	60.62	51.3	**60.97**	20.84	55.69	92.59	89.44	**95.11**
Contr.	**0.375**	0.417	1.085	0.875	0.857	**0.333**	0.417	0.681	0.833	0.875
projects	3.70	10.19	**21.99**	11.11	16.13	4.48	17.91	10.68	16.02	**21.42**
Niche	1.333	0.857	1.714	0.494	**0.239**	1.000	0.833	1.333	0.641	**0.355**
projects	1.22	6.10	7.53	12.22	**21.17**	2.13	12.77	6.451	20.20	**24.74**
Opin.	**0.714**	1.002	1.301	0.863	0.842	**0.714**	1.003	1.567	0.915	0.909
users	2.55	1.57	**11.11**	4.76	4.02	4.43	1.92	**17.65**	7.38	8.45
Black	0.938	0.529	1.267	0.443	**0.344**	0.938	0.604	1.056	0.459	**0.424**
sheep	1.99	3.65	**7.57**	4.70	5.04	4.60	8.62	**15.25**	10.49	13.71

The result of accuracy and coverage measures on users is shown in the right part of Table 4.1. From this respective, we can see that TREDs also exceed over others and the difference is larger than the respective of rating. In this measurement, every user is taken into account only once, therefore cold start users have the same influence with heavy buyers. Since TREDs do well for cold start users, this advantage is reflected by the result on users. Different from rating perspective, we can see that for controversial projects TREDs have a high user coverage. The reason is that opinions from cold start users on controversial projects contribute some consistency since there is less chance for them to buy controversial projects.

4.2 User Study

Apart from embedding our work into existing trust-aware recommender systems, we also evaluate our work in a direct manner, that is, by surveying real users in Zhongchou.com. We select some users in Zhongchou.com dataset randomly as experimental subjects. Then we use our model to recommend lists of experts to them. Finally, we survey them through questionnaires to evaluate whether they approve the experts we recommend.

Experiment Setup. We randomly pick up 140 users from Zhongchou.com and for each user we select the most relevant category as their query topic c according to their purchase history and we set the number of query experts $k = 5$, then we use the expert discovery model to recommend them lists of experts ordered by reputation respectively.

Since we are the first to discover experts in time-sensitive online shopping, there is no counterpart work to be compared. Thus we use two standard baselines: one is to pick up users randomly from dataset by the uniform distribution; the other one is to recommend similar users measured by the Pearson Correlation coefficient, which is wildly used in recommender systems.

We provide recommendation expert lists E_1, E_2 and E_3 generated from these three methods to subject i and ask him to (1) select experts he thinks in the list; (2) supplement experts, if any, by his experience; (3) rank the experts by his cognitive reasoning. The information we give to subjects to judge whether a recommended user is an expert includes two parts: (1) his demographic information including join date, location, self-description and his status in the adoption network $|\rho(u)|$, and (2) information about projects he has pledged including project name, his expenditure, comment sentiment, the amounts of sharing and liking, whether it matches presetting sale goal and whether it has purchase disputes.

We use two measures to evaluate results. First we use *precision, recall* and *F1-measure* to evaluate whether these methods generate right expert lists. Then we adopt the smooth Mean Average Precision @20(MAP@20) to evaluate whether these methods give right orders for recommended experts. Specifically, given the survey result A_i of subject i and the expert list E_j generated from a tested method, we first compute the average precision of i, donated as $ap_i@20$:

$$ap_i@20 = \frac{\sum_{k=1}^{20} \frac{1}{1-lnp(k)} \times \Delta(k)}{|A_i|} \tag{8}$$

where $p(k)$ is the precision at the cut-off k in the list E_j, and if the kth user is in A_i then $\Delta(k) = 1$, otherwise $\Delta(k) = 0$. Then we sum up all the average

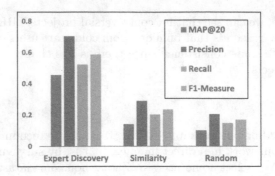

Fig. 3. The result of user study.

precisions to get the formulation of MAP@20:

$$MAP@20 = \frac{\sum_{i \in S} ap_i@20}{|S|} \qquad (9)$$

Where S is the set of subjects.

Results of Experiment. The results are shown in Fig. 3. We can see the expert discovery method is better than baselines on all measures. Specifically, it exceeds over others by 0.416, 0.321 and 0.353 on precision, recall and F1-measure respectively, which demonstrates the expert list generated by the expert discovery method matches more true experts approved by subjects. For MAP@20, our method also get better performance with the improvement of 0.315, which means our reputation computation is also effective.

5   Conclusion and Future Work

We propose a generalized framework to discover experts in time-sensitive online shopping. Firstly, it adopts the history data of users to quantify their purchase ability, influence acceptance and adoption network which captures the access pattern of users. Then it selects expert candidates by their accessibility to the query user. Finally, it computes the candidates' influence they give to community and trust they get from others as their reputation in order to filter and rank experts. The experiment results show that it can improve recommendation and achieves good performance in matching and ranking experts trusted by subjects.

In the future, we plan to further our work in following aspects:

Expert Community. In the experiments, we obverse that experts usually come up in a cluster. Thus we guess that it is more likely to find another expert near the existing one. We can apply this assumption to discover more interesting patterns such as *cluster* of experts, which may be used to improve the accuracy of discovery or accelerate the computation.

Dynamics of Purchase Ability and Trust. In practice, users' purchase ability and trust may change over time, so we want to explore whether the discovery process benefit by including their dynamics.

Acknowledgement. The work is supported by National Natural Science Foundation of China (61373022, 61532015, 71473146) and Chinese Major State Basic Research Development 973 Program (2015CB352301).

References

1. An, J., Quercia, D., Crowcroft, J.: Recommending investors for crowdfunding projects. In: Proceedings of the 23rd International Conference on World Wide Web, pp. 261–270. ACM (2014)
2. Aslay, C., Barbieri, N., Bonchi, F., Baeza-Yates, R.A.: Online topic-aware influence maximization queries. In: EDBT, pp. 295–306 (2014)
3. Barbieri, N., Bonchi, F., Manco, G.: Topic-aware social influence propagation models. Knowl. Inf. Syst. **37**(3), 555–584 (2013)
4. Breese, J.S., Heckerman, D., Kadie, C.: Empirical analysis of predictive algorithms for collaborative filtering, arXiv preprint arxiv:1301.7363 (2013)
5. Casella, G., Berger, R.L.: Statistical Inference. Oxford University Press, New York (1995). 25(98): xii, 328
6. Das, M., De Francisci Morales, G., Gionis, A., Weber, I.: Learning to question: leveraging user preferences for shopping advice. In: Proceedings of the 19th ACM SIGKDD International Conference on Knowledge Discovery and Data Mining, pp. 203–211. ACM (2013)
7. Domingos, P., Richardson, M.: Mining the network value of customers. In: Proceedings of the Seventh ACM SIGKDD International Conference on Knowledge Discovery and Data Mining, pp. 57–66. ACM (2001)
8. Guha, R., Kumar, R., Raghavan, P., Tomkins, A.: Propagation of trust and distrust. In: Proceedings of the 13th International Conference on World Wide Web, pp. 403–412. ACM (2004)
9. Jamali, M., Ester, M.: Trustwalker: a random walk model for combining trust-based and item-based recommendation. In: Proceedings of the 15th ACM SIGKDD International Conference on Knowledge Discovery and Data Mining, pp. 397–406. ACM (2009)
10. Kempe, D., Kleinberg, J., Tardos, É.: Maximizing the spread of influence through a social network. In: Proceedings of the Ninth ACM SIGKDD International Conference on Knowledge Discovery and Data Mining, pp. 137–146. ACM (2003)
11. Koren, Y., Bell, R., Volinsky, C.: Matrix factorization techniques for recommender systems. Computer **8**, 30–37 (2009)
12. Massa, P., Avesani, P.: Trust-aware recommender systems. In: Proceedings of the ACM Conference on Recommender Systems, pp. 17–24. ACM (2007)
13. O'Donovan, J., Smyth, B.: Trust in recommender systems. In: Proceedings of the 10th International Conference on Intelligent User Interfaces, pp. 167–174. ACM (2005)
14. Pavlou, P.A.: Consumer acceptance of electronic commerce: integrating trust and risk with the technology acceptance model. Int. J. Electron. Commer. **7**(3), 101–134 (2003)

15. Sarwar, B., Karypis, G., Konstan, J., Riedl, J.: Item-based collaborative filtering recommendation algorithms. In: Proceedings of the 10th International Conference on World Wide Web, pp. 285–295. ACM (2001)
16. Song, X., Tseng, B.L., Lin, C.Y., Sun, M.T.: Personalized recommendation driven by information flow. In: Proceedings of the 29th Annual International ACM SIGIR Conference on Research and Development in Information Retrieval (2006)

Temporal Recommendation via Modeling Dynamic Interests with Inverted-U-Curves

Yang Xu[1], Xiaoguang Hong[1(✉)], Zhaohui Peng[1], Guang Yang[1], and Philip S. Yu[2,3]

[1] School of Computer Science and Technology,
Shandong University, Jinan, China
zzmylq@gmail.com, {hxg,pzh}@sdu.edu.cn,
loggyt@yeah.net
[2] Department of Computer Science,
University of Illinois at Chicago, Chicago, USA
psyu@uic.edu
[3] Institute for Data Science, Tsinghua University, Beijing, China

Abstract. How to capture user interest accurately to enhance the user experience is a great practical challenge in recommender systems. Through preliminary investigation, we find that each user has his personalized interest model which may contain multiple kinds of interests, and the strength of each user interest usually has a dynamic evolution process which can be divided into two stages: rising stage and declining stage. The evolution rate of the user interests also differ from each other. Based on this finding, a recommendation framework called SimIUC is proposed, which can identify multiple user interests and adapt the inverted-U-curve to model the dynamic evolution process of user interests. Specifically, SimIUC differs from the traditional user preference based methods which use monotonously decreasing function to model user interest. It can predict the evolutionary trends of interests and make recommendations by inverted-U-interest-based collaborative filtering. We studied a large subset of data from MovieLens and netflix.com respectively. The experimental results show that our method can significantly improve the accuracy in recommendation.

Keywords: Interest modeling · Recommender systems · Inverted U curve

1 Introduction

Personalized recommendation systems that can extract information of interests from a large amount of complex data have been widely used in the case of the rapid expansion of Internet applications. User interest model plays a key role in personalized recommendation systems and it is applied in a variety of applications, e.g., microblog recommendations, product recommendations and music recommendations, etc. The key issue of personalized recommendation is how to accurately obtain the user's interests.

Temporal interaction log of a user's interaction with the system contains the user's interests. Existing studies [6, 12, 13, 15] have generally considered that the more timely the temporal information generated, the higher effectiveness in the rating prediction.

© Springer International Publishing Switzerland 2016
S.B. Navathe et al. (Eds.): DASFAA 2016, Part I, LNCS 9642, pp. 313–329, 2016.
DOI: 10.1007/978-3-319-32025-0_20

Thus, old information has lower effectiveness. Existing user interest models usually use monotonically decreasing function, e.g., linear function, exponential function, to calculate timeliness of information. However, our preliminary investigation shows that user interests have two characteristics. First, the interest pattern is personalized. For example, some of the users keep a clear interest consistently, while some others maintain multiple interests at the same time and also some ones change their interests frequently. Second, the evolution of a user interest is a nonmonotonic process which can be divided into two stages: rising stage and declining stage [1]. The evolution rate of the user interests also differs from each other. Due to these two characteristics, it is hard to accurately model a user's interest using a single monotonously decreasing function. Existing interest models cannot effectively describe interest patterns and interest evolution trends.

In this paper, we propose a new recommendation framework called SimIUC. Under SimIUC, we can effectively identify multiple user interests from his or her temporal interaction log, and adapt the inverted-U-curves to model user interest patterns to reflect the dynamic evolution process of user interests, and then we can predict user's interest evolution trend in the future and make recommendation.

What is Inverted U Curve? The inverted U curve was first advanced by economist Simon Kuznets in the 1950s and 60s, and it was used to describe the relationship between income distribution and economic development [2]. In social psychology, inverted U curve can be used to describe the evolution process of a person's interest. People interact with their interested things, and the larger the number of interactions, the higher the degree of interest [3], but the growth rate has continued to decrease [4]. When the number of interactions reaches a certain threshold, the degree of interest will drop [3]. This is a universal law in the real world, for example, you find a nice dessert in a dessert shop, thus become interested in the dessert shop. You will continue to visit the shop, and a new delicious dessert will attract you a strong interest in the shop. But as you buy more and more dessert in the shop, you may feel that the taste of the dessert in that shop is getting more and more common, while their deficiencies will also be found in this process. Thus a "very good dessert shop" may become a "good dessert shop" after a period of time, and at last, it may just become an ordinary "dessert shop". Inverted-U-curve can properly describe the dynamic evolution process of a user interest like this. To summarize, the major contributions of this paper are as follows:

- In this paper, we consider the dynamic evolution process of user interest which previous studies have rarely mentioned. A novel recommendation framework named SimIUC for modeling the dynamic evolution process of user interest is proposed. SimIUC can be used to predict user interest evolution trend and make accurate recommendation.
- In SimIUC, we design a multiple user interest identification method and an inverted-U-interest model learning algorithm to model user interest pattern and interest evolution trend.
- We systematically compare the proposed SimIUC approach with other algorithms on the dataset of Movielens and Netflix. The results confirm that our new method substantially improves the precision of recommendation.

The rest of this paper is organized as follows. Section 2 presents some notations and the algorithm framework. Section 3 introduces our novel algorithm for interest identification, and Sect. 4 details the construction of the inverted-U-interest model and the recommendation approach. Experiments and discussions are given in Sect. 5. In Sect. 6, we review the related works on temporal dynamics and user interest. Conclusions are drawn in Sect. 7.

2 Preliminaries

2.1 Notations

I is a set of all items and U is a set of all users. Our main task in this paper is modeling user interest by analyzing user temporal interaction logs and making recommendations. We first define the data model of user temporal interaction logs. $L_u = \{r_{u1}, r_{u2}, \ldots, r_{ul}\}$ denotes the temporal interaction log of user u which contains l records, and r_{ui} represents $u's$ ith interaction record expressed as a quadruple ($user, item,$ $rating, timestamp$). $L'_u = \{r'_{u1}, r'_{u2}, \ldots, r'_{uo}\}$ is the processed temporal interaction logs of user u treated by filtering out noise and labeling interest, o records the size of L'_u, and r'_{ui} is an extended record expressed as a five-tuple ($user, item, rating, interest, timestamp$). We propose algorithms to identify the user's N_u interests and learn N_u inverted-U-interest curves separately. f_{ui} represents user u's ith inverted-U-interest curve, and we denote the inverted-U-interest model of user u as $IUI_u = \{f_{u1}, f_{u2}, \ldots, f_{uN_u}\}$. Table 1 lists the frequently used symbols in this paper.

Table 1. Summary of notations

Symbol	Description
I	the set of items
U	the set of users
$L_u = \{r_{u1}, r_{u2}, \ldots, r_{ul}\}$	temporal interactive log of user u
$L'_u = \{r'_{u1}, r'_{u2}, \ldots, r'_{uo}\}$	temporal interactive log of user u with interest labeled
$IUI_u = \{f_{u1}, f_{u2}, \ldots, f_{uN_u}\}$	the set of inverted-U-interest curves of user u
HIN	the heterogeneous information network
λ	the parameter to govern the influence of feature similarity and interactive similarity
N_u	the number of user interest of user u

2.2 SimIUC Recommendation Framework

We propose a recommendation framework called SimIUC based on inverted-U-interest model. The framework is as shown in Fig. 1. In order to discover users' interests from user temporal interaction logs, SimIUC first needs to calculate the similarity matrix S between each item. In this paper, a new similarity algorithm named FIsim is proposed in SimIUC. In FIsim, we consider the similarity between different items not only based on the item features, but also the similar situation of items in the process of interactions

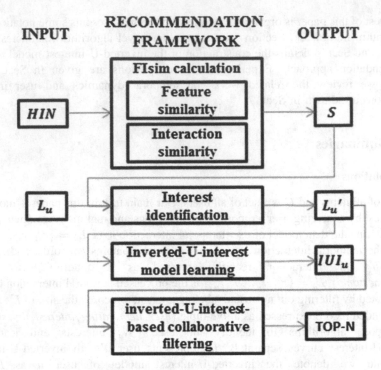

Fig. 1. SimIUC recommendation framework

between users and items. Consequently, the input information in SimIUC is all users' temporal interaction log L and items' feature information FG. Given a target user u, SimIUC uses the similarity matrix S to cluster and filter items in u's temporal interaction log L_u, getting u's N_u different interests. Then, SimIUC proposes an inverted-U-interest model learning algorithm to learn u's inverted-U-interest model IUI_u. Last, SimIUC makes accurate top-N recommendation through an inverted-U-interest-based collaborate filtering approach.

3   Interest Identification

As discussed in the previous section, in order to accurately discover users' interests from user interaction logs, the first phase of SimIUC is to calculate the similarity between each item. The proposed similarity algorithm FIsim combines the similarities of item feature and user interaction. Then we detail the interest identification algorithm.

3.1   Feature Similarity Calculation

FIsim measures the feature similarity between each item by calculating the similarity between their features in the heterogeneous information network. A heterogeneous information network(HIN) is an undirected weighted graph $G = (V, E, W)$ with an object type mapping function $\varphi : V \rightarrow \mathcal{A}$ and a link type mapping function $\phi : E \rightarrow \mathcal{R}$ and $|\mathcal{A}| > 1$, $|\mathcal{R}| > 1$. Each object $v \in V$ belongs to a particular object type $\varphi(v) \in \mathcal{A}$ and each link $e \in E$ belongs to a particular relationship $\phi(e) \in \mathcal{R}$. $W : E \rightarrow \mathbb{R}^+$ is a weight mapping from an edge $e \in E$ to a real number $\omega \in \mathbb{R}^+$. As an example, a toy IMDB network is given in Fig. 2. It is a typical heterogeneous information network, containing five types of nodes: users (U), movies (M), genres (G), directors (D), and actors (A). G, D and A are called as feature nodes in this paper.

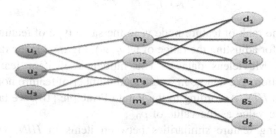

Fig. 2. An example of HIN

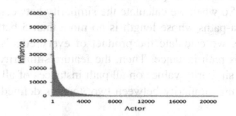

Fig. 3. Long-tailed distributions of actor influence

In G, the weight of the edges between items and features represents the influence from feature to item and depends mainly on two factors. One is the global influence of features, which is an important consideration in people's content-consumption behaviors because users tend to choose popular objects to interact. The more pro-found influence the item features has, the more possibility the user chooses the item. The other factor is association strength among features and item, which depends on the rank between the same kinds of features of the item. The higher rank the feature has, the greater association with the item it has. For example, the strength of association of the first actor is much higher than the fifth actor in a movie.

In G, the feature node v_f connects with the items which own feature f, and the set of these items denoted as I_f. The global influence of v_f is defined as:

$$p(v_f) = \left| \bigcup_{v_i \in I_f} \bigcup_{v_u \in U_i} \{v_u\} \right| \tag{1}$$

where U_i is the set of user nodes v_u which is connected with item node v_i. The definition means, given a feature node v_f, its influence is the number of users who have interacted with items which own feature f. In the example shown in Fig. 2, the influence of director d_1 is 3 and actor a_1 is 1. The weight of the edges between the item and the feature is measured as:

$$w(v_f, v_i) = e^{-\frac{r(v_f, v_i) \cdot p_{0.2}}{p(v_f)}} \tag{2}$$

where $r(v_f, v_i)$ is the rank of feature node v_f in the same type of features of item i, and $p_{0.2}$ is a parameter for adjusting the range of $w(v_f, v_i)$. Figure 3 is the distribution of the actor influence in Movielens dataset. The influence of item features presents a long-tailed distribution. We denote n_i as the total number of feature nodes belonging to the feature type i in *HIN*. According to the Pareto Principle [10], we take the $(0.2 \cdot n_i)$th maximum influence value as the value of $p_{0.2}$.

When calculating feature similarities between items in *HIN*, we introduce the concept of meta-path mentioned in [11]. $\mathbb{P}$ denotes the set of specified meta-paths. $P = (A_1 A_2 \ldots A_n)$ is a meta-path, and $p = (v_1 v_2 \ldots v_n)$ is a path instance of the meta-path. It has been shown in [11] that long meta-paths are not quite useful in calculating similarity. So when we calculate the similarity between two items, we only need to use those meta-paths whose length is no more than 5 between the two items. For one path instance, we calculate the product of every edge's weight of it as the similarity value on this path instance. Then, the feature similarity between two items i_p, i_q is the sum of the similarity values on all path instances of all meta-paths between nodes v_{i_p}, v_{i_q}. The feature similarity between two items is defined as:

$$FeatureSim(v_{i_p}, v_{i_q}) = \sum_{P \in \mathbb{P}} \sum_{p \in P} \prod_{k=1}^{n-1} w(v_k, v_{k+1}). \tag{3}$$

3.2   Interaction Similarity Calculation

FIsim considers that interactions between the user and the system are driven by interests, and the user's interests remain constantly for a period of time. Consequently, items connected with the same user have a certain similarity called interaction similarity.

The algorithm we proposed measures the interaction similarity from two dimensions: time and rating. FIsim considers that the smaller the interaction time interval or rating difference, the more similar between the two items, because the user's interest and evaluation standard remain constantly for a period of time.

The interaction similarity between the items i_p, i_q is measured as:

$$\text{sim}\left(v_{i_p}, v_{i_q}\right) = \sum_{u \in (U_p \wedge U_q)} e^{-\left[(1-\theta)\left(\frac{t_{up}-t_{uq}}{t_u}\right)^2 + \theta\left(\frac{r_{up}-r_{uq}}{\bar{r}_u}\right)^2\right]} \qquad (4)$$

$$\text{InteractSim}\left(v_{i_p}, v_{i_q}\right) = \frac{\sum_{v_a, v_b \in I} I(sim(v_a, v_b) \neq 0)}{\sum_{v_a, v_b \in I} I(sim(v_a, v_b) \neq 0) sim(v_a, v_b)} sim(v_{i_p}, v_{i_q}) \qquad (5)$$

where $I(\cdot)$ is an indicator function, U_p is a set of user nodes which connected with item p, t_{up} is the interaction time of user u with item p, t_u is the time of latest interaction, r_{up} is u's rating of p and $\bar{r}_u$ is the average rating of u. The value range of interaction similarity is limited by formula 5. λ is used to govern the influence of feature similarity and interaction similarity. The FIsim is given by the following equation:

$$\text{FIsim}\left(v_{i_p}, v_{i_q}\right) = (1 - \lambda) \cdot FeatureSim\left(v_{i_p}, v_{i_q}\right) + \lambda \cdot InteractSim(v_{i_p}, v_{i_q}). \qquad (6)$$

3.3   Interest Identification

Using the FIsim algorithm described in last section, we can get the similarities between any two items and construct similarity matrix S. The identification of user multiple interests follows a cluster filtering approach, which clusters the items in L_u and uses the principal component extraction method to filter out the noise. We treat each item in L_u as a singleton candidate cluster. For each pair of items (p, q) in L_u, we will put (p, q) in a descending sequence SEQ if the similarity of (p, q) is no smaller than a predefined threshold β. Next, we remove (p, q) from SEQ in sequence, and the two clusters which contain item p and q respectively will be merged if the similarity between them is no less than a predefined threshold γ until SEQ is empty. $sim(p, q)$ is calculated as:

$$sim(p, q) = \frac{1}{|c(i)||c(j)|} \sum_{i_p \in c(i)} \sum_{i_q \in c(j)} S(p, q) \qquad (7)$$

where $c(i)$ is the cluster which contains item i, and $S(p.q)$ is similarity between items p, q. We apply the principal component extraction method to the clustering result. The N_u maximum clusters which make $\sum_{i=1}^{N_u} size(c(i))/\sum_{i=1}^{n} size(c(i)) \geq \eta$, where η is the threshold of the first N_u principal components' cumulative contribution rate, n is the number of clusters in the clustering result and $size(c(i))$ is the number of items in cluster $c(i)$. They are selected as N_u kinds of the user interest and other clusters are considered as noise to be removed. In this paper, we set η to 0.8. Finally, we update the item records in L_u with interest tag. The pseudocode of interest identification is shown in Algorithm 1 which is inspired by Kruskal algorithm [21]. Through multiple user interest identifications, each of the remaining items in the user temporal interaction log is assigned into a corresponding interest.

Algorithm 1. Interest identification

Input : S: *the similarity matrix of all items*, L_u: *the temporal interactive* log *of the user u*,
β: *the similarity threshold of items*, γ: *the similarity threshold of clusters*
Output: L'_u: the temporal interactive log with interest labeled of the user u
  1:    **for** each item i_p, i_q in L_u **do**
  2:        **if** $FIsim(i_p, i_q) \geq \beta$ **then**
  3:            add (p, q) in Seq;
  4:    Queue $Q \leftarrow$ Seq.sort("order by $FIsim(i_p, i_q)$ descending");
  5:    each item i_k in L_u as a singleton cluster $c(i_k)$;
  6:    **while** Q is not empty **do**
  7:        $(i_p, i_q) = $ Q.top();
  8:        Calculate the similarity $sim(p, q)$ between $c(i_p)$ and $c(i_q)$ according to formula 7;
  9:        **if** $sim(p, q) \geq \gamma$ **then**
 10:            merge $c(i_p)$ and $c(i_q)$ as a new cluster;
 11:    apply the principal component extraction method to the clustering result;
 12:    select N_u maximum clusters which satisfied $\sum_{i=1}^{N_u} size(c(i))/\sum_{i=1}^{n} size(c(i)) \geq \eta$ as user's N_u interests;
 13:    $L'_u \leftarrow$ update L_u with interest tag;
 14:    return L'_u;

---

4   Learning Inverted-U-Interest Model and Making Recommendation

After interest identification, SimIUC presents a learning algorithm to learn the inverted-U-interest model and then introduces an inverted-U-interest-based collaborate filtering approach to make top-N recommendation.

4.1   Inverted-U-Interest Model

Interest Pattern. Influenced by behavior habits, every user has a personal interest pattern. There are four kinds of user interest patterns through analyzing user log, which are *specific interest pattern*, *multiple interest pattern*, *interest shifting pattern* and *random noise pattern*. We cannot make accurate predictions about future behaviors of the user with random noise pattern because when the user interacts with the system, it shows a strong randomness. So we do not consider this type of users in this paper. For the other three interest patterns, SimIUC characterizes them by building inverted-U-interest model.

Learning Algorithm of Inverted-U-Interest Model. Time is an important aspect in fitting interest curve. Because of the difference between user behavior and external environment constraints, the frequency of interaction of different users varies considerably even with the same interest. For example, two users A and B both like watching comedy movie. Due to work reasons, user A watches movies only on weekends, but user B watches movies every day. If we measure the interestingness using the same time dimension, user A's interestingness to comedy movie would be far lower than B's.

But this is not the case. Consequently, the modeling results will be greatly intervened with the interaction frequency if we use the same time dimension for all users in modeling. As illustrated in [3], the evolution of user interest relates to interaction times. Our interest modeling approach uses interaction times to build the time dimension, which eliminates the impact of interaction frequencies. The learning samples are created from L'_u, and user's inverted-U-interest model is learned by the learning algorithm SimIUC proposed. The pseudocode of learning algorithm of inverted-U-interest model is shown in Algorithm 2.

Algorithm 2. Learning inverted-U-interest model

Input : L'_u: *the temporal interactive log with interest labeled of the user u*
Output: IUI_u: *the set of inverted-U-interest models of user u*
  1:    **for** each interest k of user u **do**
  2:        Set x of sample points in interest k's learning sample are $[1,2,...,|L_u'|]$;
            Set y of sample points to the number of interactions between the user and items belong-
  3: ing
            to u's interest k when user u has interacted with system x times.
  4:        //use the improved sigmoid function and the least square method to fit sample points//
  5:        $f_{uk} \leftarrow \arg \min_{a,b,c} \Sigma_i(Y_i - \hat{Y}_i)^2 = \arg \min_{a,b,c} \Sigma_i[Y_i - (\frac{a}{1+be^{-cX_i}})]^2$;
  6:        Inverted-U-interest curve f_{uk}' is the derivative of f_{uk};
  7:        $IUI_u. add(f_{uk}')$;
  8:    return IUI_u;

Through identifying the multiple user interests, N_u different interests of target user u have been identified and each item in L'_u is assigned into the corresponding interest. In L'_u, the items belonging to u's interest k have been selected to construct the learning sample of u's interest k. A learning sample of interest k is a group of ordered pairs (x, y), in which x represents the number of interactions between user u and the system, and y is defined as the number of interactions between user u and items belonging to u's interest k. For example, there are two different interests identified for user u, and at the time of $x = 20$, which means that user u has interacted with the system for 20 times, the number of interactions between user u and items belonging to his first interest is 12 and that of the second interest is 8, then the sample point $(20, 12)$ is contained in u's first interest learning sample, and the sample point $(20, 8)$ is contained in the second one. The discrete ranges of both x and y are $[1, |L'_u|]$. Each interest learning sample is dealt with in lines 1 to 4 of Algorithm 2. In lines 5 to 6, we improve the sigmoid function as the fitting function and use the least square method to fit sample points. The prediction is given by the following equation:

$$\hat{Y} = \frac{a}{1 + be^{-cX}} \tag{8}$$

This equation involves three parameters. The least square method defines the estimate of these parameters as the values which minimize the sum of the squares between the measurements and the model. This amounts to minimizing the expression:

$$\varepsilon = \sum_i (Y_i - \hat{Y}_i)^2 = \sum_i [Y_i - (\frac{a}{1+be^{-cX_i}})]^2 \qquad (9)$$

Thus, the best parameter setting (including a, b and c) of fitting curve should be

$$\arg \min_{a,b,c} \sum_i (Y_i - \hat{Y}_i)^2 = \arg \min_{a,b,c} \sum_i [Y_i - (\frac{a}{1+be^{-cX_i}})]^2 \qquad (10)$$

N_u is the number of user interest of user u. We can get N_u fitting curves $\{f_1, f_2, ..., f_{Nu}\}$ by fitting the sample points of each interest. The inverted-U-interest curve f_k' of interest k is the derivative of f_k:

$$f_k'(x) = \left(\frac{a_k}{1+b_k e^{-c_k x}}\right)' = a_k c_k g(x)[1 - g(x)], g(x) = \frac{1}{1+b_k \cdot e^{-c_k x}} \qquad (11)$$

where a_k, b_k, c_k are the three parameters to be learned. The inverted-U-interest model of user u is the set of inverted-U-interest curves, i.e. $IUI_u = \{f_1', f_2', ..., f_{Nu}'\}$.

Case Study. The results of learning the inverted-U-interest model for the 51th user in Movielens dataset is shown in Fig. 4. The user interests are identified by interest identification algorithm. The size of L_{51}' become 36, after filtering out 4 noise records from user's temporal interaction log L_{51}, and two kinds of user interests are identified. The fitting results of two interest learning sample of 51th user is shown in Fig. 4(a), and the inverted-U-interest model has been learned is shown in Fig. 4(b).

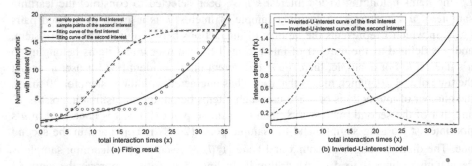

Fig. 4. The results of learning IUI model for the 51th user in Movielens data set

In Fig. 4(a), '$\times$' and '$\bigcirc$' represent sample points of the user's first interest and the second interest respectively. Each point represents that when the user has interacted with system x times, the number of interactions between him and the items belonging to his corresponding interest is y. The dotted line and the full line respectively describe the fitting result of the user's first interest and the second interest. Figure 4(a) has shown that during the first half of interactions, the user mainly interacts with the items belonging to his first interest, but in the second half, the user shifts his attention to items belonging to his second interest.

Each interest of users has its corresponding inverted-U-interest curve constructing user's inverted-U-interest model. Figure 4(b) demonstrates the inverted-U-interest model of the 51th user where x-axis represents the number of interactions between the 51th user and the system, and y-axis represents the interest strength. Figure 4(b) has shown that the interest of the 51th user has an obvious shifting in the process of interacting with the system. The old interest (dotted line in Fig. 4(b)) declines while the new interest (full line in Fig. 4(b)) rises fast. So we infer that the interest pattern of this user belongs to the "*interest shifting pattern*", in which SimIUC will focus more on the new interest in recommendation.

4.2 Inverted-U-Interest-Based Collaborative Filtering Approach

When making recommendation for x times interaction of the target user u, our basic idea is to consider u as the source to be injected with user preferences. The propagation process of user preferences is shown in Fig. 5. Preferences injected into the user u will be propagated to items i in L_u' through the corresponding interest, and tend to propagate to K nearest neighbor items i' which construct candidate item set C. For each i in C, the estimated preference $p_{u,i,x}$ of user u on item i is measured as:

$$P_{u,i,x} = \sum_{j \in L_u'} \sum_{f'k \in IUI_u} e^{f'k(x)} \cdot I(j \in interest_k) \cdot S(i,j). \tag{12}$$

where $I(\cdot)$ is an indicator function, f_k' is the inverted-U-interest curve of $interest_k$, and $S(i,j)$ is the similarity between item i and j. We sort $P_{u,i,x}$ for all no-interacted neighbor items, and then return top-N no-interacted neighbor items to user u.

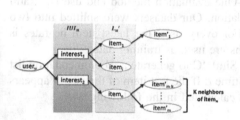

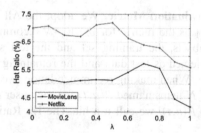

Fig. 5. The propagation process of user preferences

Fig. 6. The impact of λ on hit ratio in both datasets

5 Experiments

We have conducted a set of experiments to examine the performance of our recommendation method compared with the baselines. We begin by introducing the experimental settings, and then analyze the evaluation results.

5.1   Experimental Settings

Data Description. There are two real datasets in our experiments: Movielens [22] and Netflix [23]. Movielens and Netflix are the most widely used common datasets in recommendation research projects. The Movielens dataset contains 6,040 users who have issued 999,209 explicit ratings on a 5-point likert scale, referring to 3,883 movies.

The Netflix dataset was made available in 2006 as a part of the Netflix Prize, in which each user rated at least 20 movies, and each movie was rated by 20–250 users. It has been widely used as a large-scale data set for evaluating recommenders. As our goal is to make accurate top-N recommendation by analyzing item features and user temporal interaction logs, we extend the movie features of the Netflix dataset through using MovieLens dataset. 6000 users and related movies in Netflix dataset are randomly selected as experimental data. The global statistics of these two datasets used in our experiments are shown in Table 2.

Table 2. Characteristics of datasets

Name of dataset	MovieLens	Netflix
Number of users	6,040	6,000
Number of items	3,883	4,158
Avg. # of rated items/user	257.3	359.7
Number of ratings	999,209	2,158,471

Table 3. Disassembled algorithms

Disassembled algorithm	FIsim	IUI model
SimIUC	O	O
CosIUC	X	O
SimCF	O	X
CosCF	X	X

Evaluation Metric. We adopt the All-But-One evaluation method and use Hit Ratio [5] as the metric for the top-N recommendation. Our datasets were splitted into two subsets, the training set and the test set. For every user, the latest item he rates is selected as test data and the remaining items are used as training data.

When making recommendation, we use SimIUC to generate a recommendation list of N items named $R(u, t)$ for each user u at time t. If the test item of the user u appears in $R(u, t)$, we call it a hit. The Hit Ratio is calculated in the following way:

$$\text{Hit Ratio} = \frac{\sum_u I(T_u \in R(u, t))}{|U|} \tag{13}$$

where $I(\cdot)$ is an indicator function, $R(u, t)$ is a set of top-N items recommended to user u at time t, T_u is the test item that the user u has actually interacted with at time t.

Compared Methods. We examine the performance of the proposed SimIUC approach by comparing it with three other top-N recommendation algorithms, including ItemKNN [9], TItemKNN [6], and IFCM-IFC [12].

Item-based collaborative filtering (ItemKNN) is famous recommendation algorithm which could predict unknown item ratings for a given user by referencing item rating information from other similar items. TItemKNN is a time weighted item-based collaborative filtering method by reducing the influence of old data when predicting users' further behaviors. TItemKNN was designed for rating prediction task but it can be easily extended to top-N recommendation for temporal data. IFCM-IFC method adapts the memory forgetting curve to model the user interest for temporal recommendations.

5.2   Evaluations

Impact of Parameter λ. We first focus on analyzing the parameter λ, which governs the influence of feature similarity and interaction similarity. In the first experiment, we vary the parameter λ from 0, 0.1 to 1. The number of the nearest neighbor K is set to 100, and N is set to 20. The results of using different constant λ on both datasets are demonstrated in Fig. 6.

The results have shown that λ is important in determining the Hit Ratio, and ignoring either interaction similarity ($\lambda = 0$) or feature similarity ($\lambda = 1$) cannot generate good results. Optimal results can be gotten by combining feature similarity and interaction similarity together. The optimal value of λ is about 0.7 in MovieLens and is about 0.5 in Netflix. In the following experiments, λ is set to 0.7 in Movielens and 0.5 in Netflix.

Comparison to TItemKNN and IFCM-IFC with Different α. The decay rate α in TItemKNN and IFCM-IFC are used to control the attenuation rate of the influence of old data. In this section, we compare the Hit Ratio of SimIUC with TItemKNN and IFCM-IFC under different α on both datasets. When tuning α, K is set to 100, and N is set to 20. The results of how Hit Ratios of TItemKNN and IFCM-IFC change against α are shown in Fig. 7. Because there is no parameter α existing in SimIUC, its Hit Ratio is drawn as a straight line. The results have shown that SimIUC outperforms other algorithms when $\alpha \in [0.1,1]$ in both datasets. For TItemKNN, the optimal α is about 0.7 in Movielens, and is about 0.3 in Netflix. For IFCM-IFC, the optimal α is about 0.6 in Movielens, and in Netflix, it is about 0.2.

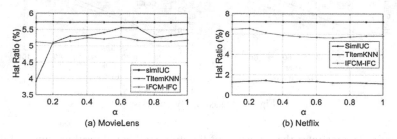

Fig. 7. The impact of α on hit ratio

Overall Accuracy Performance. In this section, we have evaluated the overall accuracy performance of SimIUC and the other three top-N recommendation algorithms, ItemKNN, IFCM-IFC, and TItemKNN. In the experiment, the number of the nearest neighbors K is set to 100, and other parameters are set to the optimal values obtained from previous experiments.

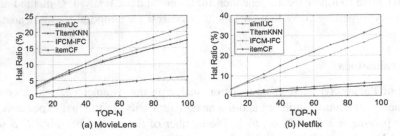

(a) MovieLens   (b) Netflix

Fig. 8. Comparisons on hit ratio of recommendation

The comparison results have been shown in Fig. 8, where SimIUC exceeds all the baselines under different size of recommendation list. By using IFCM-IFC as the benchmark, SimIUC outperforms IFCM-IFC 10.43% to 29.31% on Netflix dataset, and improves it up to 10.42% to 38.34% on MovieLens dataset. The experiment proves that the inverted-U-interest model plays an important role in improving recommendation accuracy, and SimIUC can get better accuracy of item recommendation.

Disassembly Analysis of SimIUC. The performance of each part of the SimIUC recommendation framework is analyzed in this section. In the experiment, FIsim has been replaced with cosine similarity, and the inverted-U-interest model has been removed by setting $e_k^{f'}(x)$ in formula (12) to 1. The three disassembled algorithms are shown in Table 3, in which the label "$\bigcirc$" means the corresponding part of SimIUC has been preserved, and the label "$\times$" means the corresponding part of SimIUC has been removed or replaced.

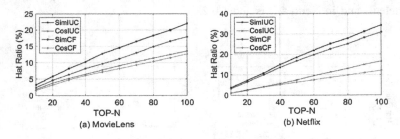

(a) MovieLens   (b) Netflix

Fig. 9. Disassembly analysis

Figure 9 demonstrates the hit ratios of the disassembled algorithms under different size of recommendation list. It has shown that SimIUC and CosIUC have achieved significant improvement in hit ratio over SimCF and CosCF by introducing inverted-U-interest model. The hit ratio of SimIUC compared with CosIUC has a significant improvement by replacing the similarity method with FIsim, and SimCF also has a notable improvement compared with CosCF on both datasets. The results lead us to conclude that both combining the feature similarity with the interaction similarity and introduction of inverted-U-interest model are necessary and effective.

6 Related Work

Temporal Dynamics in Recommendation. Temporal information is widely used in the framework of information spreading [18, 20], matrix-factorization [14], item-based CF [6] or graph [19]. Modeling temporal dynamics is indispensable when designing recommender systems.

Koren [14] models the temporal dynamics via a time-aware factorization model to predict movie ratings for Netflix. Quan [7] propose the Geographical-Temporal influences Aware Graph (GTAG) to exploit both geographical and temporal influences in time-aware POI recommendation. Chen [8] designs a Gibbs sampling algorithm to learn the receptiveness among users over time. Xiang [19] proposes a Session-based Temporal Graph (STG) to model long-term and short-term preferences and strengthen the impact of current interests via Injected Preference Fusion (IPF) method. Lathia et al. [16] and Wang [17] improve temporal diversity of recommendation across time. Compared to these work handling the temporal dynamics in different ways, SimIUC focus on modeling the dynamic evolution process of the user interests which is composed of two stages: rising stage and declining stage, and making accurate top-N recommendation based on the user interest model.

User Interest in Recommendation. In the past few years, how to capture user interest accurately has become the key issue and challenge in personalized recommendation. User interest changes over time. There are many studies using monotonously decreasing function to model user interest. Koychev [15] uses linear functions to simulate the decay of user interest. Ding et al. [6] and Andreas [13] both propose a time weighted item-based collaborative filtering method (TItemKNN) by reducing the influence of old data when predicting user's further behaviors. Chen [12] adapts the memory forgetting curve to model the human interest-forgetting curve for music recommendation. However, the method has two disadvantages. One is that the latest data is not always important while old data is not worthless all the time. The other is that users have a personalized interest model, and there are differences in the amount of interests, interest strength, interest forgetting rate and the trend of user interest evolution. It is not enough reasonable to use uniform interest model to all users.

The paper focuses on discovering user interest patterns and the trend of interest evolution by mining user interaction logs. To the best of our knowledge, this is the first work that use inverted U curves to model user interests in recommendation.

7   Conclusion and Future Work

A user's interests have a dynamic evolution process, including two stages, growth and decay. Modeling and leveraging this dynamic process for temporal recommendation poses great challenges. In this paper, we proposed a novel recommendation framework named SimIUC, which can identify multiple user interests and model the dynamic evolution process of user interests with the inverted-U-curves. Based on that, SimIUC can predict user interest evolution trend in the future and make more accurate recommendation. The experimental results have shown a significant improvement in the accuracy of our top-N recommendation method compared with the baselines. In our future works, we will try to predict the potential transfer directions of a user's interests, and acquire new interests a user just formed.

Acknowledgements. This work is supported by NSF of China (No. 61303005), 973 Program (No. 2015CB352500), NSF of Shandong, China (No. ZR2013FQ009), the Science and Technology Development Plan of Shandong, China (No. 2014GGX101047, No. 2014GGX101019). This work is also supported by US NSF grants III-1526499, and CNS-1115234.

References

1. Silvia, P.J.: Exploring the Psychology of Interest. Oxford University Press, New York (2006)
2. Stewart, B., Mark, M.: The U-curve adjustment hypothesis revisited: a review and theoretical framework. J. Int. Bus. Stud. **22**, 225–247 (1991)
3. Zajonc, R.B.: Attitudinal effects of mere exposure. J. Pers. Soc. Psychol. **19**(2), 77–78 (1968)
4. Stigler, G.J.: The adoption of the marginal utility theory. Hist. Polit. Econ. **4**(2), 571–586 (1972)
5. Karypis, G.: Evaluation of item-based top-n recommendation algorithms. In: CIKM 2001, pp. 247–254 (2001)
6. Ding, Y., Li, X.: Time weight collaborative filtering. In: CIKM 2005, pp. 485–492 (2005)
7. Quan, Y., Gao, C., Aixin, S.: Graph-based point-of-interest recommendation with geographical and temporal influences. In: CIKM 2014, pp. 659–668 (2014)
8. Chen, W., Hsu, W., Lee, M.L.: Modeling user's receptiveness over time for recommendation. In: SIGIR 2013, pp. 373–382 (2013)
9. Deshpande, M., Karypis, G.: Item-based top- N recommendation algorithms. ACM TOIS **22**(1), 143–177 (2004)
10. Newman, M.: Power laws, pareto distributions and Zipf's law. Contemp. Phys. **46**(5), 323–351 (2005)
11. Sun, Y., Han, J., Yan, X., Wu, T.: Pathsim: meta path-based top-k similarity search in heterogeneous information networks. In: VLDB 2011 (2011)
12. Chen, J., Wang, C., Wang, J.: Modeling the interest-forgetting curve for music recommendation. In: MM 2014, ACM (2014)
13. Toscher, A., Jahrer, M., Bell, R.M.: The bigchaos solution to the Netflix Grand prize (2008)
14. Koren, Y.: Collaborative filtering with temporal dynamics. In: KDD 2009, pp. 447–456 (2009)

15. Koychev, I., Schwab, I.: Adaptation to drifting user's interests. In: ECML 2000 Workshop: Machine Learning in New Information Age (2000)
16. Lathia, N., Hailes, S., Capra, L., Amatriain, X.: Temporal diversity in recommender systems. In: SIGIR 2010, pp. 210–217 (2010)
17. Wang, H., Fan, W., Yu, P.S., Han, J.: Mining concept-drifting data streams using ensemble classifiers. In: KDD 2003, pp. 226–235 (2003)
18. Senzhang, W., Xia, H., Philip, S.Y., Zhoujun, L.: MMRate: inferring multi-aspect diffusion networks with multi-pattern cascades. In: KDD 2014, pp. 1246–1255 (2014)
19. Liang, X., Quan, Y., Zhao, S., Chen, L., Zhang, X.: Temporal recommendation on graphs via long-and short-term preference fusion. In: KDD 2010, pp. 723–732 (2010)
20. Senzhang, W., Sihong, X., Xiaoming, Z., Zhoujun, L., Philip, S.Y., Xinyu, S.: Future influence ranking of scientific literature. In: SDM 2014, pp. 749–757 (2014)
21. Kruskal, J.B.: On the shortest spanning subtree of a graph and the traveling salesman problem. Am. Math. Soc. 7, 48–50 (1956)
22. MovieLens: http://grouplens.org/datasets/movielens
23. Netflix: http://www.netflixprize.com

Point-Of-Interest Recommendation Using Temporal Orientations of Users and Locations

Saeid Hosseini[✉] and Lei Thor Li

School of Information Technology and Electrical Engineering,
The University of Queensland, Brisbane, Australia
{s.hosseini,l.li3}@uq.edu.au

Abstract. Location Based Social Networks (LBSN) promotes communications among subscribers. Utilizing online check-in data supplied via LBSN, Point-Of-Interest (POI) recommendation systems propose unvisited relevant venues to the users. Various techniques have been designed for POI recommendation systems. However, diverse temporal information has not been studied adequately. From temporal perspective, as visited locations during weekday and weekend are marginally different, we choose weekly intervals to improve effectiveness of POI recommenders. However, our method is also applicable to other similar periodic intervals. People usually visit tourist and leisure spots during weekends and work related places during weekdays. Similarly, some users perform check-ins mostly during weekend, while others prefer weekday predominantly. In this paper, we define a new problem to perform recommendation, based on temporal weekly alignments of users and POIs. We argue that locations with higher popularity should be more influential. Therefore, In order to solve the problem, we develop a probabilistic model which initially detects a user's temporal orientation based on visibility weights of POIs visited by her. As a step further, we develop a recommender framework that proposes proper POIs to the user according to her temporal weekly preference. Moreover, we take succeeding POI pairs visited by the same user into consideration to develop a more efficient temporal model to handle geographical information. Extensive experimental results on two large-scale LBSN datasets verify that our method outperforms current state-of-the-art recommendation techniques.

Keywords: Point-Of-Interest recommendation · Location-Based Social Networks · Temporal influence

1 Introduction

Nowadays, pervasive applications of *Location-Based Social Networks (LBSN)* assist further communications between online users. For instance, people can easily share relevant data via LBSN check-in services (e.g. *Foursquare, Facebook places and Google+*). When a user performs a check-in at a venue, she reports the location, time and enclosed artifacts (text, photo, audio or video) [19,32]. In

S.B. Navathe et al. (Eds.): DASFAA 2016, Part I, LNCS 9642, pp. 330–347, 2016.
DOI: 10.1007/978-3-319-32025-0_21

return, LBSN data facilitates Point-Of-Interest (POI) recommenders to propose attractive locations to the users that are not yet visited by them. Therefore, POI recommendation is beneficial for people as well as industry (e.g. tourism).

The POI recommendation problem has already received increasing research attentions. Some traditional models [21,24] predict user's interest on POIs through *Random Walk and Restart* process on a graph which holds both users' social links and check-in logs. However, this model is not effective on LBSN, as friendship between users doesn't reflect similar check-in behaviours [3]. Recently, *Collaborative Filtering (CF)* methods [25,26] has achieved promising results in POI recommendation. However, the multi-facet time factor is not yet studied adequately. So, we aim to further promote *user-based collaborative filtering* through mining temporal properties of both users and POIs.

Intuitively, User-POI matrix reflects the correlation between users (rows) and POIs (columns) in LBSN. The value for each entry can be defined as binary (i.e. 1 if user visits the POI, otherwise 0). The main problem involved in POI recommendation is *data sparsity* [23]. The reason is that users only visit a small number of POIs which decreases the density of the User-POI matrix. This issue affects majority of CF methods. Therefore, additional information needs to be extracted to improve effectiveness. Ye et al. introduced the concept of *Geographical Influence(GI)* [25,26] based on the observation that people tend to visit POIs close to their own previously visited places. Later on, several studies took different kinds of *Temporal Influence* into consideration. e.g. [5,26,27] generate User-POI-time (UTP) matrix through splitting the day into time slots and learn user's feasibility to check-in at a location in each time slot. Obviously, UTP matrix is even more sparse. So, to alleviate the sparseness, [28] proposes *TICRec Framework* which applies a density estimation method and [26] computes cosine similarity based on visited locations in related time slots. Moreover, Gao et al. [5] suggest *LRT* which can be configured to adapt various temporal intervals (e.g. weekly, monthly and etc.). However, in current temporal approaches, the correlations between users and the check-ins are merely considered (i.e. User-POI or UTP matrices) and user-specific and location-specific temporal properties are not appreciated. Hence in our work, we consider such temporal alignments of both users and venues to further improve POI recommendation systems.

From temporal perspective, while certain venues are visited more during weekday or weekend, some users also show their interest to perform check-ins mainly during either weekday or weekend. We verify this through observations (Sect. 2.1). Intuitively, we can imply two facts: (a) If a user is mostly aligned toward weekday, we should offer her more from weekday oriented POIs. (b) If a group of POIs have the same rank, they should be offered to the user based on her temporal preference. We name User and POI weekly alignments as *User Act* and *POI Act* respectively.

The main challenge here is to compute the User/POI Acts through the check-in history and integrate these temporal preferences in recommendation consistently. The naïve approach is to count the number of visits during weekday/weekend and recommend POIs based on the user's temporal orientation.

Table 1. Sample of weekly oriented POIs visited more than 50 times by varied users

Weekend oriented		
POI name	Category	Weekend prob.
Downtown Los Angeles Artwalk	Museum, Arts & Entertainment,	0.99
Santa Monica Farmers Market	Shop & Service, Food	0.98
Coachella Valley Music and Arts Festival	Outdoors & Recreation, Arts & Entertainment	0.93
Social Nightclub	Nightlife Spot,	0.89
Los Angeles Memorial Coliseum	College & University, Arts & Entertainment, College Stadium,	0.88
Weekday oriented		
POI name	Category	Weekday prob.
Finnegan's Marin	Nightlife Spot, Food	0.98
Sierra College	College & University	0.91
Oviatt Library	College & University, Professional	0.88
MEVIO, Inc.	Office, Professional	0.87
Olives Gourmet Grocer	Shop & Service, Food & Drink Shop	0.83

However, not all the POIs are the same in their impacts. On the other hand, the probabilities for a user to visit different POIs are not the same. Based on [25], three factors of Collaborative Filtering, Geographical Influence and Friendship, determine the *visibility weight* for a user to check-in at a location. So, POIs should be treated differently based on their visibility weights. Our approach comprises three steps: (i) Firstly it predicts the probability for user i ($u_i \in U$) to visit any POI in her check-in history (L_i). (ii) Secondly, we use a probabilistic approach to compute u_i's act which denotes her interest toward weekday/weekend intervals. Consequently, as higher the probability of a POI is, its influence on u_i's act is considered more inflated. (iii) Finally, if the user's weekly alignment exceeds the threshold, we then apply temporal recommendation approach that initially computes the POI act for any location in primary recommendation list. According to u_i's act, it finally proposes a combined list of neutral and temporally oriented locations. We designate weekly intervals because on one hand, majority of locations visited during the weekdays and weekends are obviously divergent (Table 1) and on the other hand, the visiting pattern repeats frequently. However, our model is applicable to other similar periodic intervals.

In this paper, we also put emphasis on temporal aspect of geographical influence [25] and limit the primary observation to daily consecutive pairs. We confirm that it still follows power law distribution. This makes the cost function minimization quicker. We also employ Normal Equation (NE) instead of the Gradient Descent (GD) which obtains optimized parameters in one round. In short, the contributions of this paper are listed as follows:

- We observe the concept behind weekly oriented users and POIs. We also design a probabilistic model to compute such temporal alignments.
- We propose a recommender framework to discretise continuous stream of LBSN users and suggest them a set of POIs according to their weekly preferences. Our method outperforms the state of the art in POI recommendation.
- Finally, we have also taken subsequent POI pairs visited by the same user into consideration and using relevant temporal observation we have provided an optimized model for geographical influence.

The rest of this paper is formed as follows: Sect. 2 explains primary definitions and insightful observations. Section 3 clarifies our temporal recommendation framework. Section 4 provides empirical concepts about collaborative filtering and geographical influence. Section 5 discusses experimental results. Related research work is surveyed in Sect. 6. In Sect. 7 we close this paper with some conclusive remarks.

2   Temporal Influence

In this section, we setup two observations based on primary definitions. We verify that certain POIs and users are aligned toward either weekday or weekend.

Definition 1 (POI Act). *Given a set of POIs* $\mathbb{P} = \{p_1, p_2, \ldots, p_n\}$, *each* p_j *($\forall p_j \in \mathbb{P}$) has a POI Act denoted as p_j^a (Eq. 1), which is the margin value ($[-1, 1]$) between its probabilities to be visited during weekday (w_d) and weekend (w_e).*

$$p_j^a = \frac{W_j^d}{N_j} - \frac{W_j^e}{N_j} \tag{1}$$

Here, W_j^d and W_j^e denote the number of visits at p_j during w_d and w_e. Also N_j is its total number of visits. If p_j^a is greater than zero, it will exhibit an alignment toward w_d and if it is less than zero, it'll show that p_j is visited more during w_e. Otherwise (if $p_j^a = 0$), p_j will be neutral (not temporally aligned)

Definition 2 (User Act). *Given a set of users* $\mathbb{U} = \{u_1, u_2, \ldots, u_n\}$, *we define that each u_i ($\forall u_i \in \mathbb{U}$) has a User Act denoted as u_i^a (Eq. 2) which is the margin value ($[-1, +1]$) between probabilities of her w_d and w_e visits.*

$$u_i^a = Avg_i^d - Avg_i^e \tag{2}$$

Avg_i^d and Avg_i^e are probabilities for u_i to visit locations during w_d and w_e respectively. If u_i^a is greater than 0, it will reflect u_i's temporal preference toward w_d and if it is less than 0, it'll indicate that she is more interested in w_e.

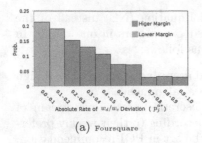
(a) Foursquare

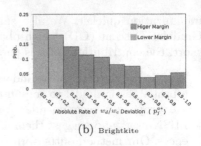
(b) Brightkite

Fig. 1. Observation on Absolute POI Act

2.1 Observations

We setup two observations to perceive that certain POIs and users can be oriented toward w_d or w_e. We use threshold T to reflect the extent of alignment. As, people visit w_e oriented places during casual Friday (e.g. they go to bar on Friday night and perform sport activities on Friday afternoon), we include Friday as weekend. Hence, w_d has one day more than w_e, and T is $\frac{1}{7} \approx 15\%$ which is consistent with uniform distribution of locations for each day in a week.

1. Absolute POI Act Observation: This observation demonstrates that many POIs are significantly used either during w_d or w_e. On the other hand, we aim to study to what extent each POI is oriented toward either w_d or w_e. Hence, for each p_j, visited by a set of users U_j, we compute p_j^{a*} (Eq. 3) as an absolute rate of temporal w_d/w_e deviation. In this inspection, we choose those locations from both datasets (Sect. 5.1) that are visited by at least 5 users.

$$p_j^{a*} = \frac{\sum_{u_i \in U_j} |p_{i,j}^d - p_{i,j}^e|}{|U_j|} \qquad (3)$$

$p_{i,j}^d$ and $p_{i,j}^e$ are the probabilities of each $u_i \in U_j$ to visit p_j during w_d and w_e (Eq. 4):

$$p_{i,j}^d = \frac{W_{i,j}^d}{W_{i,j}}, p_{i,j}^e = \frac{W_{i,j}^e}{W_{i,j}} \qquad (4)$$

Here $W_{i,j}$ is the total number of times that each $u_i \in U_j$ has visited p_j. Also, $W_{i,j}^d$ and $W_{i,j}^e$ record the visits performed exclusively during w_d and w_e.

Fig. 1a and b depict the probabilities regarding POIs' weekly deviations based on different ranges (e.g. 0.3–0.4). As highlighted in dark orange, more than 70% of the POIs in both datasets have an average absolute orientation greater than T. This means that majority of locations in both datasets are predominantly used either during the weekend or weekday.

2. Absolute User Act Observation: Similarly, for each user u_i with L_i as the check-in log, we compute u_i^{a*} (Eq. 5) as her average rate of absolute temporal w_d/w_e deviation. We select users who has visited at least 8 POIs ($\{\forall u_i \in U | |L_i| > 8\}$). Figure 2a and b illustrate relevant probabilistic bins which reflects to what extent each user is temporally oriented (disregarding the alignment toward w_d or w_e). $|p_{i,j}^a|$ is p_j's absolute POI act limited to u_i's visits.

$$u_i^{a*} = \frac{\sum_{p_j \in L_i} |p_{i,j}^a|}{|L_i|} \tag{5}$$

If u_i^{a*} is less than $T(15\%)$, we can ensure that u_i is not oriented toward w_d or w_e. However, as highlighted in dark orange (Fig. 2), 57.3% and 61.6% of users in Foursquare and Brightkite have an absolute temporal deviation more than the T. Also more than 10% of users are highly aligned toward either weekday or weekend ($u_i^{a*} > 45\%$).

Based on the observations conducted in LBSN, we can now conclude that weekly temporal influences exist for both users and POIs.

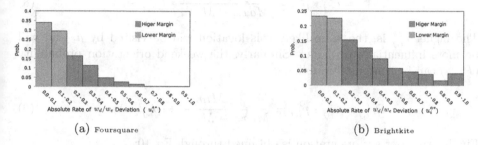

(a) Foursquare   (b) Brightkite

Fig. 2. Observation on Absolute User Act

3   Recommendation

In this section, we firstly provide an efficient approach to compute user acts and secondly we describe our recommendation framework.

3.1   User Act Efficient Model

Primary user act (Definition 2, Eq. 2) treats all POIs the same, while they differ based on POI acts and *visiting influence*. Therefore, we propose a more effective model to compute the user act. We first need to obtain user's visiting orientation toward w_d or w_e. Therefore, we compute the POI act for every location visited by u_i ($p_j \in L_i$). We use Eq. 6 to find positive or negative impacts.

$$\hat{p}_{i,j}^d = (p_{i,j}^d - \lambda), \hat{p}_{i,j}^e = (p_{i,j}^e - \lambda) \tag{6}$$

Where $\lambda \in (0,1)$ serves as a separator of w_d/w_e margins. If we assume $\lambda = 0.5$, $p_{i,j}^d = 0.75$ and $p_{i,j}^e = 0.25$, then $\hat{p}_{i,j}^d = 0.75 - 0.5 = 0.25$, which indicates that p_j has a positive impact on user $i's$ weekday act. We argue that the POI with higher probability to be visited by a user (*visiting influence*) should play a more significant role in computation of her user act. [25] comprises three influential modules of Collaboration, Friendship and Vicinity which we use to compute *visiting influence* for each location (p_j). Our modifications on baseline are described in Sect. 4. To capture *visiting influence*, we remove each p_j from L_i, we can subsequently obtain $c_{i,j}^*$ which represents the probability of u_i to visit p_j considering all three modules. We then normalize (i.e. between (0,1]) the results using Eq. 7:

$$\hat{c}_{i,j}^* = \frac{c_{i,j}^* - Min_{ci}}{Max_{ci} - Min_{ci}}, \tag{7}$$

where $Max_{ci} = arg_{max}(C_{i,k}^*), Min_{ci} = arg_{min}(C_{i,k}^*), \forall p_k \in L_i$. To get the final weekday orientation probability for each $p_j \in L_i$ we use Eq. 8:

$$Pr_{i,j}^d = \hat{c}_{i,j}^* * \hat{p}_{i,j}^d = \frac{c_{i,j}^* - Min_{ci}}{Max_{ci} - Min_{ci}} * (p_{i,j}^d - \lambda) \tag{8}$$

The higher $\hat{c}_{i,j}^*$ is, the more likely this location will be visited by u_i and will be more influential on u_i's act. Similarly, the weekend orientation probability ($Pr_{i,j}^e$) can be computed as follows:

$$Pr_{i,j}^e = \hat{c}_{i,j}^* * \hat{p}_{i,j}^e = \frac{c_{i,j}^* - Min_{ci}}{Max_{ci} - Min_{ci}} * (p_{i,j}^e - \lambda) \tag{9}$$

Finally, the user act orientation is obtained through Eq. 10:

$$\hat{u}_i^a = \left| \tilde{Avg}_i^d - \tilde{Avg}_i^e \right| \tag{10}$$

While $\tilde{Avg}_i^d$ (Eq. 11) and $\tilde{Avg}_i^e$ (12) are respective w_d/w_e average ratios.

$$\tilde{Avg}_i^d = \frac{\Sigma_{p_j \in L_i} Pr_{i,j}^d}{|L_i|} \tag{11}$$

$$\tilde{Avg}_i^e = \frac{\Sigma_{p_j \in L_i} Pr_{i,j}^e}{|L_i|} \tag{12}$$

The value regarding $\tilde{Avg}_i^d$-$\tilde{Avg}_i^e$ shows the direction. If it is greater than zero, it will indicate that user is aligned toward w_d and if it is less than zero, it will show w_e orientation.

3.2   Framework

In this section, we propose the framework which suggests a ranked list of candidate POIs for each user disregarding the extent of her temporal orientation.

As Fig. 3 depicts, we can imagine the input of a recommender system as a continuous stream of users in course of time. Utilizing check-in history, the system should suggest top @*Num* appealing locations for each user. A basic POI

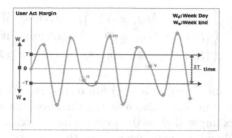

Fig. 3. Continuous stream of users is discretized by first computing relevant user acts and then utilizing of threshold T.

recommender system doesn't differentiate w_d/w_e temporal preferences, however we use threshold T to discretise input users based on their effective user acts (Sect. 3.1). If they pass the threshold, temporal method will be employed otherwise they will be treated as non-temporal users. For example, u_m and u_v are oriented to do the check-ins during w_d. However unlike u_m, u_v doesn't surpass T and is not adequately oriented toward w_d so the framework doesn't apply the temporal method for her. While the user act reflects how a user performs the check-ins in weekly cycles, POI act is used in recommendation process to suggest right POIs to the right users through utilizing of such temporal preferences.

$$f(L_i) = \begin{cases} M_{avg}(\rho, \delta) & \text{if } \hat{u}_i^a \geq T \\ usg_w & \text{otherwise} \end{cases} \tag{13}$$

As formulated in Eq. 13, the system receives u_i's check-in log (L_i). If the user act computed based on L_i exceeds threshold T, the system will utilize temporal influence. Otherwise the user will be recommended by usg_w [25] which also integrates further modifications (Sect. 4). In temporal case, ρ is the initial list of recommending POIs computed by USG and δ resembles the POI act for each item in primary recommendation list. ρ and δ are the input of M_{avg} function which performs recommendation as described in Sect. 3.3.

3.3   Temporal Act Based Recommendation

If the efficient user act is greater than threshold ($\hat{u}_i^a \geq T$), we need to follow temporal recommendation approach (M_{avg}). The method has two inputs. ρ which is the primary decently sorted recommendation list and δ which includes the acts for each of POIs in ρ. We first retrieve Top $K^*@Num$ items from ρ while $@Num$ is the number of final list (denoted as R). R is formed by three subsets of Weekday aligned(R^d), Weekend oriented(R^e) and Neutral (R^n) where $R = \{R^d, R^e, R^n\}$ and $|R| = @Num$. The final proportion for each category will follow relevant ratios from proper POIs which are computed based on efficient user act (Eq. 14).

$$M_{avg}(\rho, \delta) = \begin{cases} |R^d| = (\tilde{Avg}_i^d + \lambda - \frac{\xi}{2}) * @Num & \text{if } p_y^a > \theta \\ |R^e| = (\tilde{Avg}_i^e + \lambda - \frac{\xi}{2}) * @Num & \text{if } p_y^a < \theta \\ |R^n| = \xi * @Num & Otherwise \end{cases} \quad (14)$$

Here, $(\tilde{Avg}_i^d + \lambda - \frac{\xi}{2})$ and $(\tilde{Avg}_i^e + \lambda - \frac{\xi}{2})$ are respective w_d and w_e proportions from final recommendation list. Also θ is the threshold for detection of w_d/w_e oriented POIs. For example if $\theta = 0$, the weekday portion from the final list will comprise the POIs whose acts are greater than 0 ($\forall p_y^a \in \delta | p_y^a > 0$) and for weekend ratio the POI acts should be less than 0 ($\forall p_y^a \in \delta | p_y^a < 0$). In fact, Neutral POIs are not likely to have high scores in w_d/w_e lists. However, we still need to propose them when they gain high probabilities. Therefore we reserve a minor portion (ξ) for POIs which are not temporally aligned.

4   Utilizing Primary Influences

In this section, we provide our empirical details on two primary modules in POI recommendation which are Geographical Influence and User based Collaborative Filtering. For social influence we adapt the method used in [25].

4.1   Geographical Influence

Geographical Influence (GI) [25,26,28] declares that *individuals visit locations which are close to those they have visited previously*. Considered as another important module in location recommendation, we provide an empirical review about it. We observe two cases: (*i*) *Non-temporal:* We compute geographical distance between each POI pair in one's check-in history. (*ii*) *Temporal:* We also perform another test on consecutive check-ins within a time period. This period denotes an average distance between two subsequent POIs that a user may travel in less than a day ($T_g = 12\,\text{h}$). In order to invalidate the noise, we apply the speed condition ($\Delta_d/\Delta_t < \gamma$) in which Δ_d is the distance between two adjacent POIs and Δ_t is the time spent presumably to travel between them.

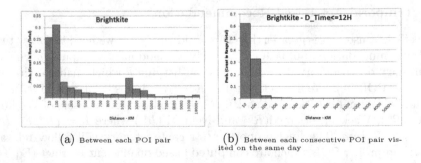

(a) Between each POI pair   (b) Between each consecutive POI pair visited on the same day

Fig. 4. Geographical influence observation: probabilities of the distance ranges

As illustrated in Fig. 4a, majority of distances between POI pairs are less than 100 km. Also, about 25 % are less than 10 kms. Notice that the scale is modified after 1000 kms and the spike shows the sum. In temporal aspect (Fig. 4b), the highest probability for distance between two daily consecutive POIs is less than 100 kms. The similar figures are seen on Foursquare. In both observations, the probability of the distance between a POI pair, follows power law distribution (*polynomial: $y = ax^m$ while x is the distance between the POI pair, y is its probability and both a and m are optimizing parameters*). However, the number of consecutive POI pairs limited to threshold T_g is merely 2.5 % of all POI pairs. This makes the cost function minimization faster while the distance probabilities will be close after applying the feature scales. Hence, we train single feature hypothesis based on the temporal set. The polynomial equation can be converted to the Linear Regression. We also employed Normal Equation (NE) instead of the Gradient Descent (GD) for minimization. NE obtains optimized parameters in one round unlike GD which applies various learning rates with multiple iterations. Indeed, NE performed well as there is only one feature in regression (distance between POI pair) and the size of the matrix was small. Here, the probability for u_i (with L_i as check-in log) to visit location l_j is the multiplication of the distance probabilities between l_j and each location $l_y \in L_i$. However, multiplying numerous decimal points will pass the minimum value and the result will be zero for many l_js. Therefore, we suggest a *log* based Eq. 15 where a and m are optimized values and d implies the distance function between a pair.

$$log(Pr[l_j|L_i]) = \sum_{l_y \in L_i} a \times d(l_y, l_j)^m. \tag{15}$$

4.2   User-Based Collaborative Filtering

In a binary logic (like [25, 26]), while L and U are respective set of locations and users, if user $u_i \in U$ has already visited $l_j \in L$, then $c_{i,j}$ will be 1 otherwise 0. Moreover, cosine similarity weight (Eq. 16) between $u_i \in U$ and $u_k \in U$ ($w_{i,k}^+$) is computed based on the number of shared POIs among ($L_i \cap L_k$). Here, $L_i \subset L$ and $L_k \subset L$ are corresponding check-in histories belonged to u_i and u_k.

$$w_{i,k}^+ = \frac{\sum_{l_j \in L_i \cap L_k} c_{i,j} c_{k,j}}{\sqrt{\sum_{l_j \in L_i} c_{i,j}^2} \sqrt{\sum_{l_j \in L_k} c_{k,j}^2}} \quad \& \quad c_{i,j}^+ = \frac{\sum_{\{\forall u_k | w_{i,k} > 0\}} w_{i,k} c_{k,j}}{\sum_{\{\forall u_k | w_{i,k} > 0\}} w_{i,k}} \tag{16}$$

In order to recommend a set of locations to u_i, we merely select items from users who share one or more POI(s) with u_i. In fact, if $L_i \cap L_k = 0$ and $l_j \in L_k$ then the check-in probability for u_i to visit l_j (Eq. 16, $c_{i,j}^+$) will be zero unless l_j is already visited by another user u_m while $L_i \cap L_m \neq 0$. Such empirical point will reduce the number of iterations during recommendation process. This is done through excluding those who does not share any location with current user. We were inspired by this experimental point to argue that influential effect of $l_j \in L_i$ on u_i's temporal preference increases by the number of visits on l_j performed by any $u_m \in U$ while $L_i \cap L_m \neq 0$.

5   Experimental Evaluation

In this section we plot and implement multiple experiments to compare our proposed method with a few alternative approaches (Sect. 5.3). Our main goal is to ensure how the concept of User/POI act can improve baseline methods which merely rely on User/POI correlations and neglect possible user and POI specific temporal influences. Nevertheless, we need to take a point into consideration that effectiveness of POI recommendation systems on LBSN datasets are always affected by low density of User-POI matrices. Hence, rather than measuring the differences using absolute values, we count on relative excellence in comparison.

Table 2. Statistics of the datasets

	Brightkite	Foursquare
Number of users	58,228	4,163
Number of locations (POIs)	772,967	121,142
Number of check-ins	4,491,143	483,813
Number of social links	214,078	32,512
Cold start ratio (less than 5 POIs)	53.36 %	14.17 %
Avg. visited POIs per user	20.93	64.66
User-POI matrix density	2.7×10^{-5}	5.33×10^{-4}

5.1   Dataset

In this paper, experiments are conducted on two large-scale real [3] LBSN datasets. Both (Foursquare[1] and Brightkite[2]) are publicly available. Relevant statistics are shown in Table 2. Furthermore, Fig. 5 depicts POI distribution of both datasets on the map. The Brightkite dataset is more extensive, however it is extremely sparse (Density: 2.7×10^{-5}) and more than 50 % of the dataset is formed by check-in history of cold start users. Similarly in Foursquare, only 8 % of user pairs share more than 5 POIs.

5.2   Evaluation Metrics

Considering top N (e.g. 5, 10 and 20) results returned by a POI recommendation system, there are two methods to evaluate its effectiveness. The first approach is survey based which employs the normalized Discounted Cumulative Gain (nDCG) [16]. The other method (used in this paper) utilizes F1-score ratios. In this method, we firstly exclude x % (default 30 %) of POIs from check-in history of any user. We then train recommendation model using rest of the POIs.

---

[1] http://www.public.asu.edu/~hgao16/.
[2] https://snap.stanford.edu/data/loc-brightkite.html.

(a) Brightkite (b) Foursquare

Fig. 5. Check-in distribution

Finally, we examine how many of excluded POIs are recovered using returned list of recommendation. As denoted in Eq. 17, $Precision@N$ is the ratio of total Number of recovered POIs (R_p) to the number of recommended POIs (N). However, $Recall@N$ would be the ratio of total Number of recovered POIs (R_p) to the number of initially excluded POIs (E_p). Indeed, Precision, Recall and F1-score metrics are computed for each test user (20 % of all dataset users) and final metrics are computed based on the total average. $F1 - score@N$ will be the final performance metric.

$$Precision@N = \frac{R_p}{N}, Recall@N = \frac{R_p}{E_p}, F1 - score@N = \frac{2 \times Precision@N \times Recall@N}{Precision@N + Recall@N}.$$
(17)

5.3 Recommendation Methods

Recommendation methods used in experiments are as follows:

User-based CF (UBCF): The primary User-based collaborative filtering.
User-based CF Temporal (UBCFT): Another version of our model which treats all the POIs the same in computation of the user act. Referring to Eq. 13, probabilities of ρ for input of M_{avg} function is calculated using Eq. 16.
USG: Denoted by USG, this method takes advantage of three modules of User-based CF, Social Influence and Geographical Influence where $0 < \alpha < 1$ and $0 < \beta < 1$ [25]. We will adapt User-Time-POI model (Baseline: [26]) in future work as we limited sparsity to User-POI matrix.
Temporal U+S+G (USGT): Model proposed in this paper (Sects. 2 and 3).

5.4 Impact of Parameters

With regard to Eq. 6, we assume $\lambda = 0.5$. to treat w_d and w_e the same. Moreover, in order to decide on the value of ξ in Eq. 14, we chose a random set of 20 % from users in both datasets and measured the rate of neutral POIs ($\{\forall p_y \in \rho | p_y^a = 0\}$) in top $K^*@Num$ items from recommendation list. As the rate was less than 10 % in both datasets, we set the value for ξ to 0.1 (e.g. 2 if @Num=20). Moreover, we set K to 10. Finally, in order to reproduce the tri-module baseline (USG:[25]), despite other automatic models in rank learning (e.g. SVM pairwise and EM), we employed tuning (Table 3). We changed the values for α and β between 0 and 1 to get the best performance @5. The optimized parameters of α and β are selected based on the best values of F1-score@5.

Table 3. USG optimised values

	F1-score @5	
	α	β
Foursquare	0.2	0.6
Brightkite	0.3	0.4

Referring to Eq. 18, $S_{i,j}$ denotes the final prediction probability for u_i to perform a check-in at location l_j. $S_{i,j}^u$ denotes user based CF probability, $S_{i,j}^s$ [25] and $S_{i,j}^g$ (Sect. 4.1) provide the values for social and geographical influence respectively. we employ feature scaling [20] to make statistical values consistent.

$$S_{i,j} = (1 - \alpha - \beta)S_{i,j}^u + \alpha S_{i,j}^s + \beta S_{i,j}^g. \tag{18}$$

5.5 Performance Comparison

Next, we discuss the results to summarize our findings. Figures 6 and 7 illustrate output of experiments for foursquare and brightkite datasets. In these figures, our proposed method (USGT) clearly outperforms other models. UBCFT exhibits a minor improvement compared to UBCF. This shows that User/POI acts must be computed based on the *visiting influence* as implemented in USGT.

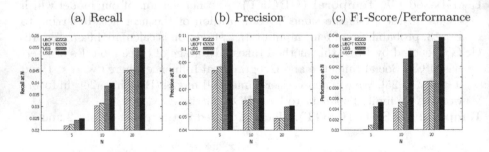

(a) Recall   (b) Precision   (c) F1-Score/Performance

Fig. 6. Comparing the methods - Foursquare dataset

Also, notice that in LBSN sparse condition when the density of User-POI matrix is extremely low, the effectiveness of the location recommenders is not inflated. For example, precision in [26] and recall in [13] are less than 4 % and for [5] both metrics are less than 3.5 %. Hence, performance evaluations are relative

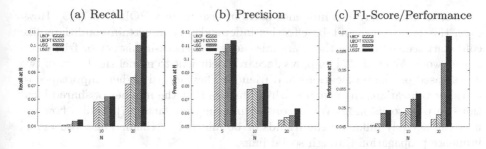

Fig. 7. Comparing the methods - Brightkite dataset

based on algorithms. Moreover, recall can be low even for active users. The reason is concealed in evaluation metrics. For instance in case of recommending @5, when all top 5 recommended POIs are recovered from initially excluded items, precision will be 100 %. However the recall value will be dependent on the number of excluded POIs. Considering a 30 % excluded from 150 POIs in a user's history, the recall will then be only $5/50 = 10\%$. Hence, if majority of active users in a dataset include numerous visited POIs in their check-in history, recall ratio for active users will still be little.

6 Related Work

As the survey [1] reports, with regard to expeditious extension of LBSN platforms (e.g. Foursquare, Gowalla and so on), POI recommendation to individuals via such mediums have become pervasive. Three modules of *Collaborative Filtering* (CF), *Social Influence* (Friendship in networks) and *Geographical Influence* are essential in such recommenders. However, integration of the time factor in any one of these modules is also considered as another multi-aspect influential parameter. This is despite the fact that some primary methods like *HIT-based* [31] and *Random Walk &Restart* [21,24] have already been used for location recommendation.

CF systems are of two types of memory and model based. Memory based approach is divided into two categories of item and user based. In user based approach the weight of similarity between certain users will be computed by employing a measurement function such as *Cosine* or *Pearson*. Subsequently, the probability for a user to visit a POI will be calculated using the rating of the similar users on the same POI [25]. We also utilize user based collaborative filtering in POI recommendation. However, in item-based approach, the co-visited POIs will be found first and then a weighted mixture of user ratings upon similar POIs will be computed [4]. CF methods in recommendation have employed various types of data such as text (e.g. [8]), GPS trajectories [9,10,30] and check-in logs [1]. Nonetheless, they have failed to gain adequate performance metrics. Therefore other components (e.g. Social and Geographical influence) have been appended to improve recommendation results.

Social links may also influence users in visiting new POIs [2,6,7,25]. However, this parameter can't be effective independently in certain conditions like cold start scenarios when the users miss adequate check-in history or friends on the network. Ye et al. [25] employs Jaccard coefficient to model similarity of two users based on shared locations and friends. They accord higher importance to the shared locations than friends which implies that the number of shared locations surpass social links. While considering the time factor [7] studies how the user can actively influence the friends or be affected by them. They also model influence propagation through social links.

Geographical Influence (GI) has been observed in several previous works [14,22,25,26,28] and states that spatial proximity between POIs matters in recommendation. On the other hand, users tend to visit the POIs which are close in distance to their previously visited locations. This has already been denoted for the POIs of the users using *Power law distribution* and *Gaussian model*. Considering the pre-assumption, GI can only be considered as an extension to the CF module as if it fails to suggest remote POIs for users even if the prediction probability might be a tiny value. We have also witnessed the GI observation (including the temporal aspect of subsequent timely ordered POIs) in our datasets. Moreover, we have employed Normal Equation to minimize the error function and exploit optimized parameters of the distribution function.

Temporal influence has also been studied from various aspects recently. Yuan et al. [26] confirm that some locations are visited in particular day/night times (e.g. library during the day or bars during the night). They prove that such day time similarity can improve location recommendation. From another perspective, [17,29] take periodicity into consideration, which is based on the intuition that some locations are visited on daily, weekly or annual intervals. [18,24] also categorize the check-ins into long-term and short-term visits. Long term determines the steady patterns witnessed all along the check-in history, while short term states otherwise. However, their approach is not applicable when the users do not visit the venues continuously. Like motif in graphs, [15,29] aim to discover repeated temporal patterns. Li et al. [11] propose a rating scheme that grants higher scores to the newly visited locations. This is based on the fact that human beings can't remember everything from long time ago.

A user's interest can be relevant to proposed POI's specifications. While we study associated temporal patterns between users and POIs, [12] takes the textual context into consideration. They use Latent dirichlet allocation (LDA) to learn a user's topics of interest and choose the POIs based on their associated topics. Current temporal models in POI recommendation including Matrix Factorization [5], Collaborative Filtering [26], Graph-based [27] and Density estimation [28] perform recommendation using user, time and POI correlations. However, we include periodic (e.g. weekly) temporal preferences of Users and Locations to improve effectiveness.

7 Conclusions

In this paper, a temporal POI recommendation system has been proposed for Location-based social networks. Relying on primary insights, we observe that certain locations are visited more during weekends while some others are aligned toward weekdays. The similar orientation is witnessed in check-in behaviour of LBSN users. Thus, an intelligent POI recommender should logically take such user/POI preferences (act) into consideration. While previous models are defined based on the correlation among users and POIs, we extend recommendation systems in a probabilistic approach to include user and POI specific temporal influences. Our framework recommends relevant POIs to users based on their weekday/weekend alignments. Furthermore, we have taken subsequent POI pairs visited by the same user into consideration and provided an optimized version of geographical influence. Proposed model in this paper outperforms current state-of-the-art recommendation techniques. Considering the importance of this temporal aspect, we plan to integrate it into model based methods in future.

Acknowledgements. The authors wish to thank Prof. Xiaofang Zhou and Prof. Shazia Sadiq from The University of Queensland for their advice. This work is supported by Australian Research Council (ARC). Grant Number DP140103171.

References

1. Bao, J., Zheng, Y., Wilkie, D., Mokbel, M.F.: A survey on recommendations in location-based social networks. Submitted to GeoInformatica (2014)
2. Cheng, C., Yang, H., King, I., Lyu, M.R.: Fused matrix factorization with geographical and social influence in location-based social networks. In: Twenty-Sixth AAAI Conference on Artificial Intelligence (2012)
3. Cho, E., Myers, S.A., Leskovec, J.: Friendship and mobility: user movement in location-based social networks. In: Proceedings of the 17th ACM SIGKDD International Conference on Knowledge Discovery and Data Mining, pp. 1082–1090. ACM (2011)
4. Ding, Y., Li, X.: Time weight collaborative filtering. In: Proceedings of the 14th ACM International Conference on Information and Knowledge Management, pp. 485–492. ACM (2005)
5. Gao, H., Tang, J., Hu, X., Liu, H.: Exploring temporal effects for location recommendation on location-based social networks. In: Proceedings of the 7th ACM Conference on Recommender Systems, pp. 93–100. ACM (2013)
6. Gao, H., Tang, J., Liu, H.: gSCorr: modeling geo-social correlations for new check-ins on location-based social networks. In: Proceedings of the 21st ACM International Conference on Information and Knowledge Management, pp. 1582–1586. ACM (2012)
7. Goyal, A., Bonchi, F., Lakshmanan, L.V.: Learning influence probabilities in social networks. In: Proceedings of the Third ACM International Conference on Web Search and Data Mining, pp. 241–250. ACM (2010)
8. Hu, B., Ester, M.: Spatial topic modeling in online social media for location recommendation. In: Proceedings of the 7th ACM Conference on Recommender Systems, pp. 25–32. ACM (2013)

9. Hung, C.-C., Peng, W.-C., Lee, W.-C.: Clustering and aggregating clues of trajectories for mining trajectory patterns and routes. VLDB J. 1–24 (2011)

10. Leung, K.W.-T., Lee, D.L., Lee, W.-C.: CLR: a collaborative location recommendation framework based on co-clustering. In: Proceedings of the 34th International ACM SIGIR Conference on Research and Development in Information Retrieval, pp. 305–314. ACM (2011)

11. Li, X., Xu, G., Chen, E., Zong, Y.: Learning recency based comparative choice towards point-of-interest recommendation. Expert Syst. Appl. **42**(9), 4274–4283 (2015)

12. Liu, B., Xiong, H.: Point-of-interest recommendation in location based social networks with topic and location awareness. SDM **13**, 396–404 (2013)

13. Liu, B., Xiong, H., Papadimitriou, S., Fu, Y., Yao, Z.: A general geographical probabilistic factor model for point of interest recommendation. IEEE Trans. Knowl. Data Eng. **27**(5), 1167–1179 (2015)

14. Liu, X., Liu, Y., Aberer, K., Miao, C.: Personalized point-of-interest recommendation by mining users' preference transition. In: Proceedings of the 22nd ACM International Conference on Conference on Information & Knowledge Management, pp. 733–738. ACM (2013)

15. Lonardi, J., Patel, P.: Finding motifs in time series. In: Proceedings of the 2nd Workshop on Temporal Data Mining, pp. 53–68 (2002)

16. Manning, C.D., Raghavan, P., Schtze, H.: Introduction to Information Retrieval, vol. 1. Cambridge University Press, Cambridge (2008)

17. Rahimi, S.M., Wang, X.: Location recommendation based on periodicity of human activities and location categories. In: Pei, J., Tseng, V.S., Cao, L., Motoda, H., Xu, G. (eds.) PAKDD 2013, Part II. LNCS, vol. 7819, pp. 377–389. Springer, Heidelberg (2013)

18. Ricci, F., Nguyen, Q.N.: Acquiring and revising preferences in a critique-based mobile recommender system. IEEE Intell. Syst. **22**(3), 22–29 (2007)

19. Symeonidis, P., Ntempos, D., Manolopoulos, Y.: Recommender Systems for Location-based Social Networks, pp. 35–38. Springer, New York (2014)

20. Tax, D.M., Duin, R.P.: Feature scaling in support vector data descriptions. Technical report (2000)

21. Tong, H., Faloutsos, C., Pan, J.-Y.: Fast random walk with restart and its applications. (2006)

22. Wang, C., Ye, M., Lee, W.-C.: From face-to-face gathering to social structure. In: Proceedings of the 21st ACM International Conference on Information and Knowledge Management, pp. 465–474. ACM (2012)

23. Wang, W., Yin, H., Chen, L., Sun, Y., Sadiq, S., Zhou, X.: Geo-sage: a geographical sparse additive generative model for spatial item recommendation, arXiv preprint, arxiv:1503.03650 (2015)

24. Xiang, L., Yuan, Q., Zhao, S., Chen, L., Zhang, X., Yang, Q., Sun, J.: Temporal recommendation on graphs via long-and short-term preference fusion. In: Proceedings of the 16th ACM SIGKDD International Conference on Knowledge Discovery and Data Mining, pp. 723–732. ACM (2010)

25. Ye, M., Yin, P., Lee, W.-C., Lee, D.-L.: Exploiting geographical influence for collaborative point-of-interest recommendation. In: Proceedings of the 34th International ACM SIGIR Conference on Research and Development in Information Retrieval, pp. 325–334. ACM (2011)

26. Yuan, Q., Cong, G., Ma, Z., Sun, A., Thalmann, N.M.: Time-aware point-of-interest recommendation. In: Proceedings of the 36th International ACM SIGIR Conference on Research and Development in Information Retrieval, pp. 363–372. ACM (2013)

27. Yuan, Q., Cong, G., Sun, A.: Graph-based point-of-interest recommendation with geographical and temporal influences. In: Proceedings of the 23rd ACM International Conference on Conference on Information and Knowledge Management, pp. 659–668. ACM (2014)

28. Zhang, J.-D., Chow, C.-Y.: TICRec: a probabilistic framework to utilize temporal influence correlations for time-aware location recommendations (2015)

29. Zhang, Y., Zhang, M., Zhang, Y., Lai, G., Liu, Y., Zhang, H., Ma, S.: Daily-aware personalized recommendation based on feature-level time series analysis. In: Proceedings of the 24th International Conference on World Wide Web, International World Wide Web Conferences Steering Committee, pp. 1373–1383 (2015)

30. Zheng, V.W., Zheng, Y., Xie, X., Yang, Q.: Collaborative location and activity recommendations with GPS history data. In: Proceedings of the 19th International Conference on World Wide Web, pp. 1029–1038. ACM (2010)

31. Zheng, Y., Zhang, L., Xie, X., Ma, W.-Y.: Mining interesting locations and travel sequences from gps trajectories. In: Proceedings of the 18th International Conference on World Wide Web, pp. 791–800. ACM (2009)

32. Zheng, Y., Zhou, X.: Computing with Spatial Trajectories. Springer, New York (2011)

TGTM: Temporal-Geographical Topic Model for Point-of-Interest Recommendation

Cong Zheng$^{(\boxtimes)}$, Haihong E, Meina Song, and Junde Song

School of Computer Science, Beijing University of Posts and Telecommunications,
No. 10, Xitucheng Road, Haidian District, Beijing, China
zhengcongbupt@163.com

Abstract. The wide spread use of location based social networks
(LBSNs) and Micro-blogging services generated large volume of users'
check-in data, which consists of user ids, textual contents, posting
timestamps, geographic information and so on. Point-of-interest (POI)
recommendation is a task to provide personalized recommendations of
interesting places to enhance the user experience in LBSNs. In this paper,
we propose 2 novel time-location-content aware POI recommendation
models which jointly integrate auxiliary temporal, textual and spatial
information to improve the performance of POI recommendation. Specifically, we utilize temporal information to partition the original user-POI
check-in frequency matrix into sub-matrices so that behavior in similar
temporal scenario can be grouped. Then, we take advantage of Latent
Dirichlet Allocation (LDA) model and spatial coordinates to infer the
POIs. Comprehensive experiments conducted using real-world datasets
demonstrate the superiority of our approach.

Keywords: Recommendation · Point of interest · Location-based social
networks · Human mobility

1 Introduction

Given the phenomenal growth rate of user population and the vast amount
of mobile devices with positioning function, location-based social networking
service has become immensely popular with hoards of online web sites, such
as Facebook Places, Google latitude, Twitter and Foursquare, etc. LBSNs now
allow users share not only their physical location coordinates and time stamps
in the form of "check-in", but also write textual opinions on the POIs they have
visited. Mining and modeling the check-in behavior using the location log data
is of great value to both users and POIs, because POI recommendation can help
people discover attractive places and may foster more potential business for the
owners of POIs.

To this end, POI recommendation has become a popular research topic in
the past few years. One of the most crucial challenges in POI recommendation
is how to cope with the extreme sparsity of user-POI check-in frequency matrix.

© Springer International Publishing Switzerland 2016
S.B. Navathe et al. (Eds.): DASFAA 2016, Part I, LNCS 9642, pp. 348–363, 2016.
DOI: 10.1007/978-3-319-32025-0_22

In addition, unlike traditional recommendations which only take the user-item rating into account, POI recommender systems is much more complex, since the mobile behavior in this scenario could be a mixture of many aspects. Particularly, when there are rich text such as comments or microblogs, the POI recommendation should be personalized, location-content aware and context depended. In light of this, diverse types of information pose another big challenge which is how to incorporate them in a unified way systematically.

Intuitively, users tend to visit restaurants within a short distance near office for lunch at noon on weekday. But on weekends, users may pay a long visit to bars at night for fun. We believe that mobile patterns can be easier found in similar temporal context. In light of this, we partition the original user-POI information into subgroups based on temporal stamp at first. Then, we employ LDA algorithm [2] to infer users' interests and the topic distributions of POIs. In [4], the authors found users' displacements (distance between 2 consecutive check-in locations) trend can be approximated by a power-law distribution. Taking into account the textual and geographical influence, on the basis of matrix factorization model, we can infer to what extent does a user prefer a POI.

In summary, the contributions of this paper are threefold:

(1) We study the relationship between users' implicit feedback check-in behavior and auxiliary information on LBSNs in terms of textual content information, geographical coordinates and temporal influence.
(2) We propose 2 novel probabilistic matrix factorization models for POI recommendation, and each of them incorporates the above three types of auxiliary information.
(3) We evaluate the presented methods by comprehensive experiments on real world LBSN datasets extracted from Twitter's API. The results demonstrate the effectiveness of our methods.

We have the usual organization: Survey, problem definition, proposed method, experiments and conclusions.

2   Related Work

2.1   Matrix Factorization

Collaborative Filtering (CF) in recommender systems can predict personalized preferences to unconsumed items [1,19]. There are two major categories of Collaborative Filtering methods. Neighborhood-based solutions predict users' potential interests by finding like-minded users [5,18], which compute similar users or items using similarity functions such as Cosine Distance or Pearson Correlation. Model-based [9,17] solutions utilize the observed ratings or tags to model the user-item interaction. A variety of successful realizations of model-based models are based on matrix factorization which decomposes the user-item rating matrix. In this paper, we resort to approaches based on matrix factorization method, so we briefly introduce it here.

Basically, matrix factorization (MF) methods factorize a rating matrix R into one user-specific matrix U and one item-specific matrix C [10]. Then the original rating matrix can be approximated by multiplying the two factorized matrices, while avoiding over-fitting using regularization terms:

$$\arg min_{U,C} \sum_{i=1}^{m} \sum_{j=1}^{n} I_{ij} \left(R_{ij} - U_i^T C_j \right)^2$$
$$+\lambda \left(\|U\|_F^2 + \|C\|_F^2 \right) \tag{1}$$

where $\|\cdot\|_F^2$ denotes the Frobenius norm, and I_{ij} is the indicator function that is equal to 1 if user i rated movie j and equal to 0 otherwise. The constant λ controls the extent of regularization in order to alleviate the over-fitting problem. The probabilistic matrix factorization (PMF) [16] gives a probabilistic explanation for the regularization.

2.2   POI Recommendation

POI recommendation was firstly studied on GPS trajectory data [23]. With the growing popularity of LBSNs, POI recommendation has drawn a lot of attention. There is a vast literature on POI recommendation exploiting geographical influence. A mutual observation has been discovered by existing studies is that people tend to visit nearby locations. User-based CF and item-based CF are studied in [11,20]. Ye et al. [20] explored social and geographical factors together under an user-based CF framework to make POI recommendation. There has been recent interest in leveraging the social and geographical properties to improve the effectiveness of POI recommendation. In [3], they proposed a matrix factorization scheme incorporating the geographical and social information by a Gaussian mixture model (GMM). In [6], authors discovered that preference derived from similar users was more important for in-town users while friendship became more important for out-of-town users.

Liu et al. [13] proposed a graph-based method which exploits temporal and geographical information in an integrated way. Yuan et al. [21] incorporated temporal influence in an user-based CF manner, and the final preference score for a candidate POI are linearly combined by the scores computed based on temporal influence and geographical influence respectively.

When it comes to the textual content information, Gao et al. [7] studied the content information on LBSNs w.r.t. POI properties, user interests, and sentiment indications under an unified POI recommendation framework. Liu and Xiong [12] studied the effect of POI-associated tags with an aggregated LDA model. Hu and Ester [8] investigated the user-interest from Twitter and Yelp according to topic modeling method. Yuan et al. [22] jointly model individual user's mobility behavior from spatial, temporal and content aspects. They captured them in a probabilistic generative model.

3 Proposed Approach

3.1 Problem Definition

The problem of POI recommendation is to recommend POIs potentially attractive to users. For ease of exposition, let $u = \{u_1, u_2, ..., u_m\}$ be the set of users and $c = \{c_1, c_2, ..., c_n\}$, be the set of POIs, where m and n denote the number of users and POIs, respectively. Each user has observable properties x_i (e.g., user's comments and check-in history). Also, each POI keeps observable properties x_j (e.g., POI's related textual description and spatial coordinates in terms of longitude and latitude). $R \in \mathbb{R}^{m \times n}$ is a check-in frequency matrix with each entry R_{ij} representing the observed check-in frequency made by u_i at c_j. Applying LDA model to all the textual descriptions related to all the users and POIs can help us obtain the topic distribution θ_i for each user u_i or π_j for POI c_j.

3.2 Time Aware User-POI Subgrouping

Subgrouping the original user-item-rating matrix based on contextual information has proven to be a promising way of improving recommender systems [15], because the generated sub-groups contain similar ratings which have higher correlations. Specifically, we use temporal information (i.e., day-of-week and hour-of-day) to partition the original user-POI check-in frequency matrix into 4 subgroups. Intuitively, people's activity may be more regular and predictable on weekday while more various and similar on weekend. So we first partition the original matrix according to day-of-week (i.e., weekday versus weekend). Next, a day (24 h) can be divided into working segment (from 08:00 am to 17:59 pm) and leisure hours (from 18:00 pm to 7:59 am of the next day). Then, the generated 2 submatrices according to day-of-week can be further derived into 4 based on hour-of-day. At last, we get 4 user-POI check-in frequency sub-matrices, in which we can make predictions by leveraging geographical and textual information.

3.3 Exploiting Textual Content and Spatial Information

LDA algorithm is known to have poor performance on short text documents such as short tweets. In this paper, we aggregate all the textual comments associated with same POI into a POI document. We also combine all the textual comments of the POIs that each user has checked in into a user document. As a consequence, we get a large document collection, and each document corresponds to one POI or user. Then LDA [2] is utilized to analyze the topic distribution of every document. Like [12], we define the matching score between user u_i and POI c_j as the similarity with regards to user's topic distribution θ_i and the topic distribution of POI π_j. We resort to Jensen-Shannon divergence to measure the similarity between above 2 multinomial topic distributions. The symmetric Jensen-Shannon divergence between them is:

$$D_{JS}(u_i, c_j) = \frac{1}{2}D(\boldsymbol{\theta_i} \parallel \boldsymbol{M}) + \frac{1}{2}D(\boldsymbol{\pi_j} \parallel \boldsymbol{M}) \qquad (2)$$

where $\boldsymbol{M} = \frac{1}{2}(\boldsymbol{\theta_i} + \boldsymbol{\pi_j})$ and $D(\cdot \parallel \cdot)$ is the Kullback-Leibler distance. The matching score is defined as:

$$S(u_i, c_j) = 1 - D_{JS}(u_i, c_j) \qquad (3)$$

Next, we exploit the spatial information. Previous work [4] investigated the distance-based displacements of consecutive check-in made by users. They found the trend of check-in frequency with regard to displacement can be approximated by a power-law distribution. To be consistent with this observed property, supposing an user is at POI c_j now, we model the probability that a user may visit another POI c_k in terms of distance as [13]:

$$p_{jk}{}^d = [Dist(j,k)]^{-1} \qquad (4)$$

where $p_{jk}{}^d$ is the probability that a user will check-in at POI c_k from POI c_j, and $Dist(j,k)$ is the distance between the 2 positions.

In practice, users may not share their position information in time, so Eq. (4) can not be simply applied in this scenario without precise spatial coordinates. Intuitively, if a user frequently check-in at a certain POI, this POI can be very influencial to his daily life because he may live or work there. While user's personal behavior is less prone to rely on the POIs which the user rarely occur. According to this phenomenon, we postulate that the attraction of POI c_j to user u_i is the weighted average of probability calculated by Eq. (4) between candidate POI c_j and the POIs in user's (user u_i) check-in log list. The weights are proportional to the user's check-in frequency at each POI in his own history. For example, suppose there are K POIs in the historical list $\boldsymbol{f_i}$ of user u_i. The geographical attractiveness of POI c_j to u_i is calculated as:

$$p_{ji}{}^d = \frac{\sum_{k=1}^{K} f_{ik} p_{jk}{}^d}{\sum_{k=1}^{K} f_{ik}} \qquad (5)$$

where f_{ik} is the probability with regard to check-in frequency of user u_i at POI c_k and $p_{jk}{}^d$ is computed by using Eq. (4). The displacement is between POI c_k and c_j.

We assume that the probability of checking in at a POI should reflect both the content and geographical influence. So we fuse these 2 factors and derive the integrated attractiveness of POI c_j to user u_i as:

$$W_{ij} = [S(u_i, c_j)]^a \times [p_{ji}{}^d]^b \qquad (6)$$

where $p_{ji}{}^d$ is calculated by Eq. (5) and $S(u_i, c_j)$ is calculated by Eq. (3). a and b are 2 decaying parameters.

3.4 Temporal-Geographical Topic Model for Personalized POI Recommendation

To jointly leverage temporal, spatial and content information, we integrate them into a probabilistic matrix factorization model (TGTM-1). Figure 1(a) shows the graphical model of TGTM-1. We define the conditional distribution of observed check-in frequencies as:

$$p(\boldsymbol{R} \mid \boldsymbol{U}, \boldsymbol{C}, \boldsymbol{W}, \sigma^2) = \prod_{i=1}^{m} \prod_{j=1}^{n} \left[\mathcal{N}(R_{ij} \mid f(\boldsymbol{U_i}, \boldsymbol{C_j}, W_{ij}), \sigma^2) \right]^{I_{ij}} \tag{7}$$

where $\mathcal{N}(x \mid \mu, \sigma^2)$ is the probabilistic density function of Gaussian distribution with mean μ and variance σ^2, and I_{ij} is the indicator function that is equal to 1 if user u_i visited POI c_j and equal to 0 otherwise. We use function $f(\boldsymbol{U_i}, \boldsymbol{C_j}, W_{ij})$ to approximate the check-in frequency of user u_i at POI c_j.

Taking spatial and content influence into consideration, we define:

$$f(\boldsymbol{U_i}, \boldsymbol{C_j}, W_{ij}) = W_{ij} \cdot \boldsymbol{U_i^T} \boldsymbol{C_j} \tag{8}$$

where W_{ij} is computed by using Eq. (6). The weighted product of user-specific and POI-specific latent feature vectors allow our model to take full advantage of all the information and make personalized POI recommendation.

We place zero-mean spherical Gaussian priors on user and POI latent feature vectors:

$$p(\boldsymbol{U} \mid \sigma_U^2) = \prod_{i=1}^{m} \mathcal{N}(\boldsymbol{U_i} \mid 0, \sigma_U^2 \boldsymbol{I}) \tag{9}$$

$$p(\boldsymbol{C} \mid \sigma_C^2) = \prod_{j=1}^{n} \mathcal{N}(\boldsymbol{C_j} \mid 0, \sigma_C^2 \boldsymbol{I}) \tag{10}$$

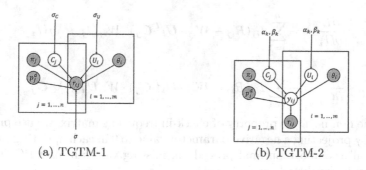

(a) TGTM-1   (b) TGTM-2

Fig. 1. Graphical model of our 2 methods

Next, through a simple Bayesian inference, the posterior distribution is given by:

$$p(\boldsymbol{U}, \boldsymbol{C} \mid \boldsymbol{R}, \sigma^2, \boldsymbol{W}, \sigma_U{}^2, \sigma_C{}^2) \propto$$
$$p(\boldsymbol{R} \mid \boldsymbol{U}, \boldsymbol{C}, \sigma^2, \boldsymbol{W}, \sigma_U{}^2, \sigma_C{}^2) p(\boldsymbol{U} \mid \sigma_U{}^2) p(\boldsymbol{C} \mid \sigma_C{}^2) =$$
$$\prod_{i=1}^{m} \prod_{j=1}^{n} [\mathcal{N}(R_{ij} \mid f(\boldsymbol{U}_i, \boldsymbol{C}_j, W_{ij}), \sigma^2)]^{I_{ij}} \prod_{i=1}^{m} \mathcal{N}(\boldsymbol{U}_i \mid 0, \sigma_U{}^2 \boldsymbol{I}) \prod_{j=1}^{n} \mathcal{N}(\boldsymbol{C}_j \mid 0, \sigma_C{}^2 \boldsymbol{I})$$

$$(11)$$

The log of the posterior distribution over the user and POI latent features is:

$$\ln p(\boldsymbol{U}, \boldsymbol{C} \mid \boldsymbol{R}, \sigma^2, \boldsymbol{W}, \sigma_U{}^2, \sigma_C{}^2) =$$
$$-\frac{1}{2\sigma^2} \sum_{i=1}^{m} \sum_{j=1}^{n} I_{ij} [R_{ij} - f(\boldsymbol{U}_i, \boldsymbol{C}_j, W_{ij})]^2 - \frac{1}{2\sigma_U{}^2} \sum_{i=1}^{m} \boldsymbol{U}_i{}^T \boldsymbol{U}_i - \frac{1}{2\sigma_C{}^2} \sum_{j=1}^{n} \boldsymbol{C}_j{}^T \boldsymbol{C}_j$$
$$-\frac{1}{2} [(\sum_{i=1}^{m} \sum_{j=1}^{n} I_{ij}) \ln \sigma^2 + md \ln \sigma_U{}^2 + nd \ln \sigma_C{}^2] + P$$

$$(12)$$

where P is a constant that does not depend on parameters and d is the number of latent features. Maximizing the log-posterior over latent vectors with fixed hyper-parameters is equivalent to minimizing the following sum-of-squared-errors objective function with quadratic regularization terms:

$$E = \frac{1}{2} \sum_{i=1}^{m} \sum_{j=1}^{n} I_{ij} (R_{ij} - W_{ij} \cdot \boldsymbol{U}_i{}^T \boldsymbol{C}_j)^2 + \frac{\lambda_U}{2} \sum_{i=1}^{m} \| \boldsymbol{U}_i \|_F^2 + \frac{\lambda_C}{2} \sum_{j=1}^{n} \| \boldsymbol{C}_j \|_F^2$$

$$(13)$$

where $\lambda_U = \sigma^2 / \sigma_U{}^2$, $\lambda_C = \sigma^2 / \sigma_C{}^2$, and $\| \cdot \|_F^2$ represents the Frobenius norm. A local minimum can be found by applying the gradient descent algorithm to the latent matrices $\boldsymbol{U}$ and $\boldsymbol{C}$ alternatively.

$$\frac{\partial E}{\partial \boldsymbol{U}_i} = - \sum_{j=1}^{n} I_{ij} (R_{ij} - W_{ij} \cdot \boldsymbol{U}_i{}^T \boldsymbol{C}_j) \cdot W_{ij} \boldsymbol{C}_j + \lambda_U \boldsymbol{U}_i \qquad (14)$$

$$\frac{\partial E}{\partial \boldsymbol{C}_j} = - \sum_{i=1}^{m} I_{ij} (R_{ij} - W_{ij} \cdot \boldsymbol{U}_i{}^T \boldsymbol{C}_j) \cdot W_{ij} \boldsymbol{U}_i + \lambda_C \boldsymbol{C}_j \qquad (15)$$

For the non-negative property of check-in frequency matrix, we use projected strategy by projecting a negative parameter value to 0 in each updating iteration.

After all the above optimal procedure, missing values in user-POI check-in matrix can be predicted as:

$$R_{ij}{}^* = \boldsymbol{U}_i{}^T \boldsymbol{C}_j \qquad (16)$$

To alleviate the potential negative values generated by Gaussian distribution, we have to use a projected strategy. We draw inspiration from [14], and propose an improved model (TGTM-2) which is more suitable for modeling nonnegative values, and it is shown in Fig. 1(b). In this model, we assume the observed check-in frequency R_{ij} in matrix $\boldsymbol{R}$ follows Poisson distribution with the mean Y_{ij} in matrix $\boldsymbol{Y}$. The elements Y_{ij} in matrix $\boldsymbol{Y}$ is defined in the same way as Eq. (8). But U_{ik} and C_{jk} are given the Gamma distribution as the empirical priors.

The generative process of every observed user-POI count R_{ij} is as follows:

1. Generate $U_{ik} \sim Gamma(\alpha_k, \beta_k), \forall k$
2. Generate $C_{jk} \sim Gamma(\alpha_k, \beta_k), \forall k$
3. Generate $Y_{ij} = W_{ij} \sum_{k=1}^{d} U_{ik} C_{jk}$
4. Generate $R_{ij} \sim Poisson(Y_{ij})$

The gamma distributions of $\boldsymbol{U}$ and $\boldsymbol{C}$ are given as follows:

$$p(\boldsymbol{U} \mid \boldsymbol{\alpha}, \boldsymbol{\beta}) = \prod_{i=1}^{m} \prod_{k=1}^{d} \frac{U_{ik}{}^{\alpha_k - 1} exp(-U_{ik}/\beta_k)}{\beta_k{}^{\alpha_k} \Gamma(\alpha_k)} \tag{17}$$

$$p(\boldsymbol{C} \mid \boldsymbol{\alpha}, \boldsymbol{\beta}) = \prod_{j=1}^{n} \prod_{k=1}^{d} \frac{C_{jk}{}^{\alpha_k - 1} exp(-C_{jk}/\beta_k)}{\beta_k{}^{\alpha_k} \Gamma(\alpha_k)} \tag{18}$$

where $\boldsymbol{\alpha} = \{\alpha_1, \cdots, \alpha_d\}$, $\boldsymbol{\beta} = \{\beta_1, \cdots, \beta_d\}$, $U_{ik} > 0, C_{jk} > 0, \alpha_k > 0$ and $\beta_k > 0$, $\Gamma(\cdot)$ is the Gamma function.

The Poisson distribution of $\boldsymbol{R}$ given $\boldsymbol{Y}$ is given by:

$$p(\boldsymbol{R} \mid \boldsymbol{Y}) = \prod_{i=1}^{m} \prod_{j=1}^{n} [\frac{Y_{ij}{}^{R_{ij}} exp(-Y_{ij})}{R_{ij}!}]^{I_{ij}} \tag{19}$$

where $Y_{ij} = W_{ij} \sum_{k=1}^{d} U_{ik} C_{jk}$, since every W_{ij} for each user-POI pair is fixed, the posterior distribution of $\boldsymbol{U}$ and $\boldsymbol{C}$ given R can be modeled as:

$$p(\boldsymbol{U}, \boldsymbol{C} \mid \boldsymbol{R}, \boldsymbol{W}, \boldsymbol{\alpha}, \boldsymbol{\beta}) \propto p(\boldsymbol{R} \mid \boldsymbol{Y})p(\boldsymbol{U} \mid \boldsymbol{\alpha}, \boldsymbol{\beta})p(\boldsymbol{C} \mid \boldsymbol{\alpha}, \boldsymbol{\beta}) \tag{20}$$

The log of the posterior distribution over the user and POI latent features is:

$$\ln p(\boldsymbol{U}, \boldsymbol{C} \mid \boldsymbol{R}, \boldsymbol{W}, \boldsymbol{\alpha}, \boldsymbol{\beta}) = \sum_{i=1}^{m} \sum_{k=1}^{d} ((\alpha_k - 1)\ln(\frac{U_{ik}}{\beta_k}) - \frac{U_{ik}}{\beta_k}) +$$
$$\sum_{j=1}^{n} \sum_{k=1}^{d} ((\alpha_k - 1)\ln(\frac{C_{jk}}{\beta_k}) - \frac{C_{jk}}{\beta_k}) + \sum_{i=1}^{m} \sum_{j=1}^{n} [I_{ij}(R_{ij}\ln Y_{ij} - Y_{ij})] + P \tag{21}$$

where P is a const. Taking derivatives of Eq. (21) with respect to $\boldsymbol{U}$ and $\boldsymbol{C}$:

$$\frac{\partial L}{\partial U_i} = \sum_{j=1}^{n} I_{ij}[W_{ij}(\frac{R_{ij}}{W_{ij}U_i C_j} - 1)C_j] + \frac{\alpha_k - 1}{U_i} - 1/\beta_k \tag{22}$$

$$\frac{\partial L}{\partial C_j} = \sum_{i=1}^{m} I_{ij}[W_{ij}(\frac{R_{ij}}{W_{ij}U_i C_j} - 1)U_i] + \frac{\alpha_k - 1}{C_j} - 1/\beta_k \tag{23}$$

Equation (16) can be used to make predictions until the optimal procedure get convergence.

4   Experimental Analysis

4.1   Description of Datasets

We use part of the datasets kindly provide by [4]. Since Twitter status messages support the inclusion of geo-tags (latitude/longitude) as well as support third-party location sharing services like Foursquare and Gowalla, the spatial coordinates data is crawled using Twitter's API, and the other data was from Foursquare, the detailed data gathering method can be found in [4]. Foursquare is a large-scale location based social network sites. It allows users to check-in at different locations writing textual comments with time stamps and spatial information. In practice, users' check-in behavior may not always rely on temporal, spatial and content influence. For example, some people may go on a business trip and the destination may be very far away from their common active area. In order to clean up the data and remove the outliers which rarely occur, we filter the users who made less than 10 check-ins, and require that each POI should be visited 10 times at least. Moreover, we also constrain that an user should visit each POI at least 3 times. In practice, a large scale POI can have slightly different geographical coordinates. In our experiment, POIs are truncated according to their unique identifier. We get 3373 unique users and 9333 unique POIs after pruning. The pruned dataset contain 31727 check-in records. The density of the user-POI check-in matrix is 0.1008 %.

4.2   Metrics

In our experiments, we split the dataset into two parts : training (80 %) and testing datasets (20 %). We use 3 metrics for evaluating the performance: Normalized Mean Absolute Error (NMAE) , Normalized Root Mean Square Error (NRMSE) [14] and $Precision@K$. $Precision@K$ is a metric designed for top-K POI recommendation. They are defined as:

$$NMAE = \frac{\sum_{r=1}^{N} | (R_r - \hat{R}_r)/R_r |}{N}$$

$$NRMSE = \sqrt{\frac{1}{N} \sum_{r=1}^{N} [(R_r - \hat{R}_r)/R_r]^2}$$

$$Precision@K = \frac{\sum_u | R(u) \cap T(u) |}{K}$$

where N is the total number of predictions, R_r is the real rating of an item and $\hat{R}_r$ is the corresponding predicted rating. $R(u)$ is the list of top-K recommended POIs and $T(u)$ are the visited locations of user u. $Precision@K$ w.r.t each user represents what percentage of POIs among the top-K recommended locations has actually been visited.

4.3   Impact of Spatial Information, Content and Temporal Influence

Since predictions are made in each subgroup independently, we can know how the models perform in each subgroup exactly. However, we do not know the overall performance with all the data in general. In light of this, after all the independent training and testing procedures in all the subgroups, we compute the metrics with all the testing datasets of all the subgroups as overall performance of the models. Take the metric $NMAE$ as an example, symbol N is the total number of predictions in all the subgroups not only one single subgroup. Other symbols can be derived in the same manner. Experimental results in Sect. 4.3 are the overall performance of the 2 proposed models.

We use grid search and fivefold cross validation to find the optimal parameters of Gamma distribution that achieve best overall experimental performance. The shape parameter of Gamma distribution $\alpha = 100$, and the scale parameter $\beta = 0.05$. The bigger β is, the wider numeric range will the Gamma distribution generate. In our experiment, we find that big β can hurt the model performance.

Parameters a and b are important because they determine to what extent do the textual contents and spatial influence affect the accuracy of prediction. Figures 2 and 3 show 3D-plots of predictive performance of our 2 models. The content parameter a balances the weight of textual contents: the bigger a is, the more we use textual content to make predictions. The spatial parameter b balances the effect of geographical influence, and the bigger b is, the more we take distance into consideration when making recommendation. We can see a clear decreasing trend from Fig. 3, the reason is that a and b affect the weighted matrix W in the exponential form. Although exponential form may generate big values generally, the performance on metric $Precision@K$ is still great. Taking all metrics into consideration, our methods gain the best predictive accuracy when $a = 10, b = 0.2$ for TGTM-1 whereas $a = 9, b = 0.3$ for TGTM-2. We also observe that the performance is more sensitive to spatial influence, slight change of parameter b can make the performance with regard to $NMAE$ and $NRMSE$ fluctuate. This indicates that both textual contents and spatial influence contribute to model performance, but the spatial information affect more. So the geographical attractiveness of POI plays a more important role in making travel decisions.

In our experiment, we find that although the performance with regards to $NMAE$ and $NRMSE$ may be poor, the metric $Precision@K$ is still relatively good (see Figs. 2 and 3). This is reasonable. Intuitively, in the POI recommendation scenario, because the check-in frequencies are in different order of magnitude, we care more about rank of recommendation results other than the accurate missing entry values. This phenomenon also demonstrates the effectiveness that we model the entries in the weighted matrix in exponential manner. Although this strategy may generate relatively big or small predictions, exponential strategy can distinguish users' wide range of check in frequency behavior well. However, this can not be simply applied to traditional rating prediction situation. Since in that situation, the numeric range of rating is fixed in 1–5, which is far more smaller than the range of check-in frequency.

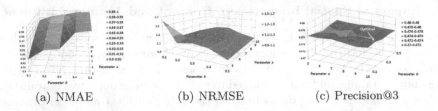

(a) NMAE   (b) NRMSE   (c) Precision@3

Fig. 2. Performance of TGTM-1 by varying content a and spatial b influence (Color figure online)

Then, we study the effect of temporal influence on our proposed model. In our proposed model, we subgroup the original user-POI check-in matrix into 4 submatrices according to time stamps (i.e., day-of-week and hour-of-day). In order to learn the temporal effect, we conducted the following experiment: using same spatial and textual parameter configuration, compare our complete model (TGTM-1 and TGTM-2) with comparisons (GTM-1 and GTM-2) regardless of the temporal influence. That means the comparisons making predictions using the original matrix without subgrouping strategy. In this experiment, using the above results, we set $a = 10, b = 0.2$ for TGTM-1 and GTM-1. We set $a = 9$, $b = 0.3$ for TGTM-2 and GTM-2. We can see from Fig. 4 that the temporal influence indeed affects the model performance a lot. Compared to the models without considering temporal influence, TGTM-2 and TGTM-1 improve the performance nearly 10 %. The Gamma priors taking time into consideration achieve better predictive performance. This result shows the effectiveness of time-based user-POI subgrouping.

4.4   Comparisons

In this subsection, we present the performance comparison between our models and some state-of-the-art POI recommendation algorithms.

BasicMF: predicts missing values according to $P \approx U^T C$, which only use the user-POI check-in matrix without any other auxiliary information (e.g., geographical, temporal information and textual contents). This one is used as baseline.

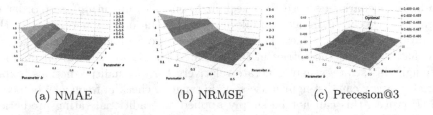

(a) NMAE   (b) NRMSE   (c) Precesion@3

Fig. 3. Performance of TGTM-2 by varying content a and spatial b influence

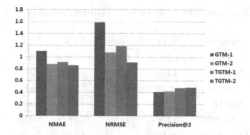

Fig. 4. Effect of temporal influence on our model

GeoCF: [20] takes into account of the spatial influence by assuming a power-law distribution and jointly integrates a user-based collaborative filtering algorithm. The recommendation is a linear combination of the spatial information and user preference.

UPT: [13] takes temporal and geographical influence into consideration, and integrates the auxiliary information into a matrix factorization model. In this approach, time stamps are also used for the purpose of subgrouping.

TLA: [12] incorporates item contents and spatial information into a matrix factorization model, thus it makes predictions based on item contents, spatial influence and check-in frequencies. In this model, spatial information is considered by using a regional level popularity factor rather than geographical coordinates.

TGTM-1 TGTM-2: [Sect. 3.4] use time stamps to do user-POI sub-grouping. In the subgroups, they makes predictions involving check-in frequencies, textual contents and spatial information into a probabilistic matrix factorization model. The user and POI latent feature matrices of TGTM-1 are assumed conforming to Gaussian priors, while the matrices of TGTM-2 are assumed conforming to Gamma priors.

In this experiment, we set $a = 10, b = 0.2$ for TGTM-1 and $a = 9, b = 0.3$ for TGTM-2. From Table 1 and Fig. 5, we can see that TGTM-2 which is applied Gamma prior indeed outperforms TGTM-1 with Gaussian prior. Table 1 also show that UPT, TGTM-1 and TGTM-2 perform different in each subgroup. The reason is that topic interests and geographical distance affect human mobile behavior slightly different when the time scenario change, especially that people tend to visit POIs which match their interests better on weekend in the expense of long distance. Because Basic MF, GeoCF and TLA do not take time information into consideration, we use their overall performance to represent the performance in each subgroup. We can also see that the performance with regard to metric $Precision@K$ is relatively high, the reason is that our data preprocessing procedure filter out the data with little relevance and all the users in testing part do not have too many check-ins. We leave large-scale comparative experiments as our future work. Although all the comparisons perform acceptable, our TGTM-2 model still improve the performance of baseline more than 30% when $K = 1$.

Table 1. Performance comparison (Dimensionality = 10, K=3)

Subgroups	Metrics	BasicMF	GeoCF	UPT	TLA	TGTM-1	TGTM-2
Weekday work time	NMAE	1.2715	0.9289	1.2834	1.2649	0.9709	0.8542
	NRMSE	3.1386	1.7517	2.6651	3.0738	1.1714	0.9107
	Precision@3	30.09 %	35.24 %	45.53 %	40.16 %	47.69 %	48.75 %
Weekday leisure time	NMAE	1.2715	0.9289	1.2946	1.2649	0.9869	0.8714
	NRMSE	3.1386	1.7517	2.6874	3.0738	1.1869	0.9139
	Precision@3	30.09 %	35.24 %	44.53 %	40.16 %	47.67 %	48.72 %
Weekend work time	NMAE	1.2715	0.9289	1.2982	1.2649	0.9846	0.87
	NRMSE	3.1386	1.7517	2.673	3.0738	1.1929	0.9288
	Precision@3	30.09 %	35.24 %	42.6 %	40.16 %	46.18 %	47.71 %
Weekend leisure time	NMAE	1.2715	0.9289	1.3006	1.2649	0.9804	0.8646
	NRMSE	3.1386	1.7517	2.6885	3.0738	1.1896	0.9189
	Precision@3	30.09 %	35.24 %	43.88 %	40.16 %	45.67 %	46.70 %

BasicMF achieves the lowest accuracy and precision, because this general model does not take any auxiliary information into consideration. For the GeoCF model, because there is no social relationship information in our experimental data, we only take the user-based collaborative filtering factor and spatial influence factor into account. We found GeoCF obtain best performance when spatial influence counts 70 %, whereas user-based collaborative filtering factor counts 30 %. This means that spatial information plays a dominating role in the POI recommendation scenario. GeoCF performs better than BasicMF model, but it is still less accurate than our methods. The reason is that although GeoCF considers more relevant data using collaborative filtering approach and take geographical influence into account, some potential important factors for POI recommendation are still lost. For example, textual content information explicitly shows whether target POI match user's interests. Moreover, temporal information intuitively explains why the user's active range is bigger on weekend. UPT utilizes temporal information to classify original data into 4 subgroups, and integrates spatial influence into matrix factorization model. The experimental results demonstrate the benefits of combining the temporal and geographical influence jointly. TLA takes advantage of textual and spatial information to make POI recommendation. UPT outperforms TLA and GeoCF, which means that the temporal information contributes more to the user's behavior than explicit textual information and inexplicit user preferences inferred using user collaborative filtering method. The probable reason is that, most people work regularly on weekday, and weekday counts more time than weekend. So the topic interest show relatively poor influence.

Table 1 and Fig. 5 also show that our proposed 2 models consistently outperform the comparisons, demonstrating that jointly taking full advantage of textual content, temporal and geographical information indeed improves the performance of POI recommendation. Although some comparisons perform well on $NMAE$ and $NRMSE$, they still can not perform better than our 2 mod-

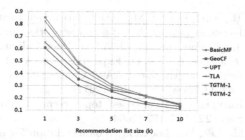

Fig. 5. *Precision@K* performance with different size of recommendation list

els on *Precision@K*. Because they can not handle the big range of check in frequency values suitably, which demonstrates that modeling the geographical influence and content information in exponential form makes our models more flexible for the big range of user-POI check-in frequencies.

5   Conclusions

In this paper, we have presented two probabilistic matrix factorization algorithms to make POI recommendation. There are several advantages of the proposed recommendation methods. First, time stamps are used to subgroup original data for the purpose of deeply understanding users' behavior patterns in different temporal scenarios. Second, the models capture the spatial influence which has been proven to be very important on user's mobile behavior. Third, deeply exploiting textual information meets the criteria of a truly personalized recommender system which is recommending POIs that match users' interests. Last but not least, the proposed approaches model the textual and geographical influence in exponential form, which is suitable for the big range of users' implicit check-in feedback data. Finally, extensive experiments on real data collected through Twitter's API validated the practical utility of our proposed methods.

Our future work includes meeting the online real time needs and conducting large scale comparative experiments to further evaluate the performance.

Acknowledgments. This work was supported by the Key Projects in the National Science and Technology Pillar Program of China under grant No. 2014BAH26F02.

References

1. Adomavicius, G., Tuzhilin, A.: Toward the next generation of recommender systems: a survey of the state-of-the-art and possible extensions. J. IEEE Trans. Knowl. Data Eng. **17**, 734–749 (2005)
2. Blei, D.M., Ng, A.Y., Jordan, M.I.: Latent Dirichlet allocation. J. Mach. Learn. Res. **3**, 993–1022 (2002)

3. Cheng, C., Yang, H., King, I., Lyu, M.: Fused matrix factorization with geographical and social influence in location-based social networks. In: 26th AAAI Conference on Artificial Intelligence, pp. 17–23. AAAI Press, California (2012)

4. Cheng, Z., Caverlee, J., Lee, K., Sui, D.: Exploring millions of footprints in location sharing services. In: 5th International AAAI Conference on Weblogs and Social Media, pp. 81–88. AAAI Press, California (2011)

5. Deshpande, M., Karypis, G.: Item-based top-n recommendation algorithms. J. ACM Trans. Inf. Syst. **22**, 143–177 (2004)

6. Ference, G., Ye, M., Lee, W.: Location recommendation for out-of-town users in location-based social networks. In: 22nd ACM International Conference on Information and Knowledge Management, pp. 721–728. ACM, New York (2013)

7. Gao, H., Tang, J., Hu, X., Liu, H.: Content-aware point of interest recommendation on location-based social networks. In: 29th AAAI Conference on Artificial Intelligence, pp. 1721–1727. AAAI Press, California (2015)

8. Hu, B., Ester, M.: Spatial topic modeling in online social media for location recommendation. In: 7th ACM Conference on Recommender Systems, pp. 25–32. ACM, New York (2013)

9. Koren, Y.: Factorization meets the neighborhood: a multifaceted collaborative filtering model. In: 14th ACM SIGKDD International Conference on Knowledge Discovery and Data Mining, pp. 426–434. ACM, New York (2008)

10. Koren, Y., Bell, R., Volinsky, C.: Matrix factorization techniques for recommender systems. J. Comput. **42**, 30–37 (2009)

11. Levandoski, J.J., Sarwat, M., Eldawy, A., Mokbel, M.F.: LARS: a location-aware recommender system. In: 28th IEEE International Conference on Data Engineering, pp. 450–461. IEEE Press, New York (2012)

12. Liu, B., Xiong, H.: Point-of-interest recommendation in location based social networks with topic and location awareness. In: 2013 SIAM International Conference on Data Mining, pp. 396–404. SIAM, Philadelphia (2013)

13. Liu, X., Liu, Y., Aberer, K., Miao, C.: Personalized point-of-interest recommendation by mining users preference transition. In: 22nd ACM International Conference on Information and Knowledge Management, pp. 733–738. ACM, New York (2013)

14. Ma, H., Liu, C., King, I., Lyu, M.R.: Probabilistic factor models for web site recommendation. In: 34th International ACM SIGIR Conference on Research and Development in Information Retrieval, pp. 265–274. ACM, New York (2011)

15. Liu, X., Aberer, K.: SoCo: a social network aided context-aware recommender system. In: 22nd International Conference on World Wide Web, pp. 781–802. W3C (2013)

16. Salakhutdinov, R., Mnih, A.: Probabilistic matrix factorization. Advances in Neural Information Processing Systems (NIPS 2007), vol. 20, pp. 1257–1264. MIT Press, Massachusetts (2007)

17. Rendle, S., Marinho, L., Nanopoulos, A., Schmidt-Thieme, L.: Learning optimal ranking with tensor factorization for tag recommendation. In: 15th ACM SIGKDD International Conference on Knowledge Discovery and Data Mining, pp. 727–736. ACM, New York (2009)

18. Resnick, P., Iacovou, N., Suchak, M., Bergstrom, P., Riedl, J.: Grouplens: an open architecture for collaborative filtering of netnews. In: 1994 ACM Conference on Computer Supported Cooperative Work, pp. 175–186. ACM, New York (1994)

19. Su, X., Khoshgoftaar, T.M.: A survey of collaborative filtering techniques. J. Adv. Artif. Intell. **2009**, 1–19 (2009)

20. Ye, M., Yin, P., Lee, W.C.: Exploiting geographical influence for collaborative point-of-interest recommendation. In: 34th International ACM SIGIR Conference on Research and Development in Information Retrieval, pp. 325–334. ACM, New York (2011)

21. Yuan, Q., Cong, G., Ma, Z., Sun, A., Thalmann, N.M.: Time-aware point-of-interest recommendation. In: 36th International ACM SIGIR Conference on Research and Development in Information Retrieval, pp. 363–372. ACM, New York (2013)

22. Yuan, Q., Cong, G., Ma, Z., Sun, A., Thalmann, N.M.: Who, where, when and what: discover spatio-temporal topics for twitter users. In: 19th ACM SIGKDD International Conference on Knowledge Discovery and Data Mining, pp. 605–613. ACM, New York (2013)

23. Zheng, Y., Zhang, L., Xie, X., Ma, W.Y.: Mining interesting locations and travel sequences from GPS trajectories. In: 18th International Conference on World Wide Web, pp. 791–800. ACM, New York (2009)

Modeling User Mobility via User Psychological and Geographical Behaviors Towards Point of-Interest Recommendation

Yan Chen[1], Xin Li[2], Lin Li[3], Guiquan Liu[1(✉)], and Guangdong Xu[4]

[1] University of Science and Technology of China, Hefei, China
{ycwustc,gqliu}@utsc.edu.cn
[2] iFlyTek Research, Hefei, China
xinli2@iflytek.com
[3] Wuhan University of Technology, Wuhan, China
cathylilin@whut.edu.cn
[4] University of Technology Sydney, Sydney, Australia
Guandong.Xu@uts.edu.au

Abstract. The pervasive employments of Location-based Social Network call for precise and personalized Point-of-Interest (POI) recommendation to predict which places the users prefer. Modeling user mobility, as an important component of understanding user preference, plays an essential role in POI recommendation. However, existing methods mainly model user mobility through analyzing the check-in data and formulating a distribution without considering why a user checks in at a specific place from psychological perspective. In this paper, we propose a POI recommendation algorithm modeling user mobility by considering check-in data and geographical information. Specifically, with check-in data, we propose a novel probabilistic latent factor model to formulate user psychological behavior from the perspective of utility theory, which could help reveal the inner information underlying the comparative choice behaviors of users. Geographical behavior of all the historical check-ins captured by a power law distribution is then combined with probabilistic latent factor model to form the POI recommendation algorithm. Extensive evaluation experiments conducted on two real-world datasets confirm the superiority of our approach over state-of-the-art methods.

Keywords: Location-based social network · Point-of-Interest recommendation · User psychological behavior · Geographical behavior · User mobility

1 Introduction

With the prevalence of GPS-enabled portable devices, Location-based Social Networks (LBSNs) have been sweeping the global world during recent years, such as Gowalla, Yelp, Facebook Places, etc. Point-of-Interests (POIs) are places that

© Springer International Publishing Switzerland 2016
S.B. Navathe et al. (Eds.): DASFAA 2016, Part I, LNCS 9642, pp. 364–380, 2016.
DOI: 10.1007/978-3-319-32025-0_23

a user may find useful information or tend to visit, e.g. restaurants or shopping malls. As an important component of LBSNs, POI has facilitated an application - POI recommendation. This interesting and useful application can not only benefit merchants by increasing their revenue through virtual marketing but also benefit customers by helping them filter out uninteresting places and reduce their decision making time thus providing satisfactory user experiences [8].

In LBSN, users could check in at a place by sharing their experiences and reviews with friends and online users. Through analyzing the check-in data, researchers can acquire a user's mobility by mining the historical check-in behavior patterns thus giving predictions to her future potential visiting places. Generally, a specific distribution is adopted to formulate a user's mobility, such as multiple Gaussian distribution [1] and power-law distribution [17]. However, such methods have an inherent flaw that they only model the user mobility from the data perspective while without considering the intrinsic reason why a user checks in at a specific place. We argue that user mobility is influenced by user psychological and geographical behaviors. The user psychological behavior can be described on two aspects. The one is when we are confronted with a series of choices, we always compare the utility of these candidates, and are more willing to choose a high utility place to visit [13]. The other is that individual differences influence the user check-in behaviors. Detailedly, a user checked in at a place for ten times and another user only visited the same place thrice. It could not indicate the former prefers the place than the latter, as the former may be a person who likes staying out or traveling and the latter may be a stay-at-home type. Besides, geographical influence matters. We may prefer a nearby place rather than a far one. Although, the distant one is a more popular and higher utility place for us.

Based on the aforementioned summarizes, we give the following assumptions:

- *Comparability:* the user check-in behavior is a comparative process, where a certain historical check-in place of a user is chosen after he compares the utility of all alternatives rather than a random decision.
- *Dissimilarity:* the frequency data or the ratings could not give actual expression of user preference. For example, a user visited a place three times while the average visiting time is 3.5, which indicates the preference to such POI is below the average. However, supposing the same POI visited by another user three times, the average is 2.5. It shows the user prefers the POI more than others. Taking a relative view to address this problem, we adopt partial order relationship deduced from frequency data rather than the frequency value of users' check-in history to learn the user preference.
- *Localization:* localization is embodied into two aspects. When users arrive at a mart or city, it is believed that they just take the places in the mart or city into account and make a decision. Hence, the candidates are localization, and it motivates us take nearest K place as our inputs. More importantly, the probability of visiting a POI is inversely proportional to the distance. From an overall perspective, users prefer a nearby place than a distant one and the check-in records are localization.

In this paper, POI recommendations are made based on the aforementioned assumptions. As demonstrated in Fig. 1, we delve into the mechanism behind the user check-in patterns and propose a probabilistic latent factor model to formulate the user psychological behavior by employing the utility theory of economics which won the Nobel Prize in 2000 [13]. The utility theory conveys that a consumer is more willing to choose those goods that have high utilities to him or her. A power-law distribution is adopted to capture all historical check-in geographical information, also called geographical behavior. To this end, we propose a novel POI recommendation algorithm by combining the user psychological and geographical behaviors to model the user mobility.

In summary, we have made several contributions in this paper:

- We propose a probabilistic latent factor model to mine users psychological behavior from a novel psychological perspective. User mobility is modeling by user psychological and geographical behaviors.
- We devise a stochastic gradient descent algorithm to learn the latent factors of both users and POIs.
- We conduct extensive experiments on two real-world datasets to evaluate our approach. The results demonstrate that our approach outperforms the state-of-the-art methods.

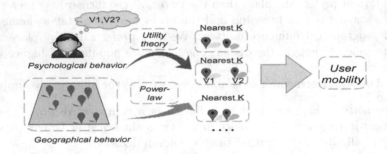

Fig. 1. The framework of the proposed algorithm.

The rest of the paper is organized as follows. We review the recent studies in Sect. 2. The problem formulation and the proposed approach are discussed in Sect. 3. Experiments are presented in Sect. 4. We conclude our work in Sect. 5.

2   Related Works

Traditional recommender systems mainly focus on exploiting explicit user-item rating matrix (usually are not available in LBSNs) via employing memory-based [9] or model-based collaborative filtering (CF) [6,7,11]. The premise behind memory-based CF is to recommend items by like-minded users of a target user and the intuition of model-based CF is to learn the latent factors e.g. by matrix factorization (MF) technology to make recommendations. Moreover, in [3,5], the authors fuse their model with contextual information like social or trust network to improve the prediction accuracy. Although explicit ratings for POIs are not available in LBSNs, traditional CF method can be applied into this filed by treating POIs as common items [16] via exploring user preference from user check-in frequency data which implicitly reflect user preference. Besides, a wide range of properties of LBSN have been extensively explored, such as geographical information [1,10,12,17,18], social connections [10,17], temporal information [4,8,12] and user preference order [8].

As an effective and popular approach to uncover users' preference, modeling user mobility plays an essential role in POI recommendation. Cheng et al. [1] assume that users tend to check in around several centers and model the distance between two locations by the same user as a Multi-center Gaussian Model (MGM) to capture the geographical influence, which is then further into MF for POI recommendation. Ye et al. [17] model user mobility by employing a power-law distribution (PD), and propose a collaborative POI recommendation algorithm based on geographical influence via naive Bayesian. Zhang et al. [18] consider that geographical influence on user mobility should be personalized when LBSNs recommend POIs to users rather than a common distribution for all users, and they model the geographical influence via using kernel density estimation (KDE). Noulas et al. [12] explore the predictive power offered by user mobility features, global mobility features and temporal features, which are finally combined in two supervised learning models. However, these works model user mobility through analyzing the check-in data and formulating the distribution without considering why a user checks in at a specific place from psychological perspective. Intuitively, a certain historical check-in place of a user is chosen under her rational choice to be granted a relatively high utility among all the nearby places.

In this paper, user mobility is captured via geographical and user psychological behaviors. Moreover, user psychological behavior is formulated by probabilistic latent factor model derived from utility theory. In [8,15], the utility theory of economics is adopted to explore the competitive process of user behaviors to make recommendation. However, Li et al. [8] adopt user-POI rating matrix to model choice behavior by adopting time window across recent visits to depict user short-term memory, and Yang et al. [15] take missing values in rating matrix as implicit feedbacks. Our approach simultaneously considers check-in frequency data and geographical information, selecting nearest K places (both visited and non-visited places) centered at each of users historical check-ins as our inputs. To the best of our knowledge, this work is the first investigation of modeling user mobility from psychological perspective in POI recommendation field.

3   The Proposed Approach

In this section, we discuss the problem formulation and propose a method for POI recommendation by modeling the user mobility considering user psychological and geographical behaviors. Firstly, we discuss the problem formulation. Then, we mine user psychological behavior using utility theory and propose a probabilistic latent factor model. A stochastic descent algorithm is devised to optimize our proposed model and then we incorporate this model with geographical behavior to formulate user mobility.

To be specific, all the users' check-in places are firstly clustered into several regions according to the clustering phenomenon of Tolbers first law of geography [2]. After that, subsets of each user's historical check-ins are chosen randomly favoring a wide coverage over all the clusters. We then randomly select the nearest K places (both visited and non-visited places) of each member in the subsets to form several collections which are the inputs of our model. Utility theory is employed to generate the matching degree of the probabilistic latent factors, by leveraging utility of a specific POI to the target user. Through the stochastic gradient descent, we could learn the latent factors of both users and POIs, which depicts the user psychological behavior. Geographical behavior of the historical check-ins of a user is captured by a power-law distribution. Finally, the information of both worlds, i.e. the psychological and geographical behaviors, is then combined thus leading to a final preference score for the future visits. By ranking those scores of a user's un-visited places, we could make recommendation from the top of the list.

3.1 Problem Formulation

Suppose that we have M users and N POIs, and let $\mathbb{U} = \{u_1, u_2, ..., u_m\}$ and $\mathbb{V} = \{v_1, v_2, ..., v_n\}$ be a set of users and POIs respectively. Each POI has a location described by longitude and latitude. $f_{u,v}$ denotes the frequency that a user u visited a POI v. All the frequency data form a user-POI check-in frequency matrix $F \in \mathbb{R}^{m \times n}$, where 0 denotes non-visited places.

As aforementioned, the nearest K places of a POI v_j (in a diversified subset chosen from multiple clusters) visited by a user u_i are chosen to form a collection, termed as a latent POI collection. C_{ij}^L denotes a latent POI collection, which means nearest K places of POI v_j visited by user u_i. In the latent POI collection, the key insight is why the user checked in at a specific place many times, while only visited other places only once or even not visited. As aforementioned, this situation is due to the user psychological behavior and higher frequency indicates more satisfaction of a user to the specific POI. Hence, all POIs in a latent POI collection can be ranged in a partial order relationship w.r.t. their frequencies. It is believed that the partial order relationship implicitly reflects the user psychological behavior.

Definition (POI Partial Order). A partial order in C_{ij}^L is shown below:

$$\succeq_{ij}^L = \{\hat{v}_1 \succeq \hat{v}_2 \succeq ... | f_{i\hat{v}_k} \geq f_{i\hat{v}_{k+1}}, \hat{v}_k \in \succeq_{ij}^L\} \tag{1}$$

where $f_{i\hat{v}_k}$ represents the frequency the user u_i visited POI $\hat{v}_k$. Figure 2 is a demonstration of POI latent collection and POI partial order collection.

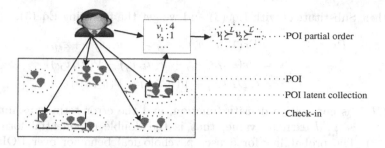

Fig. 2. A demonstration of latent POI collection and POI partial order collection.

3.2 Modeling User Psychological Behavior

Generally speaking, it is difficult to quantify the value of a POI, instead, the preference or the psychological behavior of users to the POI can be extracted from the check-in data more readily. Luckily, utility theory defines a kind of measurement of the users' preference over a set of alternatives and could help reveal the inner information underlying the comparative choice behavior of each user [13]. Hence, we use utility theory to model the user psychological behavior in our approach. In this paper, the check-in frequency data are depicted as the utility.

In order to exploit utility theory to model user psychological behavior, we define the utility function $\mathcal{U}(u,v) : v \rightarrow \mathbb{R}$ to depict the utility of POI v given by user u in a collection C^L. Naturally, high utility means more preference of a user to a POI and thus a higher frequency should be granted to the POI. As afore-mentioned, user psychological behavior implicitly reflected by the relationship between POIs in Eq. (1) becomes meaningful and can be formalized with theory utility. With the partial order in latent collection C_{ij}^L, the following Eq. (2) reveals user psychological behavior:

$$v_j \succeq v_{j'} IIF \mathcal{U}(u_i, v_j) \geq \mathcal{U}(u_i, v_{j'}) \tag{2}$$

which means $v_j \succeq v_{j'}$ if the utility of v_j for u_i is larger than that of $v_{j'}$.

According to [13], utility usually is decomposed into two parts based on random utility model. Following that is the definition:

$$\mathcal{U}(u_i, v_j) = \mathcal{V}(u_i, v_j) + \varepsilon_{ij} \tag{3}$$

where the first part is the observed utility and the second part is some uncontrollable factors such as emotion, weather or even some occasional events. In our paper, $\mathcal{V}(u_i, v_j)$ is depicted as the frequency counted in check-in data, i.e., $\mathcal{V}(u_i, v_j) = f_{ij}$. In our approach, we use $f_{ij} = U_i V_j^T$ to parameterize the observed utility [6], where $U_i \in \mathbb{R}^k$, $V_j \in \mathbb{R}^k$ are low-rank determinable latent factors for user u_i and item v_j respectively and k is the dimension of the latent factor.

The probability of user psychological behavior over alternatives can be defined using the utility of choice [14]:

$$Pr(v_j \succeq v_{j'}) = Pr(\mathcal{U}(u_i, v_j) \geq \mathcal{U}(u_i, v_{j'})) \tag{4}$$

Further, Substitute $\mathcal{U}$ with Eq. (3) and we get the following Eq. (5):

$$
\begin{aligned}
Pr(v_j \succeq v_{j'}) &= Pr(\mathcal{V}(u_i,v_j) + \varepsilon_{ij} \geq \mathcal{V}(u_i,v_{j'}) + \varepsilon_{ij'}) \\
&= Pr(\varepsilon_{ij'} \leq \varepsilon_{ij} + \mathcal{V}(u_i,v_j) - \mathcal{V}(u_i,v_{j'})) \\
&= CDF(\varepsilon_{ij} + \mathcal{V}(u_i,v_j) - \mathcal{V}(u_i,v_{j'}))
\end{aligned}
\tag{5}
$$

where CDF is cumulative density function, and the error term is assumed to satisfy $\varepsilon_{ij'} \sim i.i.d$ extreme value, that is the double exponential format as $\exp(-e^{-\varepsilon})$. The probability for a user psychological behavior over POIs in a latent POI collection is deduced as follows:

$$
Pr(v_j \succeq v_{j'}) = \frac{e^{\mathcal{V}(u_i,v_j)}}{\sum_{j'}^{|C^L|} e^{\mathcal{V}(u_i,v_{j'})}}
\tag{6}
$$

3.3   Probabilistic Latent Factor Model and Learning Algorithm

We formulate user psychological behavior for a certain historical check-in place to be granted a relatively utility among all the nearby places and propose a probabilistic latent factor model based on the utility theory. Each POI visited by a user implies a latent collection, so we can easily formulate all users' psychological behaviors and get the probability of whole observation as follows:

$$
Pr(\succeq^L) = \prod_{u \in U} \prod_{v_j \in V(u)} Pr(v_j \succeq v_{j'}) = \prod_{u \in U} \prod_{v_j \in V(u)} \frac{e^{\mathcal{V}(u_i,v_j)}}{\sum_{j'}^{|C^L|} e^{\mathcal{V}(u_i,v_{j'})}}
\tag{7}
$$

As demonstrated above, we use $f_{ij} = U_i V_j^T$ to measure the observed utility, thus assuming a spherical multivariate Gaussian prior on both U and V, that is

$$
Pr(\Omega|\Theta) = N(\Omega|0, \sigma^2 I)
\tag{8}
$$

where $\Omega = \{U, V\}$, and Θ denotes some hyper-parameters. Ω_l is a component of Ω.

By applying Bayesian theorem, the probabilistic latent factor model can be defined based on the Eqs. (7) and (8).

$$
Pr(U,V|\succeq) \propto Pr(\succeq^L |U,V) Pr(U|\Theta) Pr(V|\Theta)
\tag{9}
$$

Given a latent POI collection, to estimate the latent factors, the model can be learned by maximum the posterior (Eq. (9)). For ease of mathematical treatment, we optimize the model using the following equivalent objective function.

$$
\Omega = \arg\min_{\Omega} -[\log Pr(\succeq^L |\Omega) + \log Pr(\Omega|\Theta)]
\tag{10}
$$

where $\log Pr(\Omega|\Theta)$ can be deemed as regularizer $R(\Omega)$ to alleviate overfitting problem. Here, $R(\Omega)$ corresponds to L2 norm regularization.

To learn U and V of the probabilistic latent factor model, a stochastic gradient descent (SGD) algorithm is devised. We can carry out the gradient of above object function into two parts respectively, and the first part can be deduced using Eq. (7).

$$-\prod_{u \in U} \prod_{v_j \in V(u)} \frac{\partial \log(v_j \succeq v_{j'})}{\partial \Omega_l} = \prod_{u \in U} \prod_{v_j \in V(u)} \frac{\partial [\log(\sum_{j'}^{|C^L|} e^{\mathcal{V}(u_i, v_{j'})}) - \mathcal{V}(u_i, v_j)]}{\partial \Omega_l}$$

(11)

The second part can be viewed as:

$$\frac{\partial \log Pr(\Omega | \Theta)}{\partial \Omega_l} = \frac{\partial R(\Omega)}{\partial \Omega_l} = \beta \Omega_l$$

(12)

where β is regularization parameter.

Using $f_{ij} = U_i V_j^T$ to qualify the observed utility, the derivation of U and V can be computed as:

$$\nabla U_i = \frac{\sum_{j'}^{|C^L|} (e^{U_i V_{j'}^T} \cdot V_{j'})}{\sum_{j'}^{|C^L|} e^{U_i V_{j'}^T}} - V_j + \beta U_i$$

(13)

$$\nabla V_j = \frac{(e^{U_i V_j^T} \cdot U_i)}{\sum_{j'}^{|C^L|} e^{U_i V_{j'}^T}} - U_i + \beta V_j$$

(14)

Given a point, the model parameters are updated by $\Omega_l \leftarrow \Omega_l - \alpha \nabla \Omega_l$ and α is learning rate. Note that we only draw a batch of $\succeq^L$ to embed into our learning algorithm.

Now, we have U and V at hand after the optimization is completed above. Hence, the observe utility of a POI v_j given by a target user u_i can be estimated. Recall the definition of utility, it is a mathematical representation of user psychological behavior, and the predicted utility of a POI v_j given by user u_i can be denoted:

$$Pr[\mathcal{U}(ij)] = Pr(\mathcal{V}(ij) + \varepsilon_{ij}) = \mathcal{V}(ij) + Pr(\varepsilon_{ij}) = U_i V_j^T + C$$

(15)

where C is a constant.

3.4   Proposed Algorithm

We propose a POI recommendation algorithm by modeling user mobility via user psychological and geographical behaviors. As aforementioned, user psychological behavior is captured by utility theory. The geographical behavior measures the probability of whether a user visits a POI under the historical check-in geographical information. In our approach, a power-law distribution is adopted to formulate the geographical behavior [17]. We then formulate the user mobility by fusing the utility with the geographical behavior:

$$y_{ij} \propto Pr[\mathcal{U}(ij)] \cdot Pr(v_j | \mathbb{V}_i)$$

(16)

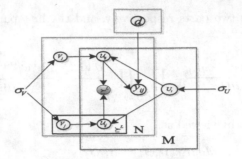

Fig. 3. The graphical representation of the proposed model.

where the $Pr(v_j|\mathbb{V}_i)$ is the likelihood probability for the user u_i to check at the POI v_j. $Pr(v_j|\mathbb{V}_i)$ is defined as:

$$Pr(v_j|\mathbb{V}_i) = \prod_{v_y \in \mathbb{V}_i} Pr[d(v_j, v_y)] \qquad (17)$$

where $d(v_j, v_y)$ denotes the distance between POI v_j to v_y and $\mathbb{V}_i$ is the historical check-in places of user u_i and $Pr[d(v_j, v_y)] = a \times d(v_j, v_y)^b$.

The graphical model is demonstrated in Fig. 3. The dark circle is the observation variable which denotes the partial order of all POI collections. σ_U and σ_V denote hyper-parameters which are placed on U and V. M and N is the size of users and POIs respectively. The blank circle of vector U_i and V_j are unknown parameters that are learnt by probabilistic latent factor model. After obtaining U_i and V_j, we can estimate the utility of POI v_j given by user u_i. With the utility of $\mathcal{U}_{ij}$ and geographical influence d, the preference of user u_i to POI v_j can be obtained, which is depicted as y_{ij}.

As aforementioned, the proposed algorithm can be shown in Algorithm 1.

Algorithm 1. MUG: Modeling User Mobility via User Psychological and Geographical Behaviors towards Point of-Interest Recommendation

---

Input: $\succeq^L$, $d(v_j, v_y)$, K, β,C
Output: y_{ui}
compute a, b; //power-law
compute $Pr(v_j|\mathbb{V}_i)$;
for *step* $= 1$ *to Max Step* **do**
$\quad$ random choose u_i from user set $\mathbb{U}$;
$\quad$ random choose v_j from the user check-in set $\mathbb{V}_i$;
$\quad$ select latent POI collections of v_j; // nearest K
$\quad$ $v_j = \max(\succeq_{ij}^L)$;
$\quad$ $U_i = U_i - \beta \nabla U_i$;
$\quad$ $V_j = V_j - \beta \nabla V_j$;

return U, V;
compute $y_{ij} = (U_i V_j^T + C) \cdot Pr(v_j|\mathbb{V}_i)$

---

4   Experiments

In this section, we compare our model with several state-of-the-art recommendation methods in terms of Information Retrieval (IR) metrics by conducting extensive experiments on two real datasets. We analyze the accuracy of prediction in our POI recommendations. To be specific, we mainly address the following questions: (1) How can we compare our proposed method with existing methods? (2) How the size of latent POI collections influences the results of the proposed method? (3) What are the performances on the cold start problem?

Table 1. Statistics of the two extracted datasets

Datasets	Statistics	User	POI
Gowalla	Max. Num of check-ins	206	308
	Avg. Num of check-ins	17.03	20.19
Brightkite	Max. Num of check-ins	140	839
	Avg. Num of check-ins	7.963	16.41

4.1   Datasets and Metrics

The proposed approach is evaluated on Gowalla[1] and Brightkite[2] datasets. Gowalla and Brightkite are location-based social networking service providers where users share their locations, activities and travel lines etc. by checking-in. We randomly extract 3000 users and 2530 POIs with 50724 check-in frequency records from the Gowalla dataset and the density of the extracted dataset is 6.68×10^{-3}. We randomly extract 5000 users and 2425 POIs with 39815 check-in frequency records from the Brightkite dataset and the density of the extracted dataset is 3.28×10^{-3}. The statistics of the two datasets are described in Table 1.

POI recommendation aims to recommend personalized top-N POIs for users, which are obtained after sorting all candidate places in ascending order according to the predicted preference. We use the following three metrics to measure the performance of all models, which are widely adopted in Information Retrieval and Document Classification.

Precision@N and Recall@N: the precision and recall of personalized top-N recommend places for a target user is defined as below:

$$Precision@N = \frac{|rec@N \bigcap rel|}{N} \tag{18}$$

$$Recall@N = \frac{|rec@N \bigcap rel|}{|rel|} \tag{19}$$

where $rec@N$ denotes the top-N recommended POIs and rel is the true visited POIs in the test data.

---

[1] http://snap.stanford.edu/data/loc-gowalla.html.
[2] http://snap.stanford.edu/data/loc-brightkite.html.

MAP: Mean Average Precision (MAP), which for a test collection is the arithmetic mean of average precision values for individual user. The definition is:

$$MAP = \frac{1}{N} \sum_{u=1}^{M} (\frac{1}{rel} \sum_{l=1}^{|rel|} Precision_u @l) \tag{20}$$

where $Precision_u @l$ denotes the precision for user u when l relevant POIs are retrieved and M is the total number of users.

4.2 Baseline and Comparison

We compare our model with the following four existing methods:

- **MF:** It is adopted in [6], which captures the user preference by factorizing the user-rating matrix to get latent factors.
- **PMF:** It is proposed in [11] and also places a spherical multivariate Gaussian prior multivariate Gaussian prior on both U and V.
- **NMF:** This method is aimed to find the non-negative matrix factors U and V [7].
- **MGM:** This algorithm models user check-in behavior via employing Multi-center Gaussian distribution in [1].
- **MUG:** This is our modeling user mobility via user psychological and geographical behaviors algorithm.

We randomly divide the two extracted datasets into training (80 %) and testing (20 %) data. The latent dimension is set as 10 for MF, PMF, NMF and our proposed algorithm, following [10]. α and β are tuned using 5 fold cross-validation grid search for all algorithms to obtain the best results. K is set as 20 and 15 for Gowalla dataset and Brightkite dataset respectively in our proposed approach. The comparative experiments repeated by three times, Tables 2, 3 and 4 report the average results of the top 5, top 10 and top 15 POIs on the ranking list, respectively. We can observe that our method outperforms all baseline methods on three metrics. PMF and MGM perform much better than MF and PMF. Moreover,the results are coincident with [10]. Note that the values of precision in Table 2 are low, because the accuracy of POI recommendation is not high on account of sparse datasets. The results are shown detailedly that:

- Compared to the best factor model NMF which does not consider geographical behavior: for instance on top 10, for Gowalla extracted dataset, our method improves the results by **33.20 %** w.r.t. precision, **48.59 %** w.r.t. recall and **28.74 %** w.r.t. MAP. As to Brightkite extracted dataset, our method improves the results by **13.12 %** w.r.t. precision, **16.84 %** w.r.t. recall and **16.47 %** w.r.t. MAP. It indicates that geographical behavior plays an important role in POI recommendation.
- Compared to the MGM which does not model user mobility from psychological perspective: on top 10, for Gowalla extracted dataset, our method improves the results by **82.09 %** on precision, and improves the results by **62.47 %** and

60.55 % on recall and MAP respectively. For Brightkite extracted dataset, our method improves the results by **181.29 %** w.r.t. precision, **40.98 %** w.r.t. recall and **93.82 %** w.r.t. MAP. MGM considers users tend to check-in around several centers and the historical check-ins follow a Gaussian distribution in every center. However, as the sparse datasets we adopted, the historical check-ins may not act up to a multi-center Gaussian distribution. Moreover, MGM is not considering latent factors of user psychological behavior. As a result, our method performs much better than MGM model.

- The results of Gowalla extracted dataset outperform Brightkite extracted dataset because Brightkite extracted dataset is sparser than Gowalla extracted dataset. There is no exception, our proposed algorithm suffers the data sparsity problem, which is one of the most challenging problems in POI recommendations.

Obviously, the proposed approach outperforms the four comparative methods. We attribute the results to the effectiveness of user mobility formulated by user psychological and geographical behaviors in POI recommendation. Besides, the estimated metrics we adopt emphasize the ranking problem which is coincided well with the learning process of our proposed probabilistic latent model. The partial order learning mechanism adopted in our approach works well with the chosen metrics as well, thus making our algorithm superior to state-of-the-art methods.

Table 2. The results of metric precision

Datasets	Pre	MF	PMF	NMF	MGM	MUG
Gowalla	@5	0.01659	0.01055	0.04892	0.02641	**0.05289**
	@10	0.01627	0.00887	0.03467	0.02536	**0.04618**
	@15	0.01088	0.00844	0.03281	0.02283	**0.03903**
Brightkite	@5	0.00740	0.00715	0.03580	0.01138	**0.04709**
	@10	0.00742	0.00459	0.03178	0.01278	**0.03595**
	@15	0.00820	0.00348	0.02362	0.01218	**0.03028**

4.3 The Cold Start Problem

The cold start problem is a potential problem in POI recommendation. It may cause the inaccuracy of prediction, as it does not gather abundant information of users or POIs, e.g. new users. To compare our method with comparative algorithms on user cold start problem, we first divide the two datasets into training (10 %) and testing (90 %) data and then measure the performances of all models in terms of recall and MAP on top 5, top 10 and top 15. The results are shown in Fig. 4. It is obvious that the performances are different from the results

Table 3. The results of metric Recall

Datasets	Rec	MF	PMF	NMF	MGM	MUG
Gowalla	@5	0.02863	0.01107	0.12630	0.08190	**0.14775**
	@10	0.03815	0.01637	0.16384	0.14984	**0.24346**
	@15	0.03904	0.02893	0.21476	0.19796	**0.25066**
Brightkite	@5	0.07724	0.02350	0.08910	0.04752	**0.10289**
	@10	0.07815	0.02122	0.12511	0.10369	**0.14619**
	@15	0.07962	0.02922	0.16398	0.13883	**0.19025**

Table 4. The results of metric MAP

Datasets	MAP	MF	PMF	NMF	MGM	MUG
Gowalla	@5	0.02144	0.01041	0.07811	0.06477	**0.11536**
	@10	0.02239	0.01252	0.10387	0.08329	**0.13373**
	@15	0.01857	0.01407	0.10893	0.09210	**0.11119**
Brightkite	@5	0.03913	0.02002	0.06572	0.03751	**0.10690**
	@10	0.03783	0.02122	0.08765	0.05267	**0.10209**
	@15	0.03968	0.02922	0.07998	0.05554	**0.11597**

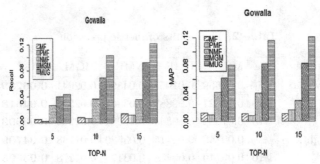

(a) Recall on cold start problem   (b) MAP on cold start problem

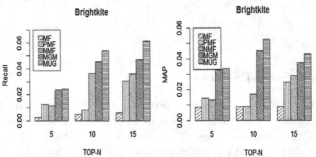

(c) Recall on cold start problem   (d) MAP on cold start problem

Fig. 4. Recall and MAP metrics on user cold start problem of both two datasets.

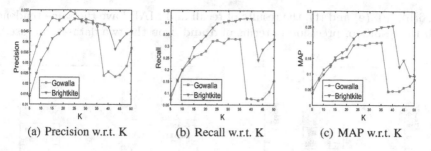

(a) Precision w.r.t. K   (b) Recall w.r.t. K   (c) MAP w.r.t. K

Fig. 5. The size of latent POI collections analysis.

of datasets divided into 80 % and 20 %. In most cases, our algorithm is superior to baseline algorithms. Moreover, MGM performs better than MF, PMF and NMF, and the results coincide with [1]. From the Fig. 4(a) and (b), on Gowalla datasets, when the check-in records are small and even many users do not have records, we observe that our MUG algorithm outperforms all baseline methods of user cold start problem on both two metrics. As to Brightkite dataset, from the Fig. 4(c) and (d), our MUG algorithm is nearly neck and neck with MGM algorithm on top 5. Luckily, on top 10 and top 15 POIs on the ranking list, MUG algorithm performs better than baseline algorithms. The reason for the good results may lie in user mobility formulated from a psychological perspective. As a whole, our MUG algorithm can improve the accuracy of recommendation on user cold start problem at a certain extent.

4.4   The Size of Latent POI Collection Analysis

Here, we analyze the influence of the size of latent POI collections. We analyze the accuracy of top 10 case on both two datasets. α and β are set as 0.08 and 0.01 for Gowalla and Brightkite datasets. From the Fig. 5(a), (b) and (c), we can observe several peaks on the line chart which show that our method is sensitive to the size of latent POI collections. For Gowalla dataset, it suggests that when $K \in [10, 35]$, our algorithm performs quite well. As to Brightkite dataset, it shows the superiority of $K \in [15, 40]$ over other numerical value intervals. The results reveal these K may be fit well to the size of our datasets we adopted.

4.5   Impact α and β

In our algorithm, the learning rate α controls how quickly the objective function descent and the regularization parameter β determines how much the regularization terms should be integrated. From the Fig. 6(a), the results demonstrate that precision negatively correlates with α and achieves the peak when $\alpha = 10^{-5}$ on Gowalla and Brightkite extracted datasets. From the Fig. 6(d), as to β, the results of precision firstly increase with smaller and smaller β, and they decay when β surpasses a certain value, i.e. 10^{-4} on our both two datasets. From

Fig. 6(b), (c), (e) and (f), the results of recall and MAP have the same tendency with the results of precision in terms of α and β on the two datasets we used.

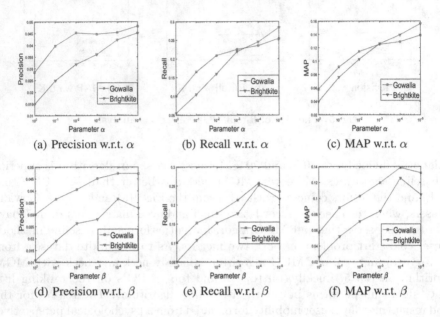

(a) Precision w.r.t. α (b) Recall w.r.t. α (c) MAP w.r.t. α

(d) Precision w.r.t. β (e) Recall w.r.t. β (f) MAP w.r.t. β

Fig. 6. Impact of parameter α and β on three metrics.

5   Conclusion and Future Work

In this paper, user psychological and geographical behaviors are combined to model user mobility for POI recommendation. The Proposed probabilistic latent factor model by adopting utility theory is used to formulate user psychological behavior and geographical behavior is captured by adopting a power-law distribution. Besides, a stochastic gradient descent algorithm is devised to learn the probabilistic latent factor model. Our approach is compared with some state-of-the-art POI recommendation algorithms on two real-world datasets and the results demonstrate that our approach achieves better recommendation performance.

For now, we only exploit information from check-in frequency data and geographical information. And data sparsity problem, which is one of the most challenging problems in real-world recommendation scenarios, also inevitably influences the performance of our method. Hence in our future work, there are two directions worthy to study: (1) How to handle with the data sparsity problem such as extremely sparse frequency data? (2) How to fuse leaving information like social networks, temporal information into our model? For future work, it is interesting to incorporate other information, e.g., temporal information, into our model to capture the user interest drift.

Acknowledgments. This research was partially supported by grants from the Science and Technology Program for Public Wellbeing of China (Grant No. 2013GS340302), National Natural Science Fund Project of China (Grant No. 61232018 and 61325010), National Social and Science Fund project of China (Grant No. 15BGL048), National 863 Plan Project of China (Grant No. 2015AA015403) and Hubei Province Support project of China (Grant No. 2015BAA072).

References

1. Cheng, C., Yang, H., King, I., Lyu, M.R.: Fused matrix factorization with geographical and social influence in location-based social networks. In: Twenty-Sixth AAAI Conference on Artificial Intelligence (2012)
2. Tobler, W.R.: A computer movie simulating urban growth in the detroit region. Econ. Geogr. **46**, 234–240 (1970)
3. Deng, S., Huang, L., Xu, G.: Social network-based service recommendation with trust enhancement. Expert Syst. Appl. **41**(18), 8075–8084 (2014)
4. Gao, H., Tang, J., Hu, X., Liu, H.: Exploring temporal effects for location recommendation on location-based social networks. In: Proceedings of the 7th ACM Conference on Recommender Systems, pp. 93–100. ACM (2013)
5. Jamali, M., Ester, M.: Trustwalker: a random walk model for combining trust-based and item-based recommendation. In: Proceedings of the 15th ACM SIGKDD International Conference on Knowledge Discovery and Data Mining, pp. 397–406. ACM (2009)
6. Koren, Y., Bell, R., Volinsky, C.: Matrix factorization techniques for recommender systems. Computer **8**, 30–37 (2009)
7. Lee, D.D., Seung, H.S.: Algorithms for non-negative matrix factorization. In: Advances in Neural Information Processing Systems, pp. 556–562 (2001)
8. Li, X., Xu, G., Chen, E., Zong, Y.: Learning recency based comparative choice towards point-of-interest recommendation. Expert Syst. Appl. **42**(9), 4274–4283 (2015)
9. Linden, G., Smith, B., York, J.: Amazon.com recommendations: item-to-item collaborative filtering. IEEE Internet Comput. **7**(1), 76–80 (2003)
10. Liu, B., Fu, Y., Yao, Z., Xiong, H.: Learning geographical preferences for point-of-interest recommendation. In: Proceedings of the 19th ACM SIGKDD International Conference on Knowledge Discovery and Data Mining, pp. 1043–1051. ACM (2013)
11. Mnih, A., Salakhutdinov, R.: Probabilistic matrix factorization. In: Advances in Neural Information Processing Systems, pp. 1257–1264 (2007)
12. Noulas, A., Scellato, S., Lathia, N., Mascolo, C.: Mining user mobility features for next place prediction in location-based services. In: 2012 IEEE 12th International Conference on Data Mining (ICDM), pp. 1038–1043. IEEE (2012)
13. Thurstone, L.L.: A law of comparative judgment. Psychol. Rev. **34**(4), 273 (1927)
14. Train, K.E.: Discrete Choice Methods with Simulation. Cambridge University Press, New York (2009)
15. Yang, S.H., Long, B., Smola, A.J., Zha, H., Zheng, Z.: Collaborative competitive filtering: learning recommender using context of user choice. In: Proceedings of the 34th International ACM SIGIR Conference on Research and Development in Information Retrieval, pp. 295–304. ACM (2011)
16. Ye, M., Yin, P., Lee, W.C.: Location recommendation for location-based social networks. In: Proceedings of the 18th SIGSPATIAL International Conference on Advances in Geographic Information Systems, pp. 458–461. ACM (2010)

17. Ye, M., Yin, P., Lee, W.C., Lee, D.L.: Exploiting geographical influence for collaborative point-of-interest recommendation. In: Proceedings of the 34th International ACM SIGIR Conference on Research and Development in Information Retrieval, pp. 325–334. ACM (2011)
18. Zhang, J.D., Chow, C.Y., Li, Y.: igeorec: A personalized and efficient geographical location recommendation framework. IEEE Trans. Serv. Comput. 8(5), 701–714 (2015)

Local Weighted Matrix Factorization for Implicit Feedback Datasets

Keqiang Wang[1], Xiaoyi Duan[1], Jiansong Ma[1], Chaofeng Sha[2(✉)],
Xiaoling Wang[1], and Aoying Zhou[1]

[1] Shanghai Key Laboratory of Trustworthy Computing,
East China Normal University, Shanghai, China
sei.wkq2008@gmail.com, duanxiaoyi@ecnu.cn, ecnumjs@163.com,
{xlwang,ayzhou}@sei.ecnu.edu.cn
[2] Shanghai Key Laboratory of Intelligent Information Processing, Fudan University,
Shanghai 200433, China
cfsha@fudan.edu.cn

Abstract. Item recommendation helps people to discover their potentially interested items among large numbers of items. One most common application is to recommend items on implicit feedback datasets (e.g., listening history, watching history or visiting history). In this paper, we assume that the implicit feedback matrix has *local* property, where the original matrix is not globally low-rank but some sub-matrices are low-rank. In this paper, we propose Local Weighted Matrix Factorization for implicit feedback (*LWMF*) by employing the kernel function to intensify *local* property and the weight function to model user preferences. The problem of sparsity can also be relieved by sub-matrix factorization in *LWMF*, since the density of sub-matrices is much higher than the original matrix. We propose a heuristic method DCGASC to select sub-matrices which approximate the original matrix well. The greedy algorithm has approximation guarantee of factor $1 - \frac{1}{e}$ to get a near-optimal solution. The experimental results on two real datasets show that the recommendation precision and recall of LWMF are both improved more than 30 % comparing with the best case of WMF.

Keywords: Recommendation systems · Local matrix factorization · Implicit feedback · Weighted matrix factorization

1 Introduction

In daily life, more and more people are going shopping online, enjoying music online and searching for restaurants' reviews before eating out. But online data are too large for people to find items that they want. So personalization recommendation can help people discover potentially interested items by analyzing user behaviors. Since they do not explicitly express user preferences. For example, ratings are explicit feedbacks which indicate users' preference, while visiting and buying history do not show users' preference directly so they are implicit

© Springer International Publishing Switzerland 2016
S.B. Navathe et al. (Eds.): DASFAA 2016, Part I, LNCS 9642, pp. 381–395, 2016.
DOI: 10.1007/978-3-319-32025-0_24

feedbacks. User *discovers* the item if her behaviors are implicit feedbacks, such as listened, watched or visited the item. Otherwise, user is unaware of the item. Different from the explicit feedback, the numerical value to describe implicit feedback is non-negative and very likely to be noisy [10].

Therefore, we assume that the implicit feedback matrix is not globally low-rank but some sub-matrices are low-rank. We call this important characteristic *local* property. Instead of decomposing the original matrix, we decompose the sub-matrix intuitively. We propose *Local Weighted Matrix Factorization (LWMF)*, integrating *LLORMA* [7] with *WMF* [10] in recommending by employing the kernel function to intensify *local* property and the weight function to intensify modeling user preference. The problem of sparsity can also be relieved by sub-matrix factorization in *LWMF*, since the density of sub-matrices is much higher than the original matrix.

The main contributions can be summarized as follows:

- We propose *LWMF* which integrates *LLORMA* with *WMF* to recommend items on implicit feedback datasets. *LWMF* utilizes the *local* property to model the matrix by dividing the original matrix into sub-matrices and relieves the sparsity problem.
- Based on kernel function, we propose *DCGASC (Discounted Cumulative Gain Anchor Point Set Cover)* to select the sub-matrices in order to approximate the original matrix better. At the same time, we conduct the theoretical submodularity analysis of the *DCGASC* objective function.
- Extensive experiments on real datasets are conducted to compare *LWMF* with state-of-the-art *WMF* algorithm. The experimental results demonstrate the effectiveness of our proposed solutions.

The rest of the paper is organized as follows. Section 2 reviews related work and Sect. 3 presents some preliminaries about *MF (Matrix factorization)*, *WMF* and *LLORMA*. Then we describe *LWMF* in Sect. 4. Section 5 illustrates the heuristic method *DCGASC* to select sub-matrices. Experimental evaluations using real datasets are given in Sect. 6. Conclusion and future work are followed in Sect. 7.

2   Related Work

In this section, we review some previous work on recommendation systems, including *K-Nearest Neighbor (KNN)*, *Matrix Factorization (MF)*, local ensemble methods, personalized ranking method and some other recommendation systems for special scenarios recommender.

One of the most traditional and popular way for recommender systems is KNN [1]. Item-based *KNN* uses the similarity techniques (e.g., cosine similarity, Jaccard similarity and Pearson correlation) between items to recommend the similar items. Then, *MF* [2–4] methods play an important role in model-based CF methods, which aim to learn latent factors on user-item matrix. *MF* usually gets better performance than *KNN*-based methods especially on rating prediction. Recently some work focuses on ensembles of sub-matrix for better approximation, such as *DFC* [5], *RBBDF* [6], *LLORMA* [7,8] and *ACCAMS* [9]. However, such methods focus on explicit feedback datasets while most of the feedbacks are implicit, such as listening times, click times and check-ins. The implicit datasets do not need users to rate the items. So Hu et al. [10] and Pan et al. [11,12] propose *Weighted Matrix Factorization (WMF)* to model implicit feedback with *Alternative Least Square (ALS)*. Our work in this paper integrates *LLORMA* with *WMF* and proposes a heuristic method to select sub-matrices. Other related work on implicit feedback datasets are ranking methods, such as *BPR* [13] and Pairwise Learning [14]. With the explosion of size of the training data, the ranking methods need use some efficient sampling techniques to reduce complexity. Finally, there are several special scenarios, such as recommending music [15], News [16], TV show [17] and POI [18,19], utilizing the additional information (e.g., POI recommender considers the geographical information) to improve prediction performance.

3   Preliminary

In this section, we present some preliminaries about *MF*, *WMF* and *LLORMA*.

3.1   Matrix Factorization

MF is a dimensionality reduction technique, which has been widely used in recommendation system especially for the rating prediction [3,4]. Due to their attractive accuracy and scalability, *MF* plays a vital role in recent recommendation system competitions, such as *Netflix Prize*[1], *KDD Cup 2011 Recommending Music Items*[2], *Alibaba Big Data Competitions*[3] and so on. Given a sparse matrix $R \in \mathbb{R}^{N \times M}$ with indicator matrix I, and latent factor number $F \ll \min\{N, M\}$. The aim of *MF* is:

$$\min_{P,Q} \sum_{u=1}^{N} \sum_{m=1}^{M} I_{um}(R_{um} - \hat{R}_{um})^2 \tag{1}$$

In order to avoid over-fitting, regularization terms are usually added to the objective function to modify the squared error. So the task is to minimize $\sum_{u=1}^{N} \sum_{m=1}^{M} I_{um}(R_{um} - \hat{R}_{um})^2 + \lambda\|P\|_F^2 + \lambda\|Q\|_F^2$. The parameter λ is used to control the magnitudes of the latent feature matrices, P and Q. Stochastic gradient descent is often used to learn the parameters [4].

3.2   Weighted Matrix Factorization

[10] argues that original *MF* is always used on explicit feedback datasets, especially for rating prediction. So Hu et al. [10] and Pan et al. [11,12] propose *Weighted Matrix Factorization* (*WMF*) to handle the cases with implicit

---

[1] http://www.netflixprize.com/.
[2] http://www.kdd.org/kdd2011/kddcup.shtml.
[3] https://102.alibaba.com/competition/addDiscovery/index.htm.

feedback. Recently, *WMF* has been widely used in TV show, music and POI (Point-of-Interests) recommendation. To utilize the undiscovered items and to distinguish between discovered and undiscovered items, a weight is added to the *MF*:

$$W_{um} = 1 + log(1 + R_{um} \times 10^{\varepsilon}) \tag{2}$$

where the constant ε is used to control the rate of increment. Considering the weights of implicit feedback, the optimization function is reformulated as follows:

$$\min_{P,Q} \sum_{u=1}^{N} \sum_{m=1}^{M} W_{um}(C_{um} - P_u Q_m^T)^2 + \lambda \|P\|_F^2 + \lambda \|Q\|_F^2 \tag{3}$$

where each entry C_{um} in the 0/1 matrix C indicates whether the user u has discovered the item m, which can be defined as $C_{um} = \begin{cases} 1 & R_{um} > 0 \\ 0 & R_{um} = 0. \end{cases}$

3.3   Low-Rank Matrix Approximation

LLORMA [7,8] is under the assumption of locally low rank instead of globally low rank. That is, limited to certain types of similar users and items, the entire rating matrix R is not low-rank but a sub-matrix R_s is low-rank. It is to say that the entire matrix R is composed by a set of low-rank sub-matrices $\mathcal{R}_s = \{R^1, R^2, ..., R^H\}$ with weight matrix set $\mathcal{T} = \{T^1, T^2, ..., T^H\}$ of sub-matrices, where T_{ij}^h indicates the sub-matrix weight of R_{ij}^h in R_h:

$$R_{um} = \frac{1}{Z_{um}} \sum_{h=1}^{H} T_{u_h m_h}^h R_{u_h m_h}^h \tag{4}$$

where $Z_{um} = \sum_{h=1}^{H} T_{u_h m_h}^h$. *LLORMA* uses the *MF* introduced in Sect. 3.1 to approximate the sub-matrix R^h. If the matrix has *local* property, we can achieve good accuracy in predicting ratings.

4   Local Weighted Matrix Factorization

In this section, we introduce our method *LWMF* for implicit datasets. Following the *LLORMA*, we first select sub-matrices from the original matrix, then each sub-matrix is decomposed by *WMF* methods as shown in Fig. 1. We propose *LWMF* which integrates *LLORMA* with *WMF* to recommend top-N items on implicit datasets. We estimate each sub-matrix R_h by *WMF* in Sect. 3.2 as follows:

$$\min_{P_h, Q_h} \sum_{u_h=1}^{N_h} \sum_{m_h=1}^{M_h} T_{u_h m_h}^h W_{u_h m_h}(R_{u_h m_h}^h - P_{u_h}^h{}^T Q_{m_h}^h)^2 + \lambda_1 \|P^h\|_F^2 + \lambda_2 \|Q^h\|_F^2 \tag{5}$$

So the original Matrix R can be approximated by the set of approximated sub-matrices $\mathcal{R}_s = \{\hat{R}^1, \hat{R}^2, ..., \hat{R}^H\}$:

$$R_{um} \approx \frac{1}{Z_{um}} \sum_{h=1}^{H} T_{u_h m_h}^h \hat{R}_{u_h m_h}^h \tag{6}$$

where $\hat{R}^h = {P^h}^T Q^h$. So there are mainly two problems: (1) How to calculate the weight matrix T^h? (2) How to select the sub-matrix set $\mathcal{R}_s$? *LLORMA* uses the kernel methods[4] to solve these two problems. *LLORMA* employs three popular smoothing kernels (i.e., the uniform kernel, the triangular kernel and the Epanechnikov kernel) to calculate the relationship between users or items. Then it randomly chooses some user-item pairs[5] from training data as the anchor point set. Each anchor point $a_h = (u_h, m_h)$ is related to other pairs (u_i, m_i) of which the kernel is greater than 0. So the related pairs and the anchor point a_h make up a sub-matrix R^h while the kernel matrix acts as the weight matrix T^h.

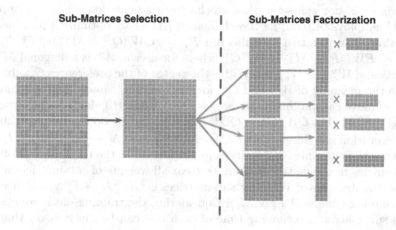

Fig. 1. Local matrix factorization

In this paper, we use the Epanechnikov kernel to calculate the relationship between two pairs (u_h, m_h) and (u_j, m_j). It is computed as the product of user Epanechnikov kernel $(E_{b_u}(u_h, u_h))$ and item Epanechnikov kernel $(E_{b_m}(m_j, m_j))$ as follows:

$$E(a_h, a_j) = E_{b_u}(u_h, u_j) \times E_{b_m}(m_h, m_j) \tag{7}$$

where

$$E_{b_u}(u_h, u_j) \propto (1 - d(u_h, u_j)^2)\, \mathbf{1}_{\{d(u_h, u_j) \le b_u\}}$$
$$E_{b_m}(m_h, m_j) \propto (1 - d(m_h, m_j)^2)\, \mathbf{1}_{\{d(m_h, m_j) \le b_m\}}$$

---

[4] https://en.wikipedia.org/wiki/Kernel_(statistics).
[5] Pair (u, m) means that the user u discovered the item m.

where b_u, b_m are the bandwidth parameters of kernel. Distance between two users or two items is the distance between two row vectors (for user kernel) or column vectors (for item kernel). The initial user latent factor and item latent factor are learned by *WMF*. Accordingly, the distance between users u_i and u_j is $d(u_i, u_j) = arccos(\frac{P_{u_i} \cdot P_{u_j}}{\|P_{u_i}\| \cdot \|P_{u_j}\|})$, where P_{u_i}, P_{u_j} are the ith and jth rows of P. The distance between items is computed in the same way.

Different from *LLORMA*, all kernel weights are set to the same value (i.e., $T^h = I^h$) in this paper. That is to say, the weights of all user-item pairs to the anchor point in the sub-matrix are identical. The intuition behind the setting is that:

1. In *LWMF*, we aim to find sub-matrices. So we just use the Epanechnikov kernel to get one sub-matrix by selecting an anchor point and do not care much about the weight;

2. Due to the weight matrix W and preference matrix C settings, the preference matrix C is not sparse so that stochastic gradient descent is not applied. *WMF* employs *Alternative Least Square (ALS)* to optimize this objective function. The iterative formulas are $P_u = (QW^u Q^T + \lambda I)^{-1} QW^u C^u$ and $Q_m = (PW^m P^T + \lambda I)^{-1} PW^m C^m$ where for user u, W^u is a diagonal $M \times M$ matrix and $W_{mm}^u = W_{um}$, and C^u is the vector of the preferences C_{um} by user u. So the meaning of W^m and C^m for item m is as same. The total running time of naive calculation for all users is $O(F^2 NM)$. [10] devises a nice trick by using the fact that $QW^u Q^T = QQ^T + Q(W^u - I)Q^T$. So the running time for each iteration is reduced to $O(F^2 \mathcal{N} + F^3 N)$, where $\mathcal{N} = \sum_{u=1}^{N} \sum_{m=1} I_{um}$. It is to say that its time complexity is in proportion to the total number of non-zero entries in the matrix R. So if we treat all weights of sub-matrices all the same, the iterative of *WMF* for sub-matrices is $O(F_s^2 \mathcal{N}_f + F_s^3 N_s)$. Otherwise, the running time is $O(F_s^2 N_s M_s)$. Considering the training data reduction of each sub-matrix, the running time of each one can be much faster than the original matrix.

3. We do some extensive experiments and find that the results of *LWMF* are almost the same whether considering kernel weight or not.

Each anchor point stands for a sub-matrix. Selecting the sub-matrix set $\mathcal{R}_s$ is in fact to select a set of anchor points. The details of selecting anchor point set is discussed in the next section.

5   Anchor Point Set Selection

Intuitively, the sub-matrix set $\mathcal{R}_s = \{R^1, R^2, ..., R^H\}$ should cover the original matrix R, that is $R = \cup_{R^h \in \mathcal{R}_s} R^h$, so these sub-matrix sets $\mathcal{R}_s$ can approximate the original matrix R better than the set that does not cover. So the anchor points selection problem can be reduced to the set cover problem. Given the candidate anchor point set $A = \{a_1, a_2, ..., a_n\}$ while every candidate point a_i can cover itself several other candidate points denoted by $A_i = \{a_i, a_{i1}, a_{i2}, ..., a_{ih}\} \subset A$, we propose the naive anchor points cover method:

Anchor Point Set Cover (ASC): returns an anchor point set $S \subset A$ such that

$$\max f(S) = |\cup_{i \in S} A_i|$$
$$s.t. |S| = K \qquad (8)$$

Here, we use all pairs (u, m) in training data as the candidate anchor points. Obviously, the ASC problem is submodular and monotone. So the greedy algorithm has $1 - \frac{1}{e}$ approximation of optimization.

However, covering all training data only once in ASC is not enough. Although performance is improved by increasing cover times, the gain is discounted, which is similar to the situation in ranking quality measures $NDCG$ (Normalized Discounted Cumulative Gain) [23] and ERR (Expected Reciprocal Rank) [22] in IR(Information Retrial). So we propose a heuristic method to model this situation as follows.

Discounted Cumulative Gain Anchor Point Set Cover (DCGASC): returns an anchor point set $S \subset A$ such that

$$\max f(S) = \sum_{a_l \in \cup_{a_i \in S} A_i} \sum_{o_l=1}^{O_l} \alpha^{o_l - 1} = \sum_{z=1}^{|S|} \sum_{a_l \in A_{iz}} \alpha^{o_{lz} - 1}$$
$$s.t. |S| = K \qquad (9)$$

where O_l denotes the covered time of a_l by itself or other selected anchor points. Below we prove that $f(\cdot)$ is also submodular and monotone.

Theorem 1. *DCGASC function is submodular and also monotone non-decreasing.*

Proof. Let $S \subseteq V \subseteq A$ and $A_h \in A \backslash V$. We have that

$$f(S \cup \{A_h\}) - f(S) = \sum_{z=1}^{|S \cup \{A_h\}|} \sum_{a_l \in A_{iz}} \alpha^{o_{lz} - 1} - \sum_{z=1}^{|S|} \sum_{a_l \in A_{iz}} \alpha^{o_{lz} - 1}$$

$$= \sum_{z=1}^{|S|} \sum_{a_l \in A_{iz}} \alpha^{o_{lz} - 1} + \sum_{a_l \in A_h} \alpha^{o_{lz}^S - 1} - \sum_{z=1}^{|S|} \sum_{a_l \in A_{lz}} \alpha^{o_{lz} - 1}$$

$$= \sum_{a_l \in A_h} \alpha^{o_{lz}^S - 1} \geq 0 \qquad (10)$$

and

$$f(S \cup \{A_h\}) - f(S) - (f(V \cup \{A_h\}) - f(V)) = \sum_{a_l \in A_h} \alpha^{o_{lz}^S - 1} - \sum_{a_l \in A_h} \alpha^{o_{lz}^V - 1}$$

$$= \sum_{a_l \in A_h} (\alpha^{o_{lz}^S - 1} - \alpha^{o_{lz}^V - 1}) = \sum_{a_l \in A_h} \alpha^{o_{lz}^S - 1}(1 - \alpha^{o_{lz}^S - o_{lz}^V}). \qquad (11)$$

Algorithm 1. *DCGASC* Greedy Algorithm

Input : Set of anchors A, anchor number K, DCGASC function f and sets A_i
              covered by each anchor a_i

Output: $S \subseteq A$ with $|S| = k$

1 $S \leftarrow \{\arg\max_{a_l \in A} |A_l|\}$;
2 **while** $|S| < K$ **do**
3 $\quad|\quad l \leftarrow \arg\max_{a'_l \in A \setminus S} f(S \cup \{a'_l\}) - f(S)$
4 $\quad|\quad S \leftarrow S \cup \{a_l\}$
5 **end**
6 **return** S

Because the number of anchor points covered satisfies that $o_{lz}^S \leqslant o_{lz}^V$ and discount parameter $\alpha \in [0, 1)$, we know that $f(S \cup \{A_h\}) - f(S) - (f(V \cup \{A_h\}) - f(V)) \geq 0$. Therefore, it is proved that the *DCGASC* function is monotone and submodular.

Due to the monotonicity and submodularity of *DCGASC* function, the greedy Algorithm 1 can provide a theoretical approximation guarantee of factor $1 - \frac{1}{e}$ as described in [24].

6   Experiments

In this section, we evaluate the method proposed in this paper using real datasets. We first introduce the datasets and experimental settings. Then we compare our method with *WMF* under specific parameter settings. We also compare results with different anchor numbers and three anchor points selection methods.

6.1   Dataset

We choose two datasets. One is the *Gowalla* from [20], one of the most popular online LBSNs datasets. Another is *YES* [21], which is a playlist dataset crawled by using the web based API[6].

The experimental dataset is chosen from Gowalla dataset within the range of latitude from 32.4892 to 41.7695 and longitude from -124.3685 to -114.5028, where locate in California and Nevada. We consider users who check in more than 10 distinct POIs and the POIs which are visited by more than 10 users. So it contains 205,509 check-ins made by 5,086 users at 7,030 locations. Finally the density of is 4.68×10^{-3} and very sparse.

YES [21] consists of radio playlists.[7] There are 431,367 playlists and 3,168 songs. A playlist corresponds to a user. Similarly, we consider users who listen in more than 10 distinct songs and the songs which are listened by more than

---

[6] http://api.yes.com.
[7] www.cs.cornell.edu/~shuochen/lme/data_page.html.

Table 1. Detail information of *Gowalla* and *YES*

Gowalla		YES	
#user	5,086	#user	40,465
#locations	7,030	#songs	2,563
#check-ins	167,404	#listens	735,367
avg.#users per loc.	23.81	avg. #user per song	286.92
avg. #loc. per user	32.91	avg. #songs per user	18.17
max #users per loc.	1,644	max #users per song	4,217
max #loc. per user	762	max #songs per user	134

10 users. So the dataset contains 40,465 users and 2,563 songs. The density is about 7.09×10^{-3}.

More details about two datasets are showed in the Table 1. We randomly select 80 % of each user's visiting locations as the training set and the 20 % as the testing set.

6.2 Setting

Next, we show the parameter values. The regularization λ is set to 0.01 and the performance of recommendation is not sensitive to this parameter. The weight parameter ε is set to 4. We set the bandwidth parameter in Epanechanikov kernel as $b_u = b_m = 0.8$. The discount α of $DCGASC$ is set to 0.7. We select 200 anchor points for Gowalla dataset and 100 anchor points for YES dataset. In the experiments, we observe that if the number of anchor points is larger, the performance is better. But the training time increases accordingly.

We employ the Precision@N and Recall@N to measure the performance. For a user u, we set $\mathcal{I}^P(u)$ as the predicted item list and $\mathcal{I}^T(u)$ as the true list in the testing dataset. So the Precision@N and Recall@N are:

$$Precision@N = \frac{1}{|U|} \sum_{u \in U} \frac{|\mathcal{I}^P(u) \bigcap \mathcal{I}^T(u)|}{N}, \; Recall@N = \frac{1}{|U|} \sum_{u \in U} \frac{|\mathcal{I}^P(u) \bigcap \mathcal{I}^T(u)|}{|\mathcal{I}^T(u)|}$$

where $|\mathcal{I}^P(u)| = N$. In our base experiments, we choose top 5, 10, 20 and 30 as evaluation metrics.

We compare two methods for implicit feedback datasets:

- *WMF:* This is the state-of-the-art method which is designed for implicit feedback [10].
- *LWMF:* This is our proposed method that takes account of two ideas of *WMF* [10] and *LLORMA* [7].

Then we compare three anchor points selection methods to study the performance of *LWMF*:

- *Random:* Sampling anchor points uniformly from training dataset as paper [7] does.

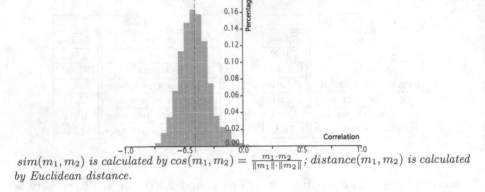

$sim(m_1, m_2)$ *is calculated by* $cos(m_1, m_2) = \frac{m_1 \cdot m_2}{\|m_1\| \cdot \|m_2\|}$*; distance*$(m_1, m_2)$ *is calculated by Euclidean distance.*

Fig. 2. Statistics of correlation between $\text{sim}(m_1, m_2)$ and $\text{distance}(m_1, m_2)$ on *Gowalla*

- *Anchor Set Cover (ASC):* Set cover method, which is submodular and monotone, and the greedy algorithm can provide a theoretical approximation guarantee of factor $1 - \frac{1}{e}$.
- *Discounted Cumulative Gain Anchor Set Cover (DCGASC):* Discounting cumulative gain of covering the points which is also submodular and monotone.

So *LWMF* can be expanded into three sub-methods *LWMF_Random*, *LWMF_ASC* and *LWMF_DCGASC*. By defualt, *LWMF* means *LWMF_DCGASC*. Each method is conducted 5 times independently. Therefore, the average score indicates the performance of the recommendation methods.

6.3 Results

In this section, we discuss the experimental results on *Gowalla* and *YES* datasets.

6.3.1 Recommendation Methods Comparison

Tables 2 and 3 list the precision and recall of methods *WMF* and *LWMF* with *DCGASC*. It shows the same result as [7] that *LORMA* outperforms *SVD*, and *LWMF* always outperforms *WMF*. With the rank r increases, both *LWMF* and *WMF* get better, but the improvements get less when r is 20 of *WMF* and 15 of *LWMF*. Noted that the results of LWMF with rank 3 are almost the same as WMF with rank $r = 20$ on both datasets. For *Gowalla* dataset, *LWMF* with rank $r = 15$ even improves the precision@5 by 35.37 % and the recall@5 by 31.37 %. For *YES* dataset, *LWMF* improves the precision@5 by 7.92 % and the recall@5 11.21 %. More obvious improvements on Gowalla is due to the local property. For example, there are some business districts in a city and business POIs are geographically close to each other within each business district. And it shows that the average cover rate of anchor points on *Gowalla* dataset is about 6.5 % while

Table 2. Precision and recall comparison on *Gowalla*

Top-N	Precision				Recall			
	5	10	20	30	5	10	20	30
WMF rank5	0.0646	0.0489	0.0363	0.0302	0.0587	0.0848	0.1229	0.1502
WMF rank10	0.0752	0.0570	0.0432	0.0359	0.0676	0.0991	0.1457	0.1786
WMF rank15	0.0819	0.0628	0.0471	0.0393	0.0748	0.1096	0.1593	0.1955
WMF rank20	0.0862	0.0659	0.0494	0.0412	0.0797	0.1165	0.1659	0.2053
LWMF rank3	0.0872	0.0662	0.0495	0.0415	0.0801	0.1152	0.1650	0.2033
	1.17%	0.34%	0.33%	0.73%	0.57%	−1.12%	−0.57%	−0.96%
LWMF rank5	0.0986	0.0750	0.0548	0.0459	0.0899	0.1292	0.1827	0.2239
	14.37%	13.67%	10.97%	11.18%	12.82%	10.93%	10.14%	9.11%
LWMF rank10	0.1101	0.0843	0.0614	0.0511	0.0992	0.1444	0.2045	0.2491
	27.70%	27.81%	24.45%	23.89%	24.48%	23.94%	23.25%	21.34%
LWMF rank15	**0.1167**	**0.0877**	**0.0638**	**0.0522**	**0.1047**	**0.1504**	**0.2131**	**0.2588**
	35.37%	33.03%	29.29%	26.49%	31.37%	29.12%	28.48%	26.08%

Table 3. Precision and recall comparison on *YES*

Top-N	Precision				Recall			
	5	10	20	30	5	10	20	30
WMF rank5	0.0793	0.0641	0.0487	0.0409	0.1030	0.2130	0.2506	0.3171
WMF rank10	0.1046	0.0842	0.0624	0.0501	0.1349	0.2765	0.3245	0.3913
WMF rank15	0.1104	0.0890	0.0655	0.0525	0.1421	0.2913	0.3393	0.4083
WMF rank20	0.1111	0.0895	0.0659	0.0530	0.1429	0.2931	0.3414	0.4118
LWMF rank3	0.1121	0.0899	0.0666	0.0542	0.1467	0.2935	0.3399	0.4116
	0.88%	0.53%	1.11%	2.29%	2.68%	0.11%	−0.45%	−0.04%
LWMF rank5	0.1167	0.0919	0.0671	0.0538	0.1517	0.3009	0.3495	0.4218
	5.02%	2.74%	1.90%	1.65%	6.18%	2.65%	2.36%	2.42%
LWMF rank10	0.1182	0.0923	**0.0684**	**0.0545**	0.1569	0.3082	0.3552	0.4245
	6.36%	3.19%	3.79%	2.84%	9.84%	5.14%	4.04%	3.09%
LWMF rank15	**0.1199**	**0.0938**	**0.0682**	**0.0545**	**0.1589**	**0.3118**	**0.3591**	**0.4282**
	7.92%	4.89%	3.49%	2.93%	11.21%	6.37%	5.17%	3.98%

it is about 11.6% on *YES* dataset. To validate the local property in *Gowalla*, we calculate the correlation between similarity of locations ($sim(m_1, m_2)$) and their geographic distance ($distance(m_1, m_2)$), where m_1 and m_2 are each pair of locations.

As shown in Fig. 2, we make statistics about correlation between similarity and geographic distance among the locations. Correlation coefficient between two locations is negative and the average is about −0.41. Therefore, two near locations are more similar than two further locations. So the *local* property leads that the user-location matrix is not globally low-rank but locally low-rank.

6.3.2   Comparison with Different Discounts for *DCGASC*

Figure 3 shows the performance of *LWMF* with different anchor numbers. For both datasets, the precision and recall of both *LWMF* and *WMF* improve while r increases and *LWMF* performances better than *WMF* with rank $r \geq 3$. For *Gowalla* dataset, *LWMF* with rank $r = 15$ and anchor number $K \geq 20$ outperforms *WMF* with rank $r = 20$. While the same performance on *YES* dataset needs $K \geq 65$ anchor points. We can see that as the number of anchor points increases, the performance gets better. Although the training time increases, the gap of running time of matrix factorization between *LWMF* and *WMF* is small. Because the running time of *WMF* is $O(F^2 N + F^3 N)$ and the sub-matrices of *LWMF* are much smaller than the original matrix (i.e., in both datasets, each sub-matrix is about 10 % of original matrix averagely). Only one sub-matrix factorization is much faster than original matrix factorization. Despite all this, *LWMF* costs more time on calculating the KDE between users and items and selecting anchor points.

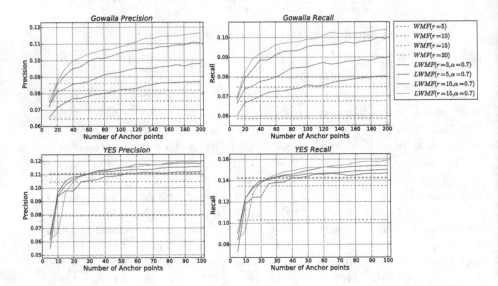

Fig. 3. Comparison with different discounts for *DCGASC* (Color figure online)

6.3.3   Anchor Point Set Selection Methods Comparison

Next, we compare the performance of *LWMF_Random*, *LWMF_ASC* and *LWMF_DCGASC* in Fig. 4. The discount parameter α is set 0.7. The method *ASC* may cover all training data before selecting K anchor points. After covering all training data, we use *Random* method to select the remaining anchor points. From Fig. 4, when the number of anchor points is small, *LWMF_DCGACS* and *LWMF_ACS* perform better in precision and recall. When the number of anchor points increases, the gap of performance among three gets less. Despite of this, *LWMF_DCGACS* and *LWMF_ACS* outperform *LWMF_Random* on both datasets and *LWMF_DCGACS* is the best.

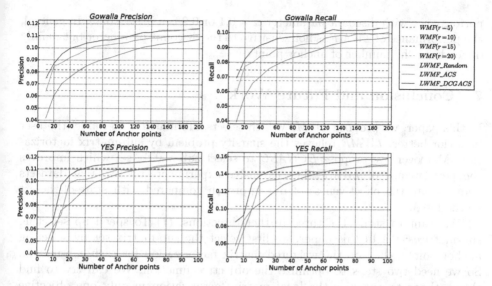

Fig. 4. Anchor point set selection methods comparison (Color figure online)

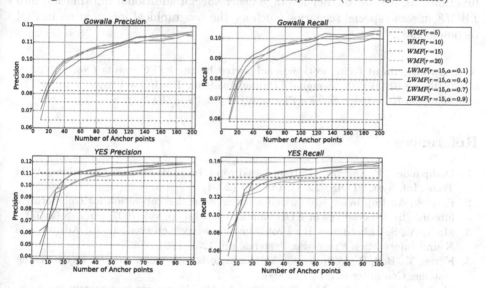

Fig. 5. Comparison with different discounts for $DCGASC$ (Color figure online)

6.3.4   Comparison with Different Discounts for $DCGASC$

Finally, we study the performance of $LWMF\_DCGACS$ with different discount parameters. For each α, we explore results obtained by varying the parameter in the range $(0, 1]$ with decimal steps. Because the results with discount parameter $\alpha \in [0.2, 0.8]$ are similar, we only plot the curves with $\alpha \in \{0.1, 0.4, 0.7, 0.9\}$ in Fig. 5. The gap of performance with four discount parameters is small. The performance with discount parameter $\alpha = 0.1$ or 0.9 is a little worse. In general, the performance of *LWMF* is not sensitive to the discount parameter but mainly depends on the number of anchor points.

7   Conclusion and Future Work

In this paper, we propose *LWMF* which selects sub-matrices to model the user behavior better. *LWMF* relieves the sparsity problem by sub-matrix factorization. Moreover, we propose *DCGASC* to select sub-matrix set which improves the performance of *LWMF*. The extensive experiments on two real datasets demonstrate the effectiveness of our approach compared with state-of-the-art method *WMF*.

We want to study the three further directions: (1) To speed up selecting sub-matrices; (2) In this paper, we first select the sub-matrix set by selecting anchor points, then do the weighted matrix factorization for each sub-matrix. So we need two steps to optimize the objective function. We can try to find the methods to optimize the local matrix factorization in only one objective function; (3) We can further leverage other special additional information into *LWMF* in some special scenarios (such as, the geographical information in POI recommender).

Acknowledgement. This work was supported by the NSFC grants (No. 61472141, 61370101 and 61021004), Shanghai Leading Academic Discipline Project (No. B412), and Shanghai Knowledge Service Platform Project (No. ZF1213).

References

1. Deshpande, M., Karypis, G.: Item-based top-n recommendation algorithms. ACM Trans. Inf. Syst. (TOIS) **22**(1), 143–177 (2004)
2. Paterek, A.: Improving regularized singular value decomposition for collaborative filtering. In: Proceedings of KDD Cup and Workshop, vol. 2007, pp. 5–8 (2007)
3. Mnih, A., Salakhutdinov, R.: Probabilistic matrix factorization. In: Advances in Neural Information Processing Systems, pp. 1257–1264 (2007)
4. Koren, Y., Bell, R., Volinsky, C.: Matrix factorization techniques for recommender systems. Computer **8**, 30–37 (2009)
5. Mackey, L.W., Jordan, M.I., Talwalkar, A.: Divide-and-conquer matrix factorization. In: Advances in Neural Information Processing Systems, pp. 1134–1142 (2009)
6. Zhang, Y., Zhang, M., Liu, Y., et al.: Localized matrix factorization for recommendation based on matrix block diagonal forms. In: Proceedings of the 22nd International Conference on World Wide Web, International World Wide Web Conferences Steering Committee, pp. 1511–1520 (2013)
7. Lee, J., Kim, S., Lebanon, G., et al.: Local low-rank matrix approximation. In: Proceedings of the 30th International Conference on Machine Learning, pp. 82–90 (2013)
8. Lee, J., Bengio, S., Kim, S., et al.: Local collaborative ranking. In: Proceedings of the 23rd International Conference on World Wide wWeb, pp. 85–96. ACM (2014)

9. Beutel, A., Ahmed, A., Smola, A.: Additive co-clustering to approximate matrices succinctly. In: Proceedings of the 24th International Conference on World Wide Web. International World Wide Web Conferences Steering Committee, pp. 119–129 (2015)
10. Hu, Y., Koren, Y., Volinsky, C.: Collaborative filtering for implicit feedback datasets. In: 8th IEEE International Conference on Data Mining: ICDM 2008, pp. 263–272. IEEE (2008)
11. Pan, R., Zhou, Y., Cao, B., et al.: One-class collaborative filtering. In: 8th IEEE International Conference on Data Mining, pp. 502–511. IEEE (2008)
12. Pan, R., Scholz, M.: Mind the gaps: weighting the unknown in large-scale one-class collaborative filtering. In: Proceedings of the 15th ACM SIGKDD International Conference on Knowledge Discovery and Data Mining, pp. 667–676. ACM (2009)
13. Rendle, S., Freudenthaler, C., Gantner, Z., et al.: BPR: Bayesian personalized ranking from implicit feedback. In: Proceedings of the 25th Conference on Uncertainty in Artificial Intelligence, pp. 452–461. AUAI Press (2009)
14. Rendle, S., Freudenthaler, C.: Improving pairwise learning for item recommendation from implicit feedback. In: Proceedings of the 7th ACM International Conference on Web Search, Data Mining, pp. 273–282. ACM (2014)
15. Yang, D., Chen, T., Zhang, W., et al.: Local implicit feedback mining for music recommendation. In: Proceedings of the 6th ACM Conference on Recommender Systems, pp. 91–98. ACM (2012)
16. Ilievski, I., Roy, S.: Personalized news recommendation based on implicit feedback. In: Proceedings of the 2013 International News Recommender Systems Workshop, Challenge, pp. 10–15. ACM (2013)
17. Zibriczky, D., Hidasi, B., Petres, Z., et al.: Personalized recommendation of linear content on interactive TV platforms: beating the cold start and noisy implicit user feedback. UMAP Workshops (2012)
18. Lian, D., Zhao, C., Xie, X., et al.: GeoMF: joint geographical modeling and matrix factorization for point-of-interest recommendation. In: Proceedings of the 20th ACM SIGKDD International Conference on Knowledge Discovery and Data Mining, pp. 831–840. ACM (2014)
19. Liu, Y., Wei, W., Sun, A., et al.: Exploiting geographical neighborhood characteristics for location recommendation. In: Proceedings of the 23rd ACM International Conference on Information, Knowledge Management, pp. 739–748. ACM (2014)
20. Cho, E., Myers, S.A., Leskovec, J.: Friendship and mobility: user movement in location-based social networks. In: Proceedings of the 17th ACM SIGKDD International Conference on Knowledge Discovery and Data Mining, pp. 1082–1090. ACM (2011)
21. Chen, S., Moore, J.L., Turnbull, D., et al.: Playlist prediction via metric embedding. In: Proceedings of the 18th ACM SIGKDD International Conference on Knowledge Discovery, Data Mining, pp. 714–722. ACM (2012)
22. Chapelle, O., Metlzer, D., Zhang, Y., et al.: Expected reciprocal rank for graded relevance. In: Proceedings of the 18th ACM Conference on Information, Knowledge Management, pp. 621–630. ACM (2009)
23. Clarke, C.L.A., Kolla, M., Cormack, G.V., et al.: Novelty, diversity in information retrieval evaluation. In: Proceedings of the 31st Annual International ACM SIGIR Conference on Research, Development in Information Retrieval, pp. 659–666. ACM (2008)
24. Nemhauser, G., Wolsey, L.A., Fisher, M.: An analysis of approximations for maximizing submodular set functions - I. Math. Program. 14(1), 265–294 (1978)

Exploring the Choice Under Conflict for Social Event Participation

Xiangyu Zhao, Tong Xu$^{(\boxtimes)}$, Qi Liu, and Hao Guo

School of Computer Science and Technology,
University of Science and Technology of China, Hefei, China
{zxy1105,tongxu,guoh916}@mail.ustc.edu.cn, qiliuql@ustc.edu.cn

Abstract. Recent years have witnessed the booming of *event driven SNS*, which allow cyber strangers to get connected in physical world. This new business model imposes challenges for event organizers to draw event plan and predict attendance. Intuitively, these services rely on the accurate estimation of users' preferences. However, due to various motivation of historical participation(i.e. attendance may not definitely indicate interests), traditional recommender techniques may fail to reveal the reliable user profiles. At the same time, motivated by the phenomenon that user may face to *conflict of invitation* (i.e. multiple invitations received simultaneously, in which only a few could be accepted), we realize that these choices may reflect real preference. Along this line, in this paper, we develop a novel conflict-choice-based model to reconstruct the decision-making process of users when facing to conflict. To be specific, in the perspective of *utility* in choice model, we formulate users' tendency with integrating content, social and cost-based factors, thus topical interests as well as latent social interactions could be both captured. Furthermore, we transfer the choice of conflict-choice triples into the pairwise ranking task, and a learning-to-rank based optimization scheme is introduced to solve the problem. Comprehensive experiments on real-world data set show that our framework could outperform the state-of-the-art baselines with significant margin, which validates the hypothesis that conflict and choice could better explain user's real preference.

Keywords: Choice model · Conflict-Choice triples · Social event · Social network

1 Introduction

Nowadays, it is commonly seen that an offline social event is organized through online social network services (SNS), in this way cyber and physical world could be connected as online strangers will now communicate face-to-face in real world. Thanks to the highly interactive experience, this new business model has become popular and attractive for millions of users all around the world, e.g., more than 9,000 groups organize new event in local communities every day at Meetup.com[1].

---

[1] http://www.meetup.com/.

© Springer International Publishing Switzerland 2016
S.B. Navathe et al. (Eds.): DASFAA 2016, Part I, LNCS 9642, pp. 396–411, 2016.
DOI: 10.1007/978-3-319-32025-0_25

This phenomenon raises new challenges for group leaders or event organizers to draw the event plan and predict the attendance, and intuitively, these analyses rely on the accurate estimation of users' preferences. Though a large amount of efforts have been made on summarizing users' historic participation, which follows the basic assumption that attendance may indicate preference. However, they may fail to describe the various motivations of users, e.g., people may attend some events only for killing leisure time, but it doesn't necessarily mean they indeed enjoy these events. Thus, new approach considering more comprehensive factors for user profiling is urgently required.

When analyzing historical event participation records, we realize that users may face to the situation of *invitation conflicts*, i.e., sometimes people may receive multiple invitation simultaneously, however, owing to the limitation, they could only select parts of them, while the rest should be rejected. Intuitively, these final decisions among conflicting invitations may better reflect users' real preference, e.g., from Meetup.com we found a programmer chose to attend a single party just on the same day with periodic iOS developing discussion, which indicates that he may be more inclined to attend such social activities. Indeed, the above example might not be occasional, and similar phenomenon could also be found in many other fields, like the alternative list in online shopping platform [10], or rating one another as "hot or not" in Facemash.com [16]. Motivated by this phenomenon, if we could precisely extract and analyze these **Conflict-Choice Triples**, i.e., pairwise conflicting choices (invitations) to be selected for one user, we could better understand users' real preference, and then effectively support related application, e.g., prediction or recommendation task.

Along this line, in this paper, we develop a novel conflict-choice-based model to reconstruct the decision-making process of users when facing pairwise invitations. To be specific, following the basic idea of choice model [18], we formulate users' tendency in the perspective of *utility* [20] with integrating users' preferences of event topics, social interaction and cost factors, thus comprehensive impacts have been captured. Furthermore, we transfer the choice of conflict-choice triples into the pairwise ranking task, thus a learning-to-rank based optimization scheme is introduced to learn users' preference. To the best of our knowledge, we are the first to discuss conflicting choice phenomenon in social event participation analysis and introduce the perspective of choice modeling for user profiling.

We conduct comprehensive experiments on real-world data set. The results show that our framework could outperform the state-of-the-art baselines with significant margin. Furthermore, to ensure the robustness and computational efficiency of our framework, we conduct parameter sensitiveness experiments and design an algorithm for optimizing model training time, which outcomes validate the hypothesis that conflicting choice could better explain user's real preference, and also confirm the application potential of our framework on social event participation analysis.

2   Conflict-Choices Model Formulation and Framework

In this section, we will introduce our novel conflict-choice-based model to reconstruct the decision-making process. According to our assumption, final decisions on conflicting choice triples could indicate users' real preference. Along this line, we first formally define the conflicting choice problem with related preliminaries, then, the conflict-choices model will be proposed with detailed technical solution. Finally, the two-stage framework will be illustrated.

2.1   Problem Statement

In this paper, we focus on the choice under conflicting invitation, which may indicate users' real preference. Specifically, we define and extract *Conflict-Choice Triples (CCT)* as follows:

Definition 1 (Conflict-Choice Triples (CCT)). *We define two events with corresponding user as a CCT if only when the following two conditions are satisfied simultaneously: (1) two invitation have been received within T days, and (2) two social events will be held within T days, where T is the periodic threshold to filter the triples. Finally, all the CCTs for a target user $\mathbf{u}$ is defined as $\mathbf{R_u}$.*

Intuitively, users' real preference could be reflected by the contrast between every event-pair from $\mathbf{R_u}$. To measure the contrast, we introduce the perspective of **Choice Utility** from *choice model*, then the problem of event participation can be defined as follows:

Definition 2 (Problem Statement). *Given a target user $\mathbf{u}$ and the set of $\mathbf{u}$'s conflict-choice triples $\mathbf{R_u} = \{r_i\}$, in which we use choice utility P_{u,e_k} to measure $\mathbf{u}$'s preference for each event $e_k \in r_i$. The problem of events participation prediction is to learn $\mathbf{u}$'s real preference by the contrast between pairwise events' choice utility from $\mathbf{R_u}$, and then utilize the real preference to analyse $\mathbf{u}$'s future participation decisions.*

In this paper, the choice utility P_{u,e_k} consists of *content-based utility C_{u,e_k}*, *social-based utility S_{u,e_k}* and *cost-based utility D_{u,e_k}*, the technical details of which will be introduced in Sect. 3.1. To define the notation of social connections, we construct social networks in which w_{uv} indicates the social influence strength from user u to user v. What should be noted is that the social influence strength w_{uv} will be trained in modeling. To describe topics distribution for each user, we exploit a vector $\mathbf{t}_u$ that will be learnt in training stage to indicate the preferences of user u, in which each dimension denotes the preference level on a specific aspect. Correspondingly, we have a vector $\mathbf{a}_k$ for each event e_k to indicate the attributes distribution, in which each dimension reflects the attribute on a specific aspect corresponding to $\mathbf{t}_u$. The mathematical notations used throughout this paper are summarized in Table 1.

Table 1. Mathematical Notations.

Symbol	Description
$\mathbf{u}$	the target user
$\mathcal{T}$	the periodic threshold
$\mathbf{E} = \{e_k\}$	the set of events
$\mathbf{R_u} = \{r_i\}$	the set of u's conflict-choice triple
$\mathbf{N_u}$	the set of u's total neighbors in the network
$\mathbf{N_{u,k}}$	the set of u's neighbors in e_k
$\mathbf{t_u}$	the profile vector for u
$\mathbf{a_k}$	the attributes vector for e_k
w_{uv}	social connection strength from user u to user v
C_{u,e_k}	u's content-based utility for e_k
S_{u,e_k}	u's social-based utility for e_k
D_{u,e_k}	u's cost-based utility for e_k
P_{u,e_k}	u's choice utility for e_k

2.2 Loss Function for Conflicting Choices

Now we turn to formulate the events participation prediction task. As mentioned above, the contrast between pairwise events' choice utility could reveal the actual preference of users, thus, we could intuitively treat this decision-making process as a pairwise ranking problem, i.e., rank the utility of two events in each conflicting event-pair. More specifically, we assume that users choosing one event of the conflicting event-pair is due to the pairwise ranking of *choice utility*, i.e., $P_{u,e_y} > P_{u,e_n}$. With the assumption above, we realize that correcting the partial ordering relation of *choice utility* in conflicting event-pairs will lead to optimal ranking results. Thus, the task of learning *choice utility* will be summarized as a pairwise ranking problem as follows:

Ranking Objective. By correcting the partial ordering relation of *choice utility* in conflicting event-pairs, we will get the appropriate *choice utility* P_{u,e_k}. To deal with this task, we formulate the loss function of pairwise ranking problem as follows:

$$\min_{w,\mathbf{t_u}} F(w, \mathbf{t_u}) = \sum_{r_i \in R_u} \sum_{e_y,e_n \in r_i} h(P_{u,e_n} - P_{u,e_y}), \quad (1)$$

where $h(x)$ is a loss function to assign a non-negative penalty according to the difference of choice utility $P_{u,e_n} - P_{u,e_y}$. Usually, we have the penalty $h(x) = 0$ when $P_{u,e_n} \leq P_{u,e_y}$. While for $P_{u,e_n} > P_{u,e_y}$, we have $h(x) > 0$ as loss. To ease the computation, here we utilize the squared loss function as follow:

$$h(x) = \max\{x, \varepsilon\}^2, \quad (2)$$

where ε presents the margin allowed for choice utility loss. To ensure accuracy of the results we set $\varepsilon = 0$, so $h(x)$ could also be rewrote as:

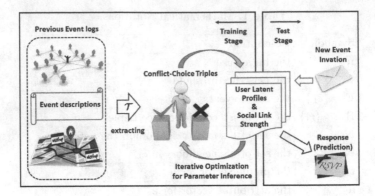

Fig. 1. Overview of our framework for event participation prediction.

$$\sum_{e_y, e_n \in r_i} h(P_{u,e_n} - P_{u,e_y}) = \sum_{e_y, e_n : P_{u,e_n} > P_{u,e_y}} (P_{u,e_n} - P_{u,e_y})^2. \qquad (3)$$

With this formulation, we could optimize the loss function to estimate social connection strength w_{uv} and users' profile vector $\mathbf{t_u}$. Simultaneously, such training stage would highlight the difference between the two events, which may be the real preference contributing to the final choice.

2.3   Two-Stage Framework

Based on the above preliminaries, now we can formally present the overview of the two-stage framework for event participation prediction. Specifically, Fig. 1 demonstrates the overview of our framework.

Training Stage. Given a target user $\mathbf{u}$ and his/her historical events $\mathbf{E_{train}} = \{e_k\}$, in which participation record (attendance/absence) sorted by time for each e_k are pre-known, so we could extract the set of u's conflict-choice triple, namely $\mathbf{R_u}$. Also, we have the event attributes $\mathbf{a}_k$ for each e_k and the connection between $\mathbf{u}$ and his/her neighbors, while the strength $\{w_{uv}\}$ are unknown. In this stage, we aim at inferring the choice utility P_{u,e_k} for each e_k of u, as well as learning the connections strength $\{w_{uv}\}$ and users' profile vector $\mathbf{t_u}$.

Test Stage. After obtaining the social connections strength $\{w_{uv}\}$ and users' profile vector $\mathbf{t_u}$ in the training stage, in the test stage, given a target user and a set of event $\mathbf{E_{test}}$ with attributes $\mathbf{a}_k$ and the corresponding social network neighbors, we aim at predicting event participation for all e_k in $\mathbf{E_{test}}$.

3   Technical Details for Prediction of Event Participation

In this section, we introduce the technical details for event participation prediction, including the detailed technical solutions for choice utility and optimization task of our framework.

3.1   Choice Utility

Here, we first introduce how to simulate a user's choice utility to an event. Intuitively, the user's choice utility should be the combination of *content-based utility* (C_{u,e_k}), *social-based utility* (S_{u,e_k}) and *cost-based utility* (D_{u,e_k}), so we choose to multiply them to formulate the user's choice utility in decision-making process. To be specific, the choice utility of $\mathbf{u}$ to e_k will be estimated as follows:

$$P_{u,e_k} = C_{u,e_k} \cdot S_{u,e_k} \cdot D_{u,e_k}. \tag{4}$$

What should be noted is that we have not set weight for each factor, but in fact the weight for each factor would redistribute spontaneously during the parameters learning in the training stage of our framework.

Content-Based Utility. Intuitively, event's description is usually an important factor of users to attend an event or not. To measure users' tendency to the events' topic, we borrow the classic *Cosine similarity* between user profile vector and event description vector to indicate the content-based utility, as users' biography and events' descriptions could be easily normalized and presented in vectors. To be specific, content-based utility will be estimated as follows:

$$C_{u,e_k} = cosine(\mathbf{t_u}, \mathbf{a_k}) = \frac{\mathbf{t_u} \bullet \mathbf{a_k}}{\|\mathbf{t_u}\|\|\mathbf{a_k}\|}, \tag{5}$$

where $\mathbf{t_u}$ is the profile vector for u that will be learnt in training stage and $\mathbf{a_k}$ is the attributes vector for e_k learnt by LDA model in our framework.

Social-Based Utility. Second, the "word-of-mouth" effect is verified that could strongly affect the decision-making process of social event participation, and at least 10%–30% of human movement could be explained by social factors [5,22,23]. So it is reasonable to investigate the social impact on social event participation, and further, the effects during conflicting choice process. To formulate the encouragement, we borrow and adapt the classic Independent Cascade (IC) model [8] for simulating the interactional influence within users, which is widely used and its effectiveness has been well proved. To be specific, *social-based utility* will be estimated as follows:

$$S_{u,e_k} = 1 - \prod_{v \in N_{u,k}} (1 - w_{vu}), \tag{6}$$

where $\mathbf{N_{u,k}}$ is the set of neighbors of $\mathbf{u}$ who attend e_k.

Cost-Based Utility. Finally, the experimental results in [11,21] inspire us to study the influence of *cost-based utility* on an individual user's event participation prediction. We apply a general nonparametric technique, known as the kernel density estimation [17] (KDE), which is widely used to estimate a probability density function of an unknown variable based on a known sample. In our case, X_u is the known sample and y is denoted as the unknown variable. The probability density function of variable y using sample X_u is given by:

$$D_{u,e_k} = \frac{1}{|X_u|\sigma} \sum_{x \in X_u} K(\frac{y-x}{\sigma}), \tag{7}$$

where $|X_u|$ is the number of sample points in X_u, σ is a smoothing parameter called bandwidth and $K(\cdot)$ is the kernel function. To ease the modeling, we apply the normal kernel, which has been widely used in related studies.

3.2   Optimization Task

As all the formulations established, finally we could discuss about the optimization task of loss function Eq. 1. To be specific, we first approach the social connection strength w by deriving the gradient of $F(\cdot)$ with respect to w_{uv} and approach the users' profile vector $\mathbf{t_u}$ by deriving the gradient of $F(\cdot)$ with respect to t_u^m, and then use a gradient based optimization method to find proper w and $\mathbf{t_u}$ that minimize $F(\cdot)$. Specially, as defining $\gamma_{e_n e_y} = P_{u,e_n} - P_{u,e_y}$, then we have the derivative as follow:

$$\frac{\partial F(w, \mathbf{t_u})}{\partial w_{vu}} = \sum_{e_y, e_n : P_{u,e_n} > P_{u,e_y}} \frac{\partial h(\gamma_{e_n e_y})}{\partial \gamma_{e_n e_y}} (\frac{\partial P_{u,e_n}}{\partial w_{vu}} - \frac{\partial P_{u,e_y}}{\partial w_{vu}}), \tag{8}$$

$$\frac{\partial F(w, \mathbf{t_u})}{\partial t_u^m} = \sum_{e_y, e_n : P_{u,e_n} > P_{u,e_y}} \frac{\partial h(\gamma_{e_n e_y})}{\partial \gamma_{e_n e_y}} (\frac{\partial P_{u,e_n}}{\partial t_u^m} - \frac{\partial P_{u,e_y}}{\partial t_u^m}), \tag{9}$$

where t_u^m is the mth dimension of $\mathbf{t_u}$ and $h'(\gamma_{e_n e_y})$ could be easily achieved as derivation of square function:

$$\frac{\partial h(\gamma_{e_n e_y})}{\partial \gamma_{e_n e_y}} = 2 \cdot (P_{u,e_n} - P_{u,e_y}). \tag{10}$$

For the social connection strength w and users' profile vector $\mathbf{t_u}$, we have:

$$\frac{\partial P_{u,e_k}}{\partial w_{vu}} = C_{u,e_k} \cdot \prod_{x \in N_{u,k}, x \neq v} (1 - w_{xu}) \cdot D_{u,e_k}, \tag{11}$$

$$\frac{\partial P_{u,e_k}}{\partial t_u^m} = \frac{a_k^m \cdot \|\mathbf{t_u}\|^2 - t_u^m \cdot \mathbf{t_u} \bullet \mathbf{a_k}}{\|\mathbf{t_u}\|^3 \|\mathbf{a_k}\|} \cdot S_{u,e_k} \cdot D_{u,e_k}, \tag{12}$$

where after each iterative round $\mathbf{t_u}$ will be normalized. To deal with the optimization task, the gradient descent methods could be exploited.

4   Experiments and Discussions

To verify our hypothesis that the choice utility affects the decision making process of potential event participants, in this section, we conduct experiments on a real-world data set to measure the event participation predicting performance with conflict-choice model. Furthermore, some representative case studies and discussion will be presented.

4.1   Experimental Setup

Data Set Pre-processing. Our experiments were conducted on the real-world data set crawled via official APIs of Meetup.com. Specially, we crawled event logs totally includes 625 groups, 50,719 social events and 99,854 related users. For details, event descriptions (e.g., location and time), participation records (attendance/absence) and user profiles are extracted.

To describe the events' attributes, we exploited the key words in the group descriptions and user profiles. 2,856 key words (or terms) with unique ID (defined by Meetup) were collected in the dictionary in total, and Latent Dirichlet Allocation (LDA) model [2] was introduced to learn the topics. Specifically, we select 20 latent topics, as Meetup system defines 34 categories of events, and majority of events focus on around 20 types which is reflected by the data set. Finally, all descriptions are presented as a 20-dimensional attribute vectors.

In offline social event scenario, we intuitively assume the distance between user's home and event location as a geographical cost factor. What should be noted is that *cost-based utility* has the potential of integrating more cost factors, e.g., weather and road condition information, by introducing multivariate kernel density estimator.

Evaluation Baselines. For more comprehensive comparisons, several state-of-the-art baselines based on different assumption are selected as follows.

(1) **Discrete Choice Model (DCM)** [20]. Discrete Choice Model (DCM) is used to predict choices between multiple discrete alternatives in economics. We utilize the DCM method as baseline, which integrates the same content and cost factors, while we utilize the number of co-occurrence members as social feature.

(2) **RankNet (RKN)** [3]. RankNet is a widely used pairwise learning-to-rank (LTR) algorithm using neural network to model underlying ranking function, which utilizes gradient descent methods for learning ranking probabilistic cost functions. As our conflict-choice model is intrinsically a pairwise ranking problem, we use RankNet as a baseline, in which we use same features with **DCM**.

(3) **LambdaMART (LAM)** [4]. LambdaMART is the boosted tree version of LambdaRank, which defines the gradient of the loss function in order to solve the problem that sorting loss function could hardly be optimized. We select it as baseline since it is among the best learning-to-rank (LTR) algorithms, in which we use same features with **DCM**.

(4) **Information Spreading** [8]. As we try to reveal latent social interactions to describe users' real preference in conflicting choices, to better validate this assumption, we conduct social-spread-based model to study whether attendance is indeed the result of "word-of-mouth" effect. Since Meetup.com ignores point-to-point connection, we construct the social connections following the common used heuristic method like in [11] that edges could be added if two people have attended the same event, and two widely studied heuristic methods are selected to set the connection strength w_{uv} as

Table 2. Overall performance of each approach.

	CCT	DCM	RKN	LAM	ISO	ISN
MAP	**0.8513**	0.7683	0.8069	0.8299	0.6980	0.6699
Improvement(%)	-	10.788	5.5003	2.5826	21.968	27.066
P-Value	-	0.0000	0.0091	0.0387	0.0000	0.0000
F1 score	**0.8016**	0.7530	0.6885	0.7050	0.6249	0.5996
Improvement(%)	-	6.4542	17.873	15.130	29.859	35.283
P-Value	-	0.0000	0.0000	0.0000	0.0000	0.0000

(1) the co-occurrence frequency (**ISO**), and (2) the Jaccard Index of common neighbors (**ISN**). Then, classic Independent Cascade (IC) model [8] will be conducted to simulate the spread process. To ensure the stable results, we repeat experiments for 500 times for each test.

4.2   Experiment Results

Due to the group-based scheme of Meetup, we treat *user-group pair* as the unit of our experiments. To be specific, for one target user in a target group, we will conduct a set of experiments, and the average results are presented as the finals.

Since we face to the severe sparse data that only less than 20 % users attended at least 5 events in a group, we assign 80 % events within one group as training samples to ensure the quality of training, while the rest 20 % are test samples. The samples are processed in time order to keep the rule of social group evolution.

As mentioned in test stage, to predict the participation, we indeed have two tasks, i.e., ranking the attendance probability with respect to their choice utility and then binary classifying to distinguish attendance/absence of participation. For each task, related metrics will be selected to measure the performance. For the ranking task, similar with the state-of-the-art learn to rank problems, **MAP** [19] is selected. For the binary classification task, typically, we select the common used **F1 score** for validation, which is a measure that combines precision and recall, namely the harmonic mean of precision and recall.

Comparison of Overall Performance. First of all, we show the overall prediction performance of our approach comparing with different baselines and the results are shown in Table 2. According to the results, we can find that our approach outperforms the other baselines with dramatic margin in **MAP** and **F1 score**, even 35 % better in some experiments. The performance highly supports our assumption that with introducing the conflicting choice utility, we could better estimate the event participation.

As expected, DCM methods performs better for binary classification, while RKN and LAM methods performs better in ranking task, which is determined by the algorithm internal mechanism. At the same time, it seems that the overall results of the DCM, RKN and LAM methods are worse than CCT. These baseline methods just make use of some statistics metrics, i.e., $|\mathbf{N}_{u,k}|$ and distance, but ignore the latent social interactions as well as probability density function of cost factors. Further, users' profile vector $\mathbf{t_u}$ are learned by LDA for these three baseline methods, which might not be enough because most people would not record all their interests in the home page. However we could train $\mathbf{t_u}$ in our CCT framework, which might be another reason.

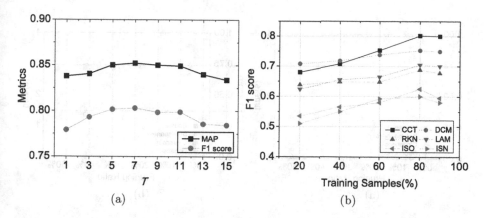

(a)   (b)

Fig. 2. Parameter Sensitiveness. (a) Prediction performance with different T (b) Prediction F1 score with different partition of training samples.

Finally, we surprisingly find that the baseline with information spreading methods, i.e., ISO and ISN, achieves the worst performance. Indeed, though preference factors are integrated between pairwise users, the information spreading methods still follows the essentially different assumption with the other three algorithms. Specifically, information spreading methods assumes the participation is mainly affected by the friends or opinion leaders' spread but not their own preference, which might not be reasonable enough. Information spreading methods ignoring *content-based utility* might be another reason. Also, the cold-start problem, which leads to insufficient pairwise interactions and sparse social network, may further impair the performance.

Evaluation on Parameter Sensitiveness. As the performance has been validated, in this subsection, we conduct the experiments for evaluating the parameter sensitiveness of our approach. In this task, there are two parameters concerned in our approach, i.e., the *periodic threshold* T, as well as the sample allocation ratio.

For the *periodic threshold* T, as mentioned in Sect. 2.1, we utilize T to describe the conflicting choice situation, thus a lower T might be better for approximation, because users face to sharper conflicting events. However, as Fig. 2(a) shows that performance achieves the peak when T is around 7 to 10 days, but not the lower the better. The reason of this phenomenon not only might be lower $\mathcal{T}$ restricts the number of conflict choice triples that covers user actual utility, but also might be the persistence of users' preference, namely users would not change their preference significantly in a short time. So even when they do not face to very sharp conflict-choice events, they also prefer to attend the events with high choice utility but reject the events with low choice utility. And this phenomenon might further indicate that most active users attend events not more than once a week.

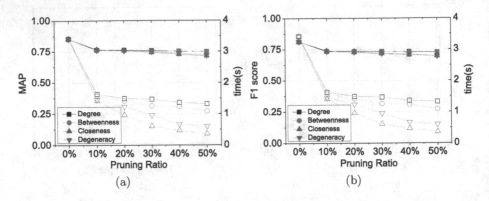

Fig. 3. Performance of Network Pruning. (a) MAP (b) F1 score.

Then, we discuss about whether the partition of training samples will influence the results, which is summarized in Fig. 2(b). We find that our framework performance improves rapidly when the partition of training samples increases, which indicates that our model is sensitive to the number of training triples. The reasons might be that we aim at predicting the events participation using social connections strength, thus it is required that most important connections strength have been trained. The methods that depend on connections strength, i.e., ISO and ISN, are sensitive to the train samples ratio for the same reason, too. On the contrary, the DCM, RKN and LAM methods keep in stable level during the train samples ratio change, since they just make use of some social statistics metrics.

Network Pruning to Optimize Training Time. As mentioned in Sect. 3.2, we use gradient descent methods to deal with the optimization task in model training stage. Specifically, we approach the social connection strength w_{uv} by first deriving the gradient of $F(\cdot)$ with respect to w_{uv}, and then use a gradient based optimization method to find proper w that minimize $F(\cdot)$, which is a time-consuming process, because the loss function iterates rounds to convergence and traverses all the connections in every round.

It is common to see that a user would not recognize all the members of every event she/he has ever attended, and the inactive neighbors of social network,

Table 3. Examples for Case Study

Precision	Sample A		Sample B		Sample C		Sample D	
	100 %		100 %		100 %		55 %	
Participation	Attend	Absent	Attend	Absent	Attend	Absent	Attend	Absent
Topic Sim	0.388	0.193	0.351	0.407	0.702	0.766	0.791	0.818
Members	16.50	17.50	7.750	2.750	11.57	11.67	51.25	45.22
Distance	5.376	5.381	11.06	10.98	1.889	5.112	18.56	11.42

such as freshers or social inactive members, are usually useless in the prediction process. So we design an algorithm for optimizing model training time by deleting the inactive neighbors of the social network. More specifically, we choose some appropriate metrics to ranking nodes in the social network, and then prune the marginal nodes. In network analysis, metrics of centrality identify the most influential persons in a social network, so we use some centrality metrics to simplify the social network by pruning nodes performing worse centrality. Here we select the widely used centrality metrics such as Degree, Betweenness, Closeness and Degeneracy centrality [1,7,14,15].

Finally, we discuss about whether the network pruning algorithm will significantly decrease the train time and how it influences the participation prediction, which is summarized in Fig. 3, in which solid symbols with solid line mean prediction performance, while hollow symbols with imaginary line mean training time. To be fair, the train time of network simplification algorithm is the sum of sorting nodes time and model training time. From the figure, we can clearly find that social network pruning could successfully improve the efficiency, while at the same time maintain relatively acceptable accuracy. And prediction performance does not degenerate when further simplify the network, the reason is that usually the actual important friends of a user are not much. Besides, we find that Closeness centrality preforms most significantly in improving efficiency.

4.3   Case Study

To better understand the performance, i.e., how the conflicting choice could reveal users' real preference, we randomly select four users as examples. Correspondingly, related social metrics of their attendance/absentee are listed. Details are shown in Table 3. Two key issues should be studied here: (1) whether conflict-choice-based model keeps working well for users with different types of utility, and (2) how the social-based utility could be summarized.

For the first issue, three types of potential participators should be carefully observed, namely the users who pay more attention to the three kinds of factors respectively. For the former three users, namely user A, B and C, we realize that user A pay more attention to content-based factors because this user prefer to attend events with higher topic similarity, and user B is a sociable user who chooses to attend events which more people attend, while user C is more likely to attend the nearby events. Besides, we find that these three typical users' participation prediction precision are 100 %, which is an intuitional evidence that our conflict-choice model is widely available.

On the contrary, for Sample D who suffers poor precision, we find that the group usually host large-scale events. With deep looking into the data, we realized that this group suffer "cold-start" problem, i.e., former members quitting and new ones coming, so social connection strength learned in training stage could not be used in participation prediction process. This phenomenon implies that stable group with strongly connections will lead to better prediction, which also supports our hypothesis of social effects.

Secondly, we discuss about the type of social-based influence. In our analysis, we set the reciprocal of attenders' amount as threshold, i.e., if connection strength passes the threshold, we treat the neighbor as *"close friend"*. We find two typically types of social-based influence, i.e., authority influence and group influence. Authority influence is the phenomenon that the target user is mostly influenced by one active member, such as event organizer. Group influence is the phenomenon that the target user is influenced by a group of people, e.g., we find a user and his 9 friends form to small community in the group, members in this community prefer to attend events with each other.

Finally, we discuss the derivative application of case studying. By illustrating the representative users above, we could find some typical patterns of all the users and events organizers can attract the right attendants and predict the attendance according to it. For instance, for users in a small community of the group, if a certain proportion members in the community accept the RSVP, we recognize that the rest of members in the community prefer to attend the event. By introducing such rules above, we could decrease predicting process time and revise the prediction results.

5   Related Work

In this section, we briefly introduce the related works of our study. In general, the related works can be mainly grouped into two categories.

The first category related to this paper is the social event recommendation, which is different with the traditional items recommendation. Specifically, some researchers focused on the conformity between users' profiles and event attributes. For example, [9] proposed a hybrid event recommender that is enriched with linked open data and content information. Furthermore, a method for recommendation by collaborative ranking of future events based on users' preferences for past events is describe in [13]. And some works focus on recommendation to a group of members, [12] proposed a personal impact topic model to enhance the group preference profile by considering the personal preferences and personal impacts of group members. Finally, there are some related works focused on other practical problems. For example, a smartphone application developed by [6] recommend events according to the users Facebook profiles.

The second category is about conflicting choice utility. In this paper, we deeply analyze events participation prediction with considering conflict choice and choice utility. Indeed, plenty efforts have been made on understanding choice model which usually predict choices between two or more discrete alternatives [18] and have been widely examined in many fields, e.g., in economics peoples choose which product to buy in online shopping platform [10]. The other topic closely related to this category is choice utility, which is a representation of preference over a set of alternatives [20]. Choice utility also usually be introduced to model the situation that users face to competitive choice, e.g., authors explored the conflicting choosing process of user behavior when facing with recommendations by adopting utility theory in [24]. However, although the works mentioned above can reappear the process of people choosing and making decision, they still may suffer some defects due to they ignore the mutual influences among people.

6 Conclusion

In this paper, we investigate how people make decisions when facing to conflicting invitations, which may reflect users' real preference. Following this assumption, we propose a novel conflict-choice-based model for better reconstructing users' decision. To be specific, we formulate users' tendency with integrating content-based utility, social-based utility and cost-based utility in the perspective of choice utility, and then transfer the choice of conflict-choice triples into the pairwise ranking task to learn the model, thus the optimization goal is formulated and solved as a ranking-based loss function. At the same time, the latent social interactions within potential attenders and their topical interests will also be revealed. Comprehensive experiments on real-world data set show that our framework could outperform the state-of-the-art baselines with significant margin, which validates the hypothesis that conflict and choice could better explain user's real preference.

Though significant performance has been achieved, as the social parameters learned might be rough to reveal latent interactions, in the future, we will target at designing more complicated scheme to describe the social-based utility, especially to extend the point-to-point interaction to the superimposed effect of multiple attenders or even little community. Also, we would like to exploit more applications of the proposed method instead of only social event participation analysis, which may further validates the applicable potential of our novel framework.

Acknowledgments. This research was partially supported by grants from the National Science Foundation for Distinguished Young Scholars of China (Grant No. 61325010), the National High Technology Research and Development Program of China (Grant No. 2014AA015203), the Natural Science Foundation of China (Grant No. 61403358), the Anhui Provincial Natural Science Foundation (Grant No. 1408085QF110) and the MOE-Microsoft Key Laboratory of USTC. Qi Liu gratefully acknowledges the support of the CCF-Tencent Open Research Fund.

References

1. Bader, G.D., Hogue, C.W.: An automated method for finding molecular complexes in large protein interaction networks. BMC Bioinformatics **4**(1), 2 (2003)
2. Blei, D.M., Ng, A.Y., Jordan, M.I.: Latent Dirichlet allocation. J. Mach. Learn. Res. **3**, 993–1022 (2003)
3. Burges, C., Shaked, T., Renshaw, E., Lazier, A., Deeds, M., Hamilton, N., Hullender, G.: Learning to rank using gradient descent. In: Proceedings of the 22nd International Conference on Machine Learning, pp. 89–96. ACM (2005)
4. Burges, C.J.: From ranknet to lambdarank to lambdamart: an overview. Learning **11**, 23–581 (2010)
5. Cho, E., Myers, S.A., Leskovec, J.: Friendship and mobility: user movement in location-based social networks. In: Proceedings of the 17th ACM SIGKDD International Conference on Knowledge Discovery and Data Mining, pp. 1082–1090. ACM (2011)
6. De Pessemier, T., Minnaert, J., Vanhecke, K., Dooms, S., Martens, L.: Social recommendations for events. In: RSWeb@ RecSys (2013)
7. Freeman, L.C.: A set of measures of centrality based on betweenness. Sociometry pp. 35–41 (1977)
8. Kempe, D., Kleinberg, J., Tardos, É.: Maximizing the spread of influence through a social network. In: Proceedings of the Ninth ACM SIGKDD International Conference on Knowledge Discovery and Data Mining, pp. 137–146. ACM (2003)
9. Khrouf, H., Troncy, R.: Hybrid event recommendation using linked data and user diversity. In: Proceedings of the 7th ACM conference on Recommender Systems, pp. 185–192. ACM (2013)
10. Liu, Q., Zeng, X., Liu, C., Zhu, H., Chen, E., Xiong, H., Xie, X.: Mining indecisiveness in customer behaviors. In: 2015 IEEE 15th International Conference on Data Mining (ICDM), pp. 281–290. IEEE (2015)
11. Liu, X., He, Q., Tian, Y., Lee, W.C., McPherson, J., Han, J.: Event-based social networks: linking the online and offline social worlds. In: Proceedings of the 18th ACM SIGKDD International Conference on Knowledge Discovery and Data Mining, pp. 1032–1040. ACM (2012)
12. Liu, X., Tian, Y., Ye, M., Lee, W.C.: Exploring personal impact for group recommendation. In: Proceedings of the 21st ACM International Conference on Information and Knowledge Management, pp. 674–683. ACM (2012)
13. Minkov, E., Charrow, B., Ledlie, J., Teller, S., Jaakkola, T.: Collaborative future event recommendation. In: Proceedings of the 19th ACM International Conference on Information and Knowledge Management, pp. 819–828. ACM (2010)
14. Newman, M.E.J.: The structure and function of complex networks. SIAM Rev. **45**(2), 167–256 (2003)
15. Sabidussi, G.: The centrality index of a graph. Psychometrika **31**(4), 581–603 (1966)
16. Schwartz, B.: Hot or not? website briefly judges looks. Harvard Crimson (2003). http://www.thecrimson.com/article.aspx
17. Silverman, B.W.: Density Estimation for Statistics and Data Analysis. CRC Press, Boca Raton (1986)
18. Train, K.E.: Discrete Choice Methods with Simulation. Cambridge University Press, Cambridge (2009)

19. Turpin, A., Scholer, F.: User performance versus precision measures for simple search tasks. In: Proceedings of the 29th Annual International ACM SIGIR Conference on Research and Development in Information Retrieval, pp. 11–18. ACM (2006)
20. Von Neumann, J., Morgenstern, O.: Theory of Games and Economic Behavior (60th Anniversary Commemorative Edition). Princeton University Press, Princeton (2007)
21. Wang, C., Ye, M., Lee, W.C.: From face-to-face gathering to social structure. In: Proceedings of the 21st ACM International Conference on Information and Knowledge Management, pp. 465–474. ACM (2012)
22. Xu, T., Liu, D., Chen, E., Cao, H., Tian, J.: Towards annotating media contents through social diffusion analysis. In: 2012 IEEE 12th International Conference on Data Mining (ICDM), pp. 1158–1163. IEEE (2012)
23. Xu, T., Zhong, H., Zhu, H., Xiong, H., Chen, E., Liu, G.: Exploring the impact of dynamic mutual influence on social event participation. In: Proceedings of 2015 SIAM International Conference on Data Mining, pp. 262–270. SIAM (2015)
24. Yang, S.H., Long, B., Smola, A.J., Zha, H., Zheng, Z.: Collaborative competitive filtering: learning recommender using context of user choice. In: Proceedings of the 34th International ACM SIGIR Conference on Research and Development in Information Retrieval, pp. 295–304. ACM (2011)

Semantics Computing and Knowledge Base

PBA: Partition and Blocking Based Alignment for Large Knowledge Bases

Yan Zhuang[1,2], Guoliang Li[1(✉)], Zhuojian Zhong[3], and Jianhua Feng[1]

[1] Tsinghua University, Beijing 100084, China
zhuang-y14@mails.tsinghua.edu.cn, {liguoliang,fengjh}@tsinghua.edu.cn
[2] PLA Navy General Hospital, Beijing 100048, China
[3] Beijing University of Posts and Telecommunications, Beijing, China
ibmzzjn@bupt.edu.cn

Abstract. The vigorous development of semantic web has enabled the creation of a growing number of large-scale knowledge bases across various domains. As different knowledge-bases contain overlapping and complementary information, automatically integrating these knowledge bases by aligning their classes and instances can improve the quality and coverage of the knowledge bases. Existing knowledge-base alignment algorithms have some limitations: (1) not scalable, (2) poor quality, (3) not fully automatic. To address these limitations, we develop a scalable partition-and-blocking based alignment framework, named PBA, which can automatically align knowledge bases with tens of millions of instances efficiently. PBA contains three steps. (1) Partition: we propose a new hierarchical agglomerative co-clustering algorithm to partition the class hierarchy of the knowledge base into multiple class partitions. (2) Blocking: we judiciously divide the instances in the same class partition into small blocks to further improve the performance. (3) Alignment: we compute the similarity of the instances in each block using a vector space model and align the instances with large similarities. Experimental results on real and synthetic datasets show that our algorithm significantly outperforms state-of-art approaches in efficiency, even by an order of magnitude, while keeping high alignment quality.

1 Introduction

With the rapid development of semantic web in the last decade, especially the promotion from the Link Open Data (*LoD*) project [4], semantic-web data has reached a considerable scale. A growing number of large-scale knowledge bases have been created across different domains (e.g., movies[1], publications[2], Biomedical Sciences[3]), thanks to the semantic web. Knowledge bases have many real-world applications, such as question answering [8], machine reading [15], knowledge support [2] and semantic search [1].

---

[1] http://www.douban.com/.
[2] http://dblp.uni-trier.de/db/.
[3] http://geneontology.org/.

© Springer International Publishing Switzerland 2016
S.B. Navathe et al. (Eds.): DASFAA 2016, Part I, LNCS 9642, pp. 415–431, 2016.
DOI: 10.1007/978-3-319-32025-0_26

There are many knowledge bases, e.g., *DBPedia* [12], *Freebase* [5], *YAGO* [20], generated from different organizations, and they contain overlapping and complementary information. It is important to integrate different knowledge bases to improve the quality and coverage of knowledge bases. There have been extensive studies in integrating them by aligning their common elements [9,10,13,19]. We can broadly classify existing studies into two categories. (1) Class Alignment. Traditional knowledge base alignment approaches focus on finding correspondences between classes, e.g., location and region, movie and film. (2) Instance Alignment. As the number of instances grows rapidly with the development of *LoD* project, the instance alignment attracts increasing attention, which finds correspondences between instances, e.g., *Beijing* and *Peking*, *Napoleon Bonaparte* and *Napoleon I*. As the number of instances (usually tens of millions) is usually larger than that of classes (usually several thousands), instance alignment is more challenging than class alignment, and recent studies focus on instance alignment. In this paper, we also study the instance alignment problem.

However, existing instance alignment methods have three limitations. (1) Not Scalable. As modern knowledge bases are becoming larger and larger, it calls for scalable algorithms. (2) Poor Quality. The integration quality is an important factor that needs to be considered in algorithm design, and efficient algorithms cannot be designed at the expense of precision and recall. (3) Not Automatic. A fully automatic algorithm is more attractive to handle large-scale knowledge bases.

To address these challenges, we propose a scalable knowledge bases alignment framework, named PBA. It digs out the main features from knowledge bases and utilizes them to enable high-quality matchings in three levels—partitions the classes into moderate size clusters, blocks the instances in each clusters by their properties, and aligns the instances in the same block. Our approach can address the three limitations and have major advantages. Firstly, PBA is scalable and can support large-scale datasets, because we only need to consider the data in the same blocks and prune those in different blocks, which significantly reduce the computation cost. Secondly, PBA has good quality, as the data that can be aligned are usually assigned into the same block. Lastly, our method is a fully automatic method.

To summarize, we make the following contributions. (1) We propose a scalable alignment framework that can automatically align millions of instances in knowledge bases. (2) We propose a new partition algorithm to divide large knowledge bases into small ones, which reduces the alignment scale and improves the performance significantly. (3) We design an efficient blocking algorithm to divide instances into different blocks. (4) We devise an alignment algorithm which aligns the data with large similarities. (5) We have conducted extensive experiments on real and synthetic datasets. Experimental results show that our algorithm significantly outperforms existing methods in efficiency, even by an order of magnitude, while keeping high alignment quality.

The rest of this paper is organized as follows. In Sect. 2, we introduce preliminaries and then present the overview of our PBA framework in Sect. 3.

Sections 4, 5 and 6 discuss the partition, blocking and alignment algorithms. Experimental results are illustrated in Sect. 7. Finally, we conclude the paper in Sect. 8.

2 Preliminaries

2.1 Problem Formulation

We use the Resource Description Framework Schema (*RDFS*[4]), which is the W3C standard for knowledge representation, to define knowledge bases. In the *RDFS* model, a fact in a knowledge base is a *RDF* triple consisting of three components: subject, predicate, object (abbreviated as *SPO*). The subject is an instance which is represented by a uniform resource identifier (*URI*) to denote a real-world entity or a class which is a category of instances with the same type. The object is a class, instance, or literal which denotes (1) the category the subject belongs to, (2) the related real-world object, or (3) a string, date or number to describe the subject. The predicate is a property or relation which is a binary relation that holds between two instances or an instance and a literal. Different classes can be connected by subclass relation which denote one class is a subset of another. The subclass relations can construct a hierarchy structure. Formally, a knowledge base can be modeled as an octuple, defined as below.

Definition 1 (Knowledge Base). *A* KNOWLEDGE BASE *(KB) is an octuple* $(\mathcal{C}, \mathcal{I}, \mathcal{L}, \mathcal{R}, \mathcal{P}, \mathcal{FR}, \mathcal{FP}, \mathcal{H})$ *where* $\mathcal{C}, \mathcal{I}, \mathcal{L}, \mathcal{R}, \mathcal{P}$ *are sets of classes, instances, literals, relations and properties respectively.* $\mathcal{FR} \subseteq \mathcal{I} \times \mathcal{R} \times \mathcal{I}$ *is a SPO triple set of relation-facts where the object is another instance called objectype property value, and* $\mathcal{FP} \subseteq \mathcal{I} \times \mathcal{P} \times \mathcal{L}$ *is a SPO triple set of property-facts where the object is a literal called datatype property value.* $\mathcal{H} \subset \mathcal{C} \times \mathcal{C}$ *denotes the subclass relation between two classes.*

Example 1. *Figure 1 is a toy example with two KBs. The circle and rectangle nodes represent instances and classes respectively. KB_2 contains a SPO* ⟨*Robert Downey,bornIn, USA*⟩ *which describes subject instance* Robert Downey *and object literal* USA *have a relationship* bornIn. *Subject instance* Robert Downey *and object instance* The Avengers II *have a relationship* actedIn. *Triple* ⟨*Robert Downey,typeof,Actor*⟩ *describes instance* Robert Downey *belongs to class* Actor. *The* subclassof *relation between classes depicts the hierarchical structure of classes.*

Because of the diversity and heterogeneity of *KBs*, two instances from different *KBs* may refer to the same real-world object. The *KB* alignment process is to find the matched instance pairs across *KBs*. We formalized the process as follows.

---

[4] http://www.w3.org/TR/rdf-primer/.

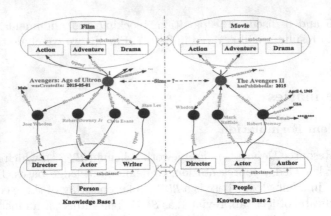

Fig. 1. A motivation example.

Definition 2 (KB ALIGNMENT). *Given two knowledge bases KB_1, KB_2 and a set of matched object pairs Γ, a KB alignment process is to find the corresponding instance pairs in KB_1 and KB_2 with certain confidence. The matching results can be represented as:*

$$\mathcal{A}_I(KB_1, KB_2, \Gamma) = \{(\mathcal{I}_1, \mathcal{I}_2, \phi) | \mathcal{I}_1 \in KB_1, \mathcal{I}_2 \in KB_2, \text{SIM}(\mathcal{I}_1, \mathcal{I}_2) > \phi, \phi \in (0, 1]\}$$

where $\mathcal{I}_1$ and $\mathcal{I}_2$ are the instances from KB_1 and KB_2, Γ is defined as prior alignment data generated by KB_1 and KB_2 automatically, ϕ is the similarity score as the alignment confidence, and the higher the value is, the more reliable the alignment is.

2.2 Related Work

Traditional *KB* alignment approaches focused mostly on aligning the classes, such as RIMOM [13], *COMA++* [7] and *Falcon* [9]. Different from our method, these approaches can only align classes and do not consider the alignment of instances.

There have been several approaches dealing with large-scale instance alignment problem of *KB* in recent years. SIGMA [10] is an iterative propagation algorithm which leverages both the relations and properties between instances to align *KB* in a greedy way, which collectively and jointly aligns the instances rather than independently. However, SIGMA needs to manually select the related relationships and properties and thus is a semi-automatic method. VMI [14] directly uses the vector space model to generate multiple vectors for different kinds of information contained in the instances, and uses a set of inverted indexes based rules to get the primary matching candidates. The similarities of matching candidates are computed as the integration of all the vector distances. However, the better result can be achieved when all the matching properties and the ways of fetching values are specified by users. ARIA [11] designs an asymmetry-resistant instance alignment framework which uses a two-phase blocking method considering concept and feature asymmetries, with a novel similarity measure overcoming structure asymmetry. Whereas, the problem is: (1) The class pair produced by the concept-based blocking is incomplete because of the different granularity in different KBs. (2) The feature-based blocking algorithm is not efficient enough for very large KBs. PARIS [19] provides a holistic probabilistic solution to align large-scale KBs. It computes alignments not only for instances, but also for classes and relations. However, it has poor efficiency and scalability. Different from existing methods, our algorithm is fully automatic and scalable while achieving high quality.

3   The PBA Framework

To address the three limitations, we should focus on the following aspects.

Scale Reduction: It is expensive to enumerate every instance pairs, especially for large knowledge bases. If we can prune most of dissimilar matching pairs, the alignment process will be accelerated greatly. To this end, we should make full use of all possible information from KBs. In the partition step, to meaningfully divide the classes into smaller partitions, we can use the class relations(*subclassof* or *typeof*) and instance identification information(e.g. *RDFS:label* property, foaf:name property or *URI*) to calculate the similarity between classes, and prune the dissimilar instances that fall in different classes. In the blocking step, to produce blocks of similar instances, we can choose datatype property values or objectype property values to block the instances with identical property values. In the alignment step, to compute the similarity between instances, we can combine the datatype property, objectype property and instance identification information together to produce the result. We will discuss how to implement the three steps later.

Matching Quality: There are several cases which prevent us to achieve better matching quality in KB alignment: (1) the same instances have different names, (2) different instances have the same name, (3) instances have different granularities, (4) the same properties have different discriminative power, and (5) entities with the same type have different number of properties. Besides, format, unit, case sensitivity, space, abbreviation and typo error etc. will all throw sand in the matching process. Our methods should consider these problems and find a perfect solution.

Prior Alignment Data: By looking through the information provided within KBs prior alignment data can be obtained in the following ways: (1) *URI*. If two instances have the same *URI*, we can take them as the same entity. (2) Owl:sameAs. This term is defined as: two *URI* references actually refer to the same thing. We can directly take an instance pair with owl:sameAs relation as a prior match. (3) Exact-string. This is done by looking for entities with the same string representation (with minimal standardization such as removing capitalization and punctuation). (4) Owl:InverseFunctionalProperty. *IFP* is the most discriminative properties which can uniquely determine its subject instance with respect to the object value in a fact, and two different instances with the same value in the *IFP* can be inferred to be the same entity. (5) *IFP*. Many *KBs* do not identify *IFP* explicitly. However, we can acquire *IFP* by the inference according to its definition. There may be absence of *URI*, Owl:sameAs, or Owl:InverseFunctionalProperty in *KBs*, but every *KB* can calculate its exact matching and *IFP* with each other. Therefore we can always obtain a number of prior alignment data. How to get enough prior data with high quality and maximize the value of these prior data to improve the matching process is another challenge. Note that prior data can be only used as initial aligned data but not the ground truth.

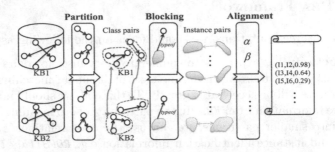

Fig. 2. Overview of the PBA framework. (The circles in the partition box represent classes and the blocks in the blocking box represent the instances belonging to the classes)

Based on these aspects, we propose a partition-and-blocking based alignment framework. The process of PBA is shown in Fig. 2.

Step 1 - Class Partition: It takes the classes of two *KBs* as inputs, and outputs a set of matched partition pairs. The process calculates the prior alignment data and applies them to divide the classes into a set of partitions, so that the similar classes are in the same partition, while classes in different partitions are dissimilar, and it also reduces the influence of name property in matching quality problems.

Step 2 - Instance Blocking: It takes the instances belonging to the same partitions derived from step 1 as inputs, and outputs a set of matched instances pairs. The process iteratively enumerates the matched property values generated by the prior alignment data to block the input instances. The blocking method can minimize the impact of the properties with different discriminative power, and mitigate the influence of different number of properties in two *KBs*.

Step 3 - Instance Alignment: It takes the instances in each block generated in step 2 as inputs, and outputs a set of matched instance pairs with confidence. The process prunes the redundant properties and relations, and designs an elaborate algorithm comprehensively considering name, properties and relations of an instance to compute the final similarity score.

Our framework divides the alignment process into three parts, which makes full use of the information in *KB* to make the alignment more efficient. The detail of the algorithm will be discussed in the following sections.

4   The Partition Algorithm

Traditional class partition methods compute the similarity between classes by textual similarity (e.g., edit distance functions of name descriptions), or structure similarity (e.g., the hierarchical structure of the classes), or the combination of the two with different weights. However, for two heterogeneous *KBs*, the structures and the names may be totally different, and thus existing methods will lose their power. In our partition process, we use the prior alignment data to address these problems. We first define the similarity measure between any two classes from different *KBs*.

Definition 3 (CLASS SIMILARITY). *Given two classes C_1 and C_2 from two KBsrespectively, we denote the number of their matched instances as $\mathbb{N}_I(C_1, C_2)$, denote the number of instances in C_1 as $|C_1|$. We use* JACCARD *to evaluate their similarity as:*

$$\text{SIM}(C_1, C_2) = \frac{\mathbb{N}_I(C_1, C_2)}{|C_1| + |C_2| - \mathbb{N}_I(C_1, C_2)}. \tag{1}$$

The class partition method considers the correlation of classes in two *KB* according to the class similarity, and partitions the classes mutually to achieve high cohesiveness within the same partitions and low relevancy between different partitions. After partitioning, instances from different partitions will be pruned and thus this method can improve the performance.

The class partition problem can be converted into weighted bipartite graph division problem. We first introduce the definition of weighted bipartite graph of *KB*.

Definition 4 (WEIGHTED BIPARTITE GRAPH). *Given two knowledge bases KB_1 and KB_2, the weighted bipartite graph of KB is represented as $BG = (X, Y, E, \mathbf{W})$, where X, Y are two sets of nodes where classes in KB_1 are the nodes in X and classes in KB_2 are the nodes in Y, $X \cap Y = \emptyset$. $E = \{(v_i, v_j)|v_i \in X, v_j \in Y\}$ is a set of edges between class nodes. $\mathbf{W}$ is a weighted edge matrix, where w_{ij} is the class similarity between two class nodes v_i and v_j calculated by Definition 3.*

The objective of *BG* division is to partition the graph into a number of subgraphs which maximizes the edge weight in each subgraph and minimizes the edge weight between subgraphs. This is similar with the goal of class partition. Hence, the method of weighted bipartite graph division can be used to solve the class partition problems.

Example 2. *The left part of Fig. 3 is a BG before partitioning derived from the motivation example. The classes* film, person, location *and etc. in the rectangle represent the nodes of BG. The gray rectangles are the nodes in X, and the white rectangles are the nodes in Y. The connection between the nodes is the edge of BG, and the thickness represents the weight. If we can divide the BG into several parts as the right part of Fig. 3 we will get the partitions of the classes.*

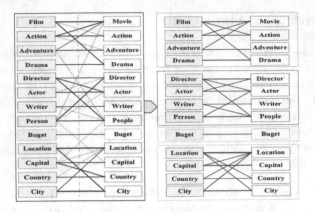

Fig. 3. A class partition example.

The *BG* division problem is NP-hard, and there have been several methods to solve it [3, 6, 18]. However, most of them have high computational complexity. As our goal is not to match each class pair, it is unnecessary to strictly follow the rule of high cohesiveness and low coupling. The conditions are met when similar classes can be found in one partition. Therefore, the partition process can be simplified. We propose a simple but scalable hierarchical agglomerative co-clustering algorithm, named HACC, based on the traditional HAC algorithm in a bottom-up fashion. We extend it for *BG* division problem with bi-directional hierarchical clustering in the same time. The data structure of a partition used in this algorithm is described as a quadruple (*id*, *first*, *second*, *totalvalue*) where *id* is a identifier of this partition; *first* is a set to save the classes in this partition; *second* is a map recording the related partitions in another *KB* whose keys are the class identifiers belonging to the related partitions and values are the similarity score; *totalvalue* is the sum of all the values of second. As shown in Algorithm 1, the proposed HACC algorithm proceeds in three stages:

Initialization: The algorithm accepts the classes of KB_1 and KB_2 and generates a similarity matrix by the prior alignment data (line 2). This stage starts from initializing a priority queue to cache the class partitions ordered by the *totalvalue* in a descending order. The partition can be generated by the *BG*, and the initialization stage constructs the bottom level of the partition hierarchy. Each node represents its own partition which produces *n* partitions in the priority queue (line 3).

Hierarchical co-clustering: This is the main stage of the HACC algorithm and is designed to mutually cluster the partition hierarchy. The algorithm iteratively merges the partitions from the priority queue according to the similarity measure θ. The process will terminate until either the number of partitions in the queue reaches 1 or there is no possibility to merge more partitions (lines 4–7).

Partition pairs generation: The final stage generates the matching partition pairs according to the merging result (line 8). As many instances in a partition

---

Algorithm 1. HACC($C_1, C_2, \Gamma, \theta, \tau$)

Input: C_1, C_2 : Classes for matching; Γ : Prior alignment data; θ : Similarity
           Threshold; τ : Partition size threshold
Output: $\mathcal{P}$: Set of matched partition pairs
1 **begin**
2     $W =$ GENERATECLASSSIMMATRIX(Γ); $BG =$ GENERATEBG(C_1, C_2, W);
3     $Q^1_{lower} =$ GENERATEQUEUE(BG, C_1); $Q^2_{lower} =$ GENERATEQUEUE(BG, C_2);
4   **while** $Q^1_{lower}.size() > 1$ *and* $Q^2_{lower}.size() > 1$ **do**
5   MERGECLASSES($Q^1_{lower}, Q^1_{lower}, \theta$); MERGECLASSES($Q^2_{lower}, Q^1_{lower}, \theta$);
6   **if** Q^1_{lower} *and* Q^2_{lower} *do not change* **then**
7   Break;
8     $\mathcal{P} = \mathcal{P} \bigcup$ GENERATEPARTITIONPAIRS(Q^1_{lower}, Q^2_{lower});
9   **for** *each partition pair* $< P_1, P_2 > \in \mathcal{P}$ **do**
10   **if** $P_1.getInstance.size() \geq \tau$ *or* $P_2.getInstance.size() \geq \tau$ **then**
11           $\theta = \theta \times \sigma$; $\mathcal{P} \setminus \langle P_1, P_2 \rangle$;
12   HACC($P_1, P_2, \mathcal{E}, \theta, \tau$);

---

will lead to higher computation costs in the following steps, these partitions will repeat the partition process in a recursive way until the size of each partition is less than the specified threshold (lines 9–12). After these steps, the final set of partition pairs will be returned.

    The computational complexity is linear with the number of classes. The complexity analysis and the details of the algorithm are provided in our technical report[5].

5 The Blocking Algorithm

The partition algorithm can effectively reduce the matching scale. However, it is still too expensive to make pairwise comparisons among all instances in the partition of very large *KBs*. *Blocking* is a common method of grouping similar instances into blocks to reduce the number of comparison. Traditionally, records are grouped together by shared properties, called blocking keys. Similarly, in *KBs* alignment problems, the instances from different *KBs* are grouped into blocks by shared values of properties or relations which indicate possible matching. These values or their variants are called blocking key values (*BKVs*). Appropriate *BKVs* will benefit the blocking process. However, it is hard to choose any fixed collection of values of properties or relations beforehand, especially when dealing with very large *KBs*. Thus, an elaborate and scalable blocking method should be introduced into the large-scale alignment problems.

    We extend the *dynamic blocking* algorithm [11,16] to overcome the difficulties by dynamically adjusting the *BKVs* at execution time rather than fixing them.

---

[5] http://dbgroup.cs.tsinghua.edu.cn/ligl/pba.pdf.

Algorithm 2. DYNAMICBLOCK($\langle I_1, I_2 \rangle$, Γ', key, t)

Input: I_1, I_2: Instance sets from partitions; Γ': Set of prior alignment pairs;
   key: Set of selected blocking key values; t: Block size threshold
Output: $\mathcal{B}$: Set of blocking pairs

```
1  begin
2  │  INITIALIZE(J);
3  │  for each feature f ∈ Γ' do
4  │  │  ⟨J₁, J₂⟩ = GENERATEBLOCKWITHBKV( f );
5  │  │  if |J₁| = 0 or |J₂| = 0 then  Γ' = Γ' \ { f } ;
6  │  └  else  J = J ∪ { ⟨f,⟨J₁, J₂⟩⟩ } ;

7  │  for each element ele ∈ J do
8  │  │  if ele.f ∉ key then
9  │  │  │  if |J₁| > t or |J₂| > t then
10 │  │  │  │  key = key ∪ { f };
11 │  │  │  │  DYNAMICBLOCK(⟨J₁, J₂⟩, Γ', key, t)
12 │  │  │  └  key = key \ { f };
13 │  └  └  else  B = B ∪ (J₁ × J₂) ;
```

The basic idea is to dynamically select different *BKVs* shared by two *KBs* to produce blocks until the size of each block is less than a specified value. Traditional dynamic blocking algorithm either requires a total order of properties or arbitrarily enumerates all possible combinations of *BKVs* to iteratively block the instances to a proper size in a recursive way. Nevertheless, in large-scale *KBs*, it is hard to get a total order of properties beforehand, and it is also intractable to enumerate all possible combinations of *BKVs* because of the high computation complexity of recursive processes. To address these limitations, we proposed a modified dynamic blocking algorithm as shown in Algorithm 2.

The algorithm first initializes an empty set *J* to store the map which uses *BKV* as key and the block generated by this *BKV* as value (line 2). Subsequently, a block is generated by each feature in the set of prior alignment pairs (lines 3–4). If the feature is not related to the instance, we will filter the feature from the prior alignment pairs (lines 5–6). Otherwise, the feature together with the block it generates will be recorded in the set *J* (lines 7–8). Then, for each *BKV* and block pairs in *J*, if the number of instances is larger than a threshold we will recursively call the dynamic blocking process with the blocking pairs and prior data generated before (lines 9–14). Finally, the blocking pairs are recorded to produce the set of blocking pairs as output (lines 15–16). In our implementation, the irrelevant *BKVs* will be deleted and the *BKV* with its block will be stored before each recursion. As the number of possible *BKVs* decreases significantly along with the blocking progress, the computation complexity will be greatly reduced.

6 The Alignment Algorithm

The vector space model (VSM) [17] can be used to align the instances in the same block [14]. However, we can not apply it directly because: (1) The number of properties of an instance in one *KB* may be much larger (or smaller) than that in another *KB*, which will lead to mismatching when we match instances with quite different number of properties. (2) The VSM model uses dot product to compute the vector similarity which ignores the relative order, whereas different properties with the same value can not be arbitrary regarded as being matched in instance alignment. For example, the *birthplace* of a person is same to the *deathplace* of another person can not be used as matching evidence. Some state-of-art methods solve the problem with human-computer interaction while we provide an automatic way based on a modified VSM method. Before introducing the details, we first give the definition of relation similarity as follows:

Definition 5 (RELATION SIMILARITY). *Given two relation sets $\mathcal{R}_1$ and $\mathcal{R}_2$ from KB_1 and KB_2 respectively. We denote the SPO triples in KB_1 as $r_1(x,y)$ where $r_1 \in \mathcal{R}_1$ and x is the subject of r_1 and y is the object of r_1, and the SPO triples in KB_2 as $r_2(x',y')$ which $r_2 \in \mathcal{R}_2$ and x' is the subject of r_2 and y' is the object of r_2. We still use* JACCARD *to define the relation similarity:*

$$\text{SIM}(r_1, r_2) = \frac{\mathbb{N}(r_1(x,y) \underset{x=x', y=y'}{\cap} r_2(x',y'))}{\mathbb{N}(r_1(x,y) \cup r_2(x',y'))}. \tag{2}$$

The numerator equals to the number of the intersection of triples from two *KBs* where both the subjects and objects are matched simultaneously according to relations r_1 and r_2. The matched subjects and objects can be directly calculated by the prior alignment data, and the matched relation pairs can be chosen by finding the maximum relation similarity in the *KB* with less relations. Similarly, if we change the object into literals with exact matching, we can define the property similarity and the matched property pairs. The relation/property pairs can be used to overcome the shortcomings of VSM, and the detailed process of the instance alignment in our framework can be represented as follows.

Relation/Property Selection. Calculate the relation and property similarity and get the relation and property pairs through prior alignment data and exact matching literals.

Vector Construction. Build three vectors to represent the instances in each block. Name vector, denoted as V_{name}, consists of terms segmented from the instance identification information and eliminates the stop words. Property vector, denoted as V_{prop}, consists of datatype properties according to the property pairs calculated by the property similarity. Relation vector, denoted as V_{rela}, consists of the objectype properties according to the relation pairs calculated by the relation similarity.

Weight Calculation. To evaluate the significance of terms in a vector, the prevalent *tf-idf* (term frequency - inverse document frequency) measure is used in this method to assign each vector component a weight according to their importance.

Candidate Selection. To reduce the computation cost, three inverted lists are used to prune dissimilar pairs.

Similarity Computation. According to the COSINE similarity measure, the similarity of name vector from two *KBs* is $\text{SIM}(V_{name}^1, V_{name}^2) = \frac{1}{W_1 W_2} \sum_{t=1}^n w_{1t} \cdot w_{2t}$. Where w_{1t} and w_{2t} are the weight of the components of the name vector, and $W_1 = \sqrt{\sum_{t=1}^n w_{1t}^2}$, $W_2 = \sqrt{\sum_{t=1}^n w_{2t}^2}$. Similarly, we can compute the similarity of property vector $\text{SIM}(V_{prop}^1, V_{prop}^2)$ and the similarity of relation vector $\text{SIM}(V_{rela}^1, V_{rela}^2)$. The final similarity consists of two parts. One is the similarity of the instance itself which is called static similarity. It is defined as $\text{SIM}_{static} = (1-\alpha)\text{SIM}(V_{name}^1, V_{name}^2) + \alpha\text{SIM}(V_{prop}^1, V_{prop}^2)$, where $\alpha \in [0,1]$ is a tuning coefficient between the name vector and property vector. The other is the similarity of its neighbors $\text{SIM}_{neighbor} = \text{SIM}(V_{rela}^1, V_{rela}^2)$. Therefore, the final similarity between instances in each blocks can be represented as $\text{SIM} = (1 - \beta)\text{SIM}_{static} + \beta\text{SIM}_{neighbor}$, where $\beta \in [0,1]$ is a tuning coefficient between the static similarity and the neighbor similarity. The final similarity shows how much we trust the matching result.

7   Experiments

In this section, we report experimental results. To evaluate the performance of our approach, we conducted two set of experiments utilizing synthetic and real-world datasets with different sizes. We compared the state-of-the-art algorithm PARIS. We also compared with exact-matching which calculated the similarity based on instance names.

7.1   Experiment Setup

Experimental Environment: All the programs were implemented in *Java* 8 and the experiments were run on the *Ubuntu* machine with Intel(R) Xeon(R) CPU E5-2670 2.60 GHz processors and 128 GB memory.

Datasets: We compare the quality of different methods using the standard metrics of precision, recall, and f-measure which can be computed based on the returned results to the ground truth. We choose the IIMB benchmark of OAEI[6] and the large-scale *Yago-DBPedia* dataset[7]. The IIMB benchmark provides *OWL/RDF* data about films, actors, and locations. *YAGO* and *DBPedia* are two famous large-scale *KBs* with a rich schema structure. Both *YAGO* [20] and *DBPedia* [12] are available as lists of triples from their respective websites. Table 1 presents the statistics information of the two datasets.

---

[6] http://oaei.ontologymatching.org/.

[7] http://webdam.inria.fr/paris/.

Table 1. Datasets

Dataset	♯ Instance	♯ Classes	♯ Properties
YAGO	3.03 M	360 K	70
DBPedia	2.49 M	0.32 K	1.2 K
IIMB	12.6 K	0.2 K	24

Table 2. Matching results (minutes for time).

Dataset	System	Precision	Recall	F-measure	Time
IIMB	PBA	98.3	92.5	95.3	0.03
	PARIS	99.6	91.6	95.4	0.06
	Exact	100	25.2	40	0.01
Yago-DBPedia	PBA (Single)	95	72.3	82.1	48
	PBA (Four)	95	72.3	82.1	23
	PARIS (Single)	93.6	71.7	81.2	748
	PARIS (Four)	93.5	72.5	81.6	499
	Exact	95.5	56.2	70.8	<1

Parameter Setting: Unless stated explicitly, parameters were set as follows by default: The prior alignment data comes from *IFP* method. The similarity threshold in the partition process $\theta = 0.01$, and the partition size threshold $\tau = 500k$. The block size threshold in the blocking process $t = 1000$. The coefficient in the alignment process $\alpha = 0.25$ and $\beta = 0.25$. The experiments in the parameter analysis section will discuss how to set the parameters.

7.2 Benchmark Test

We first run experiments on IIMB dataset. The matching result is shown in Table 2. In this experiment, the exact-matching method achieves an f-measure of 40 % in 1 s, and PARIS converges after just 3 iterations in 4 s which achieves an f-measure of 95 %. Whereas our method achieves an f-measure of 95 % in 2 s. The results show that PBA generates considerable matching results compared with PARIS but less time consuming and gets much better results than exact-matching method on the benchmark test. Note that PBA gets a slightly lower precision than the other two methods. By analyzing the false matchings in the results, we find out that some instances with few descriptions and neighboring information are modified quite a lot in their properties and relations by the OAEI, which produces more impact on the local collective approach used in our alignment process than the multiple rounds iteration algorithm in PARIS.

7.3 Real-World Dataset

Our design objective is to align comprehensive large-scale *KBs* in an efficient and automatic way. *Yago* and *DBPedia* which are located at the core position in *LoD* project can give full play to our algorithm's performance. We compare PBA with PARIS in single thread version and four-threads version, and run PARIS for 4 iterations until convergence. The precision of the instance matching is simply determined by comparing the *URIs* of the entities. To compute recall, we count the number of instances that the two *KBs* have in common. As the result, the two resources share more than 1.42 million entities, and PARIS can map them with a precision of 93.6 % and a recall of 71.7 % in 748 min in single thread version, while PBA achieves a precision of 95 % and a recall of 72.3 % in 48 min. Compared to PARIS method, PBA increased precision by 1.4 % and recall by 1.6 %, and decreased processing time by more than 15 times. Similarly, in four-thread version, PBA also outperformed PARIS in precision and recall while achieving 20 times faster. Examining the few remaining alignment errors revealed the following reasons: (1) Most errors were caused by very close related instances. Taking the mismatching *Liborio Romero* to *Perro Aguayo Jr* as an example, both of them are Mexican boxers born on the same day while PBA matches them by mistake. (2) Some errors were caused by the instances of different granularities in different *KBs*. Such as *Voorderweert* and *Sint Amands* which are the province and municipality in the same area of Belgium. PBA matches them even if they are different in *URI*. (3) Some errors were caused by the redirection references of *Wikipedia*. By crawling the website of *Wikipedia*, we get more than 16 K redirection references which ought to be correct indeed.

7.4   Tuning Parameters

Each process of PBA has a set of parameters. We analyze the performance of PBA in different parameters settings. The following analysis is made on the *Yago-DBPedia* dataset. Except the discussed parameter, the rest parameters are set to the default value.

Prior Data Switch Variable. First, we check the performance of PBA with different type of prior data computation. The result is shown in Figs. 4(a) and 5(a). It is obvious that the best one is *URI*. However, because every instance has a *URI* in our datasets and we use them as the matching standard, *URI* can not be used in our method. We notice that the precision of *URI* is not 100 %. That is because PBA can recognize redirection pairs as a match, while they are not be

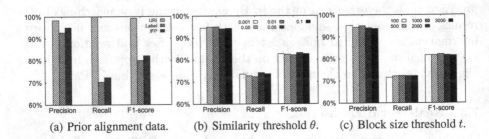

(a) Prior alignment data.   (b) Similarity threshold θ.   (c) Block size threshold t.

Fig. 4. Alignment quality in parameter experiments.

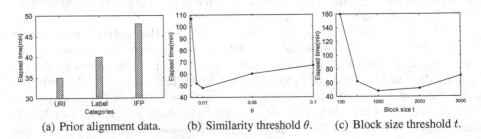

(a) Prior alignment data.   (b) Similarity threshold θ.   (c) Block size threshold t.

Fig. 5. Elapsed time in parameter experiments.

Table 3. The tradeoff coefficient.

Parameter	Precision	Recall	f-measure
$\beta = 0, \alpha = 0$	93.5	67.9	78.6
$\beta = 0, \alpha = 1$	89.4	66.6	76.3
$\beta = 1$	91.7	66.4	77

Table 4. Evaluation for partition and blocking.

Dataset	RR	PC	f-measure
IIMB	99.9	97.2	98.5
YAGO-DBPedia	99.9	89.1	94.2

included in the gold standard. Although *IFP* takes several minutes longer than exact-string, the alignment quality is much better than it. This is because in the large-scale real-world datasets, there are so many instances with duplicated names. Therefore, we adopt *IFP* as the default setting.

Similarity Threshold θ. From Fig. 4(b), we find that the matching quality is quite similar with the parameter varying from one thousandth to five percent, whereas the elapsed time is quite different in Fig. 5(b). Too small similarity threshold will lead to more class merging process and large similarity threshold will lose more useful information for matching. Both situations will cause higher computation cost. Therefore, This parameter is used to control the computational complexity, and the elapsed time should be the main consideration when choosing the value of this parameter.

Block Size Threshold t. Similar to the similarity threshold θ in the partition process, block size threshold t is another important parameter which is used to control the computational complexity in the partition blocking period according to Figs. 4(c) and 5(c). Too small block size will lead to more iteration times in the blocking algorithm, whereas larger block size will reduce the recall. We chose 1000 as the default value for the *Yago-DBPedia* dataset.

Tradeoff Coefficient α **and** β. According to Table 3, it will lead to a decline in both precision and recall when we only consider name similarity ($\beta = 0, \alpha = 0$), property similarity ($\beta = 0, \alpha = 1$), or relation similarity ($\beta = 1$) respectively. It is convinced that all three vectors constructed in the instance alignment process are very important in finding alignment pairs. The precision does not drop as much as recall because each block contains very limited amount of instances, and they are quite similar after the partition and blocking processes.

7.5 Evaluation for Partition and Blocking

We evaluate the efficiency and effectiveness of partition and blocking process. Here we use *reduction ratio* (*RR*) and *pairs completeness* (*PC*) as evaluating indicators. The reduction ratio measures the relative reduction in the comparison space of the partition and blocking process. Pairs completeness corresponds to the upper bound of recall. The blocking size threshold t in IIMB test is set to 50, while in *YAGO-DBPedia*, 500. As shown in Table 4, our partition and blocking methods show nearly perfect *RR* in both datasets. *PC* is relatively low in *YAGO-DBPedia* test, and this is because of the complexity of the class structure as well as properties and relations in the large-scale *KB*. This also explains the low recall of our PBA framework. To further improve the recall is one of our research directions.

7.6   Discussions

The experiments show that compared to PARIS, the proposed method obtains a comparable matching quality but more efficiency on the OAEI benchmark datasets and the real-world *KBs* with millions of entities. Because larger *KB* can be partitioned into proper size with negligible costs in the partition process (the partition process only take up to 2 % of the total time), the scalability is also very good especially in the parallel computation mode. As we mentioned above, the blocking process is the most time consuming part of the whole process (it will take up to 85 % of the total time). The complexity of PBA approximately can be deemed as equaling to the number of the block pairs because the computation in each block will be a constant value if we choose a fixed and small enough block size.

8   Conclusion

In this paper we present a scalable framework PBA for the alignment of large-scale knowledge bases. PBA takes full use of the prior alignment data and utilizes the comprehensive information of the *KB* to achieve high-performance matching. It reduces the matching space with elaborate partition and blocking algorithms. The experimental results on synthetic datasets from OAEI and large-scale datasets from real-world prove that our method outperforms state-of-art approaches and is very suitable as a powerful alignment tool for large-scale knowledge bases.

Acknowledgement. This work was supported by the National Grand Fundamental Research 973 Program of China (2015CB358700), the National Natural Science Foundation of China (61422205, 61373024, 61472198), Tsinghua-Tencent Joint Laboratory for Internet Innovation Technology, "NExT Research Center", Singapore (WBS:R-252-300-001-490), Huawei, Shenzhou, FDCT/116/2013/A3, MYRG105(Y1-L3)-FST13-GZ, National 863 Program of China (2012AA012600), Chinese Special Project of Science and Technology (2013zx01039-002-002) and the National Center for International Joint Research on E-Business Information Processing (2013B01035).

References

1. Abdullah, N., Ibrahim, R.: Knowledge retrieval in lexical ontology-based semantic web search engine. In: ICUIMC, Kota Kinabalu, Malaysia, 17–19 January 2013, p. 8 (2013)
2. Afacan, Y., Demirkan, H.: An ontology-based universal design knowledge support system. Knowl.-Based Syst. **24**(4), 530–541 (2011)
3. Anagnostopoulos, A., Dasgupta, A., Kumar, R.: Approximation algorithms for co-clustering. In: PODS 2008, Vancouver, BC, Canada, 9–11 June 2008, pp. 201–210 (2008)
4. Bizer, C., Heath, T., Berners-Lee, T.: Linked data - the story so far. Int. J. Semant. Web Inf. Syst. **5**(3), 1–22 (2009)

5. Bollacker, K.D., Evans, C., Paritosh, P., Sturge, T., Taylor, J.: Freebase: a collaboratively created graph database for structuring human knowledge. In: SIGMOD 2008, Vancouver, BC, Canada, 10–12 June 2008, pp. 1247–1250 (2008)
6. Dhillon, I.S.: Co-clustering documents and words using bipartite spectral graph partitioning. In: SIGKDD, San Francisco, CA, USA, 26–29 August 2001, pp. 269–274 (2001)
7. Do, H.H., Rahm, E.: Matching large schemas: approaches and evaluation. Inf. Syst. 32(6), 857–885 (2007)
8. Guo, Q., Zhang, M.: Question answering based on pervasive agent ontology and semantic web. Knowl.-Based Syst. 22(6), 443–448 (2009)
9. Hu, W., Qu, Y., Cheng, G.: Matching large ontologies: a divide-and-conquer approach. Data Knowl. Eng. 67(1), 140–160 (2008)
10. Lacoste-Julien, S., Palla, K., Davies, A., Kasneci, G., Graepel, T., Ghahramani, Z.: Sigma: simple greedy matching for aligning large knowledge bases. In: KDD 2013, Chicago, IL, USA, 11–14 August 2013, pp. 572–580 (2013)
11. Lee, S., Hwang, S.: ARIA: asymmetry resistant instance alignment. In: AAAI, Québec City, Québec, Canada, 27–31 July 2014, pp. 94–100 (2014)
12. Lehmann, J., Isele, R., Jakob, M., Jentzsch, A., Kontokostas, D., Mendes, P.N., Hellmann, S., Morsey, M., van Kleef, P., Auer, S., Bizer, C.: DBpedia - a large-scale, multilingual knowledge base extracted from wikipedia. Semant. Web 6(2), 167–195 (2015)
13. Li, J., Tang, J., Li, Y., Luo, Q.: Rimom: a dynamic multistrategy ontology alignment framework. IEEE Trans. Knowl. Data Eng. 21(8), 1218–1232 (2009)
14. Li, J., Wang, Z., Zhang, X., Tang, J.: Large scale instance matching via multiple indexes and candidate selection. Knowl.-Based Syst. 50, 112–120 (2013)
15. Lo, K.K., Lam, W.: Building knowledge base for reading from encyclopedia. In: Machine Reading, Papers from the 2007 AAAI Spring Symposium, Technical Report SS-07-06, Stanford, California, USA, 26–28 March 2007, pp. 73–78 (2007)
16. McNeill, N., Kardes, H., Borthwick, A.: Dynamic record blocking: efficient linking of massive databases in mapreduce. In: Proceedings of the 10th International Workshop on Quality in Databases (QDB) (2012)
17. Salton, G., Wong, A., Yang, C.S.: A vector space model for automatic indexing. Commun. ACM 18(11), 613–620 (1975)
18. Secer, A., Sonmez, A.C., Aydin, H.: Ontology mapping using bipartite graph. Int. J. Phys. Sci. 17, 4224–4244 (2011)
19. Suchanek, F.M., Abiteboul, S., Senellart, P.: PARIS: probabilistic alignment of relations, instances, and schema. PVLDB 5(3), 157–168 (2011)
20. Suchanek, F.M., Kasneci, G., Weikum, G.: YAGO: a large ontology from wikipedia and wordnet. J. Web Semant. 6(3), 203–217 (2008)

Knowledge Graph Completion via Local Semantic Contexts

Xiangling Zhang[1], Cuilan Du[2], Peishan Li[1], and Yangxi Li[2(✉)]

[1] School of Information, Renmin University of China, Beijing, China
{zhangxiangling,percentcent}@ruc.edu.cn
[2] National Computer Network Emergency Response Technical Team/Coordination
Center of China (CNCERT/CC), Beijing, China
liyangxi@outlook.com

Abstract. Knowledge graphs are playing an increasingly important role for many search tasks such as entity search, question answering, etc. Although there are millions of entities and thousands of relations in many existing knowledge graphs such as Freebase and DBpedia, they are still far from complete. Previous approaches to complete knowledge graphs are either factor decomposition based methods or machine learning based ones. We propose a complementary approach that estimates the likelihood of a triple existing based on similarity measure of entities and some common semantic patterns of the entities. Such a way of triple estimation is very effective which exploits the semantic contexts of entities. Experimental results demonstrate that our model achieves significant improvements on knowledge graph completion compared with the state-of-art techniques.

Keywords: Knowledge graph completion · Entity semantic similarity · Knowledge graph

1 Introduction

In recent years, a number of large-scale knowledge graphs such as DBPedia [1], Freebase [2] and YAGO2 [22] have been created. Some (e.g., Googles Knowledge Graph and Microsoft Bings Satori) have been applied in search engines to support important search tasks such as entity search and question answering. Facts in those knowledge graphs are usually expressed in the form of triple $< subject, predicate, object >$ (denoted as $< s, p, o >$ in short). Although many of these open domain knowledge graphs are very huge in terms of massive entities and relations contained, they are still incomplete on both entities and their relations. For example, 75 % persons in Freebase lack nationality information [4]. This somehow affects the wide and effective applications of knowledge graphs. Therefore, the study on knowledge graph completion is important and necessary.

Knowledge graph completion can be generally described as to estimate the probability of a triple $< s, p, o >$ which do not appear in a knowledge graph,

© Springer International Publishing Switzerland 2016
S.B. Navathe et al. (Eds.): DASFAA 2016, Part I, LNCS 9642, pp. 432–446, 2016.
DOI: 10.1007/978-3-319-32025-0_27

given s, p, and o existing individually in the knowledge graph. Traditionally, people address this problem by building models to predict a triple using the whole facts of the knowledge graph. A number of tensor-based methods have been recently proposed [5,6,9,17,18] in which the knowledge graph is modeled as a tensor. Triple prediction is then achieved through the factorization of adjacent tensor. Existing triples in knowledge graphs are treated as positive examples, and those non-existing triples are treated as negative ones according to closed-world assumption. However, it is observed [9] that the closed-world assumption is inappropriate. Krompass et al. [9] therefore propose a local closed-world assumption which improves the prediction accuracy.

Another fold of approaches for knowledge graph completion is to convert both entities and relations of the knowledge graph into low-dimensional vectors [3,11,12,21,25]. Since TransE does not work well for 1-to-N, N-to-1 and N-to-N relations , other methods such as TransH [25], TransR [12] and PTransE [11] are also proposed in this stream of work. However, both tensor-based approaches and vector-based approaches rely on the training data that may be hard to achieve (for negatives). Moreover, the implicit nature of such approaches makes it hard to debug why they do not work well for entities and relations of particular domains.

In this paper, we propose a model to effectively estimate the likelihood of a triple $< s, p, o >$ according to the semantic contexts of s and o. The intuition of the proposed solution is "If s is similar with s' and $< s', p, o >$ exists, then it is likely that $< s, p, o >$ may also exist. Similarly, if o is similar with o' and $< s, p, o' >$ exists, then it is likely that $< s, p, o >$ may also exist." The problem is then how to effectively define the similarity between entities based on their semantic contexts, and how to estimate a triple based on the similarity measures of entities. The main contributions of the paper can be summarized as follows:

- We propose a similarity measure of two entities based on their semantic contexts.
- We design an effective model to estimate the likelihood of a missing triple.
- We conduct extensive experiments on two public datasets. The results show that the proposed model significantly outperforms the state-of-the-art techniques.

The rest of the paper is organized as follows: Sect. 2 gives a related work study. Section 3 introduces the solution. Experimental study is given in Sect. 4, followed by the conclusion given in Sect. 5.

2   Related Work

There are three typical representation models for knowledge graphs: graph based model, tensor based model and low-dimension vector model. The first one simply treat the whole knowledge graph as a graph. The intuition of our model is inspired by SimRank [8]. However, SimRank does not take semantic contexts into account when evaluating the similarities of entities. It merely computes the similarity score between two objects based on the topology structure of the graph. Moreover, the computational complexity of SimRank is quite expensive. Tong et al. [24] propose an efficient algorithm for SimRank. However, the relations among entities are still ignored. There are also some related studies considering the labels of edges/paths when evaluating the similarity of entities [7,13,23]. For example, Sun et al. [23] propose a measure named PathSim to evaluate the similarity between two entities under a certain semantic path. However, users should specify the paths in advance which is not suitable for knowledge graph with the abundant types of relations and entities. Another work PRA [10] employs relation paths for inference on knowledge graphs. It has beed applied in KnowledgeVault [4].

Tensor based model regard a knowledge graph as a tensor. A score for none existing triples in a given knowledge graph could be obtained through the tensor factorization algorithms in [5,6,9,17,18]. Take RESCAL [17] as an example, knowledge graph is modelled as a tensor. Two ways refer to the entities, the other way to the relations. When a triple $< s, p, o >$ holds in knowledge graph, the value corresponding to s, p, o is 1, and 0 otherwise. RESCAL decompose the tensor into a matrix and a low-dimension core tensor. They can be used to do link prediction. Unfortunately, the computational cost is too high and the memory requirements are also high especially for knowledge graph with millions of entities and thousands of relations.

Recently, with the development of representation learning, a lot of works [3,11,12,21,25] regard both entities and relations in knowledge graph as a low-dimensional vector. All these work are inspired by [14]. Take TransE [3] as an example. The learning score function of TransE is

$$f(s, p, o) = ||\mathbf{s} + \mathbf{p} - \mathbf{o}||_2^2 \tag{1}$$

To learn vector representation for the entities and relations in knowledge graph, TransE minimize $f(s, p, o)$ if the triple $< s, p, o >$ exists, and maximize otherwise. The relations could be classified into four classes. A given relation is 1-to-1 if a subject entity can map to at most one object, such as $capital, spouse$. 1-to-N means a subject could map to many objects. N-to-1 means many subjects could map to the same one object, such as $birthplace$. N-to-N means many subjects could map to many objects, such as $starring$. TransE works well to 1-to-1 relations but is not suitable for 1-to-N, N-to-1 and N-to-N relations. The reason is that replacing subject or object to generate negative examples is valid only for 1-to-1 relations.

Both tensor-based approaches and vector-based approaches need to construct a model using the whole graph, which becomes very expensive once the graph reaches certain size. What is more, some work such as [21] requires to build a model for each relation. Moreover, the interpretability of these methods are not strong since all the reasoning mechanism is implicit.

Another family of knowledge graph completion solutions are based on inductive logic programming. Various inductive logic programming methods such as

FOIL [19], Progol [16] and Claudien [20] can be applied. For example, FOIL learns Horn clauses which cover all positive examples but none negative ones. Due to the huge search space, the computational complexity is extremely high.

3 Our Model

3.1 The Basic Idea

An knowledge graph is a directed edge-labeled graph denoted as $\mathcal{K} = \{E, U, L, \tau\}$, where (1) E is an entity set, (2) $U \subseteq E \times E$ is a set of directed edges, (3) L is a set of edge labels (predicates), and (4) $\tau : U \to L$ is a mapping function defines the mappings from the edges to the labels. Each label represents a relation between two entities. For example, $\tau(s, e) \to l$ (also can be represented as an triple $< s, l, e >$), where $s \in E$, $e \in E$, $l \in L$, $< s, e > \in U$, means that there is a relation, the predicate l, between entities s and e. We use l^{-1} to represent the reverse relation of the label l. For example, a triple $< s, l, e >$ can also be represented as another triple $< e, l^{-1}, s >$.

To evaluate the likelihood of a triple $< s, p, o >$, the basic idea is to use local semantic contexts determined by $< s, p, o >$. Specifically, to evaluate whether s satisfies a pattern $< x, p, o >$, we will compute the similarity of entities satisfying pattern $< x, p, o >$ with the entity s. The higher the similarity, the more likelihood that the fact $< s, p, o >$ holds. Similarly, to evaluate whether o satisfies a pattern $< s, p, x >$, we will compute the similarity of entities satisfying pattern $< s, p, x >$ with the entity o. To be more specific, in the running example of Fig. 1, to estimate the probability of fact $< Catch\_Me\_If\_You\_Can, starring, Tom\_Hanks >$, we compare the similarities between $Catch\_Me\_If\_You\_Can$ and the other movies where Tom_Hanks played in. On the other way, we also compare Tom_Hanks with other actors played in $Catch\_Me\_If\_You\_Can$. The greater the similarity, the more likely Tom_Hanks played in $Catch\_Me\_If\_You\_Can$.

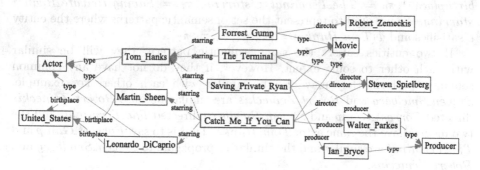

Fig. 1. A running example

To facilitate the introduction of our solution, we first give some relevant definitions as follows.

3.2 Definitions of Common Semantic Pattern

Given two entities, if they are similar, there must be a few of common features shared between them. For example, *Forrest_Gump* and *The_Terminal*, we say they are similar because they are films in which *Tom_Hanks* played. We apply the concept *semantic pattern* to define such common features.

Definition 1 (Semantic Pattern). *A semantic pattern in a knowledge graph $\mathcal{K}$ is composed of an anchor entity node e_a, and a predicate p. It is denoted as $\pi = e_a : p$.*

A semantic pattern (SP) is used to represent a set of target entities (defined in the Definition 2) having the relation p with the same anchor entity e_a. For example, to express the movies where *Tom_Hanks* played a role, we can utilize the semantic pattern $\pi_1 = Tom\_Hanks : starring^{-1}$, where *Tom_Hanks* is the anchor entity, and the predicate $starring^{-1}$ is the relation between the target entities of π_1 and the anchor entity *Tom_Hanks*. Note that $^{-1}$ indicates the direction of relation (predicate) where the anchor entity serves as an object. Another semantic pattern $\pi_2 = Movie : type^{-1}$ means a set of entities whose *type* is *Movie*. A semantic pattern exhibits the common feature of the target entities.

Definition 2 (Target entity). *If an entity e has a relation p with the anchor entity e_a. We say e is a target entity of $\pi = e_a : p$ which is denoted as $e \models \pi$.*

The set of target entities of a semantic pattern $\pi = e_a : p$ is denoted as $E(\pi) = \{e|e \models \pi\}$. For example, the set $E(\pi_1)=\{Catch\_Me\_If\_You\_Can, Saving\_Private\_Ryan, Forrest\_Gump\}$ in the running example of Fig. 1.

Actually, each entity in knowledge graphs may satisfy many semantic patterns. For example, in the Fig. 1, $Tom\_Hanks \models \pi_3$, $Tom\_Hanks \models \pi_4$, $Tom\_Hanks \models \pi_5$, $Tom\_Hanks \models \pi_6$, $Tom\_Hanks \models \pi_7$ where $\pi_3 = Forrest\_Gump : starring, \pi_4 = Actor : type^{-1}, \pi_5 = United\_States : birthplace^{-1}, \pi_6 = The\_Terminal : starring, \pi_7 = Saving\_Private\_Ryan : starring$. We use $\Phi(e)$ to represent the set of semantic patterns where the entity e satisifies and $\Phi(Tom\_Hanks) = \{\pi_3, \pi_4, \pi_5, \pi_6, \pi_7\}$.

If two entities satisfy the same semantic pattern, they will be similar with each other to some extent. However, if they do not have any common semantic pattern, they may also be similar with each other. For example, *Steven_Spielberg* and *Robert_Zemeckis* are similar because *Robert_Zemeckis* directed *Forrest_Gump* and *Steven_Spielberg* directed *The_Terminal*, and the two movies have the same actor *Tom_Hanks*. That is to say, *Forrest_Gump* and *The_Terminal* are similar and the similarity propagate to *Steven_Spielberg* and *Robert_Zemeckis*.

With the above definitions, an important issue is then how to define the similarity between entities, based on the semantic contexts determined by common semantic patterns.

To facilitate the understanding of the concepts and solution, we show some frequently used notations in Table 1.

Table 1. Frequently used notations

Notation	Description
$\mathcal{K} = \{E, U, L, \tau\}$	A knowledge graph
$\pi = e_a : p$	A semantic pattern where e_a is an anchor entity and p is a (directed) predicate
$E(\pi) = \{e \mid e \models \pi\}$	The set of target entities satisfying the semantic pattern π
$\Phi(e)$	The set of semantic patterns where the entity e satisifies
$P(e)$	The set of predicates of the entity e
$sim_s(a, b)$	The direct similarity score of two entities determined by SP
$sim_p(a, b)$	The propagated similarity score of two entities

3.3   Similarity of Entities

The semantic similarity of two entities consists of two parts: one is acquired by their common semantic patterns denoted as $sim_s(a, b)$; and the other is propagated through the same relation from the similar entities denoted as $sim_p(a, b)$.

With respect to $sim_s(a, b)$, the more common semantic patterns shared by entity a and b, the more similarity they will have. However, the weight of each common semantic pattern is different. Borrowing the idea of inverse document frequency in information retrieval, we term it as inverse semantic pattern frequency. The idea is that general semantic patterns are not as useful as non-frequent semantic patterns in computing the similarities of entities. We define the weight of an semantic pattern π as $\frac{1}{log(1+|E(\pi)|)}$ where $|E(\pi)|$ is the number of target entities of π. Note that for each semantic patterns π in the knowledge graphs, $|E(\pi)| > 1$.

$$sim_s(a, b) = \frac{\sum_{\pi \in (\Phi(a) \cap \Phi(b))} \frac{1}{log(1+|E(\pi)|)}}{\sqrt{|\Phi(a)||\Phi(b)|}} \qquad (2)$$

Besides the direct similarity determined by local semantic contexts of entities, we need also measure the propagated similarity of entities, which is defined as:

$$sim_p(a, b) = \sum_{p \in (P(a) \cap P(b))} \frac{\sum_{a' \models a:p} \sum_{b' \models b:p} sim_s(a', b')}{|E(a : p)||E(b : p)|} \qquad (3)$$

where $P(a)$ and $P(b)$ are sets of predicates of a and b respectively. According to the definition, the propagated similarity is aggregated from pairs of entities that share the same predicate to the two entities a and b.

Finally, the similarity between two entities is evaluated as the weighted sum of the two parts:

$$sim(a, b) = \alpha \cdot sim_s(a, b) + (1 - \alpha) \cdot sim_p(a, b) \qquad (4)$$

where α is the parameter tuning the weight.

3.4 The Overall Solution

Our goal is to obtain the likelihood of a triple $< s, p, o >$. The intuition of our method is to calculate the similarity between entity s and other entities s' who has a relation p with o, which means $s' \models o : p^{-1}$. We also take into account of the similarity between o and o' which has a relation p with s, which means $o' \models s : p$. Finally, we define the probability from two directions of a triple, by considering the likelihood of $s \models o : p^{-1}$, as well as that of $o \models s : p$. For evaluating the probability score of a triple $< s, p, o >$, we optimistically choose the maximal similarity score computed from the two directions.

$$score(< s, p, o >) = \beta \cdot \max_{o' \in E(s:p)} sim(o', o) + (1 - \beta) \cdot \max_{s' \in E(o:p^{-1})} sim(s', s) \quad (5)$$

which is also a weighted sum (determined by the parameter β) of the two parts. Note that the likelihood of a triplet is not normalized, which does not affect the effectiveness of triple prediction because it is a relative measure. We denote our method as LSCS for short.

4 Experiments

Our model is evaluated on two widely used knowledge graphs: WordNet [15] and Freebase [2]. We adopt three datasets (their statistics are given in Table 2) and conduct three tasks to evaluate our model.

Table 2. Statistics of the data sets

Dataset	#Relation	#Entities	#Train	#Test
FB15K	1,345	14,951	483,142	59,071
FB13	13	75,043	316,232	23,733
WN11	11	38,696	112,581	10,544

4.1 Experimental Setup

Data Sets Description

Wordnet. This knowledge graph can be seen as a combination of dictionary and thesaurus. The entities (called synsets) correspond to word senses, and the relations between entities represent lexical relations between them, such as hypernym, hyponym and meronym. We utilize WN11 used in [12, 21] which contains 11 relation types.

Freebase. Freebase is a large and growing collaborative knowledge base which provides general facts of the world. There are currently around 3.1

billion facts (triplets) and more than 80 million entities. For instance, $<$ *albert_einstein, spouse, mileva_maric* $>$ means the relation between entity *albert_einstein* and entity *mileva_maric* is *spouse*. We apply FB13 used in [12,21] and FB15K used in [3,11,12] as two data sets. For FB13, all the subjects are from people domain and 13 relations are extracted while only 7 appear in the testing data.

Baselines

We apply 5 baselines for comparison with our model. Among them, we use the code provided by the authors for TransR, TransE and PTransE. All of them consider the knowledge graph as a continuous vector space meanwhile the relations and entities are transformed into low-dimension vectors. We utilize the best configuration supplied by the paper [3,11,12]. Simultaneously, we compare with other entity similarity model including SimRank [8] and PathSim [23]. For SimRank we use a decay factor $C = 0.8$ and for PathSim we explore all possible paths.

Metrics

The metrics adopted for evaluation include: accuracy for the triple classification task, the mean rank and hits@10 for entity prediction. We consider hits@1 for relation prediction since hits@10 for approaches exceeds 95%. All these metrics are also used in [3,11,12,21].

For each test triplet, the subject of the triplet is replaced by other entities with predicate in triplet and rank these entities in descending order of scores calculated by score function in Eq. 5. We also apply the same procedure for the object of the triplet. We exhibit the mean of those predicted ranks and the hits@10 which is the proportion of correct entities ranked in the top 10. Country to expectation, the above metrics may under-estimate when some triplets already exist in the knowledge graph. In this case, before ranking we may filter out those triplets because they are true. We denote the first setting as "Raw" and the latter one as "Filter"

4.2 Experimental Results

Triple Classification

Our goal of this task is to choose the correct triple in the form of (s, p, t) in the testing set. In the testing sets of WN11 and FB13, there are triple pairs with the same subject and predicate and different objects. For each pair, the first one is positive while the other is negative. This is a binary classification task which has been investigated in [3,12,25]. All algorithms compute a score for each triple, the score could be used to determine the likelihood of each possible triple. For each pair, if the first one's score is greater, then we deem it as the positive one. The metric adopted for evaluation is *accuracy*.

Evaluation results on both WN11 and FB13 are displayed in Table 3. From the table we observe that: (1) Overall, the results on WN11 is better than FB13.

Table 3. Evaluation results of triple classification(%)

Dataset	WN11	FB13
TransE	80.24	81.09
PTransE	79.24	73.32
TransR	80.12	80.20
SimRank	98.86	82.65
PathSim	98.99	80.65
LSCS	**99.02**	**88.17**

This is because of the characteristics of data sets. Since all the subjects in FB13 are from *people* domain, we can not obtain the similarity between objects from other contexts. For instance, *Paris* and *Marseilles* are similar because both of them are the cities of *France*. Unfortunately, we could not obtain this information in FB13. In contrast, the knowledge is more comprehensive in WN11. (2) None of TransE, TransR and PTransE can outperform the other kind of methods based on entity similarity such as Pathsim on the WN11 data set. The reason behind the phenomenon maybe that methods based on entity similarity only consider the local knowledge of a triple are more precise than the model learned from the whole knowledge graph. (3) On FB13, our model significantly outperforms baseline methods. There are two reasons for this result. Firstly, local knowledge of a triple is more powerful. Secondly, our entity semantic similarity metric is more effective than SimRank and Pathsim. We not only take into account their common semantic patterns, but consider their different weights.

Figure 2 shows the accuracy of different relations on FB13 of our model. The accuracy ranges from 70.86 % (*gender*) to 97.94 % (*nationality*). We observe that using local knowledge about subjects and objects based on semantic context similarity is difficult to predict the *gender* even though there are only two values for *gender*. It is difficult to infer the *gender* from a person's local information (such as *nationality, cause_of_death, profession, ethnicity*) in intuition.

Table 4. Results of entity prediction

Metric	Mean rank		Hits@10(%)	
	Raw	Filter	Raw	Filter
TransE	147	48	52.80	74.40
PTransE	168	18	53.29	91.44
TransR	178	81	52.14	79.72
SimRank	**114**	90	55.29	70.35
PathSim	157	90	51.59	77.96
LSCS	154	**16**	**59.74**	**92.21**

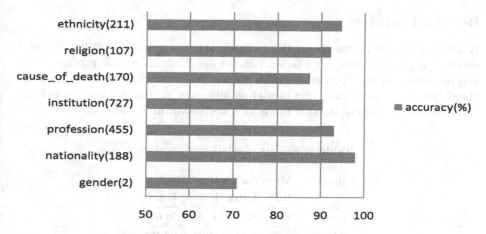

Fig. 2. Comparison of accuracy of different relations on FB13. The number in the bracket means the size of possible answer set

Entity Prediction

The goal of entity prediction is to predict the missing subject or object for a triple $< s, p, o >$ which employed in [3, 11, 12]. This task requires a set of candidate entities from the knowledge graph for the position of subject or object which is missing rather than only giving one best result.

In this paper, we utilize the FB15K dataset for this task because of the abundant relations contained by FB15K. In the testing phase, for each test triple $< s, p, o >$, we replace the subject and object entity by those which have the predicate p but not all entities in the knowledge graph. For example, to predict $< Catch\_Me\_If\_You\_Can, starring, Tom\_Hanks >$, we replace the subject $Catch\_Me\_If\_You\_Can$ with those subject entities who have a predicate $starring$, such as, $Saving\_Private\_Ryan, The\_Terminal, Forrest\_Gump$, we do not use such as $United\_States$ to replace $Catch\_Me\_If\_You\_Can$ because $United\_States$ does not have a relation $starring$ with any entity. Rank all generated triples in descending order by the prediction scores. We follow [3, 11, 12] and use two measures as our evaluation metric: mean rank and hits@10. As described in [3], the metrics are affected by the triples in knowledge graph. In other word, we also calculate a score for triples that already exist in the knowledge graph. The metrics will be under-estimated. Therefore, we filter out all those triples already hold in knowledge graph before rank and named as "Filter". The other one is "Raw". We calculate the metrics individually.

Table 4 shows the results of entity prediction. According to the results, we can observe that our model outperforms the other baselines. It implies that local information provides adequate evidences to infer a triple while the semantic context similarity is also effective.

Relation Prediction

Relation prediction aims to predict the relation between two entities. We also conduct experiment on the FB15K dataset for evaluation. We replace all of the predicates in the knowledge graph for the given two entities and then rank them. We adopt mean rank and hits@1 as measures to compare the algorithms. Experimental results of relation prediction are shown in Table 5.

Table 5. Results of relation prediction

Metric	Mean rank		Hits@1(%)	
	Raw	Filter	Raw	Filter
TransE	85	85	47.82	58.60
PTransE	2.38	2.13	68.42	**93.92**
TransR	101	100	34.71	42.35
SimRank	28	7	49.48	53.27
PathSim	31	10	52.77	59.54
LSCS	**1.86**	**1.77**	**70.42**	90.31

Generally, relation prediction is easier than entity prediction because the number of possible relations is less. Compared with the results of entity prediction in Table 4, the results of relation prediction in Table 5 are better. From the result we observe that: both PTransE and our model performs the best. Because the local knowledge plays an import role for prediction and the entities similarity metric also works of our model. Meanwhile the performance of PTransE is also very good. That is because PTransE takes into account different paths play a role for triple prediction.

Parameters Impact of LSCS

LSCS has two parameters that may affect its performance on knowledge graph completion. One is the parameter α used to balance the weight of common semantic patterns and the similarity conveyed from neighbors. Intuitively, the contribution of common semantic pattern is even greater. The other is the parameter β used to represent different weight of two directions which means one direction is to compare s with s' when $< s', p, o >$ exists in knowledge graph, the other is to compare o and o' when $< s, p, o' >$ holds in knowledge graph. We test the impacts of these two parameters on triple classification task on both WN11 and FB13 dataset. According to the results in Figs. 3 and 4, we may observe that the parameter α and β affects the performance on the two datasets differently. This is also reasonable because the characteristics are different for the two datasets.

In Fig. 3, when $\alpha = 0.95$ the performance reaches the best on FB13. When $\alpha = 1$, we notice that the performance is worse than $\alpha = 0.95$. This prove that

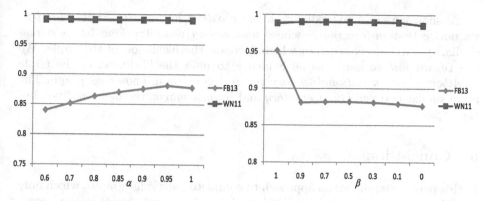

Fig. 3. The impacts of the parameter α **Fig. 4.** The impacts of the parameter β

the similarity conveyed from the similar neighbors also plays a little role. For WN11 Dataset, the value of α has little influence on the performance.

From the Fig. 4, we observe that the larger of β the performance is better for FB13 dataset. Since the subject entities in FB13 are only from *people* domain, and all the predicates are about *people*, such as *gender, nationality, cause_of_death*. The contribution of similarity between s and s' when $< s', p, o >$ holds is more important. For WN11 dataset, the value of β has little influence on the performance. Generally speaking, two directions are essential to the performance while the direction weight distribution has little effect. To sum up, the weight of two directions is equal, that is $\beta = 0.5$.

4.3 Example of Reasoning

We have observed that our model achieves good performance for knowledge graph completion through the experiments. In this part, we exhibit one example of our model.

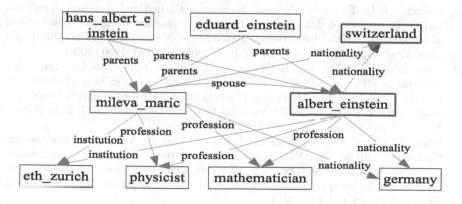

Fig. 5. A reasoning example in FB13

As shown in Fig. 5, to predict $< Albert\_einstein, nationality, Switzerland >$, we notice that $mileva\_maric$ whose $nationality$ is $switzerland$ has a strong similarity with $albert\_einstein$ which increase the likelihood of the triple. We can regard $mileva\_maric$ as an evidence to infer the likelihood of the triple $< Albert\_einstein, nationality, Switzerland >$ exists in knowledge graph. The more similar between $albert\_einstein$ and $mileva\_maric$, the more likelihood of the triple.

5   Conclusion

In this paper, we propose an approach to complete knowledge graphs, which only considers the local knowledge related to the subjects and objects in the knowledge graph. Unlike existing methods, our model is based on the similarity determined by the semantic contexts between entities to infer the likelihood of a triple. In experiments, we evaluate our models on three tasks including triple classification, entity prediction and relation prediction. Experiment results demonstrate that our model achieves significant improvements compared to TransE, TransR and PTransE as well as other entity similarity measures such as SimRank and Pathsim.

Acknowledgements. This work is supported by National Basic Research Program of China (973 Program) No. 2012CB316205, the National Science Foundation of China under grant (No. 61472426, 61170010, 61432006), the Fundamental Research Funds for the Central Universities, the Research Funds of Renmin University of China No. 14XNLQ06, and a gift by Tencent.

References

1. Auer, S., Bizer, C., Kobilarov, G., Lehmann, J., Cyganiak, R., Ives, Z.G.: DBpedia: a nucleus for a web of open data. In: Aberer, K., et al. (eds.) ASWC 2007 and ISWC 2007. LNCS, vol. 4825, pp. 722–735. Springer, Heidelberg (2007)
2. Bollacker, K.D., Evans, C., Paritosh, P., Sturge, T., Taylor, J.: Freebase: a collaboratively created graph database for structuring human knowledge. In: Proceedings of the ACM SIGMOD International Conference on Management of Data, SIGMOD 2008, Vancouver, BC, Canada, 10–12 June 2008, pp. 1247–1250 (2008)
3. Bordes, A., Usunier, N., García-Durán, A., Weston, J., Yakhnenko, O.: Translating embeddings for modeling multi-relational data. In: Advances in Neural Information Processing Systems 26: 27th Annual Conference on Neural Information Processing Systems 2013. Proceedings of a meeting held 5–8 December 2013, Lake Tahoe, Nevada, United States, pp. 2787–2795 (2013)
4. Dong, X., Gabrilovich, E., Heitz, G., Horn, W., Lao, N., Murphy, K., Strohmann, T., Sun, S., Zhang, W.: Knowledge vault: a web-scale approach to probabilistic knowledge fusion. In: The 20th ACM SIGKDD International Conference on Knowledge Discovery and Data Mining, KDD 2014, New York, NY, USA, 24–27 August 2014, pp. 601–610 (2014)

5. Drumond, L., Rendle, S., Schmidt-Thieme, L.: Predicting RDF triples in incomplete knowledge bases with tensor factorization. In: Proceedings of the ACM Symposium on Applied Computing, SAC 2012, Riva, Trento, Italy, 26–30 March 2012, pp. 326–331 (2012)

6. Franz, T., Schultz, A., Sizov, S., Staab, S.: TripleRank: ranking semantic web data by tensor decomposition. In: Bernstein, A., Karger, D.R., Heath, T., Feigenbaum, L., Maynard, D., Motta, E., Thirunarayan, K. (eds.) ISWC 2009. LNCS, vol. 5823, pp. 213–228. Springer, Heidelberg (2009)

7. Gangemi, A., Presutti, V.: Towards a pattern science for the semantic web. Semantic Web 1(1–2), 61–68 (2010)

8. Jeh, G., Widom, J.: Simrank: a measure of structural-context similarity. In: Proceedings of the Eighth ACM SIGKDD International Conference on Knowledge Discovery and Data Mining, 23–26 July 2002, Edmonton, Alberta, Canada, pp. 538–543 (2002)

9. Krompass, D., Nickel, M., Tresp, V.: Large-scale factorization of type-constrained multi-relational data. In: International Conference on Data Science and Advanced Analytics, DSAA 2014, Shanghai, China, 30 October–1 November 2014, pp. 18–24 (2014)

10. Lao, N., Mitchell, T.M., Cohen, W.W.: Random walk inference and learning in a large scale knowledge base. In: Proceedings of the 2011 Conference on Empirical Methods in Natural Language Processing, EMNLP 2011, 27–31 July 2011, John McIntyre Conference Centre, Edinburgh, UK, A meeting of SIGDAT, a Special Interest Group of the ACL, pp. 529–539 (2011)

11. Lin, Y., Liu, Z., Luan, H., Sun, M., Rao, S., Liu, S.: Modeling relation paths for representation learning of knowledge bases. In: Proceedings of the 2015 Conference on Empirical Methods in Natural Language Processing, EMNLP 2015, Lisbon, Portugal, 17–21 September 2015, pp. 705–714(2015)

12. Lin, Y., Liu, Z., Sun, M., Liu, Y., Zhu, X.: Learning entity and relation embeddings for knowledge graph completion. In: Proceedings of the Twenty-Ninth AAAI Conference on Artificial Intelligence, 25–30 January 2015, Austin, Texas, USA, pp. 2181–2187 (2015)

13. Maccatrozzo, V., Ceolin, D., Aroyo, L., Groth, P.: A semantic pattern-based recommender. In: Presutti, V., et al. (eds.) SemWebEval 2014. CCIS, vol. 475, pp. 182–187. Springer, Heidelberg (2014)

14. Mikolov, T., Sutskever, I., Chen, K., Corrado, G.S., Dean, J.: Distributed representations of words and phrases and their compositionality. In: Advances in Neural Information Processing Systems 26: 27th Annual Conference on Neural Information Processing Systems 2013. Proceedings of a meeting held December 5–8, 2013, Lake Tahoe, Nevada, United States, pp. 3111–3119 (2013)

15. Miller, G.A.: Wordnet: a lexical database for English. Commun. ACM 38(11), 39–41 (1995)

16. Muggleton, S.: Inverse entailment and progol. New Gener. Comput. 13(3&4), 245–286 (1995)

17. Nickel, M., Tresp, V., Kriegel, H.: A three-way model for collective learning on multi-relational data. In: Proceedings of the 28th International Conference on Machine Learning, ICML 2011, Bellevue, Washington, USA, 28 June–2 July 2011, pp. 809–816 (2011)

18. Nickel, M., Tresp, V., Kriegel, H.: Factorizing YAGO: scalable machine learning for linked data. In: Proceedings of the 21st World Wide Web Conference 2012, WWW 2012, Lyon, France, 16–20 April 2012, pp. 271–280 (2012)

19. Quinlan, J.R., Cameron-Jones, R.M.: FOIL: a midterm report. In: Machine Learning: ECML-93, European Conference on Machine Learning, Vienna, Austria, 5–7 April 1993, pp. 3–20 (1993)
20. Raedt, L.D., Dehaspe, L.: Clausal discovery. Mach. Learn. **26**(2–3), 99–146 (1997)
21. Socher, R., Chen, D., Manning, C.D., Ng, A.Y.: Reasoning with neural tensor networks for knowledge base completion. In: Advances in Neural Information Processing Systems 26: 27th Annual Conference on Neural Information Processing Systems 2013. Proceedings of a meeting held 5–8 December 2013, Lake Tahoe, Nevada, United States, pp. 926–934 (2013)
22. Suchanek, F.M., Kasneci, G., Weikum, G.: Yago: a core of semantic knowledge. In: Proceedings of the 16th International Conference on World Wide Web, WWW 2007, Banff, Alberta, Canada, 8–12 May 2007, pp. 697–706 (2007)
23. Sun, Y., Han, J., Yan, X., Yu, P.S., Wu, T.: Pathsim: meta path-based top-k similarity search in heterogeneous information networks. PVLDB **4**(11), 992–1003 (2011)
24. Tong, H., Faloutsos, C., Pan, J.: Fast random walk with restart and its applications. In: Proceedings of the 6th IEEE International Conference on Data Mining (ICDM 2006), 18–22 December 2006, Hong Kong, China, pp. 613–622 (2006)
25. Wang, Z., Zhang, J., Feng, J., Chen, Z.: Knowledge graph embedding by translating on hyperplanes. In: Proceedings of the Twenty-Eighth AAAI Conference on Artificial Intelligence, 27–31 July 2014, Québec City, Québec, Canada, pp. 1112–1119 (2014)

Cross-Lingual Type Inference

Bo Xu[1], Yi Zhang[1], Jiaqing Liang[1], Yanghua Xiao[1,2(✉)], Seung-won Hwang[3], and Wei Wang[1]

[1] School of Computer Science, Fudan University, Shanghai, China
{xubo,z_yi11,shawyh,weiwang1}@fudan.edu.cn, l.j.q.light@gmail.com
[2] Shanghai Internet Big Data Engineering and Technology Center, Shanghai, China
[3] Department of Computer Science, Yonsei University, Seoul, South Korea
seungwonh@yonsei.ac.kr

Abstract. Entity typing is an essential task for constructing a knowledge base. However, many non-English knowledge bases fail to type their entities due to the absence of a reasonable local hierarchical taxonomy. Since constructing a widely accepted taxonomy is a hard problem, we propose to type these non-English entities with some widely accepted taxonomies in English, such as DBpedia, Yago and Freebase. We define this problem as cross-lingual type inference. In this paper, we present CUTE to type Chinese entities with DBpedia types. First we exploit the cross-lingual entity linking between Chinese and English entities to construct the training data. Then we propose a multi-label hierarchical classification algorithm to type these Chinese entities. Experimental results show the effectiveness and efficiency of our method.

1 Introduction

With the boost of Web applications, WWW has been flooded with information on an unprecedented scale, most of which is readable only by human but not by machine. To make the machine understand the Web, great efforts have been dedicated to harvesting knowledge from the online encyclopedias, such as Wikipedia. A variety of knowledge graphs or knowledge bases thus have been constructed, such as Yago [21], DBpedia [2] and Freebase [3]. These knowledge bases contain different semantic relationships between entities and concepts (also known as types or categories).

A fundamental semantic information about entities is their types. The relationships between an entity and its types represent the instanceOf relationships. For example, William Shakespeare has the types Person, Writer, Poet, etc. Among these types, *fine-grained* types such as Writer and Poet are more important than *coarse-grained* types such as Person because they characterize

Y. Xiao—This paper was supported by the National NSFC (No. 61472085, 61171132, 61033010, U1509213), by National Key Basic Research Program of China under No. 2015CB358800, by Shanghai Municipal Science and Technology Commission foundation key project under No. 15JC1400900. Seung-won Hwang was supported by Microsoft.

S.B. Navathe et al. (Eds.): DASFAA 2016, Part I, LNCS 9642, pp. 447–462, 2016.
DOI: 10.1007/978-3-319-32025-0_28

an entity more accurately. Characterizing an entity with types especially with fine-grained types plays a more and more important role in many real applications, such as recommender systems [12], question answering [10], emerging entities discovering [11], named entity disambiguation [24], etc. We refer to this effort as *Entity Typing*. Different from traditional `Named Entity Recognition` (`NER`) task, which tends to classify entities into a small set of coarse types, such as `Person`, `Location` and `Organization` [17], `Entity Typing` task tends to assign *specific* types to their entities.

A direct solution to type an entity is assigning it to the type that the entity most likely belongs to. The likelihood can be estimated by the similarity between the entity and the entities already with the type. Clearly, this naïve solution relies on a conceptual hierarchy (which contains the *instanceOf* relation between entities and their categories) and the description about entities. However, for non-English entities, typing them is still difficult due to the lack of well-structured non-English knowledge bases, especially the *instanceOf* knowledge. Although many non-English knowledge bases contain category information about entities, it cannot serve as *instanceOf* knowledge for entity typing. We use Baidu Baike, the largest Chinese knowledge repository, as an example to illustrate their weaknesses:

- First, categories[1] in non-English knowledge bases such as Baidu Baike are actually entities' tags/topics instead of the exact types. One of the categories of entity `AK47` in Baidu Baike is 军事 (en: Military), which is the topic of the entity instead of its type.
- Second, some entities contain error categories. For example, one of the categories of entity `AK47` in Baidu Baike is 军事人物 (en: Military Person), which is wrong.
- Third, types in non-English knowledge bases can hardly be organized as a hierarchy. Even though some categories are really types of entities, many desired relationships among categories such as `subClassOf` are still sparse in non-English knowledge bases. As a result, we are blind about the granularity of a type as well as their `subClassOf` relationships, which makes the entity typing with specific types difficult.

We notice that many English knowledge bases have been available and they contain widely acknowledged `instanceOf` knowledge. Thus, we wonder *whether we can use the English knowledge bases to type an non-English entity*. We refer to this problem as *cross-lingual type inference*. In this paper, we focus on typing Chinese entities with English DBpedia knowledge bases.

---

[1] In this paper, we strictly differentiate "category" from "type". "Category" always refer to the part of the knowledge base such as Baidu Baike and Wikipedia named as "category". Most of these categories actually are only tags of an entity. Instead "type" always refer to the class that an entity can be classified into.

Motivation. We highlight that cross-lingual type inference is sufficiently motivated:

- First, reuse of types in English knowledge bases ensures the high quality of types. Many well-established knowledge bases are available in English, such as DBpedia [2] and Yago [21]. These knowledge bases provide richer and cleaner types, which are widely and successfully used in many real applications.
- Second, types from these well-established English knowledge bases form a hierarchy consisting of subClassOf relations between types. It enables specificity-aware entity typing, which has the flexibility to be adapted in different applications demanding types with different specificity.
- Third, cross-lingual type inference enables many cross-lingual search tasks. For example, suppose we are looking for all birds from all over the world. Since many birds may only appear in a local knowledge base, we need to integrate the knowledge from different knowledge repositories. However, due to knowledge bases in a specific language tend to name the type birds in its own language, such as 鸟 in Chinese. As a prerequisite, we need to type the entities in different languages with a uniform type. In our example, to retrieve all birds, we need to type the entities of birds in different local knowledge bases to the same English type birds.

Weakness of Previous Approaches. Many solutions have been proposed to type English entities with English Types. However, some of them [5,6,15] are language-dependent, and cannot be used to solve the cross-lingual type inference problem. Tipalo [5] uses natural language processing (NLP) tools to extract types from the definition, and Yago [15,21] also finds WordNet types from Wikipedia category names. For instance, they use NLP tools to find category Michael Jackson albums belonging to WordNet type album, hence all entities belonging to this category can be assigned to the type (album) in WordNet. However, with only English entities considered, it cannot be easily adapted to other languages. SDType [13] is a state-of-the-art type inference method and can be adopted to solve cross-lingual problem. However, they do not consider the type taxonomy and predict type label independently (i.e. not aware of the relationship between types), which in general will decrease the accuracy of some fine-grained type classifiers.

Challenges and Contributions. A direct solution to determine whether an entity belongs to a type is training a binary classifier for the type. However, this naive solution in general is not applicable in our setting. We still need to solve the following challenges:

- **Construction of Training Data.** To build the classifiers for each type, we need a massive amount of training data. For Chinese entities, there exist no DBpedia types. Manual labeling is costly and not applicable on a large scale of training data. We address this challenge in Sect. 4 by exploiting the cross-lingual linking between Chinese and English entities, and typing these Chinese ones with the types of their corresponding English ones.

- **Efficient-yet-effective Typing Solutions.** There exist hundreds of English types in DBpedia and tens of millions of Chinese entities to type. Classifying an entity using each classifier is obviously wasteful since an entity deserves a very small number of types. Hence, how to use the relationships between types to speed up the typing procedure while ensuring the accuracy of the typing is a challenge. We address this challenge in Sect. 5 by building hierarchical multi-label classifiers and designing a corresponding hierarchy-aware typing algorithm.
- **Sparseness of Types.** Note that *instanceOf* relations in DBpedia are still sparse. For example, *Tom Cruise* only has types *Thing, Agent* and *Person* in DBpedia and many other types such as *Artist* and *Actor* are missing. As a result, many types such as *Artist* in our example miss many members. The classifier built for *Artist* can hardly be effective due to the sparsity of its instances. To solve this problem, we first propose a type completion method as a preprocessing step to find more types for DBpedia entities in Sect. 3.2.

The rest of this paper is organized as follows. In Sect. 2, we formally define the problem of cross-lingual type inference and give an overview of our proposed system. In Sect. 3, we introduce the features we use for Chinese knowledge bases entities and the process of type completion of English DBpedia entities. In Sect. 4, we describe the construction of training data. In Sect. 5, we propose our multi-label hierarchical classification method. In Sect. 6, we present the experimental results. In Sect. 7, we review the related work and highlight the differences between our work and major existing methods. In Sect. 8, we conclude this work.

2   Overview

2.1   Problem Definition

In this section, we first formalize the cross-lingual type inference problem in Definition 1. There are many well-established knowledge bases, which have different taxonomy structures. For instance, the structure of DBpedia is a tree, while Yago is a Directed Acyclic Graph (DAG). According to [18], solutions vary from structure to structure. In this paper, we focus on typing Chinese entities with types in a tree-based taxonomy. Specifically, we type Chinese entities with DBpedia types.

Definition 1 (Cross-Lingual Type Inference). *Let $\mathcal{T}$ be the collection of all types in English knowledge base's taxonomy, and $\mathcal{E}$ be the collection of non-English knowledge base entities. Our problem is typing each non-English entity $e \in \mathcal{E}$ with a subset of types $T(e) \subset \mathcal{T}$.*

DBpedia taxonomy is a tree-based hierarchical type structure. Different from DAG structure, a node can only have one parent node. As shown in Fig. 1, types of DBpedia entities have three properties:

- First, entities have types with different granularities. For instance, William Shakespeare is not only a Person, but also a Writer.

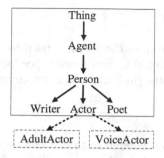

Fig. 1. Black solid square represents the types of entity `William Shakespeare` and their hierarchical relations. `AdultActor` and `VoiceActor` are the sub-types of Actor, but not the types of `William Shakespeare`.

- Second, entities may have multiple types at a certain granularity. For instance, `William Shakespeare` belongs to the types `Writer` and `Poet`, both of which are the sub-types of `Person`.
- Third, entities may not have the most specific types. For instance, `William Shakespeare` belongs to the type `Actor`, but not to any of its sub-types (`AdultActor` and `VoiceActor`).

Hence, typing Chinese entities with DBpedia types is a *tree-based, multi-label, non-mandatory leaf node, hierarchical classification problem*.

2.2 System Architecture

We introduce `CUTE`, which is short for **C**ross-ling**U**al **T**ype inf**E**rence method. Figure 2 is the system architecture of `CUTE`. We first exploit the cross-lingual entity linking between Chinese and English entities to construct the training data. Then we propose a multi-label hierarchical classification algorithm to type these Chinese entities.

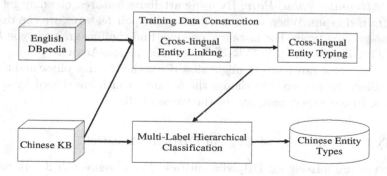

Fig. 2. System architecture of `CUTE`

3   Data

In this section, we first introduce the features used to characterize Chinese entities. Then we present the detail to find more types for DBpedia entities so that the constructed training data (in Sect. 4) is more complete.

3.1   Feature Set

To make our solution general enough so that it can be applicable to type entities in an arbitrary Chinese knowledge base, we only use features that exist in their knowledge bases. We tend to use many features in the structured data of a knowledge part (such as infobox templates, tags) instead of the features extracted from free text such as part-of-speech tags, dependency parsing results, etc. More formally, each entity $e \in \mathcal{E}$ is represented by an $|\mathcal{F}|$-dimension vector:

$$e = (f_1, f_2, ...f_j, ..., f_{|\mathcal{F}|})   \tag{1}$$

where f_j represents the j-th feature of entity e and $\mathcal{F}$ is the full feature set we used. In this paper, each feature is a binary feature, that is we use 1 and 0 to represent the existence of the j-th feature in entity e. Next, we elaborate the three types of features we used: *entity category*, *entity attribute* and *entity attribute-value pair*.

Entity Category: Categories in encyclopedia websites are used to group similar articles, which has been widely used in ontology/taxonomy construction [16,21] and type inference [1]. For instance, Chinese entity 刘德华 (en: Andy Lau) belongs to many categories, such as 歌手 (en: Singer) and 演员 (en: Actors). Using these features, we can easily infer its types.

Entity Attribute: Attributes (also called properties) of entities also play an important role in inferring their types [7,13]. For instance, one may obtain the types `agent` and `person` by 刘德华 (en: Andy Lau)'s attribute 职业 (en: Occupation).

Entity Attribute-Value Pair: By using attribute features, one may get some coarse-grained types. When using attribute-value pair features, we can discover more fine-grained types. For instance, one can find the fine-grained type `Actor` from the attribute-value pair 职业-演员 (en: Occupation-Actor).

Note that some rare features might slow down the training phase and mislead the classifiers. Hence, we only choose the features which are shared by at least β entities. In our experiment, we set the value to 10.

3.2   Type Completion of English DBpedia Entities

Many types are missing for DBpedia entities. The absence will deteriorate the quality of the training data, and in turn hurt the effectiveness of our typing solution. In this subsection, we elaborate how we solve this problem.

As we know, types of DBpedia entities are derived from their infobox template names in Wikipedia [8]. However, many of them only have general infobox templates, hence they may lack some fine-grained types. We refer to it as Type Incompleteness problem. For instance, a number of entities, such as Tom Cruise, only belong to Thing, Agent and Person in DBpedia, while more specific types such as Artist and Actor are missing. As a result of using existing DBpedia types to construct the training data, many types such as Artist and Actor in our example have quite few members. The classifiers of these types will have bad performance. As a result, entity typing with these classifiers will lead to errors.

To find more types for DBpedia entities, we exploit the category information of entities in DBpedia to complete their types. Our basic idea is discovering the subClassOf relationship between DBpedia categories and types. If a category c is a subclass of type t, then all entities in category c would belong to type t. Specifically, the type completion process consists of two steps. The first step is to estimate the probability that category c is a subclass of type t ($Pr(c \subset t)$). The second step is to compute the probability that the entity e belongs to type t ($Pr(e \in t)$).

STEP 1: $Pr(c \subseteq t)$. There are two state-of-the-art methods to estimate the probability of $Pr(c \subseteq t)$. One is Yago [21], and the other is PARIS [20]. We employ both methods to estimate the probability.

We first determine the subClassOf relations by Yago. The procedure is as follows [21]: We first segment the pre-modifier, head compound and post-modifier of the category name c. After stemming the head compound, we check whether the concatenation of pre-modifier and head compound or the head compound alone is a name of DBpedia type t. If true, we consider category c as a subclass of t. For all categories and types, if category c is a subclass of type t, we set the probability of $Pr_1(c \subset t)$ to 1. Otherwise, we set the probability to 0. Take Wikipedia category History museums in Ohio as a example, its pre-modifier, head compound and post-modifier are History, museums and in Ohio, respectively. After stemming the head compound (i.e. museum), we find that museum is a name of DBpedia type. Thus we set the probability that History museums in Ohio belongs to type museum to 1.

Then we estimate the probabilistic subClassOf relations from PARIS. As shown in Eq. 2 [20], the probability is proportional to the number of entities of category c that belong to type t:

$$Pr_2(c \subseteq t) = \frac{\#c \cap t}{\#c} \tag{2}$$

where $\#c$ is the number of entities of category c, and $\#c \cap t$ is the number of entities of category c that belong to type t.

Finally, we choose the greater one as the final probability. Because any individual one is a strong signal to suggest the subClassOf relations among categories.

$$Pr(c \subseteq t) = \max(Pr_1(c \subseteq t), Pr_2(c \subseteq t)) \tag{3}$$

STEP 2: $Pr(e \in t)$. Then, we use a Noisy-or model [19] to estimate the probability that entity e belongs to type t in Eq. 4:

$$Pr(e \in t) = 1 - \prod_{c \in C(e)} (1 - Pr(c \subseteq t)) \tag{4}$$

where $C(e)$ is the categories that entity e belongs to. If the probability of $Pr(e \in t)$ is greater than or equal to a threshold θ $(0 \leq \theta \leq 1)$, we assign the type t to entity e.

4   Training Data Construction

To effectively learn the classifiers, we need a massive amount of labelled data. For Chinese entities, there exist no English DBpedia types, and it is not practical to use human labeling. Hence, we first link Chinese entities and English entities and then use the types of English entities as the label of Chinese entities. In this way, we have many Chinese entities as well as their English types as the training data.

Cross-lingual entity linking in general is non-trivial [4,22,23]. Fortunately, our goal is just linking entities across different languages to construct the training data instead of finding as more cross-lingual entity links as possible. Thus, we are only concerned with the *precision* of the entity linking instead of *recall*. We notice that some entities in English/Chinese knowledge base may have the same Chinese label name. Hence, we only need to compare their Chinese label names of Chinese and English entities. If they share the same label name, we establish a link between them. For example, Chinese entity 威廉·莎士比亚 and English entity William Shakespeare share the same Chinese label name 威廉·莎士比亚, and hence should be linked together. Given the linked entity pairs, we directly use English entities' types to label the corresponding Chinese entities. For instance, William Shakespeare has 6 types which are used as the English types of corresponding Chinese entity 威廉·莎士比亚.

5   Multi-label Hierarchical Classification

To solve the multi-label hierarchical classification problem, we propose a local classifier per node based approach [18]. Specifically, we first train a binary classifier for each type in the hierarchy (except the root node) in the *training phase*. Then we use a top-down search to *type* the entities.

5.1   Training Phase

We first define the set of positive and negative samples for each classifier. We adopt the sibling policy proposed in [18] for this purpose. However, the generic policy is designed for *mandatory* leaf node classification problem [18] and should be customized for our *non-mandatory* leaf node classification problem. More specifically, for each type classifier $TC(t)$, we use all entities belonging to type t as positive samples, and those entities belonging to the sibling or supertype of t but not to type t as negative samples. For type `VoiceActor` in Fig. 1, entities belonging to its sibling `AdultActor` or its super type `Actor` but not to `VoiceActor` serve as negative samples. The rationality of sibling strategy is that in our hierarchical classification, we determine the membership of an entity to types from the root to specific types. That means when an entity e has been typed with type t, we need to find the best subtypes that e belongs to. Thus, we need samples from different subtypes to effectively classify entities into fine-grained types.

Then we need to train all the classifiers. Each of them is a binary classifier. We tested all possible binary classification models, such as Logistic Regression, Random Forest, and SVM. We obtained almost the same performance results by these models. As a result, considering the efficiency, we use the Logistic Regression model implemented by `scikit-learn` [14].

5.2   Typing Phase

Finally, we propose a top-down multi-label hierarchical classification algorithm to type Chinese entities with DBpedia types. The procedure is illustrated in Algorithm 1. The algorithm searches in the DBpedia taxonomy in a top-down manner. We use a queue data structure Q to store the candidate types of entities. For each entity e, we first push all types at level 1 into Q, which are the sub-types of root node of DBpedia taxonomy. Then for each candidate type t in Q, we run the classifier of t to test whether entity e belongs to the type t or not. If the result is `true`, we continue to search all its sub-types by appending its sub-types into Q. The search process ends when no more candidate types to be processed.

6   Experiments

In this section, we present the experimental results. Specifically, we verify the effectiveness of *type completion* and *cross-lingual type inference*.

6.1   Type Completion of English DBpedia Entities

We first justify type completion and determine the best threshold (θ) used for type completion. We use DBpedia2014[2] as our dataset. Specifically, we use DBpedia Ontology[3], Entity Types[4] and Entity Categories[5] as input, and return a set of new entity-type pairs. There are totally 4,191,094 entities with 14,993,020 types in DBpedia, with an average of 3.58 types per entity.

---

[2] http://oldwiki.dbpedia.org/Downloads2014.
[3] http://oldwiki.dbpedia.org/Downloads2014#dbpedia-ontology.
[4] http://oldwiki.dbpedia.org/Downloads2014#mapping-based-types.
[5] http://oldwiki.dbpedia.org/Downloads2014#articles-categories.

Algorithm 1. The Hierarchical Classification Algorithm.

Input: the feature vector $e = (f_1, f_2, ..., f_{|\mathcal{F}|})$ of entity e, all types T in DBpedia taxonomy and trained classifiers $TC(t)$ for each type $t \in T$.
Output: a collection of types $T(e)$ of entity e

1: Initialize Q to be a queue containing all types at level 1
2: **while** Q is not empty **do**
3:     $t \leftarrow Q.dequeue()$
4:     **if** $TC(t|e) == true$ **then**
5:         add t to $T(e)$
6:         **if** t has sub-types **then**
7:             add these sub-types to Q
8:         **end if**
9:     **end if**
10: **end while**

From Fig. 3(a), we can observe that the number of new entity-type pairs discovered by our approach decreased with the threshold. We also estimate how many newly extracted entity-type pairs are correct. Since no ground truth is available, we resort to using human judgement to evaluate the precision. We randomly select 1000 new entity-type pairs and ask volunteers to judge whether they are correct. From Fig. 3(b), we can see that the precision increases with the threshold. We notice that when the threshold is 0.9, our type completion approach achieves the highest precision (0.923) and finds more new entity-type pairs than threshold 1.0. Hence, we use 0.9 as the best threshold. Finally we obtain 5,463,462 new entity-type pairs.

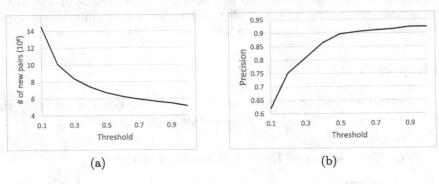

(a)   (b)

Fig. 3. Completion performance with different thresholds

6.2 Cross-Lingual Type Inference

Next we show the effectiveness and efficiency of our method. The Chinese knowledge base is from Baidu Baike[6] (the largest Chinese online encyclopedia)

---

[6] http://baike.baidu.com/.

articles which were crawled in January 2015. We obtain entity name, categories, attributes and attribute-value pairs information from each article. The English DBpedia data is from DBpedia2014, including `DBpedia Ontology`, `Entity Types`, and `Entity Chinese label names`[7]. Moreover, new entity-type pairs from type completion process are also used. There are overall 9,223,450 Chinese Baidu Baike entities and 4,191,094 English DBpedia entities. 344,576 of English entities have Chinese label names. After cross-lingual entity linking, we obtain English DBpedia types for 93,381 Chinese entities, which is large enough to train our classifiers. To evaluate the performance, we randomly select 50,000 entities for training, and 10,000 ones for test.

Metrics. Notice that our problem is hierarchical classification. Hence, we use hierarchical precision (hP), hierarchical recall (hR) and hierarchical f1-measure (hF) to evaluate the performances of our approach [18]. For an entity $e \in \mathcal{E}$, we denote the set of the ground truth types as t_e and the set of the types found by our approach as p_e. The metric we used for evaluation are defined as follows:

Hierarchical Precision (hP)

$$hP = \frac{\sum_{e \in \mathcal{E}} |t_e \cap p_e|}{\sum_{e \in \mathcal{E}} |p_e|}. \tag{5}$$

Hierarchical Recall (hR)

$$hR = \frac{\sum_{e \in \mathcal{E}} |t_e \cap p_e|}{\sum_{e \in \mathcal{E}} |t_e|}. \tag{6}$$

Hierarchical F1-measure (hF)

$$hF = \frac{2 * hP * hR}{hP + hR}. \tag{7}$$

Comparison with Baseline Methods. We compare our method with baseline methods and the state-of-the-art methods on the test data. The competitors include:

- **SDType.** SDType is a state-of-the-art method proposed by [13]. It uses the same features described in Sect. 3.1.
- **LR.** Logistic Regression Model (LR) is a strong baseline for many tasks. In this competitor, we does not consider the hierarchy structure of types, and train LR models for each type independently. The features are the same as in SDType.
- **CUTE.** The proposed method in this paper.

---

[7] http://data.dws.informatik.uni-mannheim.de/dbpedia/2014/zh/labels_en_uris_zh. nt.bz2.

Table 1. Comparison results on test data.

Method	hP	hR	hF
SDType	0.81	0.69	0.75
LR	**0.89**	0.69	0.78
CUTE	0.88	**0.71**	**0.79**

Effectiveness. The comparison result is shown in Table 1, from which we can see that our method CUTE performs sufficiently well on the test data, and achieves an hierarchical f1-measure of 0.79. SDType method has the worst performance, since it does not consider the hierarchical structure and has a poor prediction performance when an entity has multiple types at a certain granularity. Our method has a high recall and a comparable precision compared to LR. A closer look at the results reveal that our higher recall can be attributed to the sibling policy we used to define positive and negative samples. This policy helps discover more fine-grained types, leading to a higher recall. To see this, we give some examples of fine-grained types for entities found by LR and CUTE in Table 2.

Table 2. Fine-grained types for entities found by LR and CUTE.

Entity Name	LR	CUTE
长江基建	Agent, Organization	Agent, Organization, Company
功夫杀手	Work	Work, Film
林志玲	Agent, Person, Artist, Actor	Agent, Person, Artist, Actor, Model

Table 3. Running time (seconds) of LR and CUTE. *Predicting (1 K)* is the runtime to type 1 K randomly selected entities. *Predicting (10 K)* is the runtime to type all the 10 K entities in the test data set.

Method	Training	Predicting (1 K)	Predicting (10 K)
LR	3,144	40	399
CUTE	745	10	94

Efficiency. We also compare the runtime of LR with that of CUTE on test data. We only compare to LR because the above results show that it is a strong baseline. As shown in Table 3, CUTE is more efficient than LR in both model training and entity typing.

6.3   Typing Chinese Entities

Finally, we use CUTE to type all the Chinese Baidu Baike entities. In total, we get 22,725,365 types for 6,314,273 entities, with an average of 3.6 types per entity. We also estimated the precision (by Eq. 5) by manual evaluation on a sample of 1000 randomly selected entity-type pairs. The precision is 90.2 %.

Since we have typed Chinese entities with DBpedia types, it is possible to compare type distribution of Baidu Baike entities and DBpedia entities. This comparison gives us opportunities to study whether the content of a knowledge base is independent with language used to describe the knowledge. Table 4 shows the top-15 most frequent types of Baike entities. From the table we can see that, the type distributions in Baidu Baike and DBpedia are quite different. For instance, there are 1,056,106 books and 178,689 food entities in Baidu Baike, while there are only 31,029 and 6,337 counterparts, respectively, in DBpedia. This is because there are a large number of novels and food entities in Baidu Baike. Most of them are only famous in China thus are absent in DBpedia. Hence, many entities as well as their knowledge only exist in knowledge bases of local language.

Table 4. Top-15 most frequent types in Baidu Baike, and the type distribution of these types in both Baidu Baike and DBpedia.The percentage after the frequency is the proportion of entities with the type.

Types	Baike Frequency	Rank	DBpedia Frequency	Rank
dbo:Work	2,529,054 (40.1 %)	1	411,295 (9.8 %)	7
dbo:Agent	2,004,923 (31.8 %)	2	1,688,264 (40.3 %)	1
dbo:Person	1,217,988 (19.3 %)	3	1,445,104 (34.5 %)	2
dbo:Place	1,197,263 (19.0 %)	4	735,062 (17.5 %)	3
dbo:WrittenWork	1,098,019 (17.4 %)	5	56,212 (1.3 %)	4
dbo:Book	1,056,106 (16.7 %)	6	31,029 (0.7 %)	41
dbo:Organisation	790,974 (12.5 %)	7	241,286 (5.8 %)	12
dbo:PopulatedPlace	616,022 (9.8 %)	8	478,351 (11.4 %)	5
dbo:ArchitecturalStructure	492,580 (7.8 %)	9	150,254 (3.6 %)	16
dbo:Settlement	462,082 (7.3 %)	10	449,479 (10.7 %)	6
dbo:Building	454,448 (7.2 %)	11	68,582 (1.6 %)	25
dbo:Company	417,010 (6.6 %)	12	58,400 (1.4 %)	27
dbo:Species	211,536 (3.4 %)	13	252,166 (6.0 %)	10
dbo:Eukaryote	207,771 (3.3 %)	14	247,208 (5.9 %)	11
dbo:Food	178,689 (2.8 %)	15	6,337 (0.2 %)	105

7   Related Work

In this section, we review and summarize works that are most relevant to our research. These include works in named entity recognition, fine-grained entity mention recognition and fine-grained entity typing.

Named Entity Recognition. Named Entity Recognition (NER) is a widely used approach for typing problem. However, most of them only support a small set of coarse-grained types, such as Person, Location, and Organization [17]. However, our task focuses more on typing fine-grained types for entities.

Fine-Grained Entity Mention Recognition. One type of approaches for fine-grained entity mention recognition is based on Named Entity Linking (NEL). They first use NEL tools to link mentions to entities which exist in a knowledge base. Then, their types are obtained from the corresponding entity types. However, their performance mainly relies on the linking entities which exist in the knowledge base used [4]. In our setting, it is reported that there are only about 10 % (about 0.4 million) Chinese entities that have corresponding English entities [23]. Hence, it does not work to use this method alone. While in our work, we just use it to construct our training data.

Other approaches for fine-grained entity mention recognition use some hand-crafted features (such as part-of-speech tags, dependency parsing results) and external resources (such as WordNet) to build a classifier to type mentions [9, 11,24]. However, most of their features are extracted from sentences, while our work focuses on using entity-level features available in knowledge bases.

Fine-Grained Entity Typing. The state-of-the-art method for fine-grained entity typing is SDType [13]. However, it does not consider the hierarchical structure between types, and the weight value of features are fixed for all types, which is not suitable for multi-label classification problem as we empirically showed.

8   Conclusion

In this paper, we introduce CUTE for typing Chinese entities with DBpedia types. By using the cross-lingual entity linking method to construct the training data, we propose a multi-label hierarchical classification method. Extensive experiments have verified the efficiency and effectiveness of CUTE.

References

1. Palmero Aprosio, A., Giuliano, C., Lavelli, A.: Automatic expansion of DBpedia exploiting wikipedia cross-language information. In: Cimiano, P., Corcho, O., Presutti, V., Hollink, L., Rudolph, S. (eds.) ESWC 2013. LNCS, vol. 7882, pp. 397–411. Springer, Heidelberg (2013)

2. Auer, S., Bizer, C., Kobilarov, G., Lehmann, J., Cyganiak, R., Ives, Z.G.: DBpedia: a nucleus for a web of open data. In: Aberer, K., Choi, K.-S., Noy, N., Allemang, D., Lee, K.-I., Nixon, L.J.B., Golbeck, J., Mika, P., Maynard, D., Mizoguchi, R., Schreiber, G., Cudré-Mauroux, P. (eds.) ASWC 2007 and ISWC 2007. LNCS, vol. 4825, pp. 722–735. Springer, Heidelberg (2007)
3. Bollacker, K., Evans, C., Paritosh, P., Sturge, T., Taylor, J.: Freebase: a collaboratively created graph database for structuring human knowledge. In: Proceedings of the 2008 ACM SIGMOD International Conference on Management of Data, pp. 1247–1250. ACM (2008)
4. Dong, L., Wei, F., Sun, H., Zhou, M., Xu, K.: A hybrid neural model for type classification of entity mentions. In: Proceedings of the 24th International Conference on Artificial Intelligence, pp. 1243–1249. AAAI Press (2015)
5. Gangemi, A., Nuzzolese, A.G., Presutti, V., Draicchio, F., Musetti, A., Ciancarini, P.: Automatic typing of DBpedia entities. In: Cudré-Mauroux, P., Heflin, J., Sirin, E., Tudorache, T., Euzenat, J., Hauswirth, M., Parreira, J.X., Hendler, J., Schreiber, G., Bernstein, A., Blomqvist, E. (eds.) ISWC 2012, Part I. LNCS, vol. 7649, pp. 65–81. Springer, Heidelberg (2012)
6. Hoffart, J., Suchanek, F.M., Berberich, K., Weikum, G.: Yago2: a spatially and temporally enhanced knowledge base from wikipedia. Artif. Intell. **194**, 28–61 (2013)
7. Lee, T., Wang, Z., Wang, H., Hwang, S.W.: Attribute extraction and scoring: a probabilistic approach. In: 2013 IEEE 29th International Conference on Data Engineering (ICDE), pp. 194–205. IEEE (2013)
8. Lehmann, J., Isele, R., Jakob, M., Jentzsch, A., Kontokostas, D., Mendes, P.N., Hellmann, S., Morsey, M., van Kleef, P., Auer, S., et al.: DBpedia-a large-scale, multilingual knowledge base extracted from wikipedia. Semant. Web J. **5**, 1–29 (2014)
9. Ling, X., Weld, D.S.: Fine-grained entity recognition. In: AAAI. Citeseer (2012)
10. Murdock, J.W., Kalyanpur, A., Welty, C., Fan, J., Ferrucci, D.A., Gondek, D., Zhang, L., Kanayama, H.: Typing candidate answers using type coercion. IBM J. Res. Dev. **56**(3.4), 7:1–7:13 (2012)
11. Nakashole, N., Tylenda, T., Weikum, G.: Fine-grained semantic typing of emerging entities. In: ACL (1), pp. 1488–1497 (2013)
12. Passant, A.: dbrec — music recommendations using DBpedia. In: Patel-Schneider, P.F., Pan, Y., Hitzler, P., Mika, P., Zhang, L., Pan, J.Z., Horrocks, I., Glimm, B. (eds.) ISWC 2010, Part II. LNCS, vol. 6497, pp. 209–224. Springer, Heidelberg (2010)
13. Paulheim, H., Bizer, C.: Type inference on noisy RDF data. In: Alani, H., Kagal, L., Fokoue, A., Groth, P., Biemann, C., Parreira, J.X., Aroyo, L., Noy, N., Welty, C., Janowicz, K. (eds.) ISWC 2013, Part I. LNCS, vol. 8218, pp. 510–525. Springer, Heidelberg (2013)
14. Pedregosa, F., Varoquaux, G., Gramfort, A., Michel, V., Thirion, B., Grisel, O., Blondel, M., Prettenhofer, P., Weiss, R., Dubourg, V., et al.: Scikit-learn: machine learning in python. J. Mach. Learn. Res. **12**, 2825–2830 (2011)
15. Pohl, A.: Classifying the wikipedia articles into the opencyc taxonomy. In: Proceedings of the Web of Linked Entities Workshop in Conjuction with the 11th International Semantic Web Conference, vol. 5, p. 16 (2012)
16. Ponzetto, S.P., Strube, M.: Deriving a large scale taxonomy from wikipedia. In: AAAI, vol. 7, pp. 1440–1445 (2007)
17. Ritter, A., Clark, S., Etzioni, O., et al.: Named entity recognition in tweets: an experimental study. In: Proceedings of the Conference on Empirical Methods in Natural Language Processing, pp. 1524–1534 (2011)

18. Silla Jr., C.N., Freitas, A.A.: A survey of hierarchical classification across different application domains. Data Min. Knowl. Discov. **22**(1–2), 31–72 (2011)
19. Srinivas, S.: A generalization of the noisy-or model. In: Proceedings of the Ninth International Conference on Uncertainty in Artificial Intelligence, pp. 208–215. Morgan Kaufmann Publishers Inc. (1993)
20. Suchanek, F.M., Abiteboul, S., Senellart, P.: Paris: probabilistic alignment of relations, instances, and schema. Proc. VLDB endowment **5**(3), 157–168 (2011)
21. Suchanek, F.M., Kasneci, G., Weikum, G.: Yago: a core of semantic knowledge. In: Proceedings of the 16th International Conference on World Wide Web, pp. 697–706. ACM (2007)
22. Wang, Z., Li, J., Tang, J.: Boosting cross-lingual knowledge linking via concept annotation. In: Proceedings of the Twenty-Third International Joint Conference on Artificial Intelligence, pp. 2733–2739. AAAI Press (2013)
23. Wang, Z., Li, J., Wang, Z., Tang, J.: Cross-lingual knowledge linking across wiki knowledge bases. In: Proceedings of the 21st International Conference on World Wide Web, pp. 459–468. ACM (2012)
24. Yosef, M.A., Bauer, S., Hoffart, J., Spaniol, M., Weikum, G.: Hyena: hierarchical type classification for entity names (2012)

Benchmarking Semantic Capabilities
of Analogy Querying Algorithms

Christoph Lofi[(✉)], Athiq Ahamed, Pratima Kulkarni,
and Ravi Thakkar

Technische Universität Braunschweig, 38106 Brunswick, Germany
lofi@ifis.cs.tu-bs.de

Abstract. Enabling semantically rich query paradigms is one of the core challenges of current information systems research. In this context, due to their importance and ubiquity in natural language, analogy queries are of particular interest. Current developments in natural language processing and machine learning resulted in some very promising algorithms relying on deep learning neural word embeddings which might contribute to finally realizing analogy queries. However, it is still quite unclear how well these algorithms work from a semantic point of view. One of the problems is that there is no clear consensus on the intended semantics of analogy queries. Furthermore, there are no suitable benchmark dataset available respecting the semantic properties of real-life analogies. Therefore, in this, paper, we discuss the challenges of benchmarking the semantics of analogy query algorithms with a special focus on neural embeddings. We also introduce the AGS analogy benchmark dataset which rectifies many weaknesses of established datasets. Finally, our experiments evaluating state-of-the-art algorithms underline the need for further research in this promising field.

Keywords: Query processing · Human-centered information systems · Benchmarking · Analogy processing · Relational similarity · Semantics of natural language

1 Introduction

The increasing spread of the Web and its multitude of information systems call for the development of novel interaction and query paradigms in order to keep up with the ever growing amount of information. These paradigms require more sophisticated capabilities compared to established declarative SQL-style or IR-style keyword queries. Especially *human-centered* query paradigms, i.e. query paradigms which try to mimic parts of natural human communication as for example questions answering and verbose queries require sophisticated semantic processing. A central pattern in human communication which has received only little attention by the information systems research community are analogy queries [1]: in natural speech, analogies allow for communicating dense information easily and naturally by exploiting the semantic capabilities and knowledge of both communication partners. Basically, analogies can be used to map factual and behavioral properties from one (usually better known concept) to

© Springer International Publishing Switzerland 2016
S.B. Navathe et al. (Eds.): DASFAA 2016, Part I, LNCS 9642, pp. 463–478, 2016.
DOI: 10.1007/978-3-319-32025-0_29

another (usually less well known) concept by using different types of similarity assertions, therefore transferring the semantic "essence" from one to another while dropping less important differences for the sake of brevity and simplicity. This is particularly effective for explaining and teaching (e.g., "The Qur'an is the 'Islam Bible'"), but can also be used for querying when only vague domain knowledge is available ("I loved my last vacation in Hawai'i. What place would be similar in East Asia?"). Analogical thinking plays such an important role in many human cognitive abilities that it has been suggested by psychologist and linguists that analogies are the "core of cognition" [2] or even the "thing that makes us smart" [3].

However, adapting this valuable concept of natural communication into information systems proves to be very challenging: on one hand, semantics of analogies very hard to grasp algorithmically as analogy processing heavily relies on human perception, abstract inference, and common knowledge. But furthermore, algorithms which claim to be able to mimic analogical reasoning are hard to evaluate due to the lack of benchmark sets and Gold standards. In its simplest form, the core of analogy processing is measuring *relational similarity* between two pairs of words, e.g., using the example from above, one could say that the Qur'an fulfills a similar role/relation for the Islam religion as does the Bible for the Christian belief. This example also highlights one of the challenges of capturing analogy semantics: of course, the role of the Qur'an is slightly different to the role of the Bible when examined in detail, but still similar enough for explaining either concept in a general discussion. While there are several datasets which are frequently used in researching analogy semantics between word pairs, they usually ignore the *definiteness* of relationships, a second core component of analogy semantics. The definiteness directly affects the usefulness of an analogy for transferring semantics during communication (e.g., the statement 'funny is to humorous as beautiful is to attractive' has a high degree of relational similarity as both words pairs share the same relationship "is a near-synonym of", but still this analogy is not useful for describing either concept as synonymy does capture the semantic essence.) Therefore, in this paper, we discuss the challenge of benchmarking analogy algorithms in detail:

- We define and discuss different properties of analogy semantics, and highlight their importance for the benchmarking process.
- We provide a brief survey of current state-of-the-art analogy algorithms, and highlight their different base assumptions.
- We introduce a new test set for benchmarking analogy algorithms. Our test set is systematically built by expanding existing benchmark sets, and by also incorporating crowdsourcing judgements in order to capture the human aspect of analogy semantics. While we do not seek to fully replace established benchmark datasets, our dataset introduces new qualities not exhibited by previous benchmarks. Especially, we provide a balanced set of test challenges with a wide range of different analogy challenges. This allows to analyse strength and weaknesses of different algorithms in on a more fine-grained level.
- As a proof of concept, we showcase and discuss the evaluation results for two current state-of-the art algorithms using our benchmark test collection, and briefly discuss the implications of the respective results.

2 Foundations and Related Work

2.1 Analogy Semantics and Analogy Queries

Due to the ubiquity and importance of analogies in daily speech, there is long-standing interest in researching the foundations of analogy semantics in the fields of philosophy, linguistics, and in the cognitive sciences, such as [5, 6], or [7]. There have been several models for analogical reasoning in these fields, as for example very early definitions from the Greek philosophers Plato and Aristotele who propose a rather hard to formalize definition based on *shared abstractions* of two concepts [7], while the 18th century philosopher Kant defines an analogy as two pairs of concepts being connected by *identical relationship*s [8]. Other approaches see analogies as a variant of formal logics, i.e. analogy is seen as a special case of *induction* [7] or for performing *hidden inductions* [9]. Another popular model for analogies stems from the field of contemporary cognitive sciences and clarifies some of the vague concepts of Aristotle's view on analogies, and is commonly known as the *structure mapping theory* [10]. Structure mapping is assuming that knowledge is explicitly provided in form of propositional networks of nodes and predicates and claims that there is an analogy whenever large parts of the structural representation of relationships and properties of one object (the source) can be mapped to the representation of the other object (the target). This model resulted in several theoretical computational models, e.g. [11].

The aforementioned analogy definitions are rather complex and hard to grasp computationally. Therefore, most recent works on computational analogy processing rely on the simple *4-term analogy* model which is an extension of the analogy model given by Kant. Basically, a 4-term analogy is given by two sets of word pairs (the so-called *analogons*), with one pair being the source and one pair being the target. A 4-term analogy holds true if there is a high degree of *relational similarity* between those two pairs. This is denoted by $[a_1 : a_2] :: [b_1 : b_2]$, where the relation between a_1 and a_2 is similar to the relation between b_1 and b_2, as for example in $[Qur'an, Muslim] :: [Bible, Christian]$.

This model has several limitations and shortcomings, as we discuss in detail in [1]. For example, the actual semantics of "a high degree of relational similarity" from an ontological point of view is quite unclear, and many frequently used analogies cannot be mapped easily to the 4-term model (as for example the Rutherford analogy which sets a simplified model of atoms in relation to a simplified model of the solar system). Furthermore, the model ignores human perception and abstractions (e.g., the validity of analogies can change over time or even between different communication partner with different background knowledge). Still, this model for analogies is quite popular in computational analogy and linguistic research as it is easy to benchmark, and there exist several recent techniques which can approximate simple relational similarity quite well.

Therefore, in this paper, we argue for an improved interpretation of the 4-term analogy model which we introduced in [12]. The intuition underlying this model is that, basically, there can be multiple relationships between the concepts of an analogon. However, not all of them are relevant for a semantically meaningful analogy– and furthermore, some of them should even be ignored. Therefore, the model introduces the set of *defining relationships*, and an analogy holds true if the defining sets of both analogons are relational similar. For illustrating the difference and importance of this change in semantics, consider the analogy statement [*Tokyo, Japan*] :: [*Braunschweig, Germany*]. Tokyo is a city in Japan, and Braunschweig is a city in Germany, therefore both analogons contain the same "city is located in country" relationship (and this could be considered a valid analogy with respect to the simple 4-term analogy model). Still, this is a poor statement from a semantic point of view because Braunschweig is not like Tokyo at all (therefore, this statement does neither describe the essence of Tokyo nor that of Braunschweig particularly well): the defining traits (relationships) of Tokyo in Japan should at least cover that Tokyo is the single largest city in Japan, and also its capital. There are many other cities which are also located in Japan, but only Tokyo has these two defining traits. Braunschweig, however, is just a smaller and rather unknown city in Germany, and there is nothing particularly special about it (therefore, the defining relationships of both word pairs are not very similar). The closest match to a city like Tokyo in Germany should therefore be Berlin, which is also the largest city and the capital city. Understanding which relationships actually define the essence of an analogon from the viewpoint of human perception is a very challenging problem, but this understanding is crucial for judging the usefulness and value of an analogy statement. Furthermore, the definiteness may vary with different contexts (e.g., the role of Tokyo in Japan in a general discussion vs. the role of Tokyo in Japan in a discussion about fashion trends: here Hamburg/Germany might be a better match than Berlin as Hamburg is often considered the fashion capital of Germany as Tokyo is the fashion capital of Japan).

In short, there can be better or worse analogies based on two core factors (we will later encode the combined overall quality with an *analogy rating*): the *definiteness* of the relationships shared by both analogons (i.e. are the relationships shared between both analogons indeed the defining relationships which describe the *intended semantic essence*), and the *relational similarity* of the shared relationships.

To further clarify the concept of definiteness, consider the analogon [*Bordeaux, France*]. Confronted with this word pair and asked for the most obvious relationships between 'Bordeaux' and 'France', most people would answer "Bordeaux is France's city of wine". Therefore, the analogy [*Bordeaux, France*] :: [*Nappa, UnitedStates*] has a high degree of definiteness, as Nappa is also one of the most famous wine cities of the US. In contrast, [*Bordeaux, France*] :: [*Dallas, UnitedStates*] would have a low definiteness: both analogons contain the same "city in country" relationship, and even more, both are indeed the 9th largest city of their respective country, but still, these relationships would not be the ones which come to people's mind – they are not an "analogy essence".

Based on these observations, we define three basic types of analogy queries (loosely adopted from [12]) which can be used for analogy-enabled information systems:

– *Analogy confirmation* ? : [a_1, a_2] :: [b_1, b_2]
  This query checks if the given analogy statement is true, i.e. it checks if the defining relationships are similar enough (from a consensual human perspective).
– *Analogy completion* ? : [a_1, a_2] :: [$b_1, ?$]
  This query can be used to find the missing concept in a 4-term analogy. (e.g., "What is for the Islam as is the Bible for the Christians?"). This is therefore the most useful

query type in a future analogy-enabled information systems [1]. Solving this query requires identifying the set of defining relationships between a_1 and a_2, and then finding a b_2 such that the set of defining relationships between b_1 and b_2 is similar.

- *Analogon ranking* ? : $[a_1, a_2] :: ?$

  Given a single analogon, this query asks for a ranked list of potential analogons which would result in a valid analogy such that the defining relationships between both analogons is similar. This query is significantly harder from a semantic perspective than completion queries, as there is less information available with respect to the nature of the defining relationships (as there is no on second analogon given, the intended analogon essence is harder to determine.)

- *Analogon ranking multiple-choice* ? : $[a_1, a_2] :: ?\{[b_1, b_2], \ldots, [z_1, z_2]\}$

  A simpler version of the general analogon ranking query are *multiple choice ranking queries* as they are for example used in the SAT benchmark dataset (discussed below). Here, the set of potential result analogons is restricted, and an algorithm would simply need to rank the provided choices instead of freely discovering the missing analogon.

2.2 Algorithmic Analogy Processing

Unfortunately, despite the potential of analogies for querying an information system, developing analogy processing algorithms received only little attention by the database and information systems community. Few early exceptions tried to accommodate analogies in first principle-style knowledge-based systems, but used them only as fallback solutions when strict inference failed, e.g., [13]. Some other systems were based on specialized case-based reasoning techniques [14], introducing a measure of similarity into reasoning. Among early systems, most approaches relied on hand-crafted data sources as for example ontology-based approaches [15], or semi-manual structure-mapping approaches [11]. The first set of techniques which showed good performance and did not require extensive manual curation relied on different natural language processing (NLP) techniques. Especially pattern mining in large Web text collections with subsequent statistical analysis showed promising result, such as [16, 17].

A recent trend from the machine learning and computational linguistics communities is learning word embeddings using Deep Learning techniques. Word embeddings represent each word in a predefined vocabulary with a real-valued vector, i.e. words are embedded in a vector space (usually with 300–600 dimensions). Most word embeddings will directly or indirectly rely on the distributional hypothesis [18] (i.e. words frequently appearing in similar linguistic contexts will also have similar real-world semantics), and are thus particularly well-suited to measure semantic similarity and relatedness between words (which is one of the foundation of the 4-term analogy definition), e.g., see [19]. Early word embeddings were often based on dimensionality reduction techniques (like for example principal component analysis) applied to word-context-co-occurrence matrixes, e.g., [20]. However, in recent years, a new breed of neural approaches relying on Deep Learning neural networks have become popular. Early neural word embeddings trained a complex multi-layer neural network to predict the next word given a sequence of initial words of a sentence [21], or to predict a nearby word given a cue word [4]. Most of these early approaches were very slow, and it could take several months to train a single model. Recent algorithmic advancements could improve on this problem, and current approaches like the popular skip-gram negative sampling approach (SGNS) [4, 22] which uses a non-linear hidden layer neural networks can train a model in just few hours using standard desktop hardware. In this paper, we will focus exclusively on neural word embeddings in the evaluation section as they are the strongest technique available today.

The straight-forward application of word embeddings is computing similarity between two given words [19] by measuring the cosine similarity. However, many (but not all) word embeddings show some very interesting and surprising additional property: it seems that not only the cosine distance between vectors represents a measure for similarity and relatedness, but that also the difference vectors between a word pair carries analogy semantics [23]. For example, the difference between the vector for "man" and "king" seems to represent the concept of being a ruler, and the vector of "woman" plus this concept vector will result in "queen". While the full extend and reasons for this behavior is not fully understood yet, these semantics have been impressively demonstrated for several examples like countries and their capitals, countries and their currencies, or several grammatical relationships (see next section). To a certain extent, these semantics can be attributed to the distributional hypothesis: in natural speech, concepts carrying similar semantics will frequently occur in similar context. Therefore, the aforementioned concept vector should implicitly encode the defining relationships between two concepts as discussed in Sect. 2.1 (i.e.: Tokyo/Japan and Berlin/Germany will likely occur in similar contexts in natural speech, while Braunschweig/Germany will likely appear in different context and will thus have a different concept vector). This interesting property is not yet well understood, and the extends of the semantic expressiveness of neural word embeddings are still unclear.

For example, a word embedding can be used to solve *analogy completion queries* as follows [4]: Given the query $[a_1, a_2] :: [b_1, ?]$, the word embedding will provide the respective word vectors $\vec{a_1}$, $\vec{a_2}$, and $\vec{b_1}$. Then, the vector $\vec{b_2}$ representing the query's solution can be determined by finding the word vector in the trained vector space V which is closest to $\vec{a_2} - \vec{a_1} + \vec{b_1}$ (see Fig. 1) with respect to the cosine vector distance, i.e. $\vec{b_2} = \arg\max\limits_{\vec{x} \in V, \vec{x} \neq \vec{a_2}, \vec{x} \neq \vec{b_1}} \left(\vec{a_2} - \vec{a_1} + \vec{b_1} \right)^T \vec{x}.$

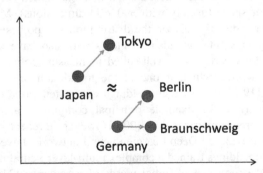

Fig. 1. Example of word embedding vectors reduced to 2-dimensions $\overrightarrow{Tokyo} - \overrightarrow{Japan} + \overrightarrow{Germany} \approx \overrightarrow{Berlin}$

2.3   Test Collections

In the following, we highlight three commonly established benchmark collection for analogy queries, and discuss their strength and weaknesses.

SAT Analogy Challenges. The SAT analogy challenges [24] deserve some special attention due to their importance with respect to previous research and its role in real-world applications. The SAT test is standardized test for general college admissions in the United States. The test features major sections on analogy challenges to assess the prospective student's vocabulary depth and general analytical skills by focusing *on multiple-choice analogon ranking queries*. The analogy challenges contained are based strictly on relational similarity between word pairs, but do not further classify the nature of this relationship or the quality of the analogy itself. As the challenge's original intent is to assess the vocabulary skills of prospective students, it contains many rare words. In order to be able to evaluate the test without dispute, there is only a single correct answer while all other answers are definitely wrong (and can't also be argued for). As a very simple example, consider this challenge from the SAT-dataset: *legend* is to *map* as is: (a) *subtitle* to *translation* (b) *bar* to *graph* (c) *figure* to *blueprint* (d) *key* to *chart* (e) *footnote* to *information*. Here, the correct answer is (d) as a key helps to interpret the symbols in a chart as does the legend with the symbols of a map. While it is easy to see that this answer is correct when the solution is provided, actually solving these challenges seems to be a quite difficult task for aspiring high school students as the correctness rates of the analogy section of SAT tests is usually reported to be around 57 %.

Unfortunately, this benchmark process measures the effectiveness of an algorithm only indirectly as an algorithm only needs to find the best answer – a task which is not too difficult as there is an unambiguous correct answer pair. In the design of our AGS dataset, we will relax this restricted design and introduce different degrees of result quality using a crowd-based analogy rating. This is rooted on the observation that analogies are usually not "correct" per se, but instead are more or less meaningful based on both definiteness and similarity of the involved relationships (see Sect. 2.1). Therefore, our dataset will also have a source word pair and multiple potentially analogous word pairs with an additional human judgement witch rates to the quality of the analogy (the *analogy rating* as discussed in Sect. 3). Depending on the strictness of the intended benchmark (i.e. by adjusting the minimal analogy rating of analogon which should be considered as being correct), our dataset can therefore support challenges with a ranked list of multiple "correct" and "incorrect" pairs.

Mikolov Benchmark Dataset. For evaluating the improved continuous Skip-gram word-embedding presented in [25], Mikolov et al. created a large test set of analogy challenges covering 14 distinct relationships. The evaluation protocol is different compared to the SAT challenge set. The dataset contains 19,558 4-term analogy tuples, and each can be assigned to one of the 14 relationships contained in the dataset. The task of an analogy algorithm to be benchmarked is to predicted the missing element of a given incomplete 4-term analogy, i.e. to solve the analogy completion query $[a, b] :: [c, ?]$.

Nine of these relationships focus on grammatical properties (like for example the relationship "is plural for a noun", e.g., [*mouse, mice*] :: [*dollar, dollars*] or "is superlative", e.g., [*easy, easiest*] :: [*lucky, luckiest*], while five relationships are of a semantic nature (i.e. "is capital city for common country" [*Athens, Greece*] :: [*Oslo, Norway*], "is capital city for uncommon country" [*Astana, Kazakhstan*] :: [*Harare, Zimbabwe*], "is currency of country", "city in state", "male-female version" [*king, queen*] :: [*brother, sister*]. The test set is generated by collecting pairs of entities which are members of the selected relationship either manually or from Wikipedia and DBpedia, and then combining these pairs into 4-term analogy tuples. For example, for the "city in state" relationship, 68 word pairs like [*Dallas, Texas*] or [*Miami, Florida*] are collected, and then combined by a cross product. Interestingly, the dataset contains only 2,467 instead of the 4,556 possible tuples. It is unclear how or why this subset was sampled that way.

A core weakness of this type of test collection is that it focuses only on rather generic relationships, as e.g., "is city in" or "is plural of". The resulting 4-term statements do usually not focus on the defining relationships between the analogon terms, and therefore most of these analogy statements are semantically weak despite high relational similarity (see discussion in Sect. 2.1). In short, this test set does not benchmark if algorithms can capture analogy semantics, but instead focuses purely on relational similarity. This is the core weakness of this dataset which we aim to rectify with our benchmark collection. Furthermore, by design, the Mikolov test set is only suitable for benchmarking analogy completion queries, and is less suitable for analogon ranking queries.

WordRep. In [26], the authors introduce the WordRep benchmark set. This dataset is based on the benchmark collection of Mikolov, and completes all missing tuples using the original word pairs. Furthermore, this test set merges some of the original categories and introduces 12 new ones, for a total of 25 categories. The word pairs for the new categories are derived automatically from both WordNet and Wikipedia. While this benchmark set is significantly larger and more complete than the Mikolov data set, it still shares the same properties, strength, and weaknesses.

3 The AGS Benchmark Collection

In the following, we will highlight the design and creation of our new AGS (Analogy Gold Standard) benchmark collection. It can be downloaded from http://www.ifis.cs.tu-bs.de/data/ags.

The semantics of real-world analogies rely on the perceived similarity and definiteness of relationships covered by the analogons from a human perspective (as discussed in Sect. 2.1 and [12]). This core insight is ignored by all established analogy test sets presented in the last section, as they simply classify statements into "correct" and "incorrect" statements with respect to relational similarity (which, in fact, renders them rather unsuitable for benchmarking analogy semantics despite their original claims, as the quality of analogies largely depends on how meaningful humans consider an analogy to be). This is the core weakness we aim to rectify with AGS by including

Table 1. Example challenge from AGS dataset

Source analogon	Target analogon	Analog rating
sushi : Japan (*food/beverages*) (*defining*)	scallops : Italy	2.57
	currywurst : Germany	4.00
	tacos : Mexico	4.67
	curry : India	4.00
	tortilla : bat	1.00
	hamburger : pen	1.33

consensual human judgements into all aspects of the collection's creation process. In order to allow for benchmarking all four query types introduced in Sect. 2.1, we designed AGS as a collection of analogy challenges. Each challenge consists of one source analogon, and a choice of potential target analogons. Each target analogon has an *analogy rating* attached, which quantifies a consensual crowd-judgement of the analogies perceived quality (from 5: very good analogy to 1: not analogous at all). This rating is an implicit measure of both relational similarity and definiteness, as each measures is hard to elicit from humans individually. As an example, consider the challenge in Table 1. Here, the defining relationship is "is a stereotypical food of the country". Most challenges have 5–6 different target analogons. We specifically took care that some of these challenges have high analogy ratings, some have very low ratings, and some have middle-ground ratings (therefore, each challenge contains a ranked mix of good and bad analogies). Therefore, this design is particularly suitable for benchmarking multiple-choice ranking queries, but in contrast to the SAT-challenges we include also ambiguous and unclear analogies. Thus, algorithms have to decide for a proper ranking instead of simply identifying a single "correct" answer. Furthermore, AGS can also be used for analogy completion queries by only considering analogons with high analogy ratings (e.g. those with an analogy rating ≥4), and using the resulting 4-term statements in a similar fashion as the statements included in the Mikolov and WordRep benchmark sets (i.e., hide one concept from the 4-term statement to be guessed by the algorithm).

Each challenge of the AGS dataset is classified by a topical domain (i.e., what is the context of the intended semantics, as for example geography, or language and grammar. For a full list of included topic domains with the number of challenges and number of resulting analogies, refer to Table 2. Domains are different from the relationship types in

Table 2. List of all topical domain categories used in AGS

Categories	#chall.	#analog.	Categories	#chall.	#analog.
Animals/plants	10	128	House/furniture/clothing	17	169
Automobiles/transportation	5	41	Humans/human relations	3	28
Electrical/electronics	4	42	Medicine/healthcare	3	28
English grammar	11	121	Movies/music	4	61
Food/beverages	11	92	People/profession	12	151
Geography/architecture	9	144	Sports	4	33

Table 3. List of additional AGS classifiers and number of challenges

Definiteness		Knowledge		Domain	
Definite relationship	Indefinite relationship	Common knowledge	Specific knowledge	Intra-domain	Inter-domain
#37	#56	#89	#4	#93	#0

the Wordrep collection: each of our challenges has an individual set of defining relationships). This allows us to drill-down benchmark results in case that certain algorithms show special strength and weaknesses based on the analogy's domain. We also classified each challenge with respect to the *definiteness of the source analogon*'s relationships. While we discussed that semantically meaningful analogies should always use defining relationships, many of the established benchmark datasets and algorithms focus only on relational similarity not caring if the relationships considered are defining or not. Therefore, in our dataset, we included a mix of "analogies" which use similar but not defining relationships (as, e.g., frequently used by the WordRep dataset), and semantically stronger analogies which use similar relationships which are also defining. Both classes are clearly marked in order to allow for experiments focusing on either subset. In addition, we introduced further classifications, as for example if a challenge focuses on *inter-domain* or *intra-domain analogies*. In our current version, the AGS dataset contains only intra-domain analogies, i.e. where both target and source are within the same topic domain. In future versions, we also expect inter-domain analogies which transfer the abstract essence of outwardly different analogons, as for example "The new X9000 tablet is the Ferrari of tablet computers" (i.e. $[X9000, tablets]$ :: $[Ferrari, cars]$, which could carry the semantics that the X9000 is extremely fast and stylish, but also very expensive.) Finally, we also classified if we expect if common knowledge is sufficient to solve an analogy challenge, or if specific domain knowledge is needed (i.e. is it likely enough to build algorithms using general corpora like Wikipedia dumps or general Web crawls, or are specialized corpora needed like for example medical publications.)

Overall, AGS contains 93 challenges classified by topic, specificity of knowledge required, definiteness of the relationships, and whether it is an intra- or inter-domain analogy (see Table 3). Each challenge includes multiple analogies with high, medium, and low analogy ratings for a total of 1040 analogies overall.

Creation of the AGS Benchmark Collection. A core goal of the design of our AGS benchmark collection is to integrate human judgements and human perception deeply into the collection's creation process. Therefore, we rely on a combination of established datasets which already include semantic human judgements, and augment this seed with additional crowdsourcing. As a starting point, we use the WordSim-353 [27] and Simlex-999 [28] benchmark sets. These established datasets are created to benchmark perceived relatedness between word pairs, and each word pair has been judged for relatedness by a large number of people. From these two datasets we selected word pairs as source word pairs for our AGS challenges based on their relatedness (assuming that such pairs will be diverse and semantically meaningful). We filtered the word pairs using expert judgements from our side with respect to the pair's potential to serve as a semantically meaningful analogon, leaving 93 pairs. For expanding a single word pair into proper AGS challenges with multiple analogies, we relied on using the CrowdFlower.com crowdsourcing platform to obtain suitable target analogons and analogy ratings. Crowdsourcing is a powerful technique to outsource small tasks requiring human intelligence to a large pool of people. However, a central challenge of crowdsourcing is controlling the result quality [29]. Therefore, according to the insights in [29], we split the rather complex task of creating our challenges in several smaller tasks which are easier to control using traditional quality control mechanisms like Gold questions, averaging, and majority voting. The first of these smaller steps was to classify each source pair into one of the 12 topic domains presented in Table 2. This is followed by a second crowdsourcing task where we ask crowd workers to provide target analogons expanding a given source pair. We used the topic domains to recruit crowd workers who felt particularly confident in that domain. For extending the source analogon, we asked specifically for some examples sharing the same essence (i.e. the same defining and similar relationships), but also for some bad analogons which are either unrelated or are related but have different essence. In a final crowdsourcing task, we ask multiple workers to assess each target analogon with respect to the analogy rating, averaging the individual judgements.

For each of the tasks described above, we only used native English speakers, and for each work package we combined the input or judgements of 5 different workers. Furthermore, each worker had to perform a quick pre-assessment task (confirming or rejecting presented analogies), and only crowd worker who could solve this task correctly were allowed to participate.

4 Benchmark Protocols and Benchmark Results

In this section, we define different benchmark protocols which can be used to benchmark a given analogy algorithm with respect to the query types identified in Sect. 2.1. Each is covered by a brief pseudocode algorithm (which focuses on a single challenge. Of course, for a full benchmark, these algorithms need to be executed for each AGS challenge). We frequently rely on selecting "correct" and "incorrect" analogies from AGS challenges. This is realized by a user defined threshold for the analogy rating allowing to adjust the strictness of the benchmark. In future, we will introduce benchmark protocols which will further differentiate result quality using numerical analogy ratings instead of working with Boolean correctness. We omitted the test protocol for open analogon ranking queries in this paper as during our experiments, none of the currently available algorithms could handle that query type in a convincingly.

Analogy Confirmation Queries $? : [a_1, a_2] :: [b_1, b_2]$. This benchmark protocol evaluates performance of a given algorithm with respect to analogy confirmation queries, checking if it can distinguish "correct" analogies from "incorrect" ones. The correctness of AGS statements is based on the analogy rating of target analogons, i.e. those with an analogy rating exceeding a minimal threshold are considered correct, and those with a rating lower than a given threshold are considered incorrect (using Table 2

and minimal correct threshold of 4, and max incorrect threshold of 2, $[sushi, Japan]$:
: $[tacos, Mexico]$ is a correct statement, $[sushi, Japan] :: [scallops : Italy]$ is excluded
from the benchmark as it is neither clearly correct nor incorrect, and $[sushi, Japan] ::$
$[hamburger, pen]$ is an incorrect statement.) The final result of this benchmark covering
multiple challenges is the percentage of correctly confirmed statements contained in a
given subset of AGS challenges.

Analogy Confirmation Benchmark(Challenge c)
Feature Required from Algorithm:

- Algorithm needs to be able to confirm or reject a given analogy statement

  Parameters:

- *min_correct*: Minimal analogy rating to consider a target analogon as "correct"
- *max_incorrect*: Maximal analogy rating to consider a target analogon as "incorrect"

  Output:

- *num_success, num_fail*: Number of correctly and incorrectly processed statements

  Protocol:

- correct statements = Combine source *c.source* with all targets $t \in c.target$ which
  have $t \geq min\_correct$
- incorrect statements = Combine *c.source* with all targets $t \in c.target$ which have
  $t < max\_incorrect$
- Check if algorithm confirms correct statements as analogies
- Check if algorithm rejects incorrect statements as analogies
- Return respective success and failure numbers

Analogy Completion Queries ? : $[a_1, a_2] :: [b_1, ?]$. In this benchmark protocol, we
check if the given algorithm can complete "correct" analogy statements. The final result
of this benchmark is the percentage of correctly completed statements contained in a
given subset of AGS challenges.

Analogy Completion Benchmark(Challenge c)
Feature Required from Algorithm:

- Algorithm needs to be able to complete an analogy statement $[a_1, a_2] :: [b_1, ?]$

  Parameters:

- *min_correct*: Minimal analogy rating to consider a target analogon as "correct"
- Output:
- *num_success, num_fail*: Number of correctly and incorrectly processed statements

  Protocol:

- correct statements = Combine source *c.source* with all targets $t \in c.target$ which
  have $t \geq min\_correct$
- From each correct statement, drop the second concept of the target analogon

- Check if the algorithm can correctly predict the dropped concept
- Return respective success and failure numbers

Analogon Ranking Multiple-Choice Queries ? : $[a_1, a_2] :: ?\{[b_1, b_2], \ldots, [z_1, z_2]\}$. In this simple ranking benchmark, we evaluate if a given analogy algorithm can select the best target analogon for a given source analogon from a limited list of candidates. In our future works, we will extend this protocol to consider rank correlation instead of focusing only on the best analogon. The final result is the percentage of correctly answered challenges.

```
Analogy Ranking Multiple Choice Benchmark(Challenge c)
```
Feature Required from Algorithm:

- Algorithm needs to be able to measure and quantify the quality of analogy between the analogon $[a_1, a_2]$ and $[b_1, b_2]$ (e.g., the relational similarity of the defining relationships of each analogon)

Output:

- *success*: Boolean result indicating if challenge was solved correctly

Protocol:

- statements = Combine source *c.source* with all targets $t \in$ *c.target*
- Measure quality of analogy for each statement
- If the statement with highest measured quality is also the statement with the highest analogy rating in AGS, return success. If not, return failure.

Benchmark Results. As a proof of concept, we present example benchmark results for two implementations of neural word embeddings in this section. We use our AGS collection with the aforementioned benchmark protocols. The algorithms under consideration are the well-known word2vec implementation by Mikolov et al. [25], and the Glove implementation by Pennington et al. [22]. Both algorithms were trained on a dump of Wikipedia, using only the implementation's predefined default parameters. Besides benchmarks covering the full extent of AGS, we also focus on different classification aspects (like only focusing either definite challenges or indefinite ones, see classification in Sect. 3).

We used the following parameters for our benchmark protocols: minimal analogy rating threshold for correct analogies of 4·0, and maximal threshold for incorrect analogies of 2·0. The results for completion and multiple-choice ranking queries are summarized in Table 4. Confirmation queries and open-rank queries are not directly supported by word-embedding based algorithms, and we therefore excluded them from this evaluation.

In general, the measured results of both word2vec and Glove on our AGS collection are rather weak. However, evaluations of the same algorithms on other benchmark sets like Wordrep showed slightly better results (see [26], around 0·25 overall accuracy). This can be explained by that fact that Wordrep uses some very limited set of relationships types which are, in comparison, rather simple in their

Table 4. Benchmark results

Protocol	Algorithm	Overall	Common knowledge	Uncommon knowledge	Definite	Indefinite
Completion	Word2Vec	0.1786	0.1781	0.1851	0.225	0.1481
	Glove	0.1960	0.1941	0.2222	0.2562	0.1563
Ranking multiple choice	Word2Vec	0.3026	0.3030	0.30	0.3125	0.2954
	Glove	0.3289	0.3181	0.30	0.3437	0.3181

semantic nature (e.g., plurals like [*fish* : *fishes*] :: [*pig*, *pigs*], or simple "city is located in" relationships). Those datasets were mostly created by mining Wikipedia, DBpedia, or Wordnet for generating a large number of example analogons using the same relationships – and the same Wikipedia texts are used to train the neural word embeddings both Glove and word2vec use. In our AGS dataset, we use analogies as provided by real humans which are inherently more complex from a semantic point of view. Therefore, our results indicate that, while algorithm relying on word embeddings might be able to deal with simple relational similarity quite well, mastering semantically rich analogy processing still requires a significant amount of future research.

5   Summary and Outlook

In this paper, we presented and discussed the challenge of benchmarking analogy processing algorithms. Such algorithms will be an important building block of future human-centered information systems trying to understand the finer semantics of natural language. Unfortunately, despite the potential importance of analogies in future query paradigms, there is still no clear definition of the intended semantics of analogy queries. Therefore, we provided several discussions focusing on that topic, and derived a set of core properties of analogy semantics. Furthermore, we highlighted basic query types and discussed how they could be benchmarked. Based on these results, we created the AGS analogy benchmark dataset, which aims to rectify several shortcomings of existing benchmark datasets. Especially, our dataset is not automatically generated from structured data sources, but instead relies on crowdsourcing and a large number of human judgements. Furthermore, we explicitly focus on semantically rich analogies instead of limiting ourselves to the significantly weaker special case of only relationally similar word pairs. Finally, we designed the dataset in such a way that all our identified query types could be benchmarked with a single dataset.

From an algorithmic point of view, the recent years have brought several impressive advancements in language understanding, and especially a new breed of deep-learning neural embedding techniques showed very impressive results for challenges related to analogy processing like measuring semantic similarity or relational similarity. Unfortunately, these algorithms do not show strong results on our new benchmark dataset, thus further emphasizing the need for continued future research in this field.

References

1. Lofi, C.: Analogy queries in information systems – a new challenge. J. Inf. Knowl. Manage. **12**, 1350021 (2013)
2. Hofstadter, D.R.: Analogy as the core of cognition. In: The Analogical Mind, pp. 499–538 (2001)
3. Gentner, D.: Why we're so smart. In: Language in Mind: Advances in the Study of Language and Thought, pp. 195–235. MIT Press (2003)
4. Mikolov, T., Yih, W., Zweig, G.: Linguistic regularities in continuous space word representations. In: Conference of the North American Chapter of the Association for Computational Linguistics: Human Language (NAACL-HLT), Atlanta, USA (2013)
5. Gentner, D., Holyoak, K.J., Kokinov, B.N. (eds.): The Analogical Mind: Perspectives from Cognitive Science. MIT Press, Cambridge (2001)
6. Itkonen, E.: Analogy as Structure and Process: Approaches in Linguistics, Cognitive Psychology and Philosophy of Science. John Benjamins Pub. Co., Amsterdam (2005)
7. Shelley, C.: Multiple Analogies in Science and Philosophy. John Benjamins Pub., Amsterdam (2003)
8. Kant, I.: Critique of Judgement. Hackett, Indianapolis (1790)
9. Juthe, A.: Argument by analogy. Argumentation **19**, 1–27 (2005)
10. Gentner, D.: Structure-mapping: a theoretical framework for analogy. Cogn. Sci. **7**, 155–170 (1983)
11. Gentner, D., Gunn, V.: Structural alignment facilitates the noticing of differences. Mem. Cogn. **29**, 565–577 (2001)
12. Lofi, C., Nieke, C.: Modeling analogies for human-centered information systems. In: Jatowt, A., et al. (eds.) SocInfo 2013. LNCS, vol. 8238, pp. 1–15. Springer, Heidelberg (2013)
13. Blythe, J., Veloso, M.: Analogical replay for efficient conditional planning. In: National Conference on Artificial Intelligence (AAAI), Providence, Rhode Island, USA (1997)
14. Leake, D.: Case-Based Reasoning: Experiences, Lessons, and Future Directions. MIT Press, Cambridge (1996)
15. Forbus, K.D., Mostek, T., Ferguson, R.: Analogy ontology for integrating analogical processing and first-principles reasoning. In: National Conference on Artificial Intelligence (AAAI), Edmonton, Alberta, Canada (2002)
16. Bollegala, D.T., Matsuo, Y., Ishizuka, M.: Measuring the similarity between implicit semantic relations from the web. In: International Conference on World Wide Web (WWW), Madrid, Spain (2009)
17. Davidov, D.: Unsupervised discovery of generic relationships using pattern clusters and its evaluation by automatically generated SAT analogy questions. In: Association for Computational Linguistics: Human Language Technologies (ACL:HLT), Columbus, Ohio, USA (2008)
18. Harris, Z.: Distributional structure. Word **10**, 146–162 (1954)
19. Lofi, C.: Measuring semantic similarity and relatedness with distributional and knowledge-based approaches. Database Soc. Jpn. J. **14**, 1–9 (2016)
20. Ştefănescu, D., Banjade, R., Rus, V.: Latent semantic analysis models on Wikipedia and TASA. In: Language Resources Evaluation Conference (LREC), Reykjavik, Island (2014)
21. Mnih, A., Hinton, G.E.: A scalable hierarchical distributed language model. Adv. Neural Inf. Process. Syst. **21**, 1081–1088 (2009)
22. Pennington, J., Socher, R., Manning, C.D.: Glove: global vectors for word representation. In: Conference on Empirical Methods on Natural Language Processing (EMNLP), Doha, Qatar (2014)

23. Collobert, R., Weston, J., Bottou, L., Karlen, M., Kavukcuoglu, K., Kuksa, P.: Natural language processing (almost) from scratch. J. Mach. Learn. Res. **12**, 2493–2537 (2011)
24. Littman, M., Turney, P.: SAT Aanalogy Challange Dataset. http://aclweb.org/aclwiki/index.php?title=SAT_Analogy_Questions_(State_of_the_art)
25. Mikolov, T., Sutskever, I., Chen, K., Corrado, G.S., Dean, J.: Distributed representations of words and phrases and their compositionality. Adv. Neural Inf. Process. Syst. **26**, 3111–3119 (2013)
26. Gao, B., Bian, J., Liu, T.-Y.: WordRep: a benchmark for research on learning word representations. In: ICML Workshop on Knowledge-Powered Deep Learning for Text Mining, Beijing, China (2014)
27. Finkelstein, L., Gabrilovich, E., Matias, Y., Rivlin, E., Solan, Z., Wolfman, G., Ruppin, E.: Placing search in context: the concept revisited. In: International Conference on World Wide Web (WWW), Hong Kong, China (2001)
28. Hill, F., Reichart, R., Korhonen, A.: SimLex-999: evaluating semantic models with (genuine) similarity estimation. Prepr. Publ. arXiv. arXiv:1408.3456 2014
29. Lofi, C., Selke, J., Balke, W.-T.: Information extraction meets crowdsourcing: a promising couple. Datenbank-Spektrum. **12**, 109–120 (2012)

Textual Data

Textual Data

Deep Learning Based Topic Identification and Categorization: Mining Diabetes-Related Topics on Chinese Health Websites

Xinhuan Chen[1]([✉]), Yong Zhang[1], Jennifer Xu[2], Chunxiao Xing[1], and Hsinchun Chen[1,3]

[1] Research Institute of Information Technology, Tsinghua National Laboratory for Information Science and Technology, Department of Computer Science and Technology, Tsinghua University, Beijing, China
xh-chen13@mails.tsinghua.edu.cn,
{zhangyong05,xingcx}@tsinghua.edu.cn, hchen@eller.arizona.edu
[2] Department of Computer Information Systems, Bentley University, Waltham, USA
jxu@bentley.edu
[3] MIS Department, University of Arizona, Tucson, USA

Abstract. As millions of people are diagnosed with diabetes every year, the demand for information about diabetes continues to increase. China is one of the countries with a large population of diabetes patients. Many Chinese health websites provide diabetes related news and articles. However, because most of the online articles are uncategorized or lack a clear topic and theme, users often cannot find their topics of interest effectively and efficiently. The problem of health topic identification and categorization on Chinese websites cannot be easily addressed by applying existing approaches and methods, which have been used for English documents, in a straightforward manner. To address this problem and meet users' demand for diabetes related information needs, we propose a deep learning based framework to identify and categorize topics related to diabetes in online Chinese articles. Our experiments using datasets with over 19,000 online articles showed that the framework achieved a higher effectiveness and accuracy in categorizing diabetes related topics than most of the state-of-the-art benchmark approaches.

Keywords: Text classification · Deep learning · Healthcare · Chinese

1 Introduction

Diabetes has become one of the most common chronic diseases that affect millions of people's lives worldwide. According to the survey by the Chinese Diabetes Society, 98.4 million people were diagnosed with diabetes and 13 million people died of diabetes in China in 2013. To meet the increasing demand for health information and knowledge many health related websites in China provide various resources and services, including health news, articles, discussion forums, and online

S.B. Navathe et al. (Eds.): DASFAA 2016, Part I, LNCS 9642, pp. 481–500, 2016.
DOI: 10.1007/978-3-319-32025-0_30

patient communities. A tremendous amount of high-quality health news and articles are contributed by healthcare professionals and experts.

However, most of the news and articles are uncategorized, making it very time-consuming and overwhelming for users to browse and search for information about specific topics. Automatically identifying topics out of uncategorized health related articles could be very useful for various types of health information users, including patients and their family members, healthcare professionals (e.g., physicians and nurses), and researchers. Especially, it can help newly diagnosed patients find valuable educational materials for self-management of their diseases and health conditions more effectively. It can also help healthcare professionals and researchers quickly learn the health and disease topics in which patients are most interested.

Many approaches and methods have been proposed for identifying and categorizing topics in English articles in various application domains, including the healthcare and medical domain [1]. Unfortunately, little work has been done in the context of health related topic identification and categorization of online Chinese articles.

The problem of health topic identification and categorization of online Chinese articles is quite challenging. It cannot be easily addressed by applying existing approaches and methods, which have been used for English articles, in a straightforward manner due to the difference between Chinese and English, the lack of Chinese medical lexicons, and the special characteristics of Chinese online health articles. First, Chinese is based on ideographic writing systems, whose structure and grammar are quite different from those of English, which is based on alphabetic. For instance, since there is no space between Chinese characters, it is more difficult to parse Chinese sentences into unambiguous word segments. Second, many prior studies adopt a standard medical knowledge base, UMLS (Unified Medical Language System), to extract medical terms and features when categorizing English articles. Unfortunately, there has not been a standard Chinese medical lexicon available. Third, there has been a lack of widely adopted standards for categorizing online Chinese health articles. Consequently, the topic categories of many online Chinese articles are wrong or misleading.

To address these challenges, we develop a Chinese domain lexicon and adopt a professional vocabulary related to diabetes and incorporate them into our topic identification and categorization framework. More importantly, our framework is based on the recently developed deep learning approach. Deep learning [2] is a promising machine learning approach based on artificial neural networks. Among many different deep learning architectures, the Deep Belief Network (DBN) has been employed in many image processing and speech recognition applications. However, this approach has not been widely employed in text categorization applications because of the information loss problem. In this research, we develop a DBN-based model to identify topics related to diabetes from Chinese online articles. We focus on articles about diabetes because it is one of the most widely studied chronic diseases and there are many diabetes patients in China. Our experiments show that our approach outperforms the state-of-the-art text categorization techniques for diabetes-related articles.

The remainder of this paper is organized as follows. We present a review of literature on deep learning and topic identification and categorization, and then describe our DBN-based deep learning framework. Next, we report on our experiments and discuss the results. Finally, we conclude this paper.

2 Related Work

In this section, we review related work about text categorization methods, especially the deep learning methods, and their application in health-related domains.

Text categorization is a set of important and well-developed classification methods for categorizing the growing number of electronic documents worldwide. Text categorization has been widely used in natural language processing and information retrieval applications, including Web page classification, spam filtering, email routing, genre classification, readability assessment, and sentiment analysis [3, 4]. Most of the existing research work in text categorization has focused on supervised machine learning methods by classifying text based on words in training documents. These approaches include Bayesian classifiers [5], Decision Tree (DT) [6], support vector machines (SVMs) [7], K-Nearest Neighbor (KNN) [8], Neural Networks [9], and combined approaches [10]. Some of these methods are selected as our benchmarks in the experiments.

Deep learning is a recently proposed classification approach. It refers to a set of machine learning techniques based on artificial neural networks, where many layers of information processing units stack up to form a hierarchical architecture. Among the various deep learning architectures, the Deep Belief Network (DBN) is widely used in the processing of images [2], audios [11], etc. Because images and sounds can be easily represented as input vectors using fixed feature sets, deep learning is a natural technique for image processing and speech recognition applications. However, only a limited number of studies have employed the deep learning techniques to categorize text documents. Liu [12] used DBN and SVM to classify text documents in a Chinese corpora. Wang et al. [13] concentrated on modeling the semantic relationship between questions and their answers using simple textual features, and presented a DBN to model the semantic relevance between questions and their answers. This is due to the difficulty of finding appropriate feature vectors to represent text. In other words, when representing text documents, the feature vectors are usually based on bag-of-words or other feature extraction methods, which often cause information loss. As a result, it remains a major challenge to use deep learning techniques in text categorization applications. In this study, we identify and combine feature vectors using different methods to represent text in DBN-based model with better categorization performance.

As a data mining and text mining approach [14, 15], classification techniques have also been used to categorize electronic medical documents (e.g., medical literatures and clinical records) and Web documents into meaningful topics in the health-related domains. These studies often rely on a standard domain lexicon during the process of document categorization. Velupillai et al. [16] reported an assertion system (pyConTextNLP) from English to Swedish (pyConTextSwe) by creating an optimized assertion lexicon for clinical Swedish. Liu et al. [17] assessed the effectiveness of several learning models (Naïve Bayes, DT, KNN, SVM) against a number of performance metrics based on a relational graph database of clinical entities. Minarro-Giménez et al. [18] applied the deep learning techniques to medical corpora to test its potential for improving the accessibility of medical knowledge by identifying the relationships. Sibunruang and Polpinij [19] leveraged an ontology of Cancer Technical Term Net to select keyword-based features, where this ontology is used as a lexicon. However, there has not been a standard medical lexicon available in Chinese, not to mention a medical ontology. This adds more difficulty to the task of categorizing medical and health related documents in Chinese.

Some researchers have included domain knowledge in the classification process to improve the performance. Sinha and Zhao [20] compared the performance of seven classification methods with and without incorporating domain knowledge. They found that incorporation of domain knowledge significantly improves classification performance. Liu et al. [21] proposed a classification method for the knowledge cards written in Japanese and Chinese using domain-specific dictionary. In this research, we develop a Chinese domain lexicon and a professional vocabulary for diabetes and incorporate them into the deep learning framework to identify topics related to diabetes from Chinese Web articles. The following section describes in detail diabetes-related topic categorization using deep learning (DBN) based classification.

3   Research Design

In this section, we propose our deep learning based topic categorization framework, which is used to find diabetes-related topics on Chinese health websites. We provide a formal definition for the topic categorization problem before presenting the framework. We then introduce the DBN-based model in detail.

3.1   Problem Definition

The topic categorization problem can be formally defined as follows:

Definition 1. Given an article collection A and a category collection C, the labeled category list for each article $a \in A$ is defined as:

$$L_a = [c_1, c_2, \ldots, c_i, \ldots, c_n], \ n \geq 1, \ c_i \in C$$

where c_1 is the primary category of each article. The output of our topic categorization framework is the primary category for each article.

3.2   The Topic Identification and Categorization Framework

Our framework includes four stages: data collection, data preprocessing, model training, and topic identification. Figure 1 presents this framework.

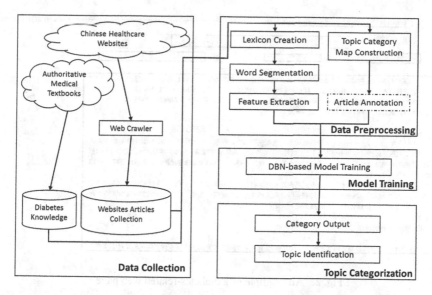

Fig. 1. The topic identification and categorization framework

Data Collection. We developed an automated web crawler to fetch and download online articles from Chinese health websites. We used the keyword "diabetes" to locate diabetes-related pages. Figure 2 presents an example of collected pages in our dataset. Text parsers were used to extract various fields from the pages including article ID, URL, title, article source, posted time, and article body as shown in Fig. 2. Navigation paths (e.g., Home Page > Diabetes Information > Child Diabetes) that may provide category information were also extracted. Note that not all websites contain navigation paths. We chose the pages with navigation paths as the training and testing data in the experiments. Additionally, some pages contain tags or keywords (e.g., "Child Diabetes", "Diabetes Complications") that highlight the focuses of the content. These tags and keywords were extracted as well. We focus on the article body of the pages and refer to them as articles in the following sections.

From all the navigation paths in the Web pages, we extracted 26 distinct diabetes-related terms, based on which we created a professional diabetes vocabulary. We also found and compiled definitions and explanations for these 26 terms based on classical books for Medicine and Diabetes in Chinese. For example, the term 低血糖 (Hypoglycemia) is defined as 低血糖反应是最常见的反应。常见于胰岛素过量、注射胰岛素后未按时进餐或活动量过大所致 (Hypoglycemia is low blood sugar. It is one of the most common symptoms of diabetes, often caused by insulin overdose, missing meals after insulin injections, or excessive physical activities.). In this study we used this professional diabetes vocabulary as part of the domain knowledge for helping the topic categorization.

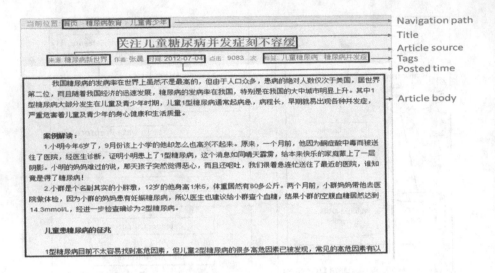

Navigation path
Title
Article source
Tags
Posted time

Article body

Fig. 2. An example of a diabetes-related web page

Data Preprocessing. Data preprocessing prepares the data for the model training stage in our framework. Data preprocessing consists of five steps: *lexicon creation, word segmentation, topic category map construction, article annotation*, and *feature extraction*.

Lexicon Creation. A domain lexicon, which contains terms in a specific domain, is usually used for feature extraction in text mining applications. Unfortunately, there has not been a standard domain lexicon in Chinese for diabetes. To build this lexicon, we combined entries from Diabetes Dictionary App [22], which was the only available commercial mobile application for Chinese diabetes patients, and the extracted tags from diabetes-related Web pages to obtain a relatively complete Chinese diabetes lexicon. The resulting lexicon contains 1,065 terms related to diabetes including its medication, treatment, care, and prevention.

Word Segmentation. We used a Chinese word segmentation tool, ICTCLAS, to remove stop words and perform word segmentation for the content bodies of Web pages.

Topic Category Map Construction. Our diabetes topic category map is a tree structure with nested levels of topic categories related to diabetes. The top level consists of main topic categories, each of which is broken down to lower-levels of sub-categories. The map was built based on both the professional vocabulary (with the 26 terms) and the navigation paths extracted from the web pages in the data collection stage. Because most navigation paths of the web pages comprised non-professional, layperson terms, we made a semantic mapping between the professional vocabulary and the extracted navigation paths with the help of domain experts. Figure 3 presents the resulting topic category map with six main categories on the top level and sub-category levels. Because of the space limit, some lower-level sub-categories are not shown on the map.

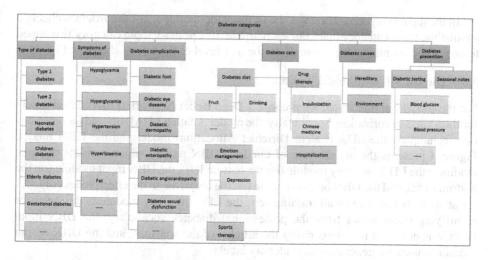

Fig. 3. Diabetes topic category map

Article Annotation. This is an optional step in this framework and can be skipped if all the collected Web pages contain navigation paths that can be used as category labels. For pages without navigation paths, they should be manually annotated based on the diabetes topic category map for the model training and testing purposes. In this research we annotated a part of our datasets (see Experiments for details about the datasets). Each article was assigned one main label but might also had one or more secondary labels.

Feature Extraction. The features used as input for the DBN-based model are then extracted from the articles. In prior studies, the bag-of-words (BOW) approach has often been employed to generate features for text data in deep learning models. In this approach, each distinct word in an article is treated as a feature. However, this approach can result in a large but sparse feature vector, significantly affecting the performance of DBN models [23]. To reduce the dimensionality of the feature vectors, we combined 1,065 diabetes related terms from our Chinese diabetes domain lexicon and 123 feature values using the TFIDF method, which is a numerical statistical method that finds weights of the most frequently used terms in an article. The first 1,065 features were binary, indicating whether the corresponding term appears in the article or not. The remaining 123 features were calculated by obtaining the 123 greatest weights from TFIDF output using a certain threshold. As a result, each article was represented as a feature vector of 1,188 dimensions. We will demonstrate the effectiveness of the feature set in the experiments.

Model Training and Topic Categorization. The DBN-based model will be described in detail in the next section. The feature vectors representing the articles serve as the inputs to the DBN input module. For each article, the DBN output module produces 26 probability values, each of which corresponds to one category node on the topic category map (see Fig. 3). At the end of this stage, the DBN-based model selects the category with the maximum probability value as the primary category for the article (as defined in Definition 1).

In the topic categorization stage, the primary topic category of the article is checked against the topic category map. For each article, this primary topic category is mapped to one of the six main category node on the first level of the tree as shown in Fig. 3.

3.3   The DBN-Based Model Training

For the model training stage, we propose a Deep Belief Network (DBN) based model for the topic categorization. Specifically, the model is called LDADBN and incorporates two techniques: the LDA (Latent Dirichlet Allocation) model and the DBN model. Figure 4 presents the architecture of our LDADBN model, which consists of three modules: the LDA auxiliary module (on the bottom left), the DBN input module (on the bottom right), and the DBN output module (on the top). Due to the effectiveness of LDA for short texts [24] and small training sets, the LDA auxiliary module is designed for identifying latent topics from the professional diabetes vocabulary; the DBN input module is designed for categorizing the articles of the websites; and the DBN output module is used for generating the category labels.

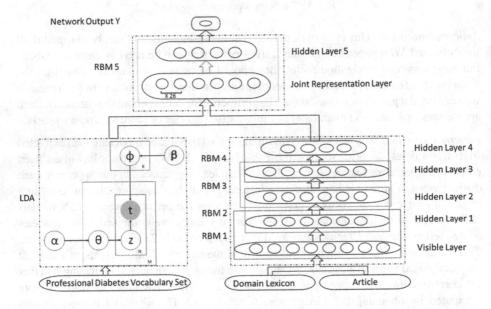

Fig. 4. The LDADBN model architecture

The LDA Auxiliary Module. The LDA model is an unsupervised machine learning technique for mining latent topics from text collections [25] and can provide supplementary information for deep learning models. Note that in the LDA model, the latent topics are different from the topic categories mentioned elsewhere in this paper. The LDA generative process can be graphically represented using the plate notation as shown in the LDA module in Fig. 4. In this module, z denotes a hidden topic; t is a word or term in a document; M is the number of documents in the data set, K is the number of latent topics; θ and ϕ are Dirichlet priors with hyper-parameters α and β respectively.

In this model, the shaded and unshaded nodes represent observed and latent variables respectively. An arrow corresponds to a conditional dependency between two variables. For example, the arrow between the Dirichlet prior θ and the hyper-parameter α represents the conditional probability of θ given α. In this figure, a box indicates repeated sampling with the number of repetitions given by the variable in the bottom of the box (e.g., N). The generative process is described as follow:

1. For each document, draw a multinomial distribution over hidden topics, $\theta \sim Dirichlet(\alpha)$;
2. For each hidden topic, draw a multinomial distribution over words, $\Phi \sim Dirichlet(\beta)$;
3. For a word t_d in a document d, sample a hidden topic z_d from θ_d, and then sample the word t_d from Φ_{z_d};

Our LDA auxiliary module takes the professional diabetes vocabulary as input. We treat each term's definition and explanation as a document d and set K = 26, which is the number of terms in the professional diabetes vocabulary, as well as the number of categories in the topic category map (see Fig. 3).

To estimate the latent parameters in the LDA module, we perform inference using the Gibbs Sampling algorithm. After LDA finishes the parameter estimation process, we obtain the parameter values for all terms and normalize the scale of each feature of each term by θ to generate feature vectors with K dimensions.

The DBN Input Module. A DBN uses a multilayered architecture that consists of one visible layer and one or more hidden layers as shown in Fig. 4. The visible layer of a DBN accepts the feature vectors of input data and delivers the data to the hidden layers [2]. DBN uses RBMs (Restricted Boltzmann Machines) as the building blocks for each layer. Several RBMs stack up to form a DBN. For example, the red box in the bottom-right of Fig. 4 represents the first RBM in this DBN. The green, blue, and purple boxes represent the second, third, and fourth RBMs, respectively. An RBM is a generative stochastic artificial neural network that consists of only two layers: the input layer and output layer as shown in Fig. 5, in which the input nodes and output nodes are denoted by I's and O's, respectively. Each layer may have one or more nodes (or neurons, units). Usually, between-layer links are learned using the training data to model the probability distributions of the inputs. However, no links are allowed to exist between nodes on the same layer, which is why this type of neural network is called restricted.

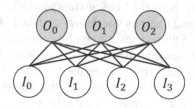

Fig. 5. The graphical representation of an RBM

The training parameters of an RBM contain the weights, w, of the links between layers and the node biases b. Each training iteration for an RBM consists of two phases. In the first phase, the states of nodes in the output layer are determined by transforming the states of nodes in the input layer using a sigmoid activation function. In the second phase, the states of the nodes in the input layer are reconstructed based on the output layer nodes' states. The process can be mathematically expressed as follow:

$$p\left(O_k = 1|I\right) = sigmoid\left(-b_k - \sum_j I_j w_{kj}\right) \tag{1}$$

$$p\left(I_j = 1|O\right) = sigmoid\left(-b_j - \sum_k O_k w_{kj}\right) \tag{2}$$

where O_k and I_j are the states for the k-th node in output layer and the j-th node in the input layer, respectively. The sigmoid activation function is defined as:

$$sigmoid(x) = 1/(1 + e^{-x}) \tag{3}$$

The nodes in the input and output layers are binary stochastic variables with values 0 or 1, corresponding to off or on of the nodes in the learning process.

After each iteration the parameters are updated based on the states of the input and output nodes. The weights w_{jk} are updated based on

$$\Delta w_{jk} = \varepsilon\left(\left\langle I_j O_k \right\rangle_{input} - \left\langle I_j O_k \right\rangle_{recon}\right) \tag{4}$$

where ε is the learning rate between 0 and 1, $\left\langle I_j O_k \right\rangle_{input}$ is the pairwise product of the state vectors for the nodes in the first phase, whereas $\left\langle I_j O_k \right\rangle_{recon}$ denotes the pairwise product of the state vectors in the second phase. The biases, b's, are updated similarly. This process is repeated until a maximum number of iterations is reached.

After the layer-wise pre-training of all RBMs in the DBN, all the parameters of the deep architecture in DBN can be fine-tuned using the back-propagation algorithm. The back-propagation considers all layers simultaneously to minimize the training error with stochastic gradient descent. The training error is calculated by comparing the network outputs and the training data category labels. The process is repeated and stops when it reaches the maximum number of iterations. After the DBN input module finishes the training process, it generates a vector for each website article.

The DBN Output Module. After the LDA auxiliary module finishes the unsupervised training using the professional diabetes vocabulary and the DBN input module finishes the supervised training using the articles, we combine the vectors representing the topic distributions of the 26 diabetes-related terms obtained from the LDA module and the feature vector representing each article from the DBN input module to form a joint representation vector. Note that the dimension of this combined vector is 26*K (the dimension of the LDA auxiliary module' output) plus the dimension of the DBN input module' output layer. This combined feature vector is used as the input to the DBN output module, which includes one or more hidden layers, to output 26 probability values. The training process of the DNB output module is the same as the DBN input module.

At the end, for each article, the LDADBN model selects the category with the maximum probability value. All parameters obtained in the LDADBN model training process are used to test the performance of the model in the experiments.

4 Experiments

We conducted a series of experiments to evaluate the performance of the LDADBN model for identifying and categorizing diabetes related topics. In this section, we report the results and findings from these experiments.

4.1 The Datasets

We collected all pages posted between July 2010 and September 2013 from two most popular Chinese health websites dedicated to diabetes related information. A summary of the two datasets is shown in the Dataset 1 column and Dataset 2 column in Table 1. Some pages contain tags such as "High Satiety" (高饱腹感) and "Diet Control" (饮食控制). There are in total 912 distinct tags in Dataset 1 and Dataset 2. We include the sample articles (and their English translations) from the two websites, respectively, as follows.

[tnbz.com] 糖尿病患者大多听说,粥是喝不得的,会使血糖迅速上升。 治疗糖尿病的医生们,也多把吃干不吃稀当成铁律告诫糖尿病患者。 肥胖者也会听说:喝粥容易消化,容易饥饿,越喝越胖。 这些说法是否完全正确呢?恐怕非常值得商榷。 一篇广为流传的文章谈到:有研究发现,等量大米煮成的米饭或粥对糖尿病患者进食后血糖有不同的影响。……

It is often said that diabetes patients should avoid eating porridge, which can cause their blood sugar level to rise rapidly. Doctors may also suggest patients with diabetes or obesity not eat porridge. The reason is that porridge is easy to digest and one may quickly get hungry again after a porridge meal. As a result, a diet with more porridge may cause one to gain weights faster. Is this really true? A widely cited study reports that using the same amount of ingredients, a rice meal and a porridge meal have different effects on the after-meal blood sugar levels of diabetes patients….

[zzcxhg.com] 一个研究报告称,在预防糖尿病心血管并发症方面,绝对素食者饮食比美国糖尿病协会推荐的饮食更有益。 大约2/3 的糖尿病患者死于心脏疾病或者是中风,因此预防心血管的发病是重中之重。 这个研究由美国医师委员会资助,该委员会倡议绝对素食者饮食……

According to a research report, vegan diet is more helpful than the diet recommended by the American Diabetes Association for preventing cardiovascular complications in diabetes patients. About two-thirds diabetes patients die of heart diseases or strokes. Thus the prevention of cardiovascular disease is a top priority. The study was funded by the U.S. Physicians Committee for Responsible Medicine, which recommended vegan diet.

Table 1. Dataset statistics summary

Dataset	Dataset 1	Dataset 2	Dataset 3	Dataset 4
Website	tnbz.com	zzcxhg.com	Combined	Combined
No. of articles	3,936	15,682	1,000	1,162
No. of all tags	6,933	7,023	–	–
No. of distinct tags	888	49	–	–

To evaluate our model's performance for uncategorized Web pages, we created Dataset 3 by randomly selecting 1,000 Web pages out of Datasets 1 and 2 (see the Dataset 3 column in Table 1). We removed the navigation paths in the original pages and manually annotated the 1,000 articles. Two graduate students with diabetes knowledge annotated the data (with 0.88 inter-coder reliability); and the remaining disagreement was resolved by in-person discussions between the two students. We found that the 26 categories in the three datasets were not uniformly distributed (see Fig. 6a–c). Because an unevenly distributed dataset may affect the performance of classification models [26], we created Dataset 4 by drawing 1,162 articles out of Datasets 1 and 2 (see the last column in Table 1) such that there were roughly an equal number of articles in each category (see Fig. 6d).

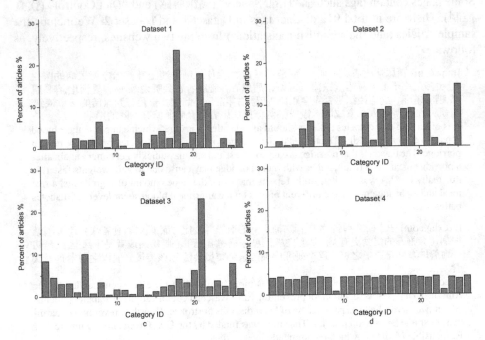

Fig. 6. The category distributions in the datasets

4.2 Evaluation Metrics and Benchmarks

The evaluation metrics of the experiments are precision, recall, F-measure and accuracy. These metrics have been widely used in data mining and machine learning studies. We selected the following five techniques as the benchmarks.

- **DBN.** This model uses only one DBN module and takes only the articles from the websites as the input. In other words, this model is equivalent to the DBN input module in our LDADBN model.
- **SVM** (Support Vector Machine). SVM is one of the most popular and effective classification techniques and been adopted in medical research. SVM constructs hyper-planes with maximum margins to divide data points with different category labels.
- **GNB** (Gaussian Naïve Bayes). GNB is a simple probabilistic classifier implementing Bayes' theorem and has been shown to perform superior in some text classification tasks.
- **PE** (Perceptron). Perceptron is one of the first artificial neural networks and suitable for supervised classification.
- **DT** (Decision Trees). DT is a non-parametric supervised learning method for classification.

4.3 Results

To compare the effectiveness of the LDADBN with the benchmarks, we conducted 10-fold cross validation on each of the four datasets. The values of the default parameters in the LDADBN model are presented in Table 2. In the training stage, we set the learning rates for the training and tuning processes to be 0.1 and 0.01, respectively, and the number of iterations to be 200.

Table 2. The default parameter values in the LDADBN model

Parameters	LDA auxiliary module	DBN input module	DBN output module
Dimensions of input	–	1,188	1,270
No. of hidden layers	–	3	4
Dimensions of hidden layers	–	594	635
Dimensions of output	26	594	26

In our implementation, the DBN input module contains three hidden layers with 594 dimensions (i.e., half of the number of input dimensions) in each of the layers. The DBN output module takes the 1,270-dimension (=26 * 26 + 594) vectors as the input and contains four hidden layers.

Table 3 summarizes the results of the performance comparison between LDADBN and the benchmarks using Dataset 1. The last four columns report the average values of metrics over the 10-fold cross validations. In Tables 3 through 6, metric values (F-measure[1] and Accuracy) that are significantly less than those of the LDADBN model are highlighted with asterisks.

Compared with the benchmark models, our LDADBN and the DBN model perform similarly and better than the benchmarks. Specifically, the LDADBN model is significantly better than all non-DBN based models in F-measure and better than GNB, PE, and DT in accuracy. Only SVM's accuracy is comparable with LDADBN. The possible reason for the non-significant difference between LDADBN and DBN is that the training data for the LDA auxiliary module in the LDADBN is too small, containing only 26 term definitions and explanations (documents). As a result, the supplementary information provided by the LDA module is marginal and not substantial enough to help enhance the performance of the DBN modules.

Table 3. Performance comparison using Dataset 1

Method	Precision %	Recall %	F-measure %	Accuracy %
LDADBN	63.02	61.28	**62.09**	**75.38**
DBN	62.67	62.93	62.73	74.78
SVM	83.15	43.74	**57.14*****	74.50
GNB	41.84	48.37	**44.86*****	**56.65*****
PE	59.40	56.82	**57.84***	**73.38****
DT	53.29	53.16	**53.20*****	**68.55*****

$*p < 0.01$; $**p < 0.005$; $***p < 0.001$

Table 4 presents the performance comparison results using Dataset 2. It shows that the performance is worse than that using Dataset 1. This is because the distribution in dataset 2 is more uneven, which results in many indistinguishable class boundaries. For example, there are no articles in 7 out of the 26 (27 %) categories in this dataset. Since the model always outputs 26 probabilities values, it may assign some articles into categories that do not exist in the dataset. However, even though the LDADBN's F-measure and accuracy are lower than those in Dataset 1, these two measures are still significantly greater than those of the non-DBN based benchmarks (except for PE in the accuracy measure). Table 5 shows the results from Dataset 3. It can be seen that when the category distributed is more even (with no missing data in any category), all classification models' performances are better; and that the LDADBN is even significantly more effective and accurate than the DBN model.

---

[1] We consider only F-measure here because it is a comprehensive metric incorporating both Precision and Recall.

Table 4. Performance comparison using Dataset 2

Method	Precision %	Recall %	F-measure %	Accuracy %
LDADBN	36.82	37.31	**38.00**	**55.29**
DBN	38.70	37.17	37.89	54.72
SVM	56.34	27.37	**36.84***	**43.14***
GNB	25.37	29.63	**27.31***	**39.45***
PE	39.85	32.79	**35.93***	54.60
DT	35.68	33.47	**34.52***	**50.61***

$*p < 0.05; ***p < 0.001$

In order to demonstrate the effect of the category distribution on model performance, we report the performance comparison results from Dataset 4 in Table 6. It shows that the LDADBN's F-measure and accuracy are improved (compared with Dataset 3 with unevenly distributed categories and similar number of articles) and significantly greater than all other non-DBN based benchmark models.

Table 5. Performance comparison using Dataset 3

Method	Precision %	Recall %	F-measure %	Accuracy %
LDADBN	53.91	53.41	**53.25**	**64.40**
DBN	50.57	51.77	**50.47***	**62.00***
SVM	68.72	29.51	**40.93****	**57.80***
GNB	34.79	40.66	**36.97***	**48.00****
PE	47.55	46.94	**47.08***	59.90
DT	40.95	44.14	**41.93***	56.20

$*p < 0.05; **p < 0.005; ***p < 0.001$

Table 6. Performance comparison using Dataset 4

Method	Precision %	Recall %	F-measure %	Accuracy %
LDADBN	65.90	63.28	**64.33**	**68.79**
DBN	64.43	61.81	62.85	67.93
SVM	67.88	53.13	**59.50***	**64.14****
GNB	55.63	54.93	**55.07****	**62.75****
PE	53.61	53.73	**53.43***	**59.65***
DT	53.95	53.51	**53.49****	**58.96***

$*p < 0.05; **p < 0.005; ***p < 0.0005$

In summary, the LDADBN model is significantly more effective and accurate in identifying and categorizing topics in diabetes related Chinese Web pages than major classification models, including SVM, GNB, PE, and DT. In most cases, the LDADBN model performs similarly to the DBN model in terms of F-measure and accuracy. For manually annotated data, the LDADBN model outperforms the DBN model.

4.4 The Effects of the Parameter Values

In order to fine tune our LDADBN model, we conducted additional experiments to examine the effects of the parameters on model performance. Especially, we focused on the effects of the number of layers and dimensions in the DBN module. We used the default values for other parameters (e.g., learning rate). Dataset 3 was used for these experiments.

Figure 7 displays the F-measures of the LDADBN as a function of the number of layers. The left of the Fig. 7 shows that as the number of layers in the LDADBN's DBN input module increases, the LDADBN's effectiveness increases until it reaches its top at the point for three layers. The right of the Fig. 7 presents the changes in LDADBN's effectiveness in response to the changes in the number of layers in the DBN output module, showing a tendency of a convex peak. There is a trade-off between the perform-ance and training time, thus we chose to use three layers in the input DBN module and four layers in the output DBN module in the LDADBN.

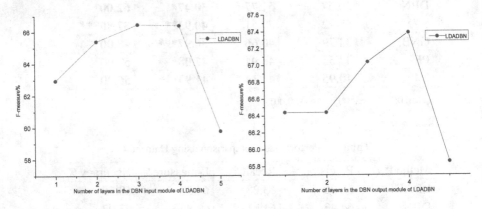

Fig. 7. The effect of the number of hidden layers in the LDADBN

Figure 8 shows the performance of the LDADBN in terms of the number of nodes (dimensions) of the hidden layers. As we can see from the Figure, as the dimension of hidden layers increases, the performance of the LDADBN goes up first and then goes down. This indicates that for the DBN-based models, having hidden units about the half of the number of input dimensions is sufficient to train the model.

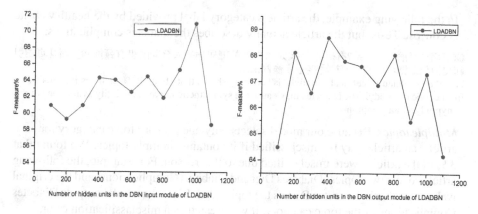

Fig. 8. The effect of the dimensions of hidden layers in the LDADBN

4.5 The Effects of the Feature Vectors

We also investigated the effects of different feature vectors on the performance of our model. In this study, we used two types of features: 1,065 binary features from the diabetes lexicon and the 123 features using the TFIDF method. We call them Binary and TFIDF, respectively. The combined vector, which was what we used in the above experiments, is called B-TFIDF. To demonstrate the value of B-TFIDF, we compared it with the Binary and TFIDF feature vectors using Dataset 1. The performance of the LDADBN model using the three different feature vectors are shown in Table 7. Our results demonstrate that the combined feature vectors, B-TFIDF, yields the best performance in terms of F-measure and accuracy.

Table 7. The performance of LDADBN using different feature vectors

Method	Precision %	Recall %	F-measure %	Accuracy %
B-TFIDF	66.03	59.14	62.40	75.5
Binary	65.46	58.47	61.77	74.7
TFIDF	59.25	20.00	29.91	59.3

4.6 Causes of Misclassification Errors

To further investigate and analyze the causes of misclassification errors, we selected some samples misclassified by the LDADBN model. We randomly selected 100 articles out of Dataset 1 and 21 % of these articles were misclassified by the LDADBN model. We conducted an in-depth analysis of the 21 articles and identified three possible causes for the misclassification errors: wrong label, multiple topics, and model error.

- *Wrong label.* We found that 10 % of the articles were "misclassified" by the LDADBN because their navigation paths, which were used as topic category labels in our experiments, were actually wrong. In other words, the LDADBN model assigned correct labels based on the articles' contents and the topic category map.

In the following example, the article's category label provided by the health website is Glucose Tests, but the article actually describes the diabetes complications.

糖尿病和颈椎病有关系?有研究显示,大约56％的糖尿病病人会同时患有颈椎病。而且可能引起类似低血糖表现的交感神经兴奋症候群……

Is diabetes related to cervical spondylosis? A study shows that about 56 % diabetes patients suffer from cervical spondylosis. Diabetes may cause the sympathetic nerve excitability syndrome similar to hypoglycemia symptoms.

- *Multiple topics.* Because our model selects only one primary topic category for each article, an article may be misclassified if it contains multiple topics. We found that 3 % of the articles were misclassified due to this reason. For example, the following article covers two topics: Diabetes Diet and Diabetes Complications; and the original website used only Diabetes Diet as the topic label. Since our model outputs Diabetes Complications as the topic category, it was treated as a misclassification error.

补充维生素C可降低视网膜病变、神经病变和肾病等糖尿病并发症危险。糖友补充维生素等营养,需要视自身病情而定……

Taking Vitamin C as a diet suppliment can reduce the risks of diabetes complications such as retinopathy,neuropathy, and kidney diseases.

- *Model error.* For the remaining 8 % of the misclassified articles, the model simply did not capture the main point of the articles. For instance, "Hyperglycemia" was mistakenly assigned by the LDADBN model as the topic category to the following article regarding glucose testing.

得了病,最忌一种心态怕麻烦;得了糖尿病更是如此。因为你不仅要精打细算地控制饮食、合理运动,按时服药、打针,还必须经常自我监测血糖……

If you have diabetes, you need to control your diet, exercise regularly, take medicine, and get injections on time. You also must often monitor and test your blood glucose level.

In short, if the training data are correctly labeled, the LDADBN model may achieve better performance in accuracy. The cases of wrong label and multiple topics show that some health websites provide some categorized articles but the category information is incorrect or misleading. There is also a lack of widely accepted standard for categorizing diabetes related topics. These problems are also the reasons that we propose this DBN-based framework. We hope to provide an effective approach (and standard) to identifying and categorizing diabetes related topics from online Chinese articles.

5 Conclusion

In this study, we propose a deep learning based framework for identifying and categorizing diabetes-related topics on Chinese health websites. This framework includes four stages: data collection, data preprocessing, DBN-based model training, and topic categorization. Our experiments using real data show that the LDADBN model outperforms several state-of-the-art benchmark categorization methods.

The contribution of our research is three-fold. First, we employ the deep learning approach to categorizing text documents, for which little research has been done to ascertain the value of the deep learning approach. To the best of our knowledge, our research framework is the first one to adopt the deep learning approach in topic identification and categorization, especially for Chinese documents in the diabetes domain. Second, to address the feature selection challenge of using the deep learning approach in text categorization applications, we propose to combine a domain lexicon based feature vector and the TFIDF based feature vector. This combined feature vector is significantly smaller than those developed using the bag-of-words approaches. Third, we incorporate domain knowledge into the process of topic categorization. We develop the professional diabetes vocabulary and use it as the input to the LDA module to model the domain knowledge related to diabetes care and treatments. The experiments show encouraging results.

One limitation of our model is that the DBN-based model with the LDA auxiliary module is not significantly better than the DBN model. The possible reason is that the size of the training data for the LDA module is too small, containing only 26 terms and their definitions and explanations. Another limitation of our model lies in the lengthy training process, which affects the model's scalability for large datasets.

Our future work will be done in three directions: (a) experiment with other types of machine learning models to formulate the domain knowledge and to help improve the performance of our DBN-based model; (b) seek more collaboration with physicians and medical professionals to refine the topic category map; and (c) extend and apply the framework to other chronicle diseases and even other domains. Moreover, for the model training process, instead of setting a fixed maximum number of iterations we would make the termination condition more adaptive and heuristic.

Acknowledgements. The research presented in this paper has been funded by grants from the National High-tech R&D Program of China (Grant No. SS2015AA020102), the US NSF grant (IIP-1417181), the 1000-Talent program, and Tsinghua University Initiative Scientific Research Program.

References

1. Yang, H., Kundakcioglu, E., Li, J., Wu, T.F., Mitchell, J.R., Hara, A.K., Pavlicek, W., Hu, L.S., Silva, A.C., Zwart, C.M., et al.: Healthcare intelligence: turning data into knowledge. IEEE Intell. Syst. **29**(3), 54–68 (2014)
2. Hinton, G.E., Osindero, S., Teh, Y.W.: A fast learning algorithm for deep belief nets. Neural Comput. **18**(7), 1527–1554 (2006)
3. Xia, R., Zong, C., Hu, X., Cambria, E.: Feature ensemble plus sample selection: domain adaptation for sentiment classification. IEEE Intell. Syst. **28**(3), 10–18 (2013)
4. Dang, Y., Zhang, Y., Chen, H.: A lexicon-enhanced method for sentiment classification: an experiment on online product reviews. IEEE Intell. Syst. **25**(4), 46–53 (2010)
5. Cheeseman, P., Kelly, J., Self, M., Stutz, J., Taylor, W., Freeman, D.: Autoclass: a bayesian classification system. In: Readings in Knowledge Acquisition and Learning. pp. 431–441. Morgan Kaufmann Publishers Inc., San Francisco (1993)
6. Quinlan, J.R.: Induction of decision trees. Mach. Learn. **1**(1), 81–106 (1986)
7. Cortes, C., Vapnik, V.: Support-vector networks. Mach. Learn. **20**(3), 273–297 (1995)
8. Yang, Y.: Expert network: effective and efficient learning from human decisions in text categorization and retrieval. In: Croft, B.W., van Rijsbergen, C.J. (eds.) Proceedings of the 17th Annual International ACM SIGIR Conference on Research and Development in Information Retrieval, pp. 13–22. Springer-Verlag New York, Inc., New York (1994)

9. Hecht-Nielsen, R.: Theory of the backpropagation neural network. In: International Joint Conference on Neural Networks, 1989 IJCNN, pp. 593–605. IEEE (1989)

10. Farid, D.M., Zhang, L., Rahman, C.M., Hossain, M., Strachan, R.: Hybrid decision tree and naive bayes classifiers for multi-class classification tasks. Expert Syst. Appl. **41**(4), 1937–1946 (2014)

11. Ghahabi, O., Hernando, J.: Deep belief networks for i-vector based speaker recognition. In: 2014 IEEE International Conference on Acoustics, Speech and Signal Processing (ICASSP), pp. 1700–1704. IEEE (2014)

12. Liu, T.: A novel text classification approach based on deep belief network. In: Wong, K.W., Mendis, B.S.U., Bouzerdoum, A. (eds.) ICONIP 2010, Part I. LNCS, vol. 6443, pp. 314–321. Springer, Heidelberg (2010)

13. Wang, B., Liu, B., Wang, X., Sun, C., Zhang, D.: Deep learning approaches to semantic relevance modeling for chinese question-answer pairs. ACM Trans. Asian Lang. Inf. Process. (TALIP) **10**(4), 21 (2011)

14. Lin, Y., Brown, R., Yang, H., Li, S., Lu, H., Chen, H.: Data mining large-scale electronic health records for clinical support. IEEE Intell. Syst. **26**(5), 87–90 (2011)

15. Klahold, A., Uhr, P., Ansari, F., Fathi, M.: Using word association to detect multi-topic structures in text documents. IEEE Intell. Syst. **29**(5), 40–46 (2014)

16. Velupillai, S., Skeppstedt, M., Kvist, M., Mowery, D., Chapman, B.E., Dalianis, H., Chapman, W.W.: Cue-based assertion classification for Swedish clinical text—developing a lexicon for pyConTextSwe. Artif. Intell. Med. **61**(3), 137–144 (2014)

17. Liu, W., Sweeney, H.J., Chung, B., Glance, D.G.: Constructing consumer-oriented medical terminology from the web a supervised classifier ensemble approach. In: Pham, D.-N., Park, S.-B. (eds.) PRICAI 2014. LNCS, vol. 8862, pp. 770–781. Springer, Heidelberg (2014)

18. Minarro-Giménez, J.A., Marín-Alonso, O., Samwald, M.: Exploring the application of deep learning techniques on medical text corpora. Stud. Health Technol. Inform. **205**, 584–588 (2013)

19. Sibunruang, C., Polpinij, J.: Ontology-based text classification for filtering cholangiocarcinoma documents from PubMed. In: Ślęzak, D., Tan, A.-H., Peters, J.F., Schwabe, L. (eds.) BIH 2014. LNCS, vol. 8609, pp. 266–277. Springer, Heidelberg (2014)

20. Sinha, A.P., Zhao, H.: Incorporating domain knowledge into data mining classifiers: an application in indirect lending. Decis. Support Syst. **46**(1), 287–299 (2008)

21. Liu, X., Cai, L., Akiyoshi, M., Komoda, N.: A classification method of knowledge cards in Japanese and Chinese by using domain-specific dictionary. In: Omatu, S., Paz Santana, J.F., González, S.R., Molina, J.M., Bernardos, A.M., Rodríguez, J.M.C. (eds.) Distributed Computing and Artificial Intelligence. AISC, vol. 151, pp. 453–460. Springer, Heidelberg (2012)

22. Omesoft: Diabetes dictionary. http://shouji.baidu.com/software/item?docid=1018036888&from=as

23. Salakhutdinov, R., Hinton, G.: Semantic hashing. Int. J. Approximate Reasoning **50**(7), 969–978 (2009)

24. Ramage, D., Dumais, S.T., Liebling, D.J.: Characterizing microblogs with topic models. ICWSM **10**, 1 (2010)

25. Blei, D.M., Ng, A.Y., Jordan, M.I.: Latent dirichlet allocation. the. J. Mach. Learn. Res. **3**, 993–1022 (2003)

26. Weiss, G.M., Provost, F.: The effect of class distribution on classifier learning: an empirical study. Rutgers University (2001)

An Adaptive Approach of Approximate Substring Matching

Jiaying Wang[1], Xiaochun Yang[1(✉)], Bin Wang[1], and Chengfei Liu[2]

[1] School of Computer Science and Engineering,
Northeastern University, Liaoning 110819, China
wangjiaying@research.neu.edu.cn, {yangxc,binwang}@mail.neu.edu.cn
[2] Department of Computer Science and Software Engineering,
Swinburne University of Technology, Melbourne, VIC 3122, Australia
cliu@swin.edu.au

Abstract. Approximate substring matching is a common problem in many applications. In this paper, we study approximate substring matching with edit distance constraints. Existing methods are very sensitive to query strings or query parameters like query length and edit distance. To address the problem, we propose a new approach using partition scheme. It first partitions a query into several segments, and finds matching substrings of these segments as candidates, then performs a bidirectional verification on these candidates to get final results. We devise an even partition scheme to efficiently find candidates, and a best partition scheme to find high quality candidates. Furthermore, through theoretical analysis, we find that the best partition scheme cannot always outperform the even partition scheme. Thus we propose an adaptive approach for selectively choosing scheme using statistic knowledge. We conduct comprehensive experiments to demonstrate the efficiency and quality of our proposed method.

Keywords: Approximate substring matching · Adaptive approach · Partition scheme · Scheme selection

1 Introduction

Approximate substring matching problem widely exists in many applications, such as text retrieval, pattern recognition, signal processing, and bioinformatics [15]. For example, spelling errors may occur when typing queries in a search engine, and typos may exist in many web pages. Approximate substring matching can help provide error tolerant high quality information retrieval. Another similar application is finding DNA subsequence in a genome database with several mutations. With the rapid growth of text data, efficiently responding approximate substring query is an important and challenging task.

In this paper, we focus on approximate substring matching with edit distance constraints, which finds all similar substrings for a given query in a string

© Springer International Publishing Switzerland 2016
S.B. Navathe et al. (Eds.): DASFAA 2016, Part I, LNCS 9642, pp. 501–516, 2016.
DOI: 10.1007/978-3-319-32025-0_31

collection, such that their edit distance with the query is within a given threshold. Existing methods can be mainly classified into two groups. The first one adopts a suffixtree-based framework [13,19], which backtracks on the suffix tree or its variant to compute the edit distance of a query with all the substrings. It saves the cost by sharing edit distance computation on same substrings. These methods are very sensitive to query's parameters. They only work well for short queries with small edit distance. Besides, the alphabet size also affects the performance, since it is more likely to have common substrings for a string collection in a small size of alphabet. The second one employs a filter and verification framework [10,17]. The idea is to utilize a filter process to find a group of substrings as candidates, and the final result can be obtained by further verifying the edit distance with the query. Since the lengths and positions of the matching are unknown beforehand, these methods can only utilize query string to do filtering. Different strings have varied influence on the efficiency. To address these problems, we propose a new method for approximate substring matching.

Recently an even partition scheme based on the pigeonhole principle shows great success in solving approximate string search and join [9,12]. In this paper, we utilize the even partition scheme as our basic method to solve the approximate substring matching problem, and utilize a variant of suffix tree called BWT to index the string collection. For a given query, we evenly partition it into several segments, and search those segments' exact matching substrings to generate a group of candidates. Then we conduct bidirectional verification on those candidates to get final results. We find that the basic method has a large room for performance improvement. The improved idea is based on an observation that a good partition scheme can help reduce the candidate size, which further reduces the verification cost. Based on this observation, we devise an improved method using a best partition scheme, which finds the minimum candidates for a query. We also propose several optimization strategies to boost the process. Furthermore, through theoretical analysis, we find that the improved method is not always a better choice, since it needs more time cost to do partition, while in some cases the even partition scheme could outperform the improved method. Therefore further improvement can be achieved if we can predict which method is more efficient. To this end, we propose an adaptive approach for approximate substring matching. In summary, we make the following contributions:

- We devise a basic method to solve approximate substring matching problem utilizing the BWT index and the even partition scheme.
- We devise an improved method using the best partition scheme, which finds the minimum candidates for a query. We also propose several optimization strategies to boost the process.
- We propose an adaptive approach by utilizing statistic knowledge to do scheme selection.
- We conduct comprehensive experiments on real data set to demonstrate the efficiency and quality of our proposed approach.

The rest of this paper is organized as follows. In Sect. 2, we review the related work. We introduce the preliminaries and formalize our problem in Sect. 3.

In Sect. 4, we introduce both the basic and improved methods for approximate substring matching. In Sect. 5, we introduce the adaptive approach for the problem. We present our experimental results in Sect. 6. Finally, we conclude the paper in Sect. 7.

2 Related Work

Given a query, finding all approximate matching strings in a string collection with edit distance constraints is a well-studied problem [15,20]. It includes two typical sub-problems. The first one aims to find complete strings similar to a query [2,4,5]. Many methods follow a filter and verification framework to solve the problem. They first find a group of candidates, then verify them to get final answers. Length filtering, position filtering, prefix filtering and several variant filtering were proposed to find candidates efficiently [1,2,8,18,21]. The second one aims to find substrings similar to a query [10,13,15]. We call it approximate substring matching. Since the lengths and positions of the matching substrings are unknown beforehand, the problem is more difficult. In this paper, we focus on solving the approximate substring matching problem.

The time complexity of the classical approach for approximate substring matching is $\mathcal{O}(nm)$ [15], in which n and m are the scale of the string collection and the query respectively. An optimal average-case algorithm solves the problem in $\mathcal{O}(n(\tau + log_\sigma m)/m)$, in which σ is the size of the alphabet, and τ is the threshold [7].

Many indexing techniques have been proposed to improve the performance. One important index method is based on suffix tree, which supports many operations on substrings. In [19], Ukkonen et al. showed how suffix tree could help boost the approximate substring matching process. Although the method is efficient, it requires much larger space than the original string collection. BWT index is a variant of suffix tree, which not only has much smaller index size but also supports many substring operations [3,6,16]. In [13], Li et al. proposed an approximate substring matching method, which utilizes BWT index to simulate the traversal of suffix tree. To allow edit operations, they enumerated all the characters in the alphabet for each character of the query string. Based on the proposed method, they developed a tool BWA [13,14], which is widely used for read alignment in bioinformatics. The method is extremely fast for small character set, such as DNA sequence.

Another important index method is based on q-gram, which is a contiguous sequence of q characters of a string. Navarro et al. utilized q-gram inverted index to address the approximate substring matching problem [17]. Yang et al. proposed variable-length grams (a.k.a. v-gram) index method to improve the performance of approximate string matching [11,23]. Kim et al. extended the v-gram index method to support approximate substring matching [10].

All these methods are very sensitive to query's parameters or contents, so the efficiency cannot be guaranteed. To address the problem, we propose a new approach which is efficient for different queries. Our idea is inspired by the pigeonhole principle (a.k.a. partition scheme), which partitions a string into several segments, then utilizes those segments to find candidates. Li *et al.* utilized even partition scheme to solve the approximate string join problem [9,12]. Yang *et al.* and Wang *et al.* utilized even partition scheme to do approximate string matching [22,25]. Yang *et al.* also utilized even partition scheme to do approximate substring search on compressed genomic data [24].

3   Preliminaries

Let Σ be a finite alphabet of characters of a string collection. String s is a sequence of characters in Σ, in which $|s|$ is the string length, and $s[i]$ is the i-th character of string s. We use $s[i,j]$ to denote a sequence (a.k.a. substring) from the i-th character to the j-th character. We call $s[i,j]$ a prefix of string s if $i = 1$, and a suffix of string s if $j = |s|$.

In this paper, we utilize edit distance as the measure of dissimilarity between two strings, which is the minimum number of edit operations (including insertion, deletion, and substitution) to transform from one string to another. Formally we use $ed(s_1, s_2)$ to denote the edit distance between string s_1 and string s_2. A query is represented as $\langle q, \tau \rangle$, in which q is the query string, and τ is the edit distance threshold. We utilize $\alpha = \frac{\tau}{|q|}$ as the error ratio of the query.

Next we define the problem of approximate substring search formally as follows. Figure 1 shows an example of the approximate substring matching problem.

Problem 1 (Approximate Substring Matching). Given a string collection S, a query string q, and an edit distance threshold τ, approximate substring matching finds all the substrings $s[i,j]$, in which $s \in S$, and $ed(s[i,j], q) \leq \tau$.

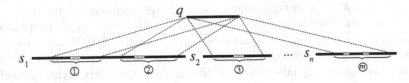

Fig. 1. An example of approximate string matching problem.

4   Approximate Substring Matching Methods

In this section, we first present a basic method for approximate substring matching, which utilizes even partition scheme. Then we show how to improve the method using a best partition scheme.

4.1 BWT and Exact Substring Matching

Burrows-Wheeler transformation (a.k.a BWT) is a common technique in data compression. Given a string s, it performs the following transformation. It first appends character $ to the end of string s, and note that $ is smaller than any character in Σ. Then it computes the suffix array SA of string s\$, which is a permutation of integers $1, 2, \ldots, |s| + 1$ such that $SA[i]$ is the start position of the i-th lexicographically smallest suffix of the string. The output of BWT is a string t such that if $SA[i] = 1$, $t[i] =$\$, otherwise $t[i] = s[SA[i] - 1]$. We show an example of the transformation in Fig. 2. The BWT result of string mississippi is ipssm$pissii.

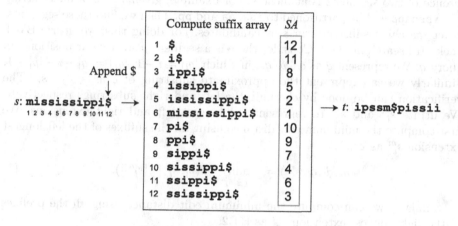

Fig. 2. A BWT example.

Now we explain how to utilize BWT to do exact substring matching. The method is called backward search, which is first proposed in [6]. Notice that any occurrence of query q in string s is a prefix of a suffix of s, since all the suffixes are lexicographically ordered, all the occurrences of query q belong to a contiguous interval of suffix array SA. The backward search method first sets the matching interval to cover all the suffix array, then scans the characters of query q from right to left. For each character it refines the matching interval to cover all the suffixes that match the current suffix of q. Take Fig. 2 as example, the process of searching query ssi is as follows. It first sets the matching interval to be $[1, 12]$, then refines the interval to be $[2, 5]$ when scans character i, and refines the interval to be $[9, 10]$ when scans character s. Finally it finds the interval $[11, 12]$, in which $SA[11] = 6$ and $SA[12] = 3$. Thus there are two exact matching substrings ssi in string s, locates in positions 3 and 6 respectively.

4.2   Basic Approximate Substring Matching Algorithm

The approximate substring matching method is based on the pigeonhole principle as depicted in Lemma 1. For simplicity, we consider searching query in one string s. We will show how to search in a string collection in Sect. 5.

Lemma 1. *Given a query string q with $\tau + 1$ segments, if a substring $s[i, j]$ is similar to q with threshold τ, $s[i, j]$ must contain a substring which exact matches a segment of q.*

Based on Lemma 1, we can solve the approximate substring matching problem by filtering and verification as follows. We first partition the query string q into $\tau + 1$ segments using an even partition scheme, in which the length difference of two segments is at most 1. For example, given $\tau = 2$, a query string $q = $ issaapp will be partitioned to iss, aa and pp. Then we find those segments' exact matching substrings in s as candidates. For doing this, we utilize BWT backward search method. Consider there is a segment p starting at position i in query q. We represent q as $q_l \cdot p \cdot q_r$, in which $|q_l| = i - 1$, $|q_r| = |q| - i - |p| + 1$. Similarly we can represent the approximate substring of q as $s_l \cdot p \cdot s_r$. The verification can be done by checking the left and right substrings respectively. We utilize s_l^m and s_r^m to represent the longest left and right extensions. We first compute the minimum edit distance using all the suffixes of the left longest extension s_l^m as Eq. 1.

$$med_l(q_l, s_l^m) = \min_{1 \le i \le |s_l^m|} ed(p_l, s_l^m[i, |s_l^m|]). \tag{1}$$

Similarly we can compute the minimum edit distance using all the prefixes of the right longest extension s_r^m as Eq. 2.

$$med_r(q_r, s_r^m) = \min_{1 \le j \le |s_r^m|} ed(p_r, s_r^m[1, j]). \tag{2}$$

Thus if there is an approximate substring $s_l \cdot p \cdot s_r$ of query q, it must satisfy $med_l(q_l, s_l^m) + med_r(q_r, s_r^m) \le \tau$. The time complexity of such verification is $\mathcal{O}(|q_l||s_l^m| + |q_r||s_r^m|)$. Since we only need to detect if the edit distance is within τ, we can improve the process to $\mathcal{O}(\tau(|q_l| + |q_r|))$, which is equal to $\mathcal{O}(\tau(|q| - |p|))$.

We give the basic approximate substring matching algorithm in Algorithm 1. It first partitions the query string evenly into $\tau + 1$ segments (line 1). For each segment, it performs a backward search using BWT index to get a matching interval (line 3). For each matching substring, it computes the minimum edit distance on the left (line 8), if the value is not greater than τ, it further computes the minimum edit distance on the right (line 10). If the summation of left and right minimum edit distance is still not greater than τ, we add it to the result.

4.3   Improved Method Based on Best Partition Scheme

In this subsection, we introduce how to improve the performance using a best partition scheme. Reconsider the example $q = $ issaapp, $\tau = 2$. For even partition, it will be partitioned into iss, aa and pp. For string $s = $ mississippi,

---

Algorithm 1. APPROXIMATESEARCH(I, q, τ)

Input: I: BWT index

       $\langle q, \tau \rangle$: A query string and its edit distance

Output: $\mathcal{A} = \{s[i,j] \mid ed(s[i,j], q) \leq \tau\}$

1 $P \leftarrow even\_partition(q, \tau + 1)$;

2 **foreach** $p \in P$ **do**

3     $[sp, ep] \leftarrow I.backward\_search(p)$;

4   **if** $sp \leq ep$ **then**

5        $\langle q_l, q_r \rangle \leftarrow split(q, p)$;

6   **for** $i \leftarrow sp$ **to** ep **do**

7           $\langle s_l^m, s_r^m \rangle \leftarrow extend(i, p)$;

8           $\tau_l \leftarrow med_l(q_l, s_l^m)$;

9   **if** $\tau_l \leq \tau$ **then**

10              $\tau_r \leftarrow med_r(q_r, s_r^m)$;

11   **if** $\tau_l + \tau_r \leq \tau$ **then**

12   add the match to $\mathcal{A}$;

13 **return** $\mathcal{A}$;

---

segment iss has two matching substrings in s, segment aa has no matching substring, and segment pp has one matching substring. So it totally needs three verifications. However, if we partition it into issa, a and pp, only segment pp has one matching substring in s, thus it only needs one verification. The formal definition of the best k-partitions is given in Definition 1.

Definition 1 (Best k-Partitions). *Given a string s, a query string q and partition number k, the best k-partitions of q is P_B, that satisfies*

$$P_B = \underset{P \in C(q,k)}{\operatorname{argmin}} \sum_{p \in G(P)} W_s(p)$$

in which $C(q, k)$ represents all the partition schemes to partition q into k segments, $G(P)$ represents a segment set which contains all segments of a partition scheme P, and $W_s(p)$ is the number of occurrences (a.k.a. weight) of a segment p in s.

A simple way to find the best $(\tau + 1)$-partitions of q is to enumerate all the partition schemes. There are totally $\binom{|q|-1}{\tau}$ schemes of partitioning query q into $\tau + 1$ segments, so it is very inefficient to do such computation.

Next we show how to improve the process using a dynamic programming algorithm. The method is based on the observation that if $P = p_1, p_2, \ldots p_{\tau+1}$ is the best $(\tau + 1)$-partitions of query q, then $P' = p_1, p_2, \ldots p_\tau$ must be the best τ-partitions of $q[1, k]$, in which $k = |p_1| + |p_2| + \ldots + |p_\tau|$. So we can divide it into a collection of simpler subproblems. More precisely, we utilize a matrix W to store the best partitions of a query, in which a cell $W[i][j]$ represents the minimum weight to partition $q[1, j]$ into i segments. So $W[\tau + 1][|q|]$ represent the minimum weight to partition q into $\tau + 1$ segments. Initially we have $W[1][j] = \mathcal{W}_s(q[1, j])$. The recursion equation is given in Eq. 3.

$$W[i][j] = \min_{i-1 \leq k \leq j-1} W[i-1][k] + \mathcal{W}_s(q[k+1, j]) \tag{3}$$

To compute $\mathcal{W}_s(q[k+1, j])$, we need to search $q[k+1, j]$ in s. Since we have built a BWT index, the value can be easily fetched by using backward search. The time complexity of computing $\mathcal{W}_s(p)$ is $\mathcal{O}(|q|)$, and the total time complexity of computing the weight for all the substrings of q is $\mathcal{O}(|q|^3)$. The process can be further improved based on the observation that we only need $\mathcal{O}(1)$ to compute $\mathcal{W}_s(q[i, j])$ if we know $\mathcal{W}_s(q[i+1, j])$, since it just needs to backward search one character $q[i]$. We can compute the weight for all the suffixes of $q[1, j]$ from right to left in $\mathcal{O}(j)$. In this way, the total time complexity is reduced to $\mathcal{O}(|q|^2)$. Based on the proposed dynamic programming algorithm, we can find the best $(\tau + 1)$-partitions of query q in $\mathcal{O}(|q|^2 \tau)$. Next we show how to further boost the process.

The first improvement is to utilize the even partition schema as an upper bound to prune all the partition schemes with larger weight. We utilize W_E to represent the weight of even partition scheme. A cell $W[i][j]$ is valid only if the value is less than W_E, and only valid cells are used to update later cells. An interesting discovery is that if cell $W[i][j]$ is valid, cell $W[i][j+1]$ must be valid. The reason is that since it partitions a longer substring into i segments, at least it can utilize the same partition scheme for the first $i-1$ segments, then the last segment is one character longer, so the weight of $W[i][j+1]$ cannot be larger than $W[i][j]$. Thus we do not need to record which cells are valid but only need to record the position of the first valid cell.

The second improvement is based on an observation: if two cell $W[i-1][k]$ and $W[i-1][k-1]$ have the same weight, we do not need to utilize $W[i-1][k]$ to update $W[i][j]$. The reason is that since we know $\mathcal{W}_s(q[k+1, j]) \geq \mathcal{W}_s q[k, j]$, it must satisfy $W[i-1][k] + \mathcal{W}_s(q[k+1, j]) \geq W[i-1][k-1] + \mathcal{W}_s(q[k, j])$. So we cannot get a smaller weight using the cell $W[i-1][k]$. To utilize the property, naively we can check if $W[i-1][k] = W[i-1][k-1]$, in which $i \leq k \leq j-1$. If that is true we can skip an unnecessary computation process.

The process can be further improved since there can be a group of continuous cells with the same value. Instead of checking those cells one by one, we can skip all of them directly. The method is based on a data structure which we call "skip link", which is a pointer from a cell $W[i-1][k]$ to the first $W[i-1][x]$, in which $x \geq k$, and $W[i-1][k] > W[i-1][x]$. The whole computation is an iterative process. We build up the skip links during the process of computing the cells in the i-th row of matrix W, then utilize them to compute the $(i+1)$-th row of W. During that computation process we can generate new skip links. The process continues until we finish the computation for all the cells in W.

After we find the best $(\tau + 1)$-partitions of query q, we can follow the same filtering and verification manner as depicted in Sect. 4.2 to find its approximate substrings. The advantage of the best partition scheme is that it finds the minimum candidates, which saves the verification cost.

5   An Adaptive Approach

In Sect. 4, we discuss two approximate substring matching methods: the basic one with even partition scheme and the improved one with best partition scheme. In this section, we first show that it is not always a good choice to utilize the improved method. Then we propose a method to adaptively select the efficient method to do approximate substring matching.

5.1   Theoretical Analysis

For a given query string q and edit distance τ, both of the methods include the following three steps.

- Partition. For the even partition scheme, the time complexity is $\mathcal{O}(\tau)$ since it can directly compute the offset for each segment. For the best partition scheme, the time complexity is $\mathcal{O}(|q|^2\tau)$. Thus the time cost of the best partition method is always larger than the cost of the even partition method.
- Search. For each segment, both of the methods need to do backward search on all the segments. The time complexity of backward search on a segment p_i is $\mathcal{O}(|p_i|)$, thus both of the methods need $\mathcal{O}(\sum_{i=1}^{\tau+1} p_i) = \mathcal{O}(|q|)$ to search all the segments.
- Verification. The cost for verifying a segment p_i is equal to its weight $\mathcal{W}(p_i)$ times single bidirectional verification cost on the segment $\mathcal{O}(\tau(|q| - |p_i|))$. So the time complexity of verifying all the segments is $\mathcal{O}(\sum_{p \in P} \tau \mathcal{W}(p_i) (|q| - |p_i|))$. Since $|p_i| < |q|$, we can simplify it $\mathcal{O}(\tau|q| \sum_{p \in P} \mathcal{W}(p_i))$. We utilize W_E and W_B to represent the total weight of even partition scheme and best partition scheme respectively, and we have $W_B \leq W_E$. The total time complexity of verification using even partition scheme and best partition scheme are $\mathcal{O}(\tau|q|W_E)$ and $\mathcal{O}(\tau|q|W_B)$ respectively.

Since the search cost of the two methods are the same, the cost difference of the two methods is $a \times (|q|^2\tau - \tau) + b \times \tau|q|(W_B - W_E)$, in which a and b are unit costs for the partition and verification process respectively. The best partition method is faster if the equation is less than 0, which satisfies $W_B < W_E - \frac{a}{b}(|q| - \frac{1}{|q|})$. In other words, we should only choose the best partition scheme when it can help reduce enough candidates.

There is a challenge to do such selection, since the value W_B can only be acquired after we compute the best partitions. But if we compute the best partitions, the time cost will also be included in the search process, so the performance cannot be better than the method using best partition scheme. The question is can we do scheme selection without computing the best partitions?

A simple solution is based on hard-coding, that is to set up a weight threshold, and if the weight of even partition method is larger than the threshold, we utilize the best partition scheme. However, it is difficult to choose such threshold, and a fixed threshold cannot be suitable for different queries. Using query length or edit distance as a threshold cannot solve the problem either.

5.2   Scheme Selection

In this subsection, we propose a method for scheme selection using statistical information. After building the index for string s, we randomly generate a group of queries with different lengths and different edit distances, then trace their time cost by utilizing the basic method and the improved method. We transform the problem to a binary classification as follows: we give each searched query a label 0 or 1, depends on which method is faster. When the basic method is faster, we set it to 0, otherwise we set it 1. Our objective is to extract features based on search query to predict which method is faster. Next we will describe the features used in this paper and the method used to do prediction.

Feature Description and Scoring. We choose a group of features to describe a query, denoted by $\vec{x}$. The choice of features is motivated by following questions: How long is the query string? How many segments will it be partitioned? What is the weight of even partition method? And how much difference do the segments have?

Based on these questions, we choose four features for our classification problem. They are the length of the query string $|q|$, the segment number $\tau + 1$, the weight of even partition method W_E, and the summation of weight difference of adjacent segments $\sum_{j=1}^{\tau} |\mathcal{W}(p_{j+1}) - \mathcal{W}(p_j)|$. The first two features are the basic information of the query. The weight of even partition method W_E is used to measure the cost of the basic method. And the summation of weight difference of adjacent segments is used to measure the probability to reduce the cost by changing the segments. Since W_E and $\sum_{j=1}^{\tau} |\mathcal{W}(p_{j+1}) - \mathcal{W}(p_j)|$ can range from very small value to very big value, we performed a logarithmic transformation on the two values, and call them e-score and d-score respectively.

Prediction. To do prediction, we utilize logistic regression, in which the probability of an example being drawn from the positive class (a.k.a. hypotheses) is $h(\vec{x})$ and

$$h(\vec{x}) = g(w_0 + \sum_{j=1}^{m} w_j x_j)$$

where $g(z) = \frac{1}{1+e^{-z}}$. We use w_j to denote the weight for j-th feature. Given the logistic regression model, we utilize randomly generated queries to fit the weight vector $\vec{w}$. And the target is to maximum the likelihood of the data, which is

$$\ell(\vec{w}) = \sum_{i=1}^{n} y_i log(h(\vec{x}_i)) + (1 - y_i) log(1 - h(\vec{x}_i))$$

in which y_i is the label of the i-th query. To achieve the target, we utilize gradient ascent method, which iteratively updates the weights to reach the maximum. The update rule is

$$w_j = w_j + \epsilon \frac{\partial}{\partial w_j} \ell(\vec{w})$$

where $\frac{\partial}{\partial w_j} \ell(\vec{w}) = \sum_{i=1}^{n}(y_i - h(\vec{w_i}))\vec{x}_{ij}$, and ϵ is the learning rate, which controls the step size of the updating.

5.3 Adaptive Approximate Substring Matching Approach

In this subsection, we present our adaptive approximate substring matching algorithms. We depict the adaptive index algorithm in Algorithm 2. It first builds a BWT index using string s (line 1), then it randomly generates a group of queries Q with different query strings and different edit distances (line 2). Each query will be used as an example to extracted features (line 4) and run both the basic and improved search methods to get its label y (line 6). Then we use logistic regression to fit the weight vector $\vec{w}$ (line 8).

We show the adaptive search algorithm in Algorithm 3. For a given query $\langle q, \tau \rangle$, we first extract its feature $\vec{x}$ (line 1), then utilize $\vec{x}$ and the weight vector $\vec{w}$ to do prediction (line 2). If the label is 0, we will utilize basic search method (line 4), otherwise we utilize the improved method (line 6). Notice that the process of generating features only takes a short time comparing to other process.

Algorithm 2. ADAPTIVEINDEX(s)

Input: s: A data string s
Output: $\langle I, \vec{w} \rangle$: BWT index and the weight vector

1   $I \leftarrow build\_bwt\_index(s)$;
2   $Q \leftarrow generate\_random\_queries()$;
3   **foreach** $\langle q, \tau \rangle \in Q$ **do**
4      $\vec{x} \leftarrow generate\_features(q, \tau)$;
5   run $basic\_search$ and $improved\_search$ method on q;
6   compute the label y;
7      $X.add(\vec{x}), Y.add(y)$;
8   logistic regression on (X, Y) to fit $\vec{w}$;
9   save $\langle I, \vec{w} \rangle$ as index;

We can easily extend the method to search queries in a string collection S by concatenating all the strings in S. We delimit them by the special character \$. And in the verification process, we drop those substrings crossing different strings.

Algorithm 3. ADAPTIVESEARCH($I, \vec{w}, \langle q, \tau \rangle$)

Input: I: BWT index
     $\vec{w}$: weight vector
     $\langle q, \tau \rangle$: A query string and its edit distance
Output: $\mathcal{A} = \{ s[i,j] \mid ed(s[i,j], q) \leq \tau \}$

1 $\vec{x} \leftarrow generate\_features(q, \tau)$;
2 Predict label y using $\vec{x}$ and $\vec{w}$;
3 **if** $y = 0$ **then**
4 $\quad \lfloor \quad \mathcal{A} \leftarrow basic\_search(q, \tau)$;
5 **else**
6 $\quad \lfloor \quad \mathcal{A} \leftarrow improved\_search(q, \tau)$;
7 **return** $\mathcal{A}$;

---

6 Experimental Study

In this section, we present our experimental study and demonstrate the efficiency and quality of our proposed methods.

6.1 Experiment Setup

All the algorithms were implemented in C++ and compiled with G++4.7 with "-O3" flags. All the experiments were run on a machine with 2.93 GHz Intel Core CPU, 8 GB main memory using Ubuntu operating system.

We utilized three real data sets to evaluate our methods, which are:

- DNA, which contains 1 GB DNA sequences taken from UCSC golden path project[1], and the alphabet size is 5.
- Protein, which contains 1 GB protein sequences taken from Swissprot project[2], and the alphabet size is 20.
- English, which contains 1 GB English text taken from Gutenberg Project[3], and the alphabet size is 236. Notice that the dataset contains not only English letters, but also punctuations and other symbols.

6.2 Evaluating Partition Methods

We first compared the candidate numbers of the even and best partition scheme. For each data set, we randomly generated 100 substrings as our searching queries. The query length varied from 10 to 200, and the error ratio varied from 1 % to 20 %. The candidate numbers of the even and best partition schemes on the three data sets are shown in Fig. 3. We can see that the candidate numbers

---

[1] https://genome.ucsc.edu/.
[2] http://www.uniprot.org/.
[3] https://www.gutenberg.org/.

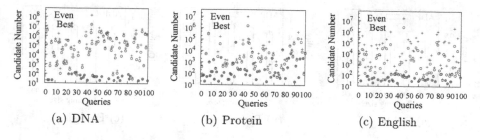

Fig. 3. Candidate numbers of the even and best partition methods.

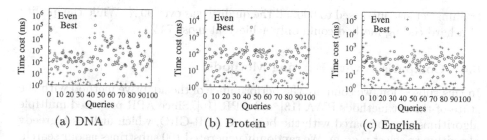

Fig. 4. Elapsed time of the basic and improved methods.

using even partition scheme were always greater than or equal to those using the best partition scheme. For some queries, the candidate numbers using the two methods had a large difference. Take English data set as an example, the maximum difference was 156 times, which means more than 99 % candidates using even partition scheme were unnecessary.

We then compared the performance of the basic method and the improved method, and the results are shown in Fig. 4. We can see that for some queries, the improved method performed much better than the basic method. While for some other queries, the basic method outperformed the improved method. The result is consistent with our early theoretical analysis. It confirms that if we could find an adaptive method to automatically select a good method, we could improve the efficiency of the searching system in average cases.

6.3 Evaluating Adaptive Method

In this subsection, we compare the quality of the proposed adaptive method. We randomly generated a group of queries as our training set. The training set size varied from 200 to 1000. We still utilized the same 100 queries as in Fig. 3 to check the quality of the adaptive method. The quality is defined as the number of times to select the faster method divided by total query number. The quality and the standard error of our method are shown in Fig. 5. With the increase of query number, the quality increased. All the quality was over 80 %. And when

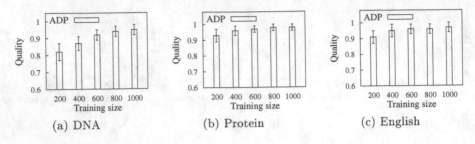

Fig. 5. Quality of the adaptive method.

training set size increased to 600, all the quality was over 90 %. While the quality of a hard coding method could only achieve at most 74 %.

6.4   Comparison with Existing Methods

In this subsection, we compare the performance of the adaptive method with the state-of-the-art methods BWA [13] and APR [10]. Since APR proposed multiple algorithms, we compared with the best one APR-GRQ, which utilizes a greedy algorithm to select v-gram. We randomly generated 100 substrings as our searching queries, varied their error ratio from 1 % to 20 %, and computed their average time cost. The result is shown in Fig. 6. On DNA data set, BWA outperformed APR. While on English dataset APR outperformed BWA. On DNA dataset, our method was close to BWA method for small error ratio. In other cases, our method was faster than the other two methods. The reason is as follows. To find the approximate matching substrings, BWA performs a depth-first search over the index. For each character of the query, it needs to enumerate all the characters in the alphabet, which is expensive for large error ratio or large alphabet. APR utilizes v-grams to find candidates. To allow larger error ratio, it must utilize small q_{min} and q_{max}, which involves a large number of candidates even for small error ratio. While our method adaptively selects a good scheme to find high quality candidates efficiently. Its performance was better than the other two methods on all the data sets.

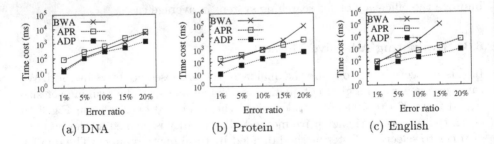

Fig. 6. Comparison with other methods.

7 Conclusion

In this paper we studied the problem for efficient approximate substring matching. We first proposed a basic method to solve the problem based on a BWT index. It utilizes an even $\tau + 1$ partition scheme to find candidates, then verifies those candidates to get the final result. To improve the quality of candidates, we then proposed an improved method, which utilizes a best $\tau + 1$ partition scheme to find candidates. Furthermore, through theoretical analysis, we find that the best partition scheme cannot always outperform the even partition scheme. To address this problem, we proposed an adaptive approach, which extracts a group of significant features for a given query, then utilizes logistic regression to do scheme selection. We conducted comprehensive experiments, which show the efficiency and quality of our proposed method.

Acknowledgements. The work was partially supported by the NSF of China for Outstanding Young Scholars under grant 61322208, the NSF of China under grants 61272178, 61572122, the NSF of China for Key Program under grant 61532021, and ARC DP140103499.

References

1. Bayardo, R.J., Ma, Y., Srikant, R.: Scaling up all pairs similarity search. In: Proceedings of the 16th International Conference on World Wide Web, pp. 131–140. ACM (2007)
2. Bocek, T., Hunt, E., Stiller, B., Hecht, F.: Fast similarity search in large dictionaries. University of Zurich, Zurich (2007)
3. Burrows, M., Wheeler, D.J.: A block-sorting lossless data compression algorithm (1994)
4. Deng, D., Li, G., Feng, J.: A pivotal prefix based filtering algorithm for string similarity search. In: Proceedings of the 2014 ACM SIGMOD International Conference on Management of Data, pp. 673–684. ACM (2014)
5. Fenz, D., Lange, D., Rheinländer, A., Naumann, F., Leser, U.: Efficient similarity search in very large string sets. In: Ailamaki, A., Bowers, S. (eds.) SSDBM 2012. LNCS, vol. 7338, pp. 262–279. Springer, Heidelberg (2012)
6. Ferragina, P., Manzini, G.: Opportunistic data structures with applications. In: Proceedings of the 41st Annual Symposium on Foundations of Computer Science, 2000, pp. 390–398. IEEE (2000)
7. Fredriksson, K., Navarro, G.: Average-optimal single and multiple approximate string matching. J. Exp. Algorithmics (JEA) **9**, 1–4 (2004)
8. Gravano, L., Ipeirotis, P.G., Jagadish, H.V., Koudas, N., Muthukrishnan, S., Srivastava, D.: Approximate string joins in a database (almost) for free. In: VLDB, pp. 491–500 (2001)
9. Jiang, Y., Deng, D., Wang, J., Li, G., Feng, J.: Efficient parallel partition-based algorithms for similarity search and join with edit distance constraints. In: Proceedings of the Joint EDBT/ICDT 2013 Workshops, pp. 341–348. ACM (2013)
10. Kim, Y., Park, H., Shim, K., Woo, K.G.: Efficient processing of substring match queries with inverted variable-length gram indexes. Inf. Sci. **244**, 119–141 (2013)

11. Li, C., Wang, B., Yang, X.: Vgram: Improving performance of approximate queries on string collections using variable-length grams. In: VLDB, pp. 303–314 (2007)
12. Li, G., Deng, D., Wang, J., Feng, J.: Pass-join: a partition-based method for similarity joins. Proc. VLDB Endowment **5**(3), 253–264 (2011)
13. Li, H., Durbin, R.: Fast and accurate short read alignment with burrows-wheeler transform. Bioinformatics **25**(14), 1754–1760 (2009)
14. Li, H., Durbin, R.: Fast and accurate long-read alignment with burrows-wheeler transform. Bioinformatics **26**(5), 589–595 (2010)
15. Navarro, G.: A guided tour to approximate string matching. ACM Comput. Surv. **33**(1), 31–88 (2001)
16. Navarro, G., Mäkinen, V.: Compressed full-text indexes. ACM Comput. Surv. (CSUR) **39**(1), 2 (2007)
17. Navarro, G., Sutinen, E., Tanninen, J., Tarhio, J.: Indexing text with approximate q-grams. In: Giancarlo, R., Sankoff, D. (eds.) CPM 2000. LNCS, vol. 1848, pp. 350–363. Springer, Heidelberg (2000)
18. Qin, J., Wang, W., Lu, Y., Xiao, C., Lin, X.: Efficient exact edit similarity query processing with the asymmetric signature scheme. In: SIGMOD Conference, pp. 1033–1044 (2011)
19. Ukkonen, E.: Approximate string-matching over suffix trees. In: Apostolico, A., Crochemore, M., Galil, Z., Manber, U. (eds.) CPM 1993. LNCS, vol. 684, pp. 228–242. Springer, Heidelberg (1993)
20. Wandelt, S., Deng, D., Gerdjikov, S., Mishra, S., Mitankin, P., Patil, M., Siragusa, E., Tiskin, A., Wang, W., Wang, J., et al.: State-of-the-art in string similarity search and join. ACM SIGMOD Rec. **43**(1), 64–76 (2014)
21. Wang, J., Li, G., Feng, J.: Can we beat the prefix filtering?: an adaptive framework for similarity join and search. In: SIGMOD Conference, pp. 85–96 (2012)
22. Wang, J., Yang, X., Wang, B.: Cache-aware parallel approximate matching and join algorithms using bwt. In: Proceedings of the Joint EDBT/ICDT 2013 Workshops, pp. 404–412. ACM (2013)
23. Yang, X., Wang, B., Li, C.: Cost-based variable-length-gram selection for string collections to support approximate queries efficiently. In: SIGMOD Conference, pp. 353–364 (2008)
24. Yang, X., Wang, B., Li, C., Wang, J., Xie, X.: Efficient direct search on compressed genomic data. In: 29th IEEE International Conference on Data Engineering, ICDE 2013, Brisbane, Australia, April 8–12, 2013, pp. 961–972 (2013)
25. Yang, X., Wang, Y., Wang, B., Wang, W.: Local filtering: Improving the performance of approximate queries on string collections. In: Proceedings of the 2015 ACM SIGMOD International Conference on Management of Data, pp. 377–392. ACM (2015)

Joint Probability Consistent Relation Analysis for Document Representation

Yang Wei[1,2]([✉]), Jinmao Wei[1,2]([✉]), Zhenglu Yang[1,2], and Yu Liu[1,2]

[1] College of Computer and Control Engineering, Nankai University,
Tianjin 300071, China
weiyang_tj@outlook.com, {weijm,yangzl}@nankai.edu.cn,
Liu_Yu@mail.nankai.edu.cn
[2] College of Software, Nankai University, Tianjin 300071, China

Abstract. Measuring the semantic similarities between documents is an important issue because it is the basis for many applications, such as document summarization, web search, text analysis, and so forth. Although many studies have explored this problem through enriching the document vectors based on the relatedness of the words involved, the performance is still far from satisfaction because of the insufficiency of data, i.e., the sparse and anomalous co-occurrences between words. The insufficient data can only generate unreliable relatedness between words. In this paper, we propose an effective approach to correct the unreliable relatedness, which keeps the joint probabilities of the co-occurrences between each word and themselves consistently equal to their occurrence probabilities throughout the generation of the relatedness. Hence the unreliable relatedness is effectively corrected by referring to the occurrence frequencies of the words, which is confirmed theoretically and experimentally. The thorough evaluation conducted on real datasets illustrates that significant improvement has been achieved on document clustering compared with the state-of-the-art methods.

Keywords: Document representation · Word relatedness · Joint probability consistency

1 Introduction

Measuring the similarities between documents is the basis for many applications, such as document summarization, web search, text analysis, and so forth, where documents are often represented as fixed-length vectors with the bag-of-words (BOW) model [9] due to its simplicity, efficiency, and good performance. However, BOW confines the similarity between two documents to the number of the common words shared by them, which incurs the problem that BOW cannot identify the similarities between documents composed of different words. As shown in Fig. 1, although all the five documents are semantically related, BOW cannot figure out the similarity between the documents d_1 and d_4.

© Springer International Publishing Switzerland 2016
S.B. Navathe et al. (Eds.): DASFAA 2016, Part I, LNCS 9642, pp. 517–532, 2016.
DOI: 10.1007/978-3-319-32025-0_32

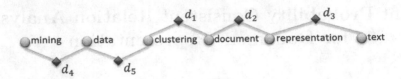

Fig. 1. An example of the relations between documents, where the diamonds stand for documents and the circles stand for words. There is a link between a diamond and a circle if the corresponding word appears in the document.

To further reveal the semantic similarities between documents, several strategies have been proposed [2,7,11,15] to enrich the document vectors with the relatedness of the words in the document collection to the original words in the corresponding documents directly. In particular, the Context Vector Model (CVM) [2] utilizes the explicit co-occurrences between words to estimate the relatedness, which is called the *explicit relatedness between words*. With CVM, d_4 will be enriched with "clustering", because its original word "data" has co-occurred with "clustering" in d_5, and d_1 will be enriched with "data" for the same reason. Then d_1 and d_4 will have two words in common, and their similarity can be discovered. Even so, the similarity between d_3 and d_4 still remains unknown for they have no explicitly related words in common.

Intuitively, it will be helpful to further exploit the *implicit relatedness* between words via their common related words. As can be seen, "representation" belonging to d_3 is implicitly related to "clustering" according to their co-occurrences with "document" in d_2 and d_1, respectively. Similarly, "data" belonging to d_4 is implicitly related to "document". By utilizing the implicit relatedness, d_3 and d_4 will have the words "clustering" and "document" in common. The Coupled term-term Relation Model (CRM) [7] has implemented this idea by defining the strength of the relative implicit relatedness between "representation" and "clustering" via "document" as the weaker strength of their explicit relatedness to "document". While taking just the arbitrary weaker strength decreases the differences among the strengths of the implicit relatedness, which further weakens the discriminativeness of the generated document vectors.

The referred document enriching methods differ from those incorporated with ontologies or dictionaries such as Wikipedia [8,10]. It's not uncommon that the external resources are mismatched or out of date for a particular document collection. Since the entities in a dictionary usually have several meanings, the referred methods are expected to help to allocate the documents to a particular one, which could be the future work of this paper.

The document enriching methods are in some sense similar to the dimensionality reduction methods, e.g., latent dirichlet allocation [3], non-negative matrix factorization [19], locally consistent concept factorization [6] and doc2vec [12], as they both utilize the relatedness between words self-contained in the document collection. The dimensionality reduction methods take this information to extract the semantic topics expressed by the documents. The similarities between documents are intuitively measured by their distributions on the topics. Significant improvements have been achieved with these methods. Nevertheless, the parameters, especially the number of topics, are often difficult to be determined.

We inherent the idea of exploiting the relatedness between words to enrich the document vectors, and argue that the fundamental issue which incurs the drawbacks of CVM and CRM is the *insufficiency of data*. The insufficiency refers to the sparse and anomalous co-occurrences between words, which make the relatedness overestimated for some words, while underestimated for some other words. The failure to find all the relatedness between words with CVM is an example of the underestimation. Specifically, due to the sparse co-occurrences, some related words haven't explicitly co-occurred with each other. CRM seems to alleviate the sparse co-occurrence problem by introducing the implicit relatedness. However, as the weaker strength of the explicit relatedness determines the implicit relatedness, CRM is very sensitive to the co-occurrence frequencies. The anomalous co-occurrences will mislead CRM to regard unrelated words as related, which is a case of the overestimation. Therefore, addressing the insufficiency problem is the key point of our method. Inspired by the common sense that the relatedness of a word to itself is one [2,7], an effective strategy is proposed. The contributions of this paper are as follow:

1. We figure out that the drawbacks of CVM and CRM are both incurred by the insufficiency of word co-occurrence frequencies, i.e., the insufficient data can only generate unreliable relatedness.
2. The proposed method effectively corrects the unreliable evaluation by referring to the occurrence frequencies of the words. In particular, the proposed method tackles the drawbacks of the existing methods by keeping the joint probability of the co-occurrence between a word and itself consistently equal to its occurrence probability, which gives a more robust evaluation of the implicit relatedness between words.
3. The experimental evaluation on document clustering demonstrates our approach significantly improves the performance compared with the state-of-the-art methods on real datasets.

The remainder of this paper is organized as follows: Sect. 2 provides some background on document representation. The novel method is proposed in Sect. 3. The experimental results are presented in Sect. 4. Finally, we draw the conclusion in Sect. 5.

2   Preliminaries

For sake of simplicity, the notions to be used are listed in Table 1.

With BOW, the m distinct words in the document collection D are used to construct the feature space. $\forall d \in D$, its document vector is represented as:

$$\Phi_{\text{bow}} : \mathbf{d} = (c_{v_1|d}, c_{v_2|d}, \cdots, c_{v_m|d})^T \in \mathbb{R}^m. \tag{1}$$

Table 1. Definition of the notations

Symbol	Definition
D	The collection of the given documents
d	A document in D
n	The number of the documents in D
m	The number of the distinct words in D
V	The collection of the distinct words in D
v_i	The i-th word in V
$c_{v_i\|D}$	The occurrence frequency of v_i in D
$c_{v_i,v_j\|D}$	The co-occurrence frequency between v_i and v_j in D
$c_{v_i\|d}$	The occurrence frequency of v_i in d
n_{v_i}	The number of the documents which contain v_i
$\mathbf{x}(i)$	The i-th value in the vector $\mathbf{x}$
$\mathbf{X}(i,j)$	The element in row i of column j of the matrix $\mathbf{X}$
$\mathbf{X}(a:b,c:d)$	The sub-matrix of $\mathbf{X}$ which contains the elements in rows a through b of columns c through d
$[\mathbf{X}_{a\times b}, \mathbf{Y}_{a\times c}]$	The conjunction of two matrices, which returns an $a \times (b+c)$ matrix $\mathbf{Z}$ with $\mathbf{Z}(i,1:b) = \mathbf{X}(i,1:b)$ and $\mathbf{Z}(i,(b+1):(b+c)) = \mathbf{Y}(i,1:c)$

Since BOW cannot figure out similar documents composed of different words, CVM [2] is proposed to reveal the meanings of documents with a set of weighted word vectors. $\forall v_i \in V$, its word vector is usually defined as [4,16]:

$$\mathbf{v}_i = (\frac{c_{v_i,v_1|D}}{c_{v_i|D}}, \frac{c_{v_i,v_2|D}}{c_{v_i|D}}, \ldots, \frac{c_{v_i,v_m|D}}{c_{v_i|D}})^T. \tag{2}$$

Generally, the meanings of words should be independent of the corpus size, so the overall frequency of v_i in the whole corpus, $c_{v_i|D}$, is introduced to give the basic context of v_i [5]. The values in $\mathbf{v}_i$ measure the relatedness of the words in the vocabulary V to v_i. Besides, if $v_j = v_i$, $\mathbf{v}_i(j) = 1$, which means the relatedness of a word to itself is one.

Together with all the word vectors, an $m \times m$ matrix is obtained, i.e., $\mathbf{V} = (\mathbf{v}_1, \mathbf{v}_2, \ldots, \mathbf{v}_m)$, which is called the *context matrix*. Then the document vector generated by CVM is:

$$\Phi_{\text{cvm}} : \mathbf{d}' = \mathbf{V}\mathbf{d} = \sum_{i=1}^{m} \mathbf{v}_i \mathbf{d}(i) \in \mathbb{R}^m. \tag{3}$$

CVM accumulates the word vectors with the weights contained by $\mathbf{d}$. The generated document vector is therefore enriched with the average relatedness of the words in V to the original words in d. In our previous work [18], an extension of the accumulation has been proposed. One main distinct contribution of this paper compared with the previous one is the estimation of the relatedness between words. As we discussed with Fig. 1, considering only the explicit relatedness hinders the enriching process to find all the related words to a document. Since our new method is developed from this new perspective, it is orthogonal to the existing one in a complementary manner.

CRM [7] introduces the relative implicit relatedness between words to discover more related words, and the *complete relatedness* between words is calculated through the weighted sum of their explicit relatedness and the average of their relative implicit relatedness via all the common related words:

$$
\mathbf{v}_i(j) = \begin{cases} \alpha \frac{c_{v_i,v_j|D}}{c_{v_i|D}} + \frac{(1-\alpha)}{|L|} \sum_{v_k \in L} \min_{a=i,j} \{ \frac{c_{v_a,v_k|D}}{c_{v_a|D}} \} & j \neq i \\ 1 & \text{else,} \end{cases}
$$
(4)

where $|L|$ denotes the number of the common related words in $L = \{ v_k | \frac{c_{v_i,v_k|D}}{c_{v_i|D}} > 0 \wedge \frac{c_{v_j,v_k|D}}{c_{v_j|D}} > 0 \}$, and α is an empirical parameter which keeps the same for all the words. With respect to Eq. (4), as the weaker strengths between the explicit relatedness of two words to their common related words are used to evaluate the strength of the implicit relatedness, it would be hard to distinguish the relatedness between an anomalously co-occurred word and a weak related word. The discriminativeness of the generated document vectors is therefore decreased.

3   Joint Probability Consistent Model

According to Eq. (2), $\forall v_i, v_j \in V$, the explicit relatedness of v_j to v_i could be regarded as the conditional probability that v_j will appear when v_i occurs:

$$
\frac{c_{v_i,v_j|D}}{c_{v_i|D}} = P(v_j|v_i).
$$
(5)

The probability view provides us the theoretical foundation to infer the implicit relatedness, and to find the relation between the complete relatedness and the occurrence frequencies of the words, which are explained in further detail below.

3.1   Definition of the Complete Relatedness

We assume the conditional probability of one word under the occurrence of another could be corrected according to their current distributions on all the words in V with the proper weights for the words.

Proposition 1. $\forall v_i, v_j \in V$, *the corrected conditional probability of v_j under the occurrence of v_i could be obtained by:*

$$
P'(v_j|v_i) = \frac{\sum_{k=1}^m \omega_k P(v_i|v_k)P(v_j|v_k)P(v_k)}{P(v_i)},
$$
(6)

where the parameters ω_k, $k = 1, \ldots, m$, are the weights for the words in V.

Proposition 1 is derived by considering the following two factors:

1st. As $P(v_j|v_i) = \frac{P(v_i,v_j)}{P(v_i)}$, re-estimating the conditional probability is equivalent to re-evaluating the joint probability $P(v_i, v_j)$. With the law of total probability[1], we have:

$$P(v_i, v_j) = \sum_{k=1}^{m} P(v_i, v_j, v_k) = \sum_{k=1}^{m} P(v_i, v_j|v_k)P(v_k), \qquad (7)$$

which means the joint probability could be further derived from the joint probabilities of the co-occurrences of v_i and v_j with each of the m distinct words. However, if the data is not sufficient to generate the joint probabilities of the co-occurrences between two words, it would be less sufficient to generate the joint probabilities of the co-occurrences among three words. To make the evaluation feasible, we argue that the co-occurrence between v_i and v_j under the occurrence of v_k mainly depends on their relatedness to v_k. $P(v_i|v_k)$ and $P(v_j|v_k)$ tend to be independent of each other, namely, $P(v_i, v_j|v_k) \simeq P(v_i|v_k)P(v_j|v_k)$. Hence we have:

$$P(v_i, v_j) \simeq \sum_{k=1}^{m} P(v_i|v_k)P(v_j|v_k)P(v_k). \qquad (8)$$

2nd. We cannot ensure that the joint probabilities could be corrected with Eq. (8) because the unreliable conditional probabilities are still involved. Meantime, as shown in Example 1, counting the co-occurrence frequency between two words through the sum of their co-occurrence frequencies with the other words often contains duplicate counts. The co-occurrences are caused by the dependencies between words. Since the words in V are not independent of each other, duplicate counts also exist when evaluating the joint probability of the co-occurrence between two words through the accumulation of the joint probabilities of the co-occurrences among three words. Therefore, we introduce the weights ω_k to smooth the potential conflict incurred by the unreliable conditional probabilities in Eq. (8) and the duplicate counts caused by the dependencies between words:

$$P(v_i, v_j) = \sum_{k=1}^{m} \omega_k P(v_i|v_k)P(v_j|v_k)P(v_k). \qquad (9)$$

Example 1. For a word segment $v_1 - v_2 - v_3 - v_4$, the sum of the co-occurrence frequencies of $v_1 - v_2 - v_3$ and $v_2 - v_3 - v_4$ is two, while the real co-occurrence frequency between v_2 and v_3 is one. Duplicate counts exist during the sum of the co-occurrence frequencies.

Dividing $P(v_i)$ from both sides of Eq. (9), we obtain Proposition 1.

From the semantic view, $P(v_i|v_k)$ and $P(v_j|v_k)$ are the explicit relatedness of v_i to v_k and v_j to v_k, respectively. Instead of taking the smaller of $P(v_i|v_k)$ and $P(v_j|v_k)$, the relative implicit relatedness between v_i and v_j via v_k is defined as $P(v_i|v_k)P(v_j|v_k)P(v_k)$. No information is abandoned.

---

[1] https://en.wikipedia.org/wiki/Law_of_total_probability.

Generally, $P(v_i|v_i)$ is forced to equal one [2,7]. Hence when $v_k = v_i$, we have:

$$\frac{P(v_i|v_i)P(v_j|v_i)P(v_i)}{P(v_i)} = \frac{P(v_j|v_i)P(v_i)}{P(v_i)} = P(v_j|v_i),$$

which is the explicit relatedness of v_j to v_i. Similarly, when $v_k = v_j$,

$$\frac{P(v_i|v_j)P(v_j|v_j)P(j)}{P(v_i)} = \frac{P(v_i|v_j)P(j)}{P(v_i)} = P(v_j|v_i).$$

So the defined conditional probability is just the complete relatedness, which combines the explicit and implicit relatedness with the weights ω_k.

3.2   Estimation of the Weights

Generally, the occurrence probabilities of the words estimated with their frequencies are more reliable than their conditional probabilities estimated with the co-occurrence frequencies because the data of the occurrence frequencies are usually more sufficient than the data of the co-occurrence frequencies. Besides, according to the Zipf's law [14], for any given document collection, the occurrence frequencies typically follow a Zipfian distribution. The occurrence probabilities of the words estimated with their frequencies are therefore thought reliable, and will be used to estimate the weights.

In accordance with the previous works [2,7], let $P(v_i|v_i) = \frac{P(v_i,v_i)}{P(v_i)} = 1$, we have $P(v_i, v_i) = P(v_i)$. With Eq. (9) we can further obtain:

$$P(v_i, v_i) = \sum_{k=1}^{m} \omega_k P(v_i|v_k)P(v_i|v_k)P(v_k) = P(v_i). \tag{10}$$

For convenience, suppose the occurrence probabilities are organized in the vector $\mathbf{p} = (P(v_1), \dots, P(v_m))^T$ and $\boldsymbol{\omega} = (\omega_1, \dots, \omega_m)^T$. By extending $\mathbf{p}$ to an $m \times m$ matrix $\mathbf{P}$ where $\mathbf{P} = (\mathbf{p}, \dots, \mathbf{p})$, Eq. (10) is equivalent to the expression:

$$\left(\mathbf{P}^T \cdot \mathbf{V} \cdot \mathbf{V}\right) \boldsymbol{\omega} = \mathbf{p}, \tag{11}$$

where $\mathbf{X}_{a \times b} \cdot \mathbf{Y}_{a \times b}$ returns an $a \times b$ matrix of which the value in row i of column k equals $\mathbf{X}(i, k)\mathbf{Y}(i, k)$. Let $\mathbf{A} = \mathbf{P}^T \cdot \mathbf{V} \cdot \mathbf{V}$, the element in row i of column k of $\mathbf{A}$ is the joint probability $P(v_i, v_i, v_k)$.

Estimating the weights is equivalent to minimizing the objective function as follows:

$$O = \|\mathbf{A}\boldsymbol{\omega} - \mathbf{p}\|^2. \tag{12}$$

The analytical solution of the objective function is:

$$\boldsymbol{\omega} = \left(\mathbf{A}^T\mathbf{A}\right)^{-1}\mathbf{A}^T\mathbf{p}. \tag{13}$$

With respect to the objective function, the potential errors incurred by the unreliable joint probabilities of the co-occurrences between the words and themselves and the duplicate counts caused by the dependencies between words are minimized by referring to the occurrence frequencies of the words. The optimal values are expected to further correct the joint probabilities of the co-occurrences between different words, because the optimal values are estimated with all the words involved, where the semantic meanings of each word are implied, and the relatedness between words depends on the semantic meanings.

3.3   Implementation of the Proposed Method

To implement the proposed method, we need to define the metric to calculate the co-occurrence frequencies between words. Among a number of alternatives [17], we follow the metric defined by CVM due to its efficiency. Formally,

$$c_{v_i,v_j|D} = \sum_{a=1}^{n} \frac{c_{v_i|d_a}}{\sum_{k=1}^{m} c_{v_k|d_a}} \cdot \frac{c_{v_j|d_a}}{\sum_{k=1}^{m} c_{v_k|d_a}}, \tag{14}$$

and $c_{v_i|D} = \sum_{j=1,j\neq i}^{m} c_{v_i,v_j|D}$ in Eq. (2), otherwise, $c_{v_i|D} = \sum_{a=1}^{n} c_{v_i|d_a}$.

In Sect. 3.2, we assume the occurrence frequencies of the words are sufficient to generate reliable occurrence probabilities. To make the assertion more plausible, we introduce the threshold TH to verify whether the occurrence frequency of a word is sufficient enough. Without loss of generality, suppose the words in V are organized in descending order according to their occurrence frequencies, and $\forall i \leq t, c_{v_i|D} > TH; \forall i > t, c_{v_i|D} \leq TH$. The estimation of the occurrence probabilities incorporated with the threshold is given as follows:

$$\forall v_i \in V, P(v_i) = \begin{cases} \frac{c_{v_i|D}}{\sum_{j=1}^{m} c_{v_j|D}} & i \leq t \\ \frac{\sum_{j=t+1}^{m} c_{v_j|D}}{(m-t)\sum_{j=1}^{m} c_{v_j|D}} & \text{else.} \end{cases} \tag{15}$$

The words with the occurrence frequencies less or equal to the threshold are called the *less frequent words*. Equation (15) gives the less frequent words the same occurrence probability.

Furthermore, we fix the weights of the less frequent words at one. As the weights are introduced to eliminate the overall duplicate counts existing in the accumulation of the joint probabilities, and the less frequent words always co-occur with some fixed words, the duplicate counts incurred by the less frequent words must be small. Besides, if two words are related to some less frequent word, they are likely to be interrelated, for the less frequent word expresses some specific meaning generally, which has more power to link the two words. So we prefer the less frequent words, and give them more weights. Consequently,

$$\boldsymbol{\omega} = [\boldsymbol{\omega}_1, \boldsymbol{\omega}_2]^T, \tag{16}$$

where $\boldsymbol{\omega}_1 = (\omega_1, \ldots, \omega_t)$ and $\boldsymbol{\omega}_2 = \underbrace{(1, \ldots, 1)}_{m-t}$.

Similarly, let

$$\mathbf{A} = [\mathbf{A}_1, \mathbf{A}_2], \tag{17}$$

where $\mathbf{A}_1$ is an $m \times t$ matrix with $\mathbf{A}_1(i, 1 : t) = \mathbf{A}(i, 1 : t)$, and $\mathbf{A}_2$ is an $m \times (m - t)$ matrix with $\mathbf{A}_2(i, 1 : (m - t)) = \mathbf{A}(i, (t + 1) : m)$.

The calculation of $\boldsymbol{w}_1$ is thereby:

$$\boldsymbol{\omega}_1 = \left(\mathbf{A}_1^T \mathbf{A}_1\right)^{-1} \mathbf{A}_1^T (\mathbf{p} - \mathbf{A}_2 \boldsymbol{\omega}_2). \tag{18}$$

$\mathbf{A}_2 \boldsymbol{\omega}_2$ returns an $m \times 1$ vector with the i-th element equal to $\sum_{k=t+1}^{m} \omega_k P(v_i, v_k)$. As $\mathbf{p}(i) = P(v_i) = \sum_{k=1}^{m} \omega_k P(v_i, v_k)$, the i-th element in $\mathbf{p} - \mathbf{A}_2 \boldsymbol{\omega}_2$ equals $\sum_{k=1}^{t} \omega_k P(v_i, v_k)$. Namely, by setting the weights of the less frequent words to one, $\mathbf{p} - \mathbf{A}_2 \boldsymbol{\omega}_2$ returns the occurrence probabilities of all the m distinct words without considering their co-occurrences with the less frequent words.

Algorithm 1. Document Vector Generation with JPCM

Input: A document vector generated with BOW $\mathbf{d}$,
        the context matrix $\mathbf{V}$,
        the weights for the words $\boldsymbol{\omega}$,
        and the occurrence probabilities of the words $\mathbf{p}$
Output: A document vector enriched with the related words
1 Let $\Omega_{m \times m} = ((\boldsymbol{\omega} \cdot \mathbf{p}), \cdots, (\boldsymbol{\omega} \cdot \mathbf{p}))$;
2 $\mathbf{V}' = \frac{\mathbf{V}(\mathbf{V}^T \cdot \Omega)}{\mathbf{P}^T}$;
3 $\mathbf{d}' = \mathbf{V}'\mathbf{d} \in \mathbb{R}^m$.

With the obtained weights, $\boldsymbol{\omega} = [\boldsymbol{\omega}_1, \boldsymbol{\omega}_2]$, the generating process of document vectors is concluded as Algorithm 1. The proposed method is called the Joint Probability Consistent Model (JPCM in short), for it keeps the joint probability of the co-occurrence between a word and itself consistently equal to its occurrence probability throughout the generation of the complete relatedness. Step 2 calculates the complete relatedness between words defined in Eq. (6), and Step 3 generates the enriched document vector according to Eq. (3), where the context matrix is replaced by the matrix $\mathbf{V}'$ fulfilled with the defined complete relatedness.

With respect to the generating process, it costs $O(m^2)$ times to calculate all the complete relatedness between words, and $O(mn)$ times to calculate the relatedness of each word to the given n documents. It also costs $m \times m$ and $m \times n$ floating space to storage the context matrix and the enriched document vectors, respectively. As the number of words is limited in practice, there is an upper bound on the complexity of calculating the relatedness between words. The overall time and space consuming for JPCM is $\max\{O(m^2), O(mn)\}$.

4   Experiments

4.1   Experimental Setup

In this section, we conduct the performance evaluation with the TDT2 and Reuters datasets[2]. The TDT2 dataset consists of data extracted from six sources,

---

[2] www.cad.zju.edu.cn/home/dengcai/Data/TextData.html.

Table 2. Statistics of the datasets

	No. words	Avg. doc. length	Avg. word freq	No. clusters	Avg. cluster size
TDT2	36771	182	51	56	179
Reuters	18933	68	30	41	200

including two newswires (APW, NYT), two radio programs (VOA, PRI) and two television programs (CNN, ABC). Those documents appearing in two or more categories are removed, and the categories with more than 10 documents are kept, thus leaving us with 10,021 documents in total. The Reuters dataset contains 21,578 documents which are grouped into 135 classes. Those documents with multiple category labels are discarded, and the categories with more than 10 documents are selected. This leaves us with 8,213 documents in total. Table 2 provides the statistics of the datasets.

The performance of the pairwise similarity evaluation is an important criterion to verify the qualities of the representations for documents. Generally, with good representation, the similarities between semantically related documents should obtain high scores, while the similarities between unrelated documents should obtain low scores. This is consistent with the purpose of the clustering task that similar documents are organized into the same group, while dissimilar documents are organized into different groups. Therefore, we evaluate the document representation methods on document clustering problem with the k-means clustering algorithm. The cosine of document vectors is chosen to be the similarity measure. The evaluation of the similarities between documents directly affects the results of k-means, thus can reflect the qualities of the representation methods for documents. The methods to be compared include:

1. The popular BOW method is used as the baseline to evaluate JPCM.
2. The dimensionality reduction methods, namely, the Non-negative Matrix Factorization based document representation (NMF) method [19] and the Locally Consistent Concept Factorization method (LCCF) [6], which were once tested with the above datasets for document clustering, are also used as the baselines.
3. CVM and CRM are performed to verify the improvement of JPCM.
4. The proposed method JPCM is tested.

With respect to the referred methods, the occurrence frequencies of the words in each document are smoothed with the tf-idf weighting scheme.[3] While for JPCM, the raw frequencies are used because we assume the occurrence frequencies of the words are sufficient. For NMF, the dimension of the feature space is predefined as the number of the classes in the document collection, each dimension corresponds to a topic, and the cluster label of a document is the topic belonging to the dimension with the maximum value, which is claimed in [19]. For the other methods, the generated document vectors are used as the input of the k-means clustering algorithm.

---

[3] https://en.wikipedia.org/wiki/Tf-idf.

For consistency with NMF and LCCF, the accuracy (AC) [19] and the normalized mutual information (NMI) [1,19] are adopted to evaluate the performance of document clustering. Given a document vector d_a, let r_a and l_a be the cluster label and the label provided by the document collection, respectively. AC is defined as follows:

$$AC = \frac{\sum_{a=1}^{n} \delta(l_a, map(r_a))}{n}, \tag{19}$$

where n denotes the total number of documents in the test, $\delta(l_a, map(r_a))$ is the delta function that equals one if $l_a = map(r_a)$ and equals zero otherwise, and $map(r_a)$ is the mapping function that maps each cluster r_a to the equivalent label from the document collection. The best mapping can be found by using the Kuhn-Munkres algorithm [13].

Let n_l be the number of documents in class l, n_r be the number of samples in cluster r, and $n_{l,r}$ be the number of samples in class l and cluster r, then:

$$NMI = \frac{2\sum_{l,r} n_{l,r} log \frac{n n_{l,r}}{n_l n_r}}{\sum_l n_l log \frac{n_l}{n} + \sum_r n_r log \frac{n_r}{n}}. \tag{20}$$

It is easy to check that NMI ranges from 0 to 1. Since the cluster number is predefined as the number of the classes in the document collection, NMI $= 1$ if cluster r is identical to class l, and NMI $= 0$ if the two sets are independent.

4.2   Performance Evaluations

The clustering results shown in Table 3 are obtained with TH equivalent to 600 for TDT2 and 40 for Reuters, respectively. For each given class number, 50 test runs were performed on different randomly chosen classes and the average performance is reported in the table. Hence the evaluation was conducted on 450 subsets for both of the datasets. Furthermore, the paired-samples t-test [20] was used to analyze the significance of the improvements achieved by the proposed method. The results are shown in Table 4, where each value is the ratio of the improvement of JPCM compared with one of the other methods with the probability of 0.9 on the 450 paired samples. These experiments reveal a number of interesting points:

(1) CVM and the proposed method both perform better than BOW, which illustrates the power of enriching the generated document vectors with the relatedness of the words involved. The bad performance of CRM agrees with our analysis that CRM gives an improper evaluation of the implicit relatedness between words. Such evaluation brings unrelated words into the document vectors in practice, which decreases the discriminativeness of the document vectors.

(2) JPCM achieves significant performance improvement, specifically, an average of 6.5 % performance improvement has been achieved compared with CVM, the best of the other methods. The improvement indicates the necessity of correcting the unreliable relatedness generated with the insufficient data and also

Table 3. Clustering results on the TDT2 and reuters datasets

No. class	2	3	4	5	6	7	8	9	10	Avg.
TDT2-AC
BOW	0.904	0.849	0.784	0.731	0.714	0.699	0.646	0.653	0.615	0.733
NMF	0.924	0.873	0.838	0.770	0.751	0.744	0.663	0.655	0.620	0.760
LCCF	0.958	0.890	**0.897**	0.853	0.860	0.842	0.771	0.764	0.730	0.841
CVM	**0.964**	0.904	0.861	0.800	0.823	0.765	0.728	0.729	0.691	0.807
CRM	0.924	0.835	0.789	0.726	0.716	0.672	0.630	0.625	0.597	0.724
JPCM	0.936	**0.911**	0.895	**0.864**	**0.870**	**0.865**	**0.826**	**0.833**	**0.817**	**0.869**
TDT2-NMI
BOW	0.720	0.767	0.745	0.733	0.759	0.754	0.727	0.739	0.720	0.740
NMF	0.779	0.780	0.750	0.722	0.743	0.737	0.694	0.698	0.689	0.733
LCCF	0.867	0.799	0.803	0.787	0.812	0.808	0.770	0.774	0.758	0.798
CVM	0.895	0.847	0.821	0.785	0.820	0.797	0.778	0.784	**0.770**	0.811
CRM	0.777	0.737	0.719	0.682	0.710	0.685	0.665	0.677	0.662	0.702
JPCM	**0.982**	**0.955**	**0.928**	**0.889**	**0.881**	**0.870**	**0.791**	**0.801**	0.763	**0.873**
Reuters-AC
BOW	0.780	0.655	0.613	0.557	0.556	0.517	0.439	0.449	0.492	0.562
NMF	0.768	0.690	0.663	0.600	0.589	0.552	0.483	0.485	0.502	0.592
LCCF	0.805	0.665	0.657	0.618	0.575	0.567	**0.534**	**0.514**	0.517	0.606
CVM	0.825	0.693	0.675	0.601	0.603	0.553	0.478	0.489	0.528	0.605
CRM	0.841	0.707	0.657	0.593	0.591	0.539	0.468	0.460	0.492	0.594
JPCM	**0.879**	**0.779**	**0.765**	**0.655**	**0.631**	**0.583**	0.497	0.493	**0.526**	**0.645**
Reuters-NMI
BOW	0.362	0.413	0.486	0.448	0.522	0.510	0.437	0.467	0.556	0.467
NMF	0.276	0.368	0.444	0.406	0.475	0.456	0.381	0.420	0.487	0.413
LCCF	0.327	0.365	0.466	0.405	0.466	0.475	0.421	0.439	0.496	0.429
CVM	0.451	0.459	0.538	0.485	0.548	0.533	0.462	0.490	0.573	0.504
CRM	0.470	0.422	0.495	0.426	0.481	0.461	0.386	0.403	0.486	0.448
JPCM	**0.547**	**0.559**	**0.613**	**0.518**	**0.572**	**0.558**	**0.489**	**0.500**	**0.581**	**0.549**

Table 4. Performance improvements of JPCM (%)

	AC					NMI				
	BOW	NMF	LCCF	CVM	CRM	BOW	NMF	LCCF	CVM	CRM
TDT2	17.9	13.8	3.1	7.2	19.2	15.7	16.9	7.9	6.2	22.2
Reuters	13.3	7.1	4.4	5.5	7.2	15.2	29.9	25.0	6.9	20.1

demonstrates the effectiveness of using the occurrence frequencies of the words on addressing the insufficiency problem. The superiority of JPCM results in the following two practical aspects: 1st, it finds more related words to enrich the document vectors compared with CVM. 2nd, it generates more robust relatedness than CRM.

(3) JPCM also performs better than NMF or LCCF. Although we haven't compared all the dimensionality reduction methods, the following conclusion
could still be derived. The dimensionality reduction methods extract the
latent topics implied by the document collection, and assign the documents
to these topics, which incurs the issue of determining the number of topics.
In particular, the worse performance of NMF or LCCF with the NMI test
may attribute to the too few number of the predefined topics, which forces
the documents belonging to different topics assigned to the same predefined
topic (NMI tests the purity of each cluster).

4.3   Parameter Selection

The value of TH is essential to our JPCM method. Figure 2 shows how the performance of JPCM varies with the values of TH. The horizontal axis is the value
of TH, and the vertical axis denotes the corresponding average performance. It
shows that JPCM achieves consistent better performance compared with CVM
and CRM with the values of TH ranging from 2 to 400. So we can choose the
value of TH in a wide range.

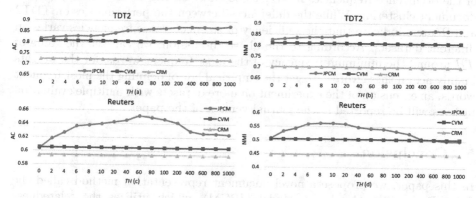

Fig. 2. The performances of JPCM with the different values of TH.

The performance of JPCM has been improved sustainedly until the value of
TH comes to about 600 on TDT2 and 20 on Reuters, where 99 % and 93 % of
the words are regarded as the less frequent words according to Fig. 3. In this
sense, only a small number of the weights for the most frequent words need to
be estimated. As the weights are affected by the duplicate counts existing in
the accumulation defined by Proposition 1, the phenomenon is consistent with
the common sense that the most frequent words are more likely to co-occur
with other words. Tuning the weights of the most frequent words therefore eliminates the duplicate counts. On the other hand, the weights of the words are
also determined by the reliability of the conditional probabilities of the words.
The occurrence frequencies of the most frequent words are usually sufficient to
be referred. Intuitively, the most frequent words are just the so called stop words.
Although the stop words are meaningless independently, they are used as the

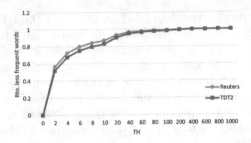

Fig. 3. The average proportions of the less frequent words in all the subsets with the different values of TH.

glue to link words to express some specific meanings in practice. The stop words are perhaps the best supporters to express the meanings of the other words. Hence tuning the weights of the most frequent words is plausible.

The variations of the performances of JPCM with different evaluation metrics on the same dataset are very similar, which demonstrates that the sufficiency of the occurrence frequencies has the same effect on the accuracy and purity of document clustering. While the differences between the performances on TDT2 and Reuters indicate that the sufficiency of the occurrence frequencies varies on different datasets. According to the above discussion, we can set the value of TH around the minimum frequency of the most frequent words, for the weights mainly act on these words. To get the empirical law of defining the most frequent words, an extension of the experiment on more datasets with multiple evaluation metrics will be reported in the journal version of the paper.

5   Conclusion

In this paper, we propose a novel document representation method called the Joint Probability Consistent Model (JPCM), which utilizes the relatedness between words to enrich the generated document vectors. We figure out that the existing methods with the same purpose generate unreliable relatedness between words because of the insufficiency of data. JPCM corrects the unreliable relatedness by referring to the occurrence frequencies of each word. Practically, the well-defined relatedness between words with JPCM gives a comprehensive consideration of the explicit and implicit relatedness between words, the weights of which are automatically estimated with the proposed optimization method. In this sense, JPCM is expected to generate more reliable relatedness between words than the existing methods which utilize part of the explicit and implicit relatedness only. The superiority of JPCM to the state-of-the-art methods has also been demonstrated by the thorough evaluation on document clustering. As a result, JPCM has more power on revealing the semantic similarities between documents than the existing methods.

Acknowledgments. This work was supported by the National Natural Science Foundation of China under grant 61070089, the Science Foundation of TianJin under grant 14JCYBJC15700, and the National Science Foundation of China under grant 11431006.

References

1. Andrews, N.O., Fox, E.A.: Recent developments in document clustering. Technical report, Computer Science, Virginia Tech (2007)
2. Billhardt, H., Borrajo, D., Maojo, V.: A context vector model for information retrieval. J. Am. Soc. Inf. Sci. Technol. **53**(3), 236–249 (2002)
3. Blei, D.M., Ng, A.Y., Jordan, M.I.: Latent Dirichlet allocation. J. Mach. Learn. Res. **3**, 993–1022 (2003)
4. Blunsom, P., Grefenstette, E., Hermann, K.M., et al.: New directions in vector space models of meaning. In: Proceedings of the 52nd Annual Meeting of the Association for Computational Linguistics (2014)
5. Bullinaria, J.A., Levy, J.P.: Extracting semantic representations from word co-occurrence statistics: a computational study. Behav. Res. Methods **39**(3), 510–526 (2007)
6. Cai, D., He, X., Han, J.: Locally consistent concept factorization for document clustering. IEEE Trans. Knowl. Data Eng. **23**(6), 902–913 (2011)
7. Cheng, X., Miao, D., Wang, C., Cao, L.: Coupled term-term relation analysis for document clustering. In: The 2013 International Joint Conference on Neural Networks (IJCNN), pp. 1–8. IEEE (2013)
8. Gabrilovich, E., Markovitch, S.: Overcoming the brittleness bottleneck using Wikipedia: enhancing text categorization with encyclopedic knowledge. In: AAAI, vol. 6, pp. 1301–1306 (2006)
9. Harris, Z.S.: Distributional structure. Word (1954)
10. Hu, X., Zhang, X., Lu, C., Park, E.K., Zhou, X.: Exploiting wikipedia as external knowledge for document clustering. In: Proceedings of the 15th ACM SIGKDD International Conference on Knowledge Discovery and Data Mining, pp. 389–396. ACM (2009)
11. Kalogeratos, A., Likas, A.: Text document clustering using global term context vectors. Knowl. Inf. Syst. **31**(3), 455–474 (2012)
12. Le, Q.V., Mikolov, T.: Distributed representations of sentences and documents. In: Proceedings of the 31st International Conference on Machine Learning. W&CP, vol. 32 (JMLR) (2014)
13. Lovász, L., Plummer, M.: Matching theory. Annals of Discrete Mathematics, vol. 29. Amsterdam Publishing, Amsterdam (1986)
14. Powers, D.M.: Applications and explanations of Zipf's law. In: Proceedings of the Joint Conferences on New Methods in Language Processing and Computational Natural Language Learning, pp. 151–160. Association for Computational Linguistics (1998)
15. Rungsawang, A.: DSIR: the first TREC-7 attempt. In: TREC, pp. 366–372 (Citeseer) (1998)
16. Turney, P.D., Pantel, P., et al.: From frequency to meaning: vector space models of semantics. J. Artif. Intell. Res. **37**(1), 141–188 (2010)
17. Wei, Y., Wei, J., Xu, H.: Context vector model for document representation: a computational study. In: Li, J., Ji, H., Zhao, D., Feng, Y. (eds.) NLPCC 2015. LNCS, pp. 194–206. Springer, Switzerland (2015)

18. Wei, Y., Wei, J., Yang, Z.: Enriching document representation with the deviations of word co-occurrence frequencies. In: Wang, G., Zomaya, A., Perez, G.M., Li, K. (eds.) ICA3PP 2015. LNCS, pp. 241–254. Springer, Switzerland (2015)
19. Xu, W., Liu, X., Gong, Y.: Document clustering based on non-negative matrix factorization. In: Proceedings of the 26th Annual International ACM SIGIR Conference on Research and Development in Informaion Retrieval, pp. 267–273. ACM (2003)
20. Zimmerman, D.W.: Teacher's corner: a note on interpretation of the paired-samples t test. J. Educ. Behav. Stat. **22**(3), 349–360 (1997)

Automated Table Understanding
Using Stub Patterns

Roya Rastan[1]([✉]), Hye-young Paik[1], John Shepherd[1], and Armin Haller[2]

[1] The University of New South Wales, Sydney, NSW, Australia
{rrastan,hpaik,jas}@cse.unsw.edu.au
[2] Australian National University, Canberra, ACT, Australia
armin.haller@anu.edu.au

Abstract. Tables in documents are a rich source of information, but not yet well-utilised computationally because of the difficulty of extracting their structure and data automatically. In this paper, we progress the state-of-the-art in automatic table extraction by identifying common patterns in table headers to develop rules and heuristics for determining table structure. We describe and evaluate a table understanding system using these patterns and rules.

Keywords: Table understanding · Table logical structure · Table stub analysis · Table categories · Category hierarchy

1 Introduction

Tables are a widely-available source of inter-related data, and there has been significant effort dedicated to automatically extracting and manipulating their structure and data [3,15]. The task of *Automatic Table Understanding* exploits the same features that humans use to try to extract the relationships amongst table cells and ultimately obtain a representation of the table data which is independent of its layout.

Much prior table understanding work utilised external knowledge sources such as domain ontologies. However, recent work such as [7,13] has aimed to be more generic by using only the extant features of the table (e.g., layout, cell content) as a basis for designing table understanding systems. The aim of such work is to automatically transform the table data into a generic data structure and create a basis for further data analysis (e.g. semantic relation detection, query answering).

Considering the diversity of table structures, it may seem challenging to formulate a one-fits-all solution for understanding tables. However, it is generally accepted that tables can be *understood* if one can detect the hierarchical structure of table headers (both row headers and column headers) properly and determine how each table data cell can be uniquely accessed through them.

A range of different approaches have been taken to deal with this complexity. These include, for instance, only processing specific types of table such as *Well-Formed Tables* [6] and *Multi-Dimensional Tables* [1], or involving a human in the

© Springer International Publishing Switzerland 2016
S.B. Navathe et al. (Eds.): DASFAA 2016, Part I, LNCS 9642, pp. 533–548, 2016.
DOI: 10.1007/978-3-319-32025-0_33

header detection process [10], or removing layout features and analysing tables based on the cell arrangement of their headers [13].

While these approaches all contributed to advancing table understanding techniques, they did not consider additional header features that enable more accurate discovery of the hierarchical structure of table headers.

In this paper, we identify patterns of layout features that commonly occur in the table stubs, and which enable us to detect the *header hierarchy* in many cases that would be missed by existing methods.

The patterns give us a concrete basis for building automatic table understanding algorithms that can be effectively applied to a wider range of tables than existing methods. Specifically, we do not assume that tables are *well-formed* (in the sense of [6]), and we do not require human intervention during the process.

We simply require a segmented table as input and determine the header hierarchy, and all *access paths* for each data cell from it. We evaluate the performance of our methods against the state-of-the-art on three well-known public datasets.

This work makes the following contributions: (i) we provide a classification of layout features in table stubs and an interpretation of the header hierarchy they imply; (ii) we present algorithms for automatic table understanding that make fewer assumptions about the source tables than existing algorithms, (iii) we demonstrate how our approach leads to a straight-forward transformation of source tables to the well-known Wang Abstract Table [14] model.

2   Preliminaries

In this section, we briefly explain the concepts and terminology used throughout the paper; more detailed descriptions can be found, if required, in [11].

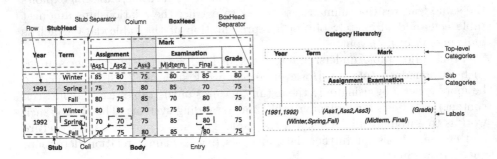

Fig. 1. Wang Table Terminology

Table Regions. In discussing table structures, we follow Wang's table terminology [14] which is widely used in the literature.

As shown in Fig. 1, Wang divides a table into four regions: *stub*, *stubhead*, *boxhead* and *body*. The regions are delineated by a *stub separator* and a *boxhead separator* which are frequently, but not always, shown as physical lines.

The body (lower-right) contains the basic data (called *entries*). The rest of the table (stub, stubhead, boxhead) contains *labels* which are used to locate the basic data. The boxhead (upper-right) consists of labels whose values are used to access individual columns. The stubhead (upper-left) and stub (lower-left) contain labels whose values are used to access individual rows.

Categories, Labels and Access Paths. The above discussion focuses on the physical features of tables. Now we introduce the logical features of tables that are relevant to Table Understanding.

Categories and Labels: A table is a two-dimensional representation of a multi-dimensional space. Labels are arranged in a hierarchy that maps the multi-dimensional space onto the table's two-dimensional data grid. In Wang's work, a table dimension is referred to as a *category*. We can view the category hierarchy as a set of trees, with one tree for each category, and with each node labelled (see e.g. Fig. 1). The roots of the trees in the hierarchy are called *top-level categories*, intermediate nodes are *sub-categories*, and the leaf nodes are simply *labels*.

Access Paths: An *access path* is a sequence of labels that leads from a top-level category to a row or column of entries. An entry can be *uniquely accessed* via a combination of the access paths created from *each* of the top-level categories. One or more of these access paths starts from the boxhead region; one or more of these starts from the stub region.

As an example, consider the table in Fig. 1. The *entries* are the average marks of the assignments and examinations in a course. *Year* is a top-level category that consists of the labels 1991 and 1992. *Term* is a top-level category that consists of the labels Winter, Spring, and Fall. *Mark* is a top-level category that consists of two sub-categories (*Assignment* and *Examination*) and a label Grade. The entry 85 at the top-left corner of the body can be accessed via a combination of three access paths: Year.1991, Term.Winter and Mark.Assignment.Ass1.

In our work, the task of *Table Understanding* focuses on detecting the categories of a table and the access paths to its entries. The output of *Table Understanding* consists of a Wang *Abstract Table* model [14], components of which we have described above. Since an Abstract Table represents the logical relationships of the cells in a table, it is considered to be a 'presentation/layout independent' representation of the table. Such a representation has been found to be useful by many applications that utilise the extracted data [1,7].

Tasks in End-to-End Table Processing. Table Understanding is the second phase of the complete end-to-end table processing task, originally defined in [2]. The first phase consists of *Table Extraction*, which deals with locating tables in a document and segmenting them into individual cells, rows and columns.

In earlier work [11], we describe the design and implementation of an end-to-end table processing system (TEXUS) in which we provide a concrete set of well-specified tasks that fit together to form a processing pipeline (Fig. 2). The pipeline takes a document as input and produces an *Abstract Table* model for each table located in the input document. We define the following tasks: (1) Document Converting: converts the input document to our own document model, (2) Locating: finds the outer boundaries of tables, (3) Segmenting: recognises the inner boundaries in a table (cells, rows and columns), (4) Functional Analysis: identifies the role of each cell in a segmented table (Header, Access, Data), and (5) Structural Analysis: detects categories and access paths.

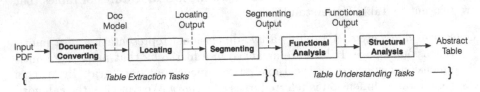

Fig. 2. An end-to-end table processing pipeline in TEXUS

In concrete terms, we consider *Table Understanding* to be a combination of functional analysis and structural analysis. Later in the paper, we present the implementation of our *automatic table understanding* algorithms in terms of these two tasks. For the remainder of the paper, we assume that tables are processed up to segmenting, which means we can refer to individual cells, rows and columns.

Note that the term *Table Interpretation* is used to refer to tasks that map table data to a specific set of domain entities and relationships. It is thus dependent on the application domain. While the Abstract Tables produced by our system provide a basis for table interpretation, detailed discussion of this is outside the scope of this paper. Note also that our methods assume that the table and cell boundaries have already been identified, so we do not discuss the work on *Table Extraction* [9] further in this paper.

3   Related Work

In this section, we focus our discussion on other work on the *Table Understanding* task and summarise how our work extends existing approaches.

The work by Fang et al. [4] employs machine learning techniques to classify tables into different types (e.g. complex, long, folded) using layout features in the headers. However, they do not attempt to extract categories. Seth et al. [12] describe a taxonomy of table column headers and propose a model to convert the geometric structure of the headers to a representation equivalent to Wang notation. However, their work is applicable only to column headers.

The most cited work in automated *Table Understanding* comes from research groups led by George Nagy and David Embley. They began with a semi-automated approach where human assistance is required during the category detection phase. This approach is used in many tools like WNT [5], TAT [10], and VeriClick [8]. To move one step further towards automation of this process, they developed a learning-based classification algorithm to detect the four *critical cells* in a table that help to distinguish the header/column cells (labels) from data cells (entries) [6]. In their recent work, they propose an algorithmic solution to index table columns and rows so that detection of the critical cells can be automated [13].

Much of the above work has focused on discovering categories in the boxhead, with less attention paid to the stub (in particular, similar processing strategies are used for both). This paper focuses on extracting category information from the stub. Specifically, we note that table designers often use layout and styling conventions (e.g. bullet points, indentation) in the stub to encode the category hierarchy. This information has been disregarded in previous work, but we believe that it can be exploited to accurately extract category hierarchies from the stub. Doing so extends the variety of tables that can be automatically mapped to Wang's Abstract Table model.

4   Stub Analysis

In this section, we describe our approach to *Stub Analysis*. As noted above, the *stub* region consists of one or more table columns and extends from the leftmost column to the start of the body region. The labels in the stub define categories which are used to build access paths to individual table rows. The aim of stub analysis is to identify the category hierarchy implied by the arrangement of labels in the stub, and we achieve this by analysing patterns in these labels.

4.1   Stub Patterns

In defining the patterns, we draw on from the work of Fang [4] and Wang [14]. Wang identified layout styling rules for the stub, and Fang investigated different structural organisations commonly applied in arranging labels. We extend their work based on observations from a variety of tables in public table datasets.
We use the following kinds of features:

- Formatting Features: including font face, font size and colour
- Layout Features: including indentation, bullets, numbering and spanned cells
- Content-based Features: including repetition of labels and empty cells

In each pattern, there is an implied hierarchy among the labels in that pattern. We use the terms "enclosing label" to indicate that one label is the 'parent' of some other label and "enclosed label" to indicate that one label is subordinate to another. If A is the enclosing label of B, then B is an enclosed label of A.

Downward Expansion Patterns. This kind of pattern occurs when a single column is used to encode the entire category hierarchy. A column containing this pattern will be the last column in the stub (i.e. we do not need to consider more columns to discover the category hierarchy after finding this pattern). Examples of such patterns are shown in Fig. 3.

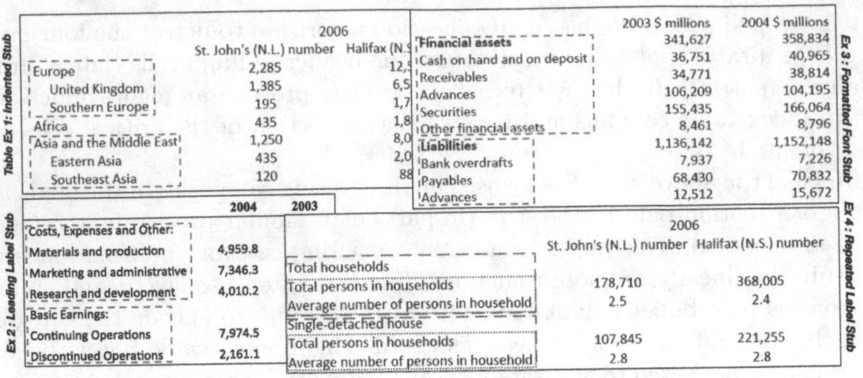

Table Ex 1: Indented Stub

2006	St. John's (N.L.) number	Halifax (N.S...
Europe	2,285	12,...
United Kingdom	1,385	6,5...
Southern Europe	195	1,7...
Africa	435	1,8...
Asia and the Middle East	1,250	8,0...
Eastern Asia	435	2,0...
Southeast Asia	120	88...

Ex 3: Formatted Font Stub

	2003 $ millions	2004 $ millions
Financial assets	341,627	358,834
Cash on hand and on deposit	36,751	40,965
Receivables	34,771	38,814
Advances	106,209	104,195
Securities	155,435	166,064
Other financial assets	8,461	8,796
Liabilities	1,136,142	1,152,148
Bank overdrafts	7,937	7,226
Payables	68,430	70,832
Advances	12,512	15,672

Ex 2 : Leading Label Stub

	2004	2003
Costs, Expenses and Other:		
Materials and production	4,959.8	
Marketing and administrative	7,346.3	
Research and development	4,010.2	
Basic Earnings:		
Continuing Operations	7,974.5	
Discontinued Operations	2,161.1	

Ex 4 : Repeated Label Stub

	2006	
	St. John's (N.L.) number	Halifax (N.S.) number
Total households		
Total persons in households	178,710	368,005
Average number of persons in household	2.5	2.4
Single-detached house		
Total persons in households	107,845	221,255
Average number of persons in household	2.8	2.8

Fig. 3. Examples of Downward Expansion Patterns

Downward expansion patterns can be recognised by the occurrence of the following features:

Indented Stub: Uses indentation (e.g. white spaces, tabs, bullets, or numbering) to indicate the hierarchy. Indented labels are enclosed by the nearest non-indented label above them.

Leading Label Stub: Uses special characters like ':', '=', ':-' at the end of the enclosing label ("leading label") to indicate that subsequent labels without this character are enclosed.

Formatted Font Stub: Uses font formatting features (e.g., modifications of font appearances - bold, italic, larger size, underlines, colour) on the enclosing label. The formatted label encloses all the following plain format labels.

Repeated Label Stub: Indicated by an identical set of labels being repeated consistently in the column. The set of repeating labels are enclosed by the nearest non-repeated label above them.

Forward Expansion Patterns. This kind of pattern occurs when the category hierarchy is encoded across multiple columns. In terms of the category hierarchy, a label in the first column is a parent of one or more labels in the second column, a label in the second column is a parent of one or more labels in the third column, and so on, until the first data column is reached. Examples of such patterns are shown in Fig. 4.

Spanned Cell Stub: Uses spanned cells in column j to indicate the association with several cells in column $j + 1$. The label in column j encloses all labels in column $j + 1$ whose cells are adjacent to the spanned cell.

Empty Cell Stub: Uses empty label cells in column j to indicate that the adjacent cell in column $j + 1$ is enclosed by the closest non-empty label cell above the empty cell in column j. The non-empty cell and its following empty cells are treated like a spanned cell.

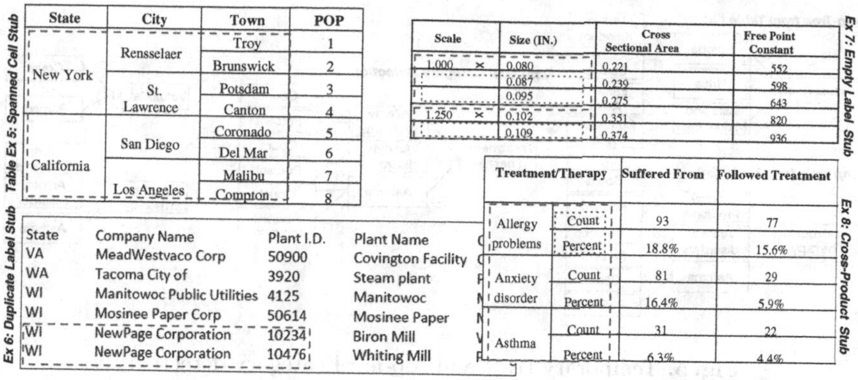

Table Ex 5: Spanned Cell Stub

State	City	Town	POP
New York	Rensselaer	Troy	1
		Brunswick	2
	St. Lawrence	Potsdam	3
		Canton	4
California	San Diego	Coronado	5
		Del Mar	6
	Los Angeles	Malibu	7
		Compton	8

Ex 6: Duplicate Label Stub

State	Company Name	Plant I.D.	Plant Name
VA	MeadWestvaco Corp	50900	Covington Facility
WA	Tacoma City of	3920	Steam plant
WI	Manitowoc Public Utilities	4125	Manitowoc
WI	Mosinee Paper Corp	50614	Mosinee Paper
WI	NewPage Corporation	10234	Biron Mill
WI	NewPage Corporation	10476	Whiting Mill

Ex 7: Empty Label Stub

Scale	Size (IN.)	Cross Sectional Area	Free Point Constant
1.000 ×	0.080	0.221	552
	0.087	0.239	598
	0.095	0.275	643
1.250 ×	0.102	0.351	820
	0.109	0.374	936

Ex 8: Cross-Product Stub

Treatment/Therapy		Suffered From	Followed Treatment
Allergy problems	Count	93	77
	Percent	18.8%	15.6%
Anxiety disorder	Count	81	29
	Percent	16.4%	5.9%
Asthma	Count	31	22
	Percent	6.3%	4.4%

Fig. 4. Examples of Forward Expansion Patterns

Duplicate Label Stub: Uses duplicate labels in column j to indicate that the label is associated with multiple labels in column $j + 1$. A duplicated label in column j encloses all of the associated labels in column $j + 1$. The set of duplicated label cells could be treated like a single spanned cell.

Cross-Product Stub: In this pattern, a set of labels in column $j + 1$ are consistently repeated for each spanned cell in column j. The labels in column $j + 1$ are enclosed by the corresponding spanned cell label in column j.

4.2 Forming a Temp Tree for Category Hierarchy from Patterns

In this section, we explain how we interpret the patterns to form a temporary tree of labels in the stub region, which will be used later on to determine the top-level categories and the category hierarchy.

First, we take the stub head as the root of the tree. If the stub head is empty, we create a virtual root. In a downward expansion pattern, the enclosing labels form the first level of the tree (children of root). Any corresponding enclosed labels become the children of the enclosing label. In a forward expanding pattern, starting from the left, the first column becomes the first level of the tree, the second column becomes the second level of the tree and so on. Figure 5 shows the temporary trees derived from Table Ex1 and Table Ex8 respectively.

4.3 Top-Level Categories and Access Paths

After building the temporary trees, we analyse them to detect the top-level categories, which become the roots of the trees in the category hierarchy. The paths in the category hierarchy, from root to leaf nodes (labels) form the access paths to the rows of data cells.

The top level of the category hierarchy is generally provided by the leftmost stub head. However, repeated sub-trees can be an indication of the existence of

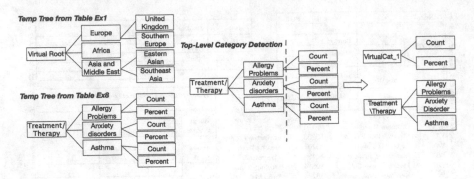

Fig. 5. Temporary trees and top-level category detection

another top-level category. Therefore, we analyse the tree level-by-level checking for a repeated label set on one level. If found, a separate tree is created to represent the new top-level category. The repeated labels are attached as children of this category. The right side of the Fig. 5 shows this type of category detection for Table Ex8 (which has {count, percent} consistently repeated).

If there is an appropriate label in the stub head for the new top-level category, it will be considered as the label for the top-level category, otherwise we use VirtualCat$_i$ (virtual category) as the label (see e.g. Fig. 5).

We have focused on category trees formed in the stub region in detail, but a similar process can be used to determine the category hierarchy in the boxhead region. One difference between the boxhead and stub is that the layout patterns in the boxhead are mostly forward expanding patterns. The same logic for detecting enclosing and enclosed labels and temporary tree formation is applied, except that in the boxhead, we expand the patterns from top-to-bottom rather than left-to-right (as in the stub).

The category hierarchy determined in this way forms the basis for the Abstract Table representation of tables. Each data entry in the table is associated with a set of access paths, where an access path originates from one of the top-level categories identified through the table understanding process.

5   Implementation

In this section, we present the overall design of TEXUS[1], focusing on the implementation of the Table Understanding sub-system. The details of the other major sub-system in TEXUS (Table Extraction) is explained in [11].

One major goal of TEXUS is to facilitate the systematic development and reuse of concepts and implementations in table processing systems. Thus, TEXUS implements the table processing tasks as a set of REST-based service components which can be executed individually or as a pipe-line of components.

---

[1] http://tate.srvr.cse.unsw.edu.au/tate2/TATE.html.

Individual components deal with the following tasks: document conversion, table location, segmentation, functional analysis, structural analysis.

Each component takes two sources of data: an input data model instance and a set of configuration parameters. The outputs are an output data model instance and a set of properties (e.g., header type or error codes if any). To store data model instances, we have chosen to use XML, because (i) it is suitable for describing structured textual information, (ii) it is a platform-independent open standard, and (iii) it is easily transformable into different formats when necessary. For example, we can visualise the output of any components using a simple XSLT[2] script.

In our current implementation, we receive a PDF document as input. The Document Conversion component converts it to an XML document according to our document model. Executing the processing pipeline up to the Segmenting component produces an XML file which contains detected cells in `<td>...</td>` and places the text chunks from each table row in `<tr>..</tr>`. This becomes the input to the next sub-system *Table Understanding* which consists of the Functional and Structural Analysis components.

Note that the algorithms below give a high-level view of the implementation. The procedures for finding patterns are described above, and so we simply mention them by name in the algorithms.

5.1   Functional Analysis

Regardless of the source of the input (i.e., by our own Table Extraction subsystem or an external system), we assume that the following information is accessible from the input: (i) coordinates of the cells, rows and columns, (ii) content of the individual cells in the table, (iii) spanned and merged cells, (iv) formatting styles of cells (e.g., alignment, font face, font color).

The goal of the Functional Analysis component is to assign one of three different functional roles (Data, Access, Header) to each table cell. After detecting the stub and boxhead regions, the cells in those regions are assigned the role of *Access* and *Header* respectively. Algorithm 1 shows how we determine the boundaries of these two regions. The remaining cells outside of these regions are considered *Data* cells (i.e., the entries).

To ensure the accuracy of Functional Analysis we have an extra procedure to check if the data region is recognised correctly. First, we assume the bottom right corner cell in the table boundary is a data cell. Then its neighbour cells are analysed based on the coordinates of the cell and alignment of the cell content. The content type is one of: numeric, alphabetic, alphanumeric, percent, date, currency, plain text and blank cell. A *type attribute*, which is the concatenation of all types found in the cells of a row, is assigned to each row. The algorithm scans table lines one by one towards the top of the table. The final meeting point of this bottom-up scanning process and top-down pattern-based detection is considered the boxhead separator.

---

[2] XSLT, http://www.w3.org/TR/xslt.

Algorithm 1. Functional Analysis (Segmented Table ST)

```
1: boxHead [Rows]= ST.getFirstRow()
2: Stub [Columns] = ST.getFirstColumn()
3: ncols = ST.getRowSize(); nrows = ST.getColumnSize()
4: for (i=1, i ≤ ncols, i++) do
5:     if findForwardExpansionPattern(i) then
6:         boxHead.add(ST.getRow(i+1))
7:     else
8:         for (j=1, j ≤ nrows, j++) do
9:             if findForwardExpansionPattern(j) then
10:                stub.add (ST.getColumn(j+1))
11:            else
12:                if (boxHead.size()=ncols) Or (stub.size()= nrows) then
13:                    print("Invalid Table for Functional Analysis")
14:                else
15:                    return (boxHead, Stub)
16:                end if
17:            end if
18:        end for
19:    end if
20: end for
```

5.2   Structural Analysis

The purpose of Structural Analysis is to detect the top-level categories and the category hierarchies. The access paths to each data entry are automatically mapped from these categories to form the Abstract Table representation.

First, we scan the stub and boxhead to find downward and forward expansion patterns (see Algorithm 2).

Algorithm 2. Structural Analysing (Functioned Table FT)

```
1: for (i=1, i ≤ FT.getStubSize(), i++) do
2:     downPattern [Pattern] = findDownwardExpansionPatterns (i)
3:     forwardPattern [Pattern] = findForwardExpansionPatterns (i)
4: end for
5: for (j=1, j ≤ FT.getBoxHeadSize(), j++) do
6:     headerPattern [Pattern]= findForwardExpansionPatterns (j)
7: end for
8: prefixEnclosingCells(downPattern,forwardPattern,headerPattern,FT)
9: findStubCategories(FT)
10: findHeaderCategories(FT)
11: for (i=boxHead.size(), i ≤ FT.getRowSize(), i++) do
12:     for (j=stubRegion.size, j ≤ FT.getColumnSize(), j++) do
13:         findAccessPaths(FT(i,j))
14:     end for
15: end for
```

For every pattern, a set of enclosing and enclosed cells will be detected and each enclosed cell will be prefixed by an enclosing one. After that there will be one path from the boxhead and one from the stub for every data cell. At this point, each data cell is indexed by an access cell in the same row and a header cell in the same column. In order to detect the exact number of top-level categories, we analyse the paths, searching for repeated sub-paths which indicate the existence of another top-level category, after which the paths will be updated to include an access path for each top-level category.

5.3 Annotated Table Metadata

Generally, the table metadata information is extracted directly from the tables or, in some cases supplied by a user. In our system, *Descriptive Metadata* records table attributes based on layout formatting, and locations. Examples of such data include document type, page numbers, caption, etc. This type of metadata is captured in our extraction tool and expressed in XML (according to a purpose-built XML schema). *Structural Metadata* captures primarily the Function and Structure views of the table. Examples of such data include labels and headers on rows and columns, and the categories and their hierarchies.

Both, the descriptive and structural metadata are currently only captured syntactically within our XML document. In future work we aim to express the structural metadata with the Metadata Vocabulary for Tabular Data currently under development in the W3C, while we also aim to analyse and map the content of arbitrary tables through a semantic analysis.

6 Evaluation

We evaluate our system in two ways: first, we benchmark our results against the latest work by [7,13] using the same dataset. Second, we analyse the overall performance of our system against two other well-known datasets.

6.1 Benchmarking with the DocLab Dataset

In the first part of the evaluation, we compare our results to the latest work in table understanding by Seth [13] and Nagy [7] whose evaluations were based on the DocLab dataset[3]. We used the same dataset as used in [7,13][4] which contained 200 HTML tables, pre-processed and converted to CSV formats, and ground truths. Table 1 shows the characteristics of the tables in the dataset in terms of the number of top-level categories, the number of columns in the stub regions and the stub patterns. We note that the pre-processing of the HTML to CSV has removed much of the layout and formatting features. We manually traced some of the features such as indentation and bullets, but font formatting was not traceable.

---

[3] http://www.iapr tc11.org/mediawiki/index.php/The_DocLab_Dataset_for_Evalua ting_Table_Interpretation_Methods.

[4] We are grateful to the authors of work for sharing the dataset.

Table 1. The DocLab dataset

By Table Type	# of Tbls	By Stub Patterns	# of Tbls
2-Category	177	**Downward Expansion**	**61**
Muti-Category	21	Layout and Formatting	43
One-column Stub	194	Content-based	18
Multi-Column Stub	4	**Forward Expansion**	**4**
		Layout and Formatting	0
		Content-based	4

In [13], the ground truth of the 200 tables was provided in the form of *4 critical cells* which determines the boundary of the stub and entries (i.e., body region that contains the data). CC1 and CC2 represents the the top-left and bottom-right cells of the stub head and CC3 and CC4 are indicators of the top-left and bottom-right cells of the data-cell region. To be able to compare our results with these, we transformed the XML output of our functional analysis component to highlight the locations of the cells that correspond to these 4 cells.

To compare the structural analysis part, we count the number of top-level categories detected by our system both from the stub and boxhead regions, and compare the numbers with the comparable work presented in [7]. Since we provide a more detailed hierarchy for each category, we can only compare the top-level categories across the different methods.

Results. Seth et al. [13] reported 100 % accuracy in their system for functional analysis (excluding two erroneous tables). We also achieve the same accuracy, in that our system correctly segments and functionally analyses all tables (except for the same two erroneous tables).

Seth et al. [13] also reports 3 seconds processing time for detecting critical cells for all 200 tables. In our case, the pipeline processing takes one table at a time. So, the whole process on average, took 4 seconds per table.

In terms of the structural analysis, Nagy et al. [7] reported that their system correctly detected all 21 cases in the dataset that contain tables with more than two top-level categories (multi-categories). We also detected all 21 cases correctly and the number of top-level categories in each table was also 100 % correct.

However, by recognising patterns on the stub region, we have detected more top-level categories than other approaches. In particular, we detect the hierarchy completely in one-column stubs. Figure 6 shows an example of this. From the layout structure, we can understand that there are relationships between "Renewable Total" and all its enclosed labels, and also between "Waste" and its enclosed labels. By correctly detecting this through our analysis, further understanding of the relationship (e.g. 'kind of', 'is member of') through semantic interpretation becomes easier.

7	Source	2003	2004	2005	2006	2007
8	Total	948,446	962,942	978,020	986,215	994,888
9	Renewable Total	96,847	96,357	98,746	101,934	107,954
10	Biomass	9,628	9,711	9,802	10,100	10,839
11	Waste	3,758	3,529	3,609	3,727	4,134
12	Landfill Gas	863	859	887	978	1,319
13	MSW1	2,442	2,196	2,167	2,188	2,218
14	Other Biomass2	453	474	554	561	598
15	Wood and Derived Fuels3	5,871	6,182	6,193	6,372	6,704
16	Geothermal	2,133	2,152	2,285	2,274	2,214
17	Hydroelectric Conventional	78,694	77,641	77,541	77,821	77,885
18	Solar/PV	397	398	411	411	502
19	Wind	5,995	6,456	8,706	11,329	16,515
20	Nonrenewable Total	851,599	866,585	879,274	884,281	886,934

Fig. 6. Example of more detailed sub-category

6.2 Performance on ICDAR and PDF-Trex Datasets

With the first part of the evaluation, we have demonstrated that our system can perform just as effectively as the best known work in table understanding, with the advantage of producing more detailed category hierarchies from the stub.

In the second part of the evaluation, we investigate the performance of our system on a wider range of table types. We have compared our performance against two public datasets well-known to the table processing research community: *the ICDAR competition dataset*[5] containing 67 PDF documents with 156 tables and *PDF-TREX*[6] containing 100 PDF documents with 164 tables, in Italian and English.

As these datasets do not have ground truth for table understanding[7], we have manually created ground-truthed datasets based on two human judges. The human judges were asked to nominate the top-level categories and complete categories with their hierarchical structure for each table.

Table 2 shows the characteristics of the tables in the dataset in terms of the number of categories, the number of columns in the stub regions and the stub patterns. As shown, most of the tables have only one column in the stub which strengthens our argument that a detailed analysis on the stub region, such as layout formatting in a single column, should provide more accurate table understanding. The summary also highlights that the layout and formatting in download expansion patterns is the most common pattern in the datasets, in which the indented stub (using whitespace, bullets and numbering) is the most dominant.

For the analysis, we used our table extraction sub-system in TEXUS, to produce the correct segmenting output of the tables to be fed into our table understanding sub-system.

---

[5] http://www.tamirhassan.com/dataset/.

[6] http://staff.icar.cnr.it/ruffolo/files/PDF-TREX-Dataset.zip.

[7] The ICDAR dataset only has ground truth for table extraction (locating and segmenting).

Table 2. The PDF-Trex and ICDAR datasets

By Tbl Type	PDF Trex	ICDAR	Tot	By Stub Patterns	PDF Trex	ICDAR	Tot
2-Category	136	124	260	**Down'd Expan**	104	44	148
Muti-Category	28	32	60	Layout/Format	59	25	84
One-col Stub	143	139	282	Content	45	19	64
Multi-Col Stub	21	17	38	**Forw'd Expan**	21	17	38
				Layout/Format	12	13	25
				Content	9	4	13

Results. We have correctly performed functional analysis on 96 % of the tables in the dataset (309 tables/320 tables). The unsuccessful cases were:

- 2 folded tables (i.e., when the stub itself is repeated in the table),
- 7 tables with repeated header rows in the middle of the table,
- one long table extending across two pages,
- one table with vertical text direction in the header rows (we only process horizontal text direction).

The result of structural analysis is show in two parts: First, the top part of Table 3 shows the performance of detecting stub patterns by type using the precision and recall measures. Second, on the bottom part of the table shows the performance on detecting the top-level categories and category hierarchy on 309 valid tables using the $Total_{sim}$ measure defined below.

In order to measure the effectiveness of our approach to detecting the top-level categories and the category hierarchies, we created an XML representation of the *category hierarchies* for each table in the ground-truth datasets and compared this against the output of our structural analysis module. There were 309 valid tables considered: 249 two-category, and 60 multi-category.

We measured the similarity of hierarchy trees (in XML files) based on the tree edit distance [16]. We defined three edit operations, *Insert, Delete* and *Rename* at node level with the same cost function of 1. Then the overall similarity for the two trees were calculated as follow:

$$Total_{Sim}(T_0, T_G) = \frac{match(T_0, T_G)}{diff(T_0, T_G) + match(T_0, T_G)} \tag{1}$$

where T_G is the ground-truth, T_0 is the output of the system, $diff(T_0, T_G)$ is the number of nodes to be edited to map the T_0 to T_G, and $match(T_0, T_G)$ is the number of nodes that remain unchanged.

We report the average performance over all valid tables in the dataset. As can be seen in Table 3, we detected all top-level categories correctly in both table types. The differences in category hierarchies were mainly due to false detection of layout features, in particular changes in font colours and sizes.

Table 3. Results of table understanding performance on ICDAR and PDF-Trex

Pattern		Recall	Precision	F-measure
Downward expansion	Layout/format	.97	.91	.94
	Content	.94	.93	.94
Forward expansion	Layout/format	1	1	1
	Content	1	1	1

	Top-Lvl category	Category hierarchy
2-category	1	.87
Multi-category	1	.92

7 Conclusion and Future Work

In this paper, we presented a novel approach to automated table understanding, based on an analysis of commonly-occurring patterns in the stub region of tables. An important aspect of our approach is that the use of patterns means that we do not rely on the use of external knowledge. Our approach is different to prior work on discovering category hierarchies, which focussed on the boxhead region, by also considering layout patterns in the *stub* region.

We incorporated the stub layout patterns defined above into the functional and structural analysis components of our TEXUS system, and analysed its performance over widely-used public datasets. Our system performs at least as well as the latest work in the area, and surpasses it by extracting more accurate category hierarchies for certain kinds of tables. Accurate category hierarchies are important in allowing the output of the system to be more effectively use for further table analysis such as semantic interpretation applications and table similarity detection.

References

1. Alrayes, N., Luk, W.-S.: Automatic transformation of multi-dimensional web tables into data cubes. Data Warehousing and Knowledge Discovery. LNCS, vol. 7448, pp. 81–92. Springer, Heidelberg (2012)
2. e Silva, A.C., Jorge, A., Torgo, L.: Design of an end-to-end method to extract information from tables. IJDAR **82**(2–3), 144–171 (2006)
3. Embley, D.W., Hurst, M., Lopresti, D., Nagy, G.: Table-processing paradigms: a research survey. IJDAR **8**(2–3), 66–86 (2006)
4. Fang, J., Mitra, P., Tang, Z., Giles, C.L.: Table header detection and classification. In: AAAI (2012)
5. Jha, P., Nagy, G.: Wang notation tool: layout independent representation of tables. In: ICPR, pp. 1–4. IEEE (2008)
6. Nagy, G.: Learning the characteristics of critical cells from web tables. In: ICPR, pp. 1554–1557. IEEE (2012)

7. Nagy, G., Seth, S., Embley, D.W.: End-to-end conversion of html tables for populating a relational database. In: DAS, pp. 222–226. IEEE (2014)

8. Nagy, G., Tamhankar, M.: Vericlick: an efficient tool for table format verification. In: IS&T/SPIE Electronic Imaging, pp. 1–9 (2012)

9. Oro, E., Ruffolo, M.: PDF-TREX: an approach for recognizing and extracting tables from pdf documents. In: ICDAR, pp. 906–910. IEEE (2009)

10. Padmanabhan, R.K.: Table abstraction tool. PhD thesis, Citeseer (2009)

11. Rastan, R., Paik, H.-Y., Shepherd, J.: TEXUS: a task-based approach for table extraction and understanding. In: DocEng2015, pp. 25–34 (2015)

12. Seth, S., Jandhyala, R., Krishnamoorthy, M., Nagy, G.: Analysis and taxonomy of column header categories for web tables. In: IAPR, pp. 81–88. ACM (2010)

13. Seth, S., Nagy, G.: Segmenting tables via indexing of value cells by table headers. In: ICDAR, pp. 887–891. IEEE (2013)

14. Wang, X.: Tabular abstraction, editing, and formatting. PhD thesis, University of Waterloo (1996)

15. Zanibbi, R., Blostein, D., Cordy, J.R.: A survey of table recognition. Doc. Anal. Recogn. **7**(1), 1–16 (2004)

16. Zhang, K., Shasha, D.: Simple fast algorithms for the editing distance between trees and related problems. SIAM J. Comput. **18**(6), 1245–1262 (1989)

Author Index